Birkhäuser

Pseudo-Differential Operators
Theory and Applications
Vol. 8

Pseudo-Differential Operators: Theory and Applications is a series of moderately priced graduate-level textbooks and monographs appealing to students and experts alike. Pseudo-differential operators are understood in a very broad sense and include such topics as harmonic analysis, PDE, geometry, mathematical physics, microlocal analysis, time-frequency analysis, imaging and computations. Modern trends and novel applications in mathematics, natural sciences, medicine, scientific computing, and engineering are highlighted.

André Unterberger

Pseudodifferential Analysis, Automorphic Distributions in the Plane and Modular Forms

 Birkhäuser

André Unterberger
Mathématiques
U.F.R. des Sciences
Université de Reims
Moulin de la Housse, B.P. 1039
51687 Reims Cedex 2
France
andre.unterberger@gmail.com, andre.unterberger@math.cnrs.fr

2010 Mathematics Subject Classification: 11F37, 11F72, 47G30

ISBN 978-3-0348-0165-2   e-ISBN 978-3-0348-0166-9
DOI 10.1007/978-3-0348-0166-9

Library of Congress Control Number: 2011935042

Cover design: SPi Publisher Services

Printed on acid-free paper

Springer Basel AG is part of Springer Science+Business Media

www.birkhauser-science.com

To the memory of Paul Malliavin

Contents

Introduction

Most practitioners of pseudo-differential analysis and of analytic number theory would probably regard the two fields as being as far apart as conceivable. However, we wish here to convey the idea that, if deepened in its appropriate aspects, pseudo-differential analysis (mostly, but not only, one-dimensional pseudo-differential analysis in this book) may find a place in the bag of tools of modular form theory. To our PDE colleagues, we shall simply offer the apology that doing some export cannot hurt: more seriously, we have written this book under the assumption that some readers with very little, or no previous knowledge of automorphic function theory, might wish to find a reason to approach this fascinating domain, in which exact formulas of much aesthetic appeal are often the reward of spectral-theoretic questions. Analysts may also find, in the first chapter, aspects of pseudo-differential analysis unknown to them.

Few things in mathematics are duller than a linear form, such as the action of testing a distribution on functions. However, if you can make an operator from your distribution, you may then test it against pairs of functions, endowing it as a result with a more interesting hermitian structure. If you are lucky, you will obtain an explicit sum of squares: examples will occur in Chapter 7. How to make in a useful way, from a distribution in $\mathbb{R}^2$, an operator on functions on the real line, say from Schwartz's space $\mathcal{S}(\mathbb{R})$ to $\mathcal{S}'(\mathbb{R})$, is the starting point of pseudo-differential analysis. The simplest, and most successful way to do so, is the so-called Weyl calculus, or symbolic calculus, of operators: the symbol of an operator is the distribution it is built from.

This will lead to one first reason to let pseudo-differential analysis enter modular form theory which, or so we hope, number theorists may find compelling. When considering eigenfunctions, possibly generalized (Eisenstein series or Maass forms), of the modular Laplacian Δ, it is always the pair of arguments $\pm\nu$ (pure imaginary numbers in the second case), rather than the eigenvalue $\frac{1-\nu^2}{4}$, that enters functions on the spectrum. Of course, functions of $\frac{1-\nu^2}{4}$ are just the same as even functions of ν, but the fact remains that, more often than not, one has to deal with products of a function of ν by the same function taken at $-\nu$. This is especially clear when dealing with such composite objects as functions of type $L(s, f \times g)$ [4, p. 72].

Now, there is a very natural transformation $\mathfrak{S} \mapsto f_0$ (Theorem 1.1.3) from distributions on $\mathbb{R}^2$ to functions in the hyperbolic half-plane $\Pi = SL(2,\mathbb{R})/SO(2)$ for which the operator $\Delta - \frac{1}{4}$ appears as the image of the square of a first-order differential operator, to wit the Euler operator in $\mathbb{R}^2$. Such a transformation is best defined in terms of pseudo-differential analysis, since it is none other than the result of testing the operator with Weyl symbol $\mathfrak{S}$ on a diagonal pair of (Gaussian) functions on the line canonically parametrized by $z \in \Pi$. There is another way to let this transformation, or associates of it, enter the picture: one can link it to the so-called dual Radon transformation from one homogeneous space of $G = SL(2,\mathbb{R})$ (the space G/MN with the standard notation for the Iwasawa decomposition $G = NAK$, and $M = \{\pm I\}$) to the homogeneous space G/K. Needless to say, the Weyl calculus benefits from all desirable so-called covariance properties, so that the transformation under examination does not destroy automorphy properties relative to any arithmetic group (we here limit ourselves to the case of $\Gamma = SL(2,\mathbb{Z})$) one may have in mind. Automorphic objects in $\Pi = G/K$ are automorphic functions of the usual kind, i.e., functions invariant under the group of fractional-linear transformations (in the complex coordinate) of the hyperbolic half-plane associated to matrices in Γ, while automorphic distributions in $\mathbb{R}^2$ are by definition distributions invariant under the linear changes of real coordinates associated to the same group of matrices.

Automorphic pseudo-differential analysis is just pseudo-differential analysis, in which one restricts one's interest in symbols which are automorphic distributions. Using Weil's metaplectic representation [66], it amounts to the same to say that one considers only operators from $\mathcal{S}(\mathbb{R})$ to $\mathcal{S}'(\mathbb{R})$ which commute with all operators from the metaplectic representation lying above elements of Γ: to make this explicit, it means that they commute with the operator of multiplication of a function of x by $e^{i\pi x^2}$, as well as with the Fourier transformation. This pseudo-differential analysis [61] has considerable specificity: on one hand, its structure relies on most features from the spectral theory of the modular Laplacian, including Hecke's theory and L-functions, sometimes of a composite kind; on the other hand, since automorphic distributions are very singular, usual methods of pseudo-differential analysis, for instance boundedness theorems or composition theorems of the usual species, are not applicable. The development of composition formulas in automorphic pseudo-differential analysis leads to an original approach towards the analysis of bilinear operations in non-holomorphic modular form theory. We shall not review automorphic pseudo-differential analysis in depth in the present book — though we shall give more than a few hints — but a byproduct of this approach, going beyond known results regarding series of Kloosterman sums, will play a crucial role in the construction and study, in Chapter 4, of a certain class of automorphic functions.

Pseudo-differential analysis will have an obviously central role in the last two chapters of the book. In a greater part of the book, it will be felt, in an indirect way, by the fact that automorphic distribution theory (on $\mathbb{R}^2$) will have the upper hand in comparison to automorphic function theory (in Π): this will have considerable advantages, for instance, in Chapter 5. In each of the chapters 4 to 7 (which contain the main results of possible arithmetic significance), the zeros of the zeta and, possibly, other L-functions, or those lying on any given parallel to the critical line, occur in some important role: we do not believe that the results obtained constitute a new approach towards the major questions relative to these functions. Each of these chapters would call for complements and generalisations, or deepening: this is especially true of the adelic Chapter 7, which is yet mostly an outlook towards future developments.

We now turn to a more detailed description of the structure of the book. Chapter 1 provides the necessary background about the Weyl calculus. It differs in an essential way from other introductions to pseudo-differential analysis, and it is tailored not for the needs in PDE, but for those in number theory. On one hand, there is much emphasis on representation-theoretic properties; on the other hand, the popular (Moyal-type) rule of composition of symbols bears no relation to the one useful here. Theorem 1.1.3 gives the main properties of the map $\mathfrak{S} \mapsto f_0$ from distributions in $\mathbb{R}^2$ to functions on Π alluded to before. With more detail, this map is defined by the equation $f_0(z) = (\phi_z^0|\mathrm{Op}(\mathfrak{S})\phi_z^0)$, where $\mathrm{Op}(\mathfrak{S})$ is the operator with symbol $\mathfrak{S}$, and $\phi_{-\frac{1}{z}}^0$ is a normalized version of the function $x \mapsto e^{-i\pi\bar{z}x^2}$. The map $\mathfrak{S} \mapsto f_0$ intertwines the two actions of $SL(2,\mathbb{R})$ in both spaces of distributions or functions, by linear or fractional-linear changes of coordinates: this makes it possible to restrict it to automorphic distributions, getting automorphic functions as a result. The other fundamental property of this map is that it transfers the operator $\pi^2\mathcal{E}^2$ in the plane, where $2i\pi\mathcal{E} = x\frac{\partial}{\partial x} + \xi\frac{\partial}{\partial \xi} + 1$, to the operator $\Delta - \frac{1}{4} = (z-\bar{z})^2\frac{\partial^2}{\partial z\partial\bar{z}} - \frac{1}{4}$. We consider only even distributions here (i.e., distributions invariant under the map $(x,\xi) \mapsto (-x,-\xi)$). Such a distribution $\mathfrak{S}$, automorphic or not, is not characterized by its image f_0: what is needed to this effect is to complete f_0 into a pair $\Theta\mathfrak{S} = (f_0, f_1)$, where the function f_1 is defined in a comparable way, using in place of ϕ_z^0 the next simplest (odd) function ϕ_z^1 attached to z. An important operator acting on symbols, to be denoted as $\mathcal{G}$, is defined by the fact that, given any distribution $\mathfrak{S} \in \mathcal{S}'(\mathbb{R}^2)$, the pseudo-differential operator with symbol $\mathcal{G}\mathfrak{S}$ is the composition of the operator $\mathrm{Op}(\mathfrak{S})$ with the operator $u \mapsto \check{u}$, with $\check{u}(x) = u(-x)$. The operator $\mathcal{G}$, a simple rescaling of the symplectic Fourier transformation in $\mathbb{R}^2$, plays a surprisingly central role, even in p-adic pseudo-differential analysis. A distribution $\mathfrak{S}$ is characterized by its image f_0 (rather than by the pair (f_0, f_1)) if and only if it is $\mathcal{G}$-invariant. From the relation between the Euler operator $2i\pi\mathcal{E}$ and the hyperbolic Laplacian Δ, one sees in particular that automorphic distribution theory (in $\mathbb{R}^2$) is slightly subtler than automorphic function theory (in Π), since every eigenfunction f (possibly generalized) of Δ for the eigenvalue $\frac{1-\nu^2}{4}$ is "covered" by exactly two distributions, automorphic if f is, homogeneous of degrees $-1-\nu$ and $-1+\nu$, the images of each other under $\mathcal{G}$.

The Weyl calculus benefits from two distinct covariance properties, in connection with the Heisenberg representation and the metaplectic representation: to each of these, one can associate a composition formula. Only the first (Moyal-type) one is generally known — it is the one useful for applications of pseudo-differential analysis to PDE — but it is the second one, introduced in [61, section 17], that is more important for number-theoretic applications: both types will be considered here. The last section of Chapter 1 deals with the totally radial Weyl calculus: this is obtained when only operators in $\mathcal{S}(\mathbb{R}^n)$ commuting with the action of rotations, and sending every function to a radial distribution, are considered. Looking for an efficient symbolic calculus of such operators, one is led in a natural way, again, to using the hyperbolic half-plane, at least as a first step.

Associates of the map $\mathfrak{S} \mapsto f_0$ are obtained as a result of composing this map by functions, in the spectral-theoretic sense, of Δ on the left, or by functions of $2i\pi\mathcal{E}$ on the right: note that an operation of the second kind is identical to one of the first kind if and only if it involves an even function of $2i\pi\mathcal{E}$. A standard such associate, defined with the help of considerations of harmonic analysis only, is the dual Radon transformation already alluded to from even functions on the plane or, what amounts to the same, functions on the homogeneous space G/MN, to functions on the half-plane G/K. Chapter 2 is devoted to a study of the Radon transformation: in particular, we make the relation between the map $\mathfrak{S} \mapsto f_0$ and the dual Radon transformation explicit, obtaining at the same time a few formulas useful in the sequel. The latter half of this chapter is concerned with the analysis of a certain function $\chi_{\rho,\nu}$ of one variable, built with the help of the hypergeometric function. The results obtained will be applied in Chapter 4, and our reasons for studying the function $\chi_{\rho,\nu}$ will be given presently with more profit.

In Chapter 3, we provide the necessary background regarding automorphic functions, recalling such notions as Eisenstein series, Hecke eigenforms, L-functions ... (in the case of the full modular group only, for simplicity): the first two notions have analogues which are automorphic distributions, the map $\mathfrak{S} \mapsto f_0$ defined before providing the correspondence, together with an obvious terminology. We spend some more time on Roelcke-Selberg expansions, and on matters related to Kloosterman sums. The analytic continuation of certain series of Kloosterman sums has been much studied, as a consequence of investigations (in particular [21]) originating with Selberg's work [45]. It is necessary for our purposes, however, to go beyond these results, finding when $\operatorname{Re} s$ and $\operatorname{Re} t$ are positive and $|\operatorname{Re}(s - t)| < 1$ the analytic continuation of the Dirichlet series in two variables defined when $\operatorname{Re} s > 1, \operatorname{Re} t > 1$ by the equation, in which $k \in \mathbb{Z}$,

$$\zeta_k(s,t) = \frac{1}{4} \sum_{\substack{m_1 m_2 \neq 0 \\ (m_1, m_2) = 1}} |m_1|^{-s} |m_2|^{-t} \exp\left(2i\pi k \frac{\overline{m_2}}{m_1}\right), \qquad (0.1)$$

with $m_2 \overline{m_2} \equiv 1 \bmod m_1$. The solution to this problem is rather lengthy: its main features will be expounded in Section 3.6. It was obtained in some previous work [60], as a byproduct of the spectral analysis of the pointwise product of two Eisenstein series, which will be detailed in Section 3.5. This latter problem was mainly solved as a preparation towards the following basic problem of automorphic pseudo-differential analysis [61]: compute the symbol of the composition (the definition of which requires some care) of two operators the symbols of which are Eisenstein distributions. Though we shall not come back to this problem in any detail, we shall give a short survey, in Section 3.4, of some of its features.

In Chapter 4, we analyze the properties of a new class of automorphic functions. Such functions are formally easy to construct, quite generally, with the help of Poincaré series, starting from a function ψ in Π, and setting $f = \sum_{g \in \Gamma/\Gamma_\psi} \psi \circ g^{-1}$, where Γ_ψ is the subgroup of Γ under which ψ remains invariant: summing over the quotient set prevents one from repeating infinitely many times the same term, in the case when Γ_ψ is infinite. Two examples, in which the group Γ_ψ is the group $\Gamma \cap N$ consisting of matrices $\left(\begin{smallmatrix} 1 & b \\ 0 & 1 \end{smallmatrix}\right)$ with $b \in \mathbb{Z}$, are well-known. In the first one, one takes $\psi(z) = (\operatorname{Im} z)^{\frac{1-\nu}{2}}$ with $\operatorname{Re} \nu < -1$, getting as a result the Eisenstein series $f = E_{\frac{1-\nu}{2}}$. In the second one, one takes $\psi(z) = (\operatorname{Im} z)^{\frac{1-\nu}{2}} e^{2i\pi k z}$ for some $k \in \mathbb{Z}$, obtaining a special case of Selberg's series [45]. In Chapter 4, we shall start, instead of an N-invariant function ψ, from an A-invariant function (where A is the set of matrices $\left(\begin{smallmatrix} a^{\frac{1}{2}} & 0 \\ 0 & a^{-\frac{1}{2}} \end{smallmatrix}\right)$ with $a > 0$), or more generally from a function such that $\psi(az) = a^{\frac{\rho-1}{2}} \psi(z)$ for some fixed number ρ and every $a > 0$. In general, Γ_ψ then reduces to $\{\pm \left(\begin{smallmatrix} 1 & 0 \\ 0 & 1 \end{smallmatrix}\right)\}$, so that new convergence problems arise. Poincaré-style series can also be built in the realm of automorphic distribution (in the plane) theory: one such example will be given in Section 4.1, as its study will demand proving a few geometric estimates needed in the sequel.

The function $(\operatorname{Im} z)^{\frac{1-\nu}{2}}$ which gives rise, under the Poincaré summation process, to the Eisenstein series $E_{\frac{1-\nu}{2}}$, is the image, up to multiplication by a constant, of the function $(x, \xi) \mapsto |\xi|^{\nu-1}$ by the dual Radon transformation. It is therefore tempting to generalize Eisenstein series by starting, instead of a function as simple as $|\xi|^{\nu-1}$, from a function of the two variables x, ξ separately homogeneous in each, to wit a function such as

$$\hom_{\rho,\nu}(x, \xi) = |x|^{\frac{\rho+\nu-2}{2}} |\xi|^{\frac{\nu-\rho}{2}} : \tag{0.2}$$

this function will also occur quite naturally from our study of composition formulas in the Weyl calculus. The role of the two parameters ρ, ν is quite distinct. The second one refers to the global degree of homogeneity $\nu - 1$ of $\hom_{\rho,\nu}$, i.e., to the fact that this is an eigenfunction of $2i\pi\mathcal{E}$ for the eigenvalue ν: the corresponding "spectral line" is defined by $\operatorname{Re} \nu = 0$, since $\mathcal{E}$ is formally self-adjoint in $L^2(\mathbb{R}^2)$. The parameter $\rho - 1$ corresponds to an eigenvalue of the operator (the product of which by i is formally self-adjoint) $x\frac{\partial}{\partial x} - \xi\frac{\partial}{\partial \xi}$, and the appropriate spectral line is defined by $\operatorname{Re} \rho = 1$: we chose ρ, rather than $\rho - 1$, as a parameter, to help not making any confusion between ρ and ν; also, $\frac{\rho}{2}$ will have to move throughout the critical strip for the zeta function, and the spectral line $\operatorname{Re}\rho = 1$ will correspond to the critical line. If one starts from a distribution in $\mathbb{R}^2$ homogeneous of degree $-1-\nu$, or from an eigenfunction of Δ (in Π) for the eigenvalue $\frac{1-\nu^2}{4}$, the Poincaré-type series (in the automorphic distribution environment), assumed to be convergent, built from such an object, will have the same property. Nothing comparable can hold relative to the parameter ρ: but, as will be seen, something important will remain from it after the summation has been performed.

The case when $\rho = 1$ is a special one on several accounts. One can verify that the image, under the dual Radon transform, of the function $\mathrm{hom}_{1,\nu}$, is a multiple of the function $\psi(z) = \mathfrak{P}_{\frac{\nu-1}{2}}\left(-i\frac{\operatorname{Re} z}{\operatorname{Im} z}\right) + \mathfrak{P}_{\frac{\nu-1}{2}}\left(i\frac{\operatorname{Re} z}{\operatorname{Im} z}\right)$ involving Legendre functions. There is nothing one can do with the Poincaré series $\frac{1}{2}\sum_{g\in\Gamma}\psi(g.z)$, as it converges for no value of ν. One can trace the reason for this as lying in the invariance of the function ψ under the change $\nu \mapsto -\nu$: to recover convergence, it is necessary to break the function ψ into two parts, the transforms of each other under the symmetry $\nu \mapsto -\nu$.

No longer specializing the parameter ρ, but assuming that $0 < \operatorname{Re}\rho < 2$, we define with the help of the hypergeometric function, if $\nu \notin \mathbb{Z}$ and $\rho \pm \nu \notin 2\mathbb{Z}$, a certain function $\chi_{\rho,\nu}$ of one real variable (2.3.31) (already alluded to when discussing Chapter 2); we consider then the function

$$\psi_{\rho,\nu}(z) = (\operatorname{Im} z)^{\frac{\rho-1}{2}} \chi_{\rho,\nu}^{\text{even}}\left(\frac{\operatorname{Re} z}{\operatorname{Im} z}\right), \tag{0.3}$$

where $\chi_{\rho,\nu}^{\text{even}}$ is the even part of $\chi_{\rho,\nu}$, and make the following observations. First, and most important, the dual Radon transform of $\mathrm{hom}_{\rho,\nu}$ is a multiple of the sum $\psi_{\rho,\nu} + \psi_{\rho,-\nu}$. From the two eigenvalue equations expressing the bihomogeneity of $\mathrm{hom}_{\rho,\nu}$, it follows by general properties that its dual Radon transform undergoes a multiplication by $a^{\frac{\rho-1}{2}}$ under any change of variable $z \mapsto az$ with $a > 0$, and that it is an eigenfunction of Δ for the eigenvalue $\frac{1-\nu^2}{4}$. The first of these two properties is also, obviously, satisfied by the function $\psi_{\rho,\nu}$. So far as the second eigenvalue equation is concerned, it is still satisfied by $\psi_{\rho,\nu}$ in the complement of the hyperbolic line from 0 to $i\infty$: however, the $\frac{\partial}{\partial x}$-derivative of this function (which is continuous in Π) has a jump at points on this line. Making this discontinuity explicit, one obtains that, in the distribution sense, one has

$$\left(\Delta - \frac{1-\nu^2}{4}\right)\psi_{\rho,\nu} = C(\rho,\nu)(\operatorname{Im} z)^{\frac{\rho-1}{2}}\delta_{(0,i\infty)}, \tag{0.4}$$

where $C(\rho,\nu)$ is an explicit constant important in the theory and $\delta_{(0,i\infty)}$ is the measure supported in the line under consideration, coinciding with $\frac{dy}{y}$ in terms of the coordinate $y = \operatorname{Im} z$. Chapter 2 ends with an intrinsic distinction, in spectral-theoretic terms, between the functions $\psi_{\rho,\nu}$ and $\psi_{\rho,-\nu}$ when $\operatorname{Re}\nu \neq 0$.

Chapter 4 is concerned, for the essential, with the construction and analysis of the series

$$f_{\rho,\nu}(z) = \frac{1}{2} \sum_{g \in \Gamma} \psi_{\rho,\nu}(g.z). \tag{0.5}$$

It converges when $\operatorname{Re}\nu < -1 - |\operatorname{Re}\rho - 1|$, though this is somewhat more difficult to establish than the corresponding convergence of Eisenstein series: indeed, there are "more" terms since one cannot divide here the group Γ by any subgroup larger than $\{\pm(\begin{smallmatrix} 1 & 0 \\ 0 & 1 \end{smallmatrix})\}$, and geometric estimates are more involved. After having defined $f_{\rho,\nu}$ in the initial domain $\operatorname{Re}\nu < -1 - |\operatorname{Re}\rho - 1|$, we need to continue it analytically to the domain $\operatorname{Re}\nu < 1 - |\operatorname{Re}\rho - 1|$. All the difficulties concentrate on the continuation, in this latter domain, of the function

$$z \mapsto \frac{1}{2} \sum_{\substack{m \in \mathbb{Z}^{\times},\, n \in \mathbb{Z} \\ m | n(n+1)}} \left|\frac{m_1}{m_2}\right|^{\frac{1-\rho}{2}} |m|^{\frac{1-\nu}{2}} \left(\frac{|n + \frac{1}{2} - mz|^2}{|m|y}\right)^{\frac{\nu-1}{2}}, \tag{0.6}$$

where the pair m_1, m_2 is characterized by the conditions $m = m_1 m_2$ and $1 \leq m_1 | n + 1, m_2 | n$. A Fourier expansion substitutes for this problem the equivalent one, already mentioned, of continuing analytically the function $\zeta_k(s,t)$ in (0.1).

A summary of the main results regarding $f_{\rho,\nu}$ is as follows. Fixing ρ with $\frac{\rho}{2}$ in the critical strip, the function $f_{\rho,\nu}$ extends as a meromorphic function of ν for $\operatorname{Re}\nu < 1 - |\operatorname{Re}\rho - 1|$, with the following poles: the non-trivial zeros of zeta, and the points $i\lambda_p$, with $\frac{1+\lambda_p^2}{4}$ in the even part of the discrete spectrum of Δ, to wit the part for which there exist cusp-forms invariant under the symmetry $z \mapsto -\bar{z}$. Next, one has the equation

$$f_{\rho,\nu} + f_{\rho,-\nu} = -\frac{C(\rho,\nu)}{\nu} \frac{\zeta^*(\frac{\rho-\nu}{2})\zeta^*(\frac{\rho+\nu}{2})}{\zeta^*(\nu)} E_{\frac{1+\nu}{2}}, \tag{0.7}$$

involving the Eisenstein series $E_{\frac{1+\nu}{2}}$ and the "full zeta function" $\zeta^*(s) = \pi^{-\frac{s}{2}}\Gamma(\frac{s}{2})$ $\cdot \zeta(s)$. On the other hand, the function $(C(\rho,\nu))^{-1} f_{\rho,\nu}$ is invariant under the symmetry $\rho \mapsto 2 - \rho$. Finding the asymptotic expansion, as $\operatorname{Im} z \to \infty$, of $f_{\rho,\nu}(z)$, makes it possible, finally, to obtain the complete (Roelcke-Selberg) spectral decomposition of this function: it does not lie in $L^2(\Gamma\backslash\Pi)$, but it does so after one has subtracted from it a certain linear combination of the Eisenstein series $E_{\frac{1+\rho}{2}}$ and $E_{\frac{3-\rho}{2}}$. An essential property, a consequence of (0.4), is the following.

Denote as Σ the one-dimensional subset of Π consisting of the (disjoint) union of all lines congruent, under elements of Γ, to the hyperbolic line from 0 to $i\infty$: making from the measure $\frac{1}{2}\left[(\operatorname{Im} z)^{\frac{\rho-1}{2}} + (\operatorname{Im} z)^{\frac{1-\rho}{2}}\right]\delta_{(0,i\infty)}$, in an obvious way, an automorphic measure $ds_\Sigma^{(\rho)}$ supported in Σ, one has the identity, in the distribution sense,

$$\left(\Delta - \frac{1 - \nu^2}{4}\right) f_{\rho,\nu} = 2C(\rho,\nu)ds_\Sigma^{(\rho)}. \tag{0.8}$$

Making a careful (not quite standard) analysis of the resolvent of Δ, one obtains the more precise result

$$(2C(\rho,\nu))^{-1} f_{\rho,\nu} = \left(\Delta - \frac{1 - \nu^2}{4}\right)^{-1} ds_\Sigma^{(\rho)} \quad \text{if } \mathrm{Re}\,\nu < 0, \tag{0.9}$$

which makes a clear distinction between $f_{\rho,\nu}$ and $f_{\rho,-\nu}$ possible. It is just as well to give, in place of the Roelcke-Selberg expansion of $f_{\rho,\nu}$, that of the one-dimensional automorphic object $ds_\Sigma^{(\rho)}$, which is of course to be regarded in some appropriate weak sense,

$$ds_\Sigma^{(\rho)} = \frac{1}{2}\left(E_{\frac{1+\rho}{2}} + E_{\frac{3-\rho}{2}}\right) + \frac{1}{16\pi}\int_{-\infty}^\infty \frac{\zeta^*(\frac{\rho - i\lambda}{2})\zeta^*(\frac{\rho + i\lambda}{2})}{\zeta^*(1 + i\lambda)} E_{\frac{1 - i\lambda}{2}}\,d\lambda$$

$$+ \frac{1}{4}\sum_{p,j \text{ even}} L^*(\frac{\rho}{2}, \mathcal{M}_{p,j})\mathcal{M}_{p,j} : \tag{0.10}$$

the functions $\mathcal{M}_{p,j}$ are Hecke eigenforms (only the ones invariant under the symmetry $z \mapsto -\bar{z}$ are to be considered here) and, again, the "full L-series" $L^*(s, \mathcal{M})$ is the L-series $L(s, \mathcal{M})$ completed by the Archimedean factor (a product of two Gamma functions) which makes its functional equation simple. In view of Dunford's integral formula, one may also consider the images of the measure $ds_\Sigma^{(\rho)}$ under all operators of the kind $H\left(2\sqrt{\Delta - \frac{1}{4}}\right)$ where we assume that $H = H(\mu)$ is an even holomorphic function in some strip $|\mathrm{Im}\,\mu| < \beta_0$, such that $\int_{\mathrm{Im}\,\mu=\beta} |\mu|^2|H(\mu)|^2 d\mu < \infty$ for every β with $|\beta| < \beta_0$. The spectral density of every automorphic function so defined is a C^∞ function of λ: within any appropriate subspace of $L^2(\Gamma\backslash\Pi)$ making this extra condition valid, let us consider the closure S_ρ of the linear space of all functions $H\left(2\sqrt{\Delta - \frac{1}{4}}\right) ds_\Sigma^{(\rho)}$. It is clear that from the knowledge of S_ρ, one can determine whether $\mathrm{Re}\,\rho = 1$ or not and, if such is the case, the value of ρ. However, whether S_ρ is independent of ρ when $\mathrm{Re}\,\rho \neq 1$ cannot be answered at present: the continuous part (relative to the spectral decomposition of the modular Laplacian) of this space is so if and only if the Riemann hypothesis is true for zeta, while the discrete part is independent of ρ if and only if the Riemann hypothesis is true for all L-functions attached to cusp-forms of even type (relative to the symmetry $z \mapsto -\bar{z}$), and all eigenvalues of the even part of Δ are simple. Needless to say, even though (0.10) gives some interpretation of the zeros of zeta lying on any given parallel to the critical line, we do not believe that this indicates any possible line of attack on any of these deep conjectures.

At the end of Section 4.7, we raise questions regarding possible generalizations of the function $f_{\rho,\nu}$, as one would come to when replacing Σ by another set of lines, and refer to links, tenuous or strong, with previous work on quadratic extensions of the rationals, by Hejhal [16], Zagier[69] and ourselves [60, Sec. 19–20].

Automorphic distribution theory is again crucial in Chapter 5, another central chapter of the book. Let us denote as $\mathfrak{E}_\nu$ "the" automorphic distribution which is the analogue of the Eisenstein series $E_{\frac{1-\nu}{2}}$, and as $\mathfrak{M}_{p,j}$ "the" analogue, in the same spirit, of the Maass-Hecke form $\mathcal{M}_{p,j}$. Actually, as already mentioned, automorphic distribution theory (on $\mathbb{R}^2$) is more precise than automorphic function theory in the upper half-plane (this is why an automorphic distribution is characterized by *a pair* of automorphic functions) and, as will be seen in Section 3.2, there are two modular distributions, one the image of the other under the symplectic Fourier transformation, corresponding to just one non-holomorphic modular form. Given any pair h, f of functions lying in the image of $\mathcal{S}_{\text{even}}(\mathbb{R}^2)$ under the operator $2i\pi\mathcal{E}(1 + 2i\pi\mathcal{E})$, the series

$$\langle \mathfrak{P}, h \otimes f \rangle = \sum_{g \in \Gamma} \int_{\mathbb{R}^2} (h \circ g)(x, \xi) f(x, \xi) dx d\xi \qquad (0.11)$$

is convergent. The main result of Chapter 5 is the identity

$$\langle \mathfrak{P}, \bar{h} \otimes h \rangle = \frac{1}{2\pi} \int_{-\infty}^{\infty} |\langle \mathfrak{E}_{i\lambda}, h \rangle|^2 |\zeta(i\lambda)|^{-2} d\lambda + 2 \sum_{p \neq 0} \sum_{j} |\Gamma\left(\frac{i\lambda_p}{2}\right)|^2 |\langle \mathfrak{M}_{p,j}, h \rangle|^2 : \qquad (0.12)$$

note that it is not a priori obvious, from its definition, that the left-hand side is non-negative. The proof of this identity is quite delicate: besides automorphic distribution theory, it relies on the theory of series of Kloosterman sums, especially the version, based on the automorphic Green's operator for the modular Laplacian, as developed by Iwaniec [21].

Another part of the book (Chapter 6) deals with arithmetic questions involved in connection with the totally radial Weyl calculus. A comparison with automorphic pseudo-differential analysis, in which non-holomorphic modular form theory enters the structure of the symbols under study, may clarify things at this point: here, arithmetic enters, instead, the functions, or rather discretely supported measures, operators are applied to; besides, it is an extension of holomorphic modular form theory that is now relevant. For the main part, Chapter 6 consists in a quotation of arithmetic results from a recent book [63] of ours: but the symbolic calculus, or quantization process (called the "soft" calculus), which has to be used is introduced here in a quite natural way, while in the quoted work it appeared as a branch in a forest of assorted symbolic calculi, in which the reader had probably no desire to venture. From an arithmetic point of view, the main features of this chapter consist in a necessary extension of the Rankin-Selberg unfolding method, and in an application of puzzling results of Shimura [46] and Iwaniec [22], involving the critical zeros of zeta, to operator theory.

The last part of the book (Chapter 7), still in a quite underdeveloped stage, raises the question whether pseudo-differential analysis should be generalized to an adelic setting: what we have in mind, here, is a pseudo-differential analysis of operators acting on *complex-valued* functions on adeles. Our main point, in this direction, is the following. Starting from a certain problem in automorphic distribution theory, one is led to asking for a version of pseudo-differential analysis in which Planck's constant would depend on the prime p under consideration. This is not possible while staying within Archimedean analysis: at this point, calling for an adelic substitute seems to be required.

Let us stress that we certainly do not consider the present chapter (half of which deals with Archimedean analysis anyway) as an introduction to adelic pseudo-differential analysis: this, on the number-theoretic side of the question, would require another author. We have been satisfied, here, with recalling some concepts of p-adic analysis, and making a few calculations, in this context, specifically related to the problem in automorphic distribution theory we started with. Even though the modification demanded, for each p, by the change of Planck's constant, has been addressed, it is unclear, to us, how the various p-adic pseudo-differential analyses so defined should be pieced together: the usual restricted direct product machinery does not seem to provide the right answer. On the other hand, which may be some justification for this chapter or so we hope, the Archimedean developments which precede in this book — the Weyl calculus, the Radon transformation, elements of representation theory — may be helpful in providing some guidelines for a future possible adelic theory: but one should certainly not limit oneself to adeles of the field $\mathbb{Q}$, and one reason not to do so has been indicated in Remark 7.2.b.(iv).

The chapter starts with the construction of a certain automorphic distribution (in the Archimedean sense) $\mathfrak{T}_\infty - \mathcal{G}\mathfrak{M}_\infty$, where $\mathfrak{T}_\infty$ and $\mathfrak{M}_\infty$ have a purely arithmetic character in that their definition involves series with arithmetic coefficients but no analytic factor such as a Gamma function: the essential property of this distribution is that it coincides with a certain series of Eisenstein distributions $\mathfrak{E}_{-\mu}$ and, possibly, some of their $\frac{d}{d\mu}$-derivatives, taken over the set of non-trivial zeros μ of zeta. If one agrees with the point of view regarding pseudo-differential analysis expounded in the very beginning of this introduction, one is led to the conviction that part of the deeper structure of the automorphic distribution $\mathfrak{T}_\infty$ may be hidden in that of the operator of which it is a symbol. While the operator with symbol $\mathcal{G}\mathfrak{M}_\infty$ can be perfectly understood in a classical distribution setting (7.1.40), and turns out to have an interesting structure, truly understanding the operator $\mathrm{Op}\,(\mathfrak{T}_\infty)$ seems to be a quite difficult task. The distribution $\mathfrak{T}_\infty$ is the limit, as the integer N goes to ∞ while absorbing all primes, of a sequence $(\mathfrak{T}_N)$ with the following property. If not the operator with symbol $\mathfrak{T}_N$, that with the rescaled symbol $N^{i\pi\mathcal{E}}\mathfrak{T}_N$ has a completely clear structure: it is a finite-rank operator from $\mathcal{S}(\mathbb{R})$ to $\mathcal{S}'(\mathbb{R})$ associated to a finite family, depending on N, of discretely supported measures $\mathfrak{d}_\rho$ on the line (here, $\rho \in (\mathbb{Z}/N\mathbb{Z})^\times$ has nothing to do with the number ρ from Chapters 1 and 2), with interesting properties. As a consequence of this, the first component f_0 of the Θ-transform (*cf. supra*) of $\mathfrak{T}_N$ can be expressed as a nice sum of squares with, however, one crucial minus sign (Theorem 7.2.1).

The necessity to rescale the symbol $\mathfrak{T}_N$ with the help of the operation $N^{i\pi\mathcal{E}}$ prevents one from finding a good interpretation of the operator with such a symbol, even more so of the operator with symbol $\mathfrak{T}_\infty$. This takes us to the question we started the current discussion with, of making Planck's constant depend on p. The adelic point of view, even when insufficiently developed, might have good heuristic value in the search for fundamentally new useful Hilbert space structures on appropriate spaces of automorphic distributions: this may, or not, help understanding the automorphic distribution, a series of Eisenstein distributions, which was the starting point of this last chapter.

We dedicate this book to the memory of Paul Malliavin. It is thanks to him that, almost half a century ago, we had our first contact both with singular integral operators (which were soon to become pseudo-differential operators) and with (holomorphic) modular form theory. We have never ceased, in the intervening decades, marveling at his mathematical accomplishments.

Chapter 1

The Weyl calculus

We start with a description of the basic features of the Weyl pseudo-differential analysis, to be used throughout the book: the emphasis is on group-theoretic properties, which is what is needed in the sequel. In the first section of the chapter, we shall show in which way the analysis of operators from the Weyl calculus by means of their diagonal matrix elements against appropriate families, parametrized by points of the hyperbolic half-plane Π, of functions of Gaussian type, establishes a link between function theory on the plane and on the hyperbolic half-plane. This Θ-transformation will appear in most parts of the book. Later, in Chapter 3, it will specialize as a correspondence from automorphic distribution theory in the plane to automorphic function theory in Π. In Section 1.2, we discuss two quite different composition formulas, meaning by this two analyses of the (partially defined) bilinear map $\#$ such that the composition $\text{Op}(h_1)\text{Op}(h_2)$ of the operators with symbols h_1 and h_2 should agree with the operator $\text{Op}(h_1 \# h_2)$: most practitioners of pseudo-differential analysis will only be familiar with the first one. In Section 1.3, we derive from a restriction of the n-dimensional Weyl calculus an efficient symbolic calculus of totally radial operators in $\mathbb{R}^n$: remarkably, this demands that the appropriate species of symbols should live, again, on Π.

1.1 An introduction to the usual Weyl calculus

A symbolic calculus of operators is a linear one-to-one way of associating operators, say on $L^2(\mathbb{R}^n)$, to functions of $2n$ variables: with the exception of the totally radial calculus, we shall be mostly concerned, in this book, with the one-dimensional case. One of the best-known ways of doing this, that which consists in associating with an operator its integral kernel, fails on two major accounts. The first one has to do with the fact that, under such a correspondence, the composition of operators has nothing to do whatever, even on an approximate level, with the pointwise multiplication of integral kernels; the second one is that this correspondence does not benefit from a large group of *visible* symmetries, in a sense to be made clear shortly.

The function h on $\mathbb{R}^{2n}$ associated to some operator A under a given symbolic calculus is called the *symbol* of A while, in the reverse direction, one generally writes $A = \text{Op}(h)$. Since operators on $L^2(\mathbb{R}^n)$ very seldom commute, the composition of operators can never correspond, under any symbolic calculus, to the pointwise multiplication of functions, which is a commutative operation. Still, considering for instance two differential operators A_1 and A_2 of orders m_1 and m_2 respectively, the top-order part of $A_1 A_2$, which can be defined as the equivalence class of the product when differential operators of order $\leq m_1 + m_2 - 1$ are neglected, is the same as the top-order part of $A_2 A_1$. Pseudo-differential analysis was developed, towards the needs of partial differential equations, as a symbolic calculus (several possibilities have been considered), in which the two bilinear operations under consideration, to wit the composition of operators and the pointwise multiplication of symbols, would roughly correspond to each other, modulo error terms of "lower order". We shall not, here, approach this domain of applications, in which hundreds of papers and a few major books [56, 53, 30, 19, 47, 35] have been written. So as to prevent misunderstanding, let us make it clear, however, that PDE people are certainly not interested in symbolic calculi of *differential* operators: the point is that good symbolic calculi (e.g. Weyl's) make it possible to construct auxiliary operators needed for the solution of PDE problems; the simplest instance concerns the construction of parametrices, i.e., approximate inverses, of elliptic operators.

Possibly the most obvious pair of non-commuting bounded operators on $L^2(\mathbb{R})$ is the pair $\tau_{y,0}, \tau_{0,\eta}$, defined by the equations

$$(\tau_{y,0}u)(x) = u(x - y) \quad \text{and} \quad (\tau_{0,\eta}u)(x) = u(x)e^{2i\pi\eta x}. \tag{1.1.1}$$

One can combine these two operations into an operation $\tau_{y,\eta}$ (almost a product of the two): coming back to the n-dimensional case, we assume that y and η lie in $\mathbb{R}^n$ and set

$$(\tau_{y,\eta}u)(x) = u(x - y)e^{2i\pi<x-\frac{y}{2},\eta>}. \tag{1.1.2}$$

If one introduces an extra real parameter t, one notes the identity

$$\frac{1}{2i\pi}\frac{d}{dt}\left(\tau_{ty,t\eta}u\right)(x)$$

$$= \left[-\frac{1}{2i\pi}\sum y_j u_j'(x - ty) + \langle x - ty, \eta\rangle u(x - ty)\right]e^{2i\pi\langle x-\frac{ty}{2},t\eta\rangle}$$

$$= \left(\sum \eta_j x_j - \frac{1}{2i\pi}y_k\frac{\partial}{\partial x_k}\right)\left(\tau_{ty,t\eta}u\right)(x). \tag{1.1.3}$$

It is thus natural to think of the operator $\tau_{ty,t\eta}$ as being the exponential $\exp(2i\pi tD)$, where D is the differential operator (on functions of x) $D = \sum\left(\eta_j x_j - \frac{1}{2i\pi}y_k\frac{\partial}{\partial x_k}\right)$. Setting

$$(Q_j u)(x) = x_j u(x), \quad (P_j u)(x) = \frac{1}{2i\pi} \frac{\partial u}{\partial x_j}, \tag{1.1.4}$$

we shall write, assuming without loss of generality that $t = 1$,

$$T_{y,\eta} = \exp(2i\pi(\langle \eta, Q \rangle - \langle y, P \rangle)). \tag{1.1.5}$$

To give this equation a more than formal meaning, we must refer to Stone's theorem on one-parameter groups of unitary operators, to be found in many places, for instance [42]. On many occasions, we shall be dealing with *explicit* one-parameter groups $(U_t)_{t \in \mathbb{R}}$ of unitary operators in some Hilbert space H, and we shall only need the easy part of Stone's theorem, to wit that such a group has a well-defined self-adjoint generator

$$D = \frac{1}{2i\pi} \frac{d}{dt} \Big|_{t=0} U_t : \tag{1.1.6}$$

recall that the operator D is not, generally, a bounded operator in H, and that the notion of self-adjoint operator has a precise meaning, which demands defining its domain; in this case, it is just the set of vectors u in H such that $t^{-1}(U_t u - u)$ has a limit in H as $t \to 0$.

Given two pairs (y, η) and (y', η'), one has the formula, of immediate verification,

$$T_{y,\eta} T_{y',\eta'} = e^{i\pi[(y,\eta),(y',\eta')]} T_{y+y',\eta+\eta'} \tag{1.1.7}$$

if one introduces the *symplectic form* $[,]$ on $\mathbb{R}^n \times \mathbb{R}^n$, by definition the (alternate) bilinear form such that

$$[(y, \eta), (y', \eta')] = -\langle y, \eta' \rangle + \langle y', \eta \rangle. \tag{1.1.8}$$

We assume that the reader is familiar with the basic language of representation theory. From (1.1.7), it is easy to define with the help of the symplectic form a group structure on the set-theoretic product $\mathbb{R}^n \times \mathbb{R}^n \times \mathbb{R}$, then a unitary representation π of the group obtained in $H = L^2(\mathbb{R}^n)$, such that $\pi(y, \eta; 0) = T_{y\eta}$. The group and representation so defined are called the Heisenberg group and Heisenberg representation. Alternatively, one can weaken the notion of representation to that of projective representation, which consists, given a topological group G and a Hilbert space H, in defining, for every $g \in G$, the operator $\pi(g)$ only up to multiplication by an indeterminate constant $\omega(g) \in \mathbb{C}$ of absolute value 1 (such an indeterminate factor will be called, generally, a *phase factor*), weakening of necessity the basic property of a representation to the relation $\pi(g)\pi(g') = \omega(g, g')\pi(gg')$. Then, (1.1.7) shows that the map $(y, \eta) \mapsto T_{y,\eta}$ is a projective representation of the additive group $\mathbb{R}^n \times \mathbb{R}^n$ in $L^2(\mathbb{R}^n)$.

Of course, all concepts or proofs based on infinitesimal elements will disappear from the more arithmetic parts of the book, in particular the sections devoted to extending the Weyl calculus to a *p*-adic setting. Coming back to our present environment, the symplectic form on $\mathbb{R}^n \times \mathbb{R}^n$ gives rise to the *symplectic group* $\mathrm{Sp}(n, \mathbb{R})$, by definition the group of linear automorphisms g of $\mathbb{R}^n \times \mathbb{R}^n$ preserving the symplectic form: this means, given any two vectors $Y = (y, \eta)$ and $Y' = (y', \eta')$, that one has the identity

$$[gY, gY] = [Y, Y']. \tag{1.1.9}$$

When $n = 1$ (the case that will occur most frequently here), the symplectic group coincides with $SL(2, \mathbb{R})$. There is a notion of *symplectic* Fourier transformation $\mathcal{F}^{\mathrm{symp}}$ on $L^2(\mathbb{R}^n \times \mathbb{R}^n)$ (no such notion exists on odd-dimensional spaces, and in the one-dimensional case $\mathcal{F}$ will denote the usual Fourier transform, normalized in the way comparable to (1.1.11) below), defined by the equation

$$(\mathcal{F}^{\mathrm{symp}} h)(y, \eta) = \int_{\mathbb{R}^n \times \mathbb{R}^n} h(x, \xi) e^{2i\pi(\langle y, \xi\rangle - \langle x, \eta\rangle)} dx d\xi : \tag{1.1.10}$$

it may look not very different from the usual Euclidean Fourier transformation

$$(\mathcal{F}^{\mathrm{euc}} h)(y, \eta) = \int_{\mathbb{R}^n \times \mathbb{R}^n} h(x, \xi) e^{-2i\pi(\langle x, y\rangle + \langle \xi, \eta\rangle)} dx d\xi, \tag{1.1.11}$$

but it has the fundamental property (the verification of which is trivial) that it commutes with all transformations $h \mapsto h \circ g^{-1}$ with $g \in \mathrm{Sp}(n, \mathbb{R})$, whereas the Euclidean Fourier transformation commutes with such transformations for g in the orthogonal group of $\mathbb{R}^{2n}$: these two groups cannot be compared generally, but when $n = 1$, the group $SO(2)$ is a proper subgroup of $SL(2, \mathbb{R})$. Since $(\mathcal{F}^{\mathrm{euc}})^2$ is the operator which transforms a symbol h into the symbol $(x, \xi) \mapsto h(-x, -\xi)$, the symplectic Fourier transformation is an involution, i.e., $(\mathcal{F}^{\mathrm{symp}})^2 = I$.

One of several (equivalent) ways of introducing the Weyl calculus Op is based on this property, and leads to the definition

$$\mathrm{Op}(h) = \int_{\mathbb{R}^n \times \mathbb{R}^n} (\mathcal{F}^{\mathrm{symp}} h)(y, \eta) \exp\left(2i\pi(\langle \eta, Q\rangle - \langle y, P\rangle)\right) dy d\eta \tag{1.1.12}$$

if $h \in \mathcal{S}(\mathbb{R}^n \times \mathbb{R}^n)$, the Schwartz space of C^∞ functions on $\mathbb{R}^{2n}$ rapidly decreasing at infinity: in this way, it is immediate that, for any function $\tilde{h} \in \mathcal{S}(\mathbb{R}^{2n})$, the operator with symbol $(x, \xi) \mapsto \int \tilde{h}(y, \eta) e^{2i\pi(\langle \eta, x\rangle - \langle y, \xi\rangle)} dy d\eta$ is the operator $\int \tilde{h}(y, \eta) \exp\left(2i\pi(\langle \eta, Q\rangle - \langle y, P\rangle)\right) dy d\eta$. Thus, in one sense, the Weyl symbolic calculus is the correspondence obtained when substituting the pair of (vector-valued) operators (Q, P) to the pair of $\mathbb{R}^n$-valued functions (x, ξ): but this (which could not make sense for arbitrary functions of (x, ξ) for reasons of non-commutativity) is only true after the symbol has been expanded as a superposition of exponentials with linear exponents.

From (1.1.5) and (1.1.2), the integral kernel of the operator $\exp(2i\pi(\langle \eta, Q\rangle - \langle y, P\rangle))$ is the function $(x_1, y_1) \mapsto \delta(y_1 - x_1 + y) e^{2i\pi\langle x_1 - \frac{y}{2}, \eta\rangle}$, from which it follows that the integral kernel k of the operator $\mathrm{Op}(h)$ is the function

$$k(x, y) = (\mathcal{F}_2^{-1}h)\left(\frac{x+y}{2}, x-y\right), \qquad (1.1.13)$$

where $\mathcal{F}_2^{-1}h$ denotes the inverse Fourier transform of h with respect to the second variable in $\mathbb{R}^n$, i.e., the function defined by the equation $(\mathcal{F}_2^{-1}h)(x, z) = \int h(x, \xi)e^{2i\pi\langle z,\xi\rangle}d\xi$. From this equation, it follows that, just as the map which associates an operator to its integral kernel, the map Op extends as an isometry from $L^2(\mathbb{R}^{2n})$ onto the Hilbert space of all Hilbert-Schmidt endomorphisms of $L^2(\mathbb{R}^n)$. Equation (1.1.13) leads to the more traditional way of defining the Weyl calculus, by means of the equation

$$(\mathrm{Op}(h)u)(x) = \int_{\mathbb{R}^n \times \mathbb{R}^n} h\left(\frac{x+y}{2}, \eta\right) e^{2i\pi\langle x-y,\eta\rangle}u(y)dyd\eta. \qquad (1.1.14)$$

We now come to the all-important concept of *Wigner function*. Given a pair (u, v) of functions in $L^2(\mathbb{R}^n)$, their Wigner function $W(v, u)$ is the function on $\mathbb{R}^n \times \mathbb{R}^n$ which makes the identity

$$(v|\mathrm{Op}(h)u) = \int_{\mathbb{R}^n \times \mathbb{R}^n} h(x, \xi)W(v, u)(x, \xi)dxd\xi \qquad (1.1.15)$$

valid for every symbol $h \in L^2(\mathbb{R}^{2n})$: note that we define the scalar product $(|)$ on $L^2(\mathbb{R}^n)$ by the equation

$$(v|u) = \int_{\mathbb{R}^n} \bar{v}(x)u(x)dx, \qquad (1.1.16)$$

as an object antilinear with respect to the variable on the left side. The function $W(v, u)$ can be obtained for instance by a computation of the transpose of the map $h \mapsto k$ in (1.1.13), applying the result to the function $u \otimes \bar{v}$: we obtain

$$W(v, u)(x, \xi) = 2^n \int_{\mathbb{R}^n} \bar{v}(x+t)u(x-t)e^{4i\pi\langle t,\xi\rangle}dt. \qquad (1.1.17)$$

It is immediate, with the help of an integration by parts in order to treat extra powers of ξ, that this function lies in $\mathcal{S}(\mathbb{R}^n \times \mathbb{R}^n)$ if both u and v lie in $\mathcal{S}(\mathbb{R}^n)$. A consequence, using (1.1.15), is that the operator $\mathrm{Op}(h)$ still makes sense as a linear operator from $\mathcal{S}(\mathbb{R}^n)$ to its dual space $\mathcal{S}'(\mathbb{R}^n)$ as soon as $h \in \mathcal{S}'(\mathbb{R}^{2n})$. On the other hand, if $h \in \mathcal{S}(\mathbb{R}^{2n})$, the operator $\mathrm{Op}(h)$ extends as a linear operator from the whole of $\mathcal{S}'(\mathbb{R}^n)$ to $\mathcal{S}(\mathbb{R}^n)$: the simplest way to see this is to observe that, in view of (1.1.13), the integral kernel of $\mathrm{Op}(h)$ also lies in $\mathcal{S}(\mathbb{R}^{2n})$ in this case. Spaces resembling the spaces $\mathcal{S}$ or $\mathcal{S}'$, on the line or on $\mathbb{R}^2$, will play an important role everywhere: the consideration of singular species of symbols, or the application of operators with smooth symbols to rather general measures on the line, is essential in applications of pseudo-differential analysis to arithmetic.

The Wigner function has another, dual, role: given ϕ, ψ in $L^2(\mathbb{R}^n)$, the Wigner function $W(\psi, \phi)$ is the Weyl symbol of the rank-one operator $u \mapsto (\psi|u)\phi$. This could be seen immediately from (1.1.13), since the integral kernel of the operator under consideration is the function $k(x, y) = \bar{\psi}(y)\phi(x)$. However, we prefer, since this is a general phenomenon, to remark that the fact that the same concept of Wigner function plays the two roles under consideration is a consequence of the isometry property (from $L^2(\mathbb{R}^{2n})$ to the space of Hilbert-Schmidt operators) of the Weyl calculus: it suffices indeed to apply to two rank-one operators $A_j = \mathrm{Op}(h_j)$ the polarized version $\mathrm{Tr}(A_1^* A_2) = \int_{\mathbb{R}^{2n}} \bar{h}_1(x, \xi)h_2(x, \xi)dxd\xi$ of the isometry property of the calculus, concluding with the help of the fact that a total subspace of the space of Hilbert-Schmidt operators consists of all rank-one operators.

The Weyl symbolic calculus benefits from two species of symmetries, or more precisely *covariance* properties. The first one is expressed in the formula

$$\tau_{y,\eta}\mathrm{Op}(h)\tau_{y,\eta}^{-1} = \mathrm{Op}((x, \xi) \mapsto h(x - y, \xi - \eta)), \quad (y, \eta) \in \mathbb{R}^n \times \mathbb{R}^n : \quad (1.1.18)$$

it is valid whenever $h \in \mathcal{S}'(\mathbb{R}^{2n})$, after it has been observed, of course, that the operators $\tau_{y,\eta}$ preserve both the space $\mathcal{S}(\mathbb{R}^n)$ and the space $\mathcal{S}'(\mathbb{R}^n)$. The proof is immediate, with the help of (1.1.14) and (1.1.7), together with (1.1.5).

This formula is of constant use, even though, for applications to modular form theory, the second covariance property, associated to the *metaplectic* representation, is more fundamental. This representation, defined in full generality (including the p-adic and adelic situations) in [66], is in the present Archimedean environment a genuine (as opposed to projective only) unitary representation Met in $L^2(\mathbb{R}^n)$ of the *metaplectic group*, by definition the twofold cover of $\mathrm{Sp}(n, \mathbb{R})$ (a connected group, the fundamental group of which is $\mathbb{Z}$). It is linked (*loc.cit.*) to the Heisenberg representation by the formula

$$\mathrm{Met}(\tilde{g}) \exp(2i\pi(\langle \eta, Q \rangle - \langle y, P \rangle))\mathrm{Met}(\tilde{g})^{-1} = \exp(2i\pi(\langle \eta', Q \rangle - \langle y', P \rangle)), \quad (1.1.19)$$

in which $\tilde{g}$ is an arbitrary element of the metaplectic group, the canonical image of which in $\mathrm{Sp}(n, \mathbb{R})$ is g (one then says that $\tilde{g}$ lies *above* g), and the vectors $\begin{pmatrix} y' \\ \eta' \end{pmatrix}$ and $\begin{pmatrix} y \\ \eta \end{pmatrix}$ are linked by the relation $\begin{pmatrix} y' \\ \eta' \end{pmatrix} = g\begin{pmatrix} y \\ \eta \end{pmatrix}$. Using this equation together with the definition (1.1.14) of the operator $\mathrm{Op}(h)$, it is immediate that one has, for every $h \in \mathcal{S}'(\mathbb{R}^{2n})$, the covariance formula

$$\mathrm{Met}(\tilde{g})\mathrm{Op}(h)\mathrm{Met}(\tilde{g})^{-1}) = \mathrm{Op}(h \circ g^{-1}) : \quad (1.1.20)$$

the symbol $h \circ g^{-1}$ is of course the one obtained from h after one has applied it the linear change of coordinates on $\mathbb{R}^{2n}$ associated with g^{-1}. Again, the left-hand side of (1.1.20) only makes sense after it has been observed, as was done in [66], that operators in the image of the metaplectic representation preserve the space $\mathcal{S}(\mathbb{R}^n)$ and extend as automorphisms of the dual space $\mathcal{S}'(\mathbb{R}^n)$.

For our present purposes, it will be sufficient to make the metaplectic representation explicit up to an indeterminacy factor ± 1 (in particular, this will make it well-defined as a projective representation): it then becomes possible to regard it as defined on the group $Sp(n, \mathbb{R})$, rather than on the twofold cover of that group. In this sense, one can list the unitary transformations $\operatorname{Met}(g)$ for g lying in an appropriate set of generators of $Sp(n, \mathbb{R})$, as follows: (i) if $g = \begin{pmatrix} A & 0 \\ 0 & A'^{-1} \end{pmatrix}$ with $A \in GL^+(n, \mathbb{R})$ and A' denoting the transpose of A, $\operatorname{Met}(g)$ is (plus or minus) the transform $u \mapsto v$ with $v(x) = (\det A)^{-\frac{1}{2}} u(A^{-1}x)$; (ii) if $g = \begin{pmatrix} I & 0 \\ C & I \end{pmatrix}$, where C is a symmetric $(n \times n)$-matrix, the same holds with $v(x) = u(x)e^{i\pi\langle Cx, x\rangle}$; (iii) finally, if $g = \begin{pmatrix} 0 & I \\ -I & 0 \end{pmatrix}$, the same holds with $v = e^{-\frac{i\pi n}{4}} \mathcal{F}^{\text{euc}}$.

Note that the covariance equation (1.1.20) makes sense even if $\operatorname{Met}(\tilde{g})$ is only defined up to an arbitrary phase factor. From its (almost) explicit definition on generators, the metaplectic representation is not irreducible, but acts within $L^2_{\text{even}}(\mathbb{R}^n)$ and $L^2_{\text{odd}}(\mathbb{R}^n)$ separately: the two terms can then be shown to be acted upon in an irreducible way. This puts forward the role of the involution ch on $\mathcal{S}'(\mathbb{R}^n)$ such that

$$(\operatorname{ch} u)(x) = \check{u}(x) = u(-x), \quad x \in \mathbb{R}^n, \tag{1.1.21}$$

and of the two operations on symbols which correspond to composing an operator on both sides, or on one side only, with the operator ch. The first identity, to wit

$$\operatorname{ch} \operatorname{Op}(h)\operatorname{ch} = \operatorname{Op}((x, \xi) \mapsto h(-x, -\xi)), \tag{1.1.22}$$

though trivial to verify in a direct way, can also be regarded as a consequence of item (iii) in the presentation above of the metaplectic representation: using the relation $(\mathcal{F}^{\text{euc}})^2 = \operatorname{ch}$, so that the operator $\pm e^{-\frac{i\pi n}{2}} \operatorname{ch}$ is an element of the metaplectic representation lying above the matrix $\begin{pmatrix} 0 & I \\ -I & 0 \end{pmatrix}^2 = \begin{pmatrix} -I & 0 \\ 0 & -I \end{pmatrix}$, one can derive (1.1.22) from (1.1.20). This equation implies, in particular, that only even symbols must be used if one is interested only in operators on $\mathcal{S}(\mathbb{R}^n)$ preserving the parity of functions. The second identity, not a covariance equation, demands that we should compute the operation $\mathcal{G}$ on symbols making the identity

$$\operatorname{Op}(h)\operatorname{ch} = \operatorname{Op}(\mathcal{G}h), \tag{1.1.23}$$

or $\operatorname{Op}(\mathcal{G}h)u = \operatorname{Op}(h)\check{u}$, valid. It is of course easy to make this computation, for instance by using the link (1.1.13) between the Weyl symbol h and the integral kernel k of the same operator, together with the fact that if k is the integral kernel of an operator A, that of A ch is the function $(x, y) \mapsto k(x, -y)$. One obtains the formula, valid in any dimension,

$$(\mathcal{G}h)(x, \xi) = 2^n \int_{\mathbb{R}^{2n}} h(y, \eta)e^{4i\pi(\langle x, \eta\rangle - \langle y, \xi\rangle)} \, dy d\eta. \tag{1.1.24}$$

Note that $\mathcal{G}$ is just a rescaled version (by a factor 2) of the symplectic Fourier transformation (1.1.10): when we have defined the Euler operator $2i\pi\mathcal{E}$ (1.1.39), we can connect the two transformations by the equation $\mathcal{G} = 2^{2i\pi\mathcal{E}}\mathcal{F}^{\text{symp}}$ (*cf.* (7.1.1)). The operator $\mathcal{G}$ and its *p*-adic variants will play an important role in applications of pseudo-differential analysis to number theory. In particular, we shall often use the fact that the symbol of the operator $u \mapsto \check{u}$ is $2^{-n}\delta$, where δ is the unit mass at $0 \in \mathbb{R}^{2n}$.

One can break the space of operators in $L^2(\mathbb{R}^n)$ into four parts (even-even, even-odd, odd-even and odd-odd), a self-explaining notion after we have made it clear that even-odd operators, for instance, are those which send even functions to odd ones and kill odd functions. The corresponding symbols are characterized as those being even and $\mathcal{G}$-invariant, odd and $\mathcal{G}$-invariant, odd and $\mathcal{G}$-anti-invariant, finally even and $\mathcal{G}$-anti-invariant. A remark pertinent to quantization theory as well as to applications of pseudo-differential analysis to modular form theory is that the Weyl calculus has a much nicer behaviour than any of its four parts. For instance, the Weyl symbol of an operator on the line as simple as the multiplication by x^2 is just, as will be seen shortly, the function $h(x, \xi) = x^2$, while that of the *even-even* part of this operator is the complicated distribution

$$\frac{1}{2}(h + \mathcal{G}h)(x, \xi) = \frac{1}{2}[x^2 - \frac{1}{16\pi^2}\delta(x)\delta''(\xi)]. \tag{1.1.25}$$

Facing this situation will have consequences throughout the book: in particular, it will explain why, from a certain point of view, it is better to let symbols live on the homogeneous space $G/N \sim \mathbb{R}^2\backslash\{0\}$ of $G = SL(2, \mathbb{R})$ (with $N = \{(\begin{smallmatrix} 1 & b \\ 0 & 1 \end{smallmatrix}),$ $b \in \mathbb{R}\}$) than on the space G/K (with $K = \{(\begin{smallmatrix} \cos\theta & \sin\theta \\ -\sin\theta & \cos\theta \end{smallmatrix}), \theta \bmod 2\pi\}$), a model of which is the hyperbolic half-space, or Poincaré half-space $\Pi = \{z \in \mathbb{C}: \operatorname{Im} z > 0\}$. It will also explain why dealing with appropriate pairs of non-holomorphic modular forms, rather than individual ones, in a way connected to the Lax-Phillips scattering theory for the automorphic wave equation [34], has important advantages.

Specializing from now on in this section in the one-dimensional case, let us explain our last comment, relying for this on the notion of *family of coherent states*. Forgetting the reason, having to do with Physics, which gave this notion its name, we only retain the representation-theoretic part of it: given a topological group G and a unitary representation π of G in some Hilbert space H, a related family of coherent states will be just a family of elements of H making up a total subset, permuted with one another, up to phase factors, under any operator $\pi(g)$, $g \in G$.

In order to build useful families of coherent states for each of the two irreducible parts of the metaplectic representation, we start — number theorists will remember at this point the usual Poincaré construction of modular forms — from a function already invariant, up to phase factors, under all operators $\text{Met}(\tilde{g})$ for $\tilde{g}$ above any element of some "large" subgroup of $SL(2, \mathbb{R})$, in the present case the subgroup $K = SO(2)$. We first show that the pair of (normalized) functions

$$\phi_i^0(x) = 2^{\frac{1}{4}}e^{-\pi x^2}, \quad \phi_i^1(x) = 2^{\frac{3}{4}}\pi^{\frac{1}{2}}xe^{-\pi x^2} \tag{1.1.26}$$

satisfies the required invariance property (note that the superscript 0 or 1 refers to parity, and that, as will be apparent when generalized later, the subscript i denotes the base-point of Π). This demands considering the all-important harmonic oscillator

$$L = \pi(Q^2 + P^2) = \pi x^2 - \frac{1}{4\pi}\frac{d^2}{dx^2}. \qquad (1.1.27)$$

This operator is consistently treated in elementary Physics textbooks, with the help of the so-called creation and annihilation operators: we shall not give reminders here, and a full set of eigenfunctions of L will only be needed in Section 7.2. Let us just recall that L has a purely discrete spectrum without multiplicity, which implies that its full spectral resolution is caught in the list of its square-integrable eigenfunctions: the spectrum is the set $\frac{1}{2} + \mathbb{N} = \{\frac{1}{2}, \frac{3}{2}, \ldots\}$, and the eigenfunctions corresponding to the two lowest eigenvalues $\frac{1}{2}$ and $\frac{3}{2}$ are the functions ϕ_i^0 and ϕ_i^1. One does not even need Stone's theorem in order to define the unitary group $t \mapsto \exp(-itL)$: one has in particular

$$\exp(-itL)\phi_i^0 = e^{-\frac{it}{2}}\phi_i^0, \quad \exp(-itL)\phi_i^1 = e^{-\frac{3it}{2}}\phi_i^1. \qquad (1.1.28)$$

Now, it was found by Mehler (the complete reference seems, unfortunately, to have disappeared from the contemporary literature; we shall give, in a moment, a proof of the formula based on the Weyl calculus) that the operator $\exp(-itL)$ has, for $0 < t < \pi$, an explicit integral kernel k_t, given as

$$k_t(x, y) = e^{-\frac{i\pi}{4}}(\sin t)^{-\frac{1}{2}}\exp\left(\frac{i\pi}{\sin t}[(x^2 + y^2)\cos t - 2xy]\right): \qquad (1.1.29)$$

looking at $k_{\frac{\pi}{2}}$, one obtains the equation $\exp(-\frac{i\pi}{2}L) = e^{-\frac{i\pi}{4}}\mathcal{F}$, from which the way to extend the map $t \mapsto k_t$ as a group homomorphism follows. Using the list of metaplectic unitaries given between (1.1.20) and (1.1.21), one sees after a change of variable $y \mapsto y\sin t$ that the operator with integral kernel $(x, y) \mapsto e^{-\frac{i\pi}{4}}(\sin t)^{-\frac{1}{2}}\exp\left(-\frac{2i\pi xy}{\sin t}\right)$ is one of the two metaplectic operators (one the negative of the other) lying above the matrix $\left(\begin{smallmatrix} 0 & 1 \\ -1 & 0 \end{smallmatrix}\right)\left(\begin{smallmatrix} \frac{1}{\sin t} & 0 \\ 0 & \sin t \end{smallmatrix}\right) = \left(\begin{smallmatrix} 0 & \sin t \\ -\frac{1}{\sin t} & 0 \end{smallmatrix}\right)$; using again the case (ii) of the same list, one obtains that the operator $\exp(-itL)$ is (up to multiplication by ± 1 again) a metaplectic operator lying above the matrix $\left(\begin{smallmatrix} 1 & 0 \\ \frac{1}{\tan t} & 1 \end{smallmatrix}\right)\left(\begin{smallmatrix} 0 & \sin t \\ -\frac{1}{\sin t} & 0 \end{smallmatrix}\right)\left(\begin{smallmatrix} 1 & 0 \\ \frac{1}{\tan t} & 1 \end{smallmatrix}\right) = \left(\begin{smallmatrix} \cos t & \sin t \\ -\sin t & \cos t \end{smallmatrix}\right)$. Since such a family of matrices, taken for $0 < t < \pi$, generates the group $K = SO(2)$, it follows that the functions ϕ_i^0 and ϕ_i^1 are indeed invariant, up to phase factors, under all metaplectic unitary transformations above $SO(2)$.

We have considered, on several occasions, the linear action $\left(\begin{smallmatrix} x \\ \xi \end{smallmatrix}\right) \mapsto g\left(\begin{smallmatrix} x \\ \xi \end{smallmatrix}\right)$ of $g \in SL(2, \mathbb{R})$ on $\mathbb{R}^2$: we consider now the action of $g = \left(\begin{smallmatrix} a & b \\ c & d \end{smallmatrix}\right)$ on Π, to be denoted as $z \mapsto g.z$, by means of fractional-linear transformations, defined in the usual way as $z \mapsto \frac{az+b}{cz+d}$. It follows from what precedes that, up to multiplication by phase factors, the function $\mathrm{Met}(\tilde{g})\phi_i^j$ (with $j = 0, 1$) only depends, if $\tilde{g}$ lies above

$g \in SL(2, \mathbb{R})$, on the class gK: as is well-known and immediate, the knowledge of this class is equivalent to that of the point $z = g.i$. To make the computation simple, we choose $g = \left(\begin{smallmatrix} a & 0 \\ c & a^{-1} \end{smallmatrix} \right)$, with $a > 0$, if $z = \frac{ai}{ci+a^{-1}}$, in other words if $-z^{-1} = a^{-2}i - ca^{-1}$: then, the list of metaplectic unitaries between (1.1.20) and (1.1.21) gives

$$(\mathrm{Met}(\tilde{g})\phi_i^j)(x) = 2^{\frac{i\pi cx^2}{a}} a^{-\frac{1}{2}} \phi_i^j(a^{-1}x), \tag{1.1.30}$$

hence

$$(\mathrm{Met}(\tilde{g})\phi_i^0)(x) = a^{-\frac{1}{2}} e^{\frac{i\pi cx^2}{a}} \phi_i^0(a^{-1}x)$$

$$= 2^{\frac{1}{4}} \left(\mathrm{Im}\,(-z^{-1}) \right)^{\frac{1}{4}} \exp \frac{i\pi x^2}{\bar{z}}. \tag{1.1.31}$$

Doing the same, starting this time from the function ϕ_i^1, we are led to introducing the pair of functions which occur in the following theorem.

Theorem 1.1.1. *Given $z \in \Pi$, set*

$$\phi_z^0(x) = 2^{\frac{1}{4}} \left(\mathrm{Im}\,(-z^{-1}) \right)^{\frac{1}{4}} \exp \frac{i\pi x^2}{\bar{z}},$$

$$\phi_z^1(x) = 2^{\frac{3}{4}} \pi^{\frac{1}{2}} \left(\mathrm{Im}\,(-z^{-1}) \right)^{\frac{3}{4}} x \exp \frac{i\pi x^2}{\bar{z}}. \tag{1.1.32}$$

Given $g = \left(\begin{smallmatrix} a & b \\ c & d \end{smallmatrix} \right) \in SL(2, \mathbb{R})$ and any $\tilde{g}$ lying above g in the metaplectic group, one has for some phase factors ω_0, ω_1 depending on z, g the equations

$$\mathrm{Met}(\tilde{g})\phi_z^0 = \omega_0 \phi_{\frac{az+b}{cz+d}}^0, \quad \mathrm{Met}(\tilde{g})\phi_z^1 = \omega_1 \phi_{\frac{az+b}{cz+d}}^1. \tag{1.1.33}$$

The set $\{\phi_z^0 : z \in \Pi\}$ (resp. $\{\phi_z^1 : z \in \Pi\}$) is total in $L_{\mathrm{even}}^2(\mathbb{R})$ (resp. $L_{\mathrm{odd}}^2(\mathbb{R})$). Any even distribution $h \in \mathcal{S}'(\mathbb{R}^2)$ is characterized by the pair of functions

$$f_0(z) = (\phi_z^0 | \mathrm{Op}(h)\phi_z^0), \quad f_1(z) = (\phi_z^1 | \mathrm{Op}(h)\phi_z^1): \tag{1.1.34}$$

the pair (f_0, f_1) will be called the Θ-transform of h.

Proof. Equations (1.1.33) follow from the definition of ϕ_z^j as the image of ϕ_i^j under $\mathrm{Met}(\tilde{g})$ for some $\tilde{g}$ with $g.i = z$. The density claims are then a consequence of Schur's lemma and of the irreducibility of the two components of the representation Met. Actually, for the odd case only, one has a more precise result, to wit the formula, the proof of which is straightforward (a simplification will occur from the measure-preserving change $z \mapsto -z^{-1}$ on the left-hand side)

$$(8\pi)^{-1} \int_\Pi |(\phi_z^1|u)|^2 dm(z) = \|u\|^2, \quad u \in L_{\mathrm{odd}}^2(\mathbb{R}), \tag{1.1.35}$$

in which $dm(z) = (\mathrm{Im}\,z)^{-2} d\mathrm{Re}\,z\, d\mathrm{Im}\,z$ is the usual measure on Π invariant under all transformations $z \mapsto g.z$. Approximating functions in $\mathcal{S}(\mathbb{R})$ with the appropriate parity by linear combinations of functions ϕ_z^j with $j = 0, 1$ and $z \in \Pi$, one sees, since an operator such as $\text{Op}(h)$ with h even preserves the parity of functions, that $h \in \mathcal{S}'_{\text{even}}(\mathbb{R}^2)$ is characterized by the pair of functions $(w, z) \mapsto (\phi_w^j | \text{Op}(h)\phi_z^j)$ on $\Pi \times \Pi$. Calling such a function a matrix element of the operator $\text{Op}(h)$ against the pair (ϕ_w^j, ϕ_z^j), the remaining problem consists in showing that $\text{Op}(h)$ is already characterized by its set of diagonal matrix elements. To do so, we simply remark, according to (1.1.32), that the function

$$(w, z) \mapsto (\text{Im}\,(-w^{-1})\text{Im}\,(-z^{-1}))^{-\frac{2j+1}{4}} (\phi_w^j | \text{Op}(h)\phi_z^j) \qquad (1.1.36)$$

is sesquiholomorphic, i.e., holomorphic with respect to w and antiholomorphic with respect to z. As such, it is certainly characterized by its restriction to the diagonal of $\Pi \times \Pi$. $\qquad\square$

One has $W(\phi_z^j, \phi_z^j) = W(\phi_i^j, \phi_i^j) \circ g^{-1}$ if $z = g.i$: indeed, this follows from the interpretation of $W(\psi, \phi)$ as the Weyl symbol of the rank-one operator $u \mapsto (\psi | u | \phi)$ together with the covariance formula (1.1.20). It is an easy task, using (1.1.17), to obtain the pair of formulas

$$W(\phi_i^0, \phi_i^0)(x, \xi) = 2e^{-2\pi(x^2+\xi^2)}, \quad W(\phi_i^1, \phi_i^1) = 2[4\pi(x^2 + \xi^2) - 1]e^{-2\pi(x^2+\xi^2)}, \qquad (1.1.37)$$

from which one thus has, more generally,

$$W(\phi_z^0, \phi_z^0)(x, \xi) = 2 \exp\left(-\frac{2\pi}{\text{Im}\,z}|x - z\xi|^2\right),$$

$$W(\phi_z^1, \phi_z^1)(x, \xi) = 2\left[\frac{4\pi}{\text{Im}\,z}|x - z\xi|^2 - 1\right] \exp\left(-\frac{2\pi}{\text{Im}\,z}|x - z\xi|^2\right). \qquad (1.1.38)$$

Remark 1.1.a. Despite the fact that the symbol h is characterized by the pair (f_0, f_1), it carries more immediately usable information. For instance, if h is a real non-negative (locally integrable) function, the function f_0 is non-negative, while the converse is far from being true. Also note that if the operator $\text{Op}(h)$ is bounded in $L^2(\mathbb{R})$, self-adjoint (i.e., $h = \bar{h}$) and semi-definite positive, both functions f_0 and f_1 are non-negative: but there is no converse, and there is no implication in any direction between the positivity of a real symbol and that of the associated operator. Approximate versions of these inexistent implications do exist, however, and can be taken [57] as the basis of some way to analyze the continuity properties of pseudo-differential operators.

Definition 1.1.2. In the plane $\mathbb{R}^2$, we introduce the Euler operator

$$2i\pi\mathcal{E} = x\frac{\partial}{\partial x} + \xi\frac{\partial}{\partial \xi} + 1; \qquad (1.1.39)$$

on the half-plane Π, we consider the usual hyperbolic Laplacian

$$\Delta = -4(\text{Im}\,z)^2 \frac{\partial^2}{\partial z \partial \bar{z}} \qquad (1.1.40)$$

(the same as $-y^2 \left(\frac{\partial^2}{\partial x^2} + \frac{\partial^2}{\partial y^2} \right)$, but we are at present prevented, since we use coordinates x, ξ on $\mathbb{R}^2$, from setting $z = x + iy$).

Initially considered as an operator from $C_0^\infty(\Pi)$ (the space of C^∞ functions in Π with compact support) to $L^2(\Pi) = L^2(\Pi, dm)$, the operator Δ is symmetric, or formally self-adjoint, i.e., satisfies the identity $(f_2|\Delta f_1)_{L^2(\Pi)} = (\Delta f_2|f_1)_{L^2(\Pi)}$ for any pair of functions in $C_0^\infty(\Pi)$: with the help of some elementary argument using the fact that Δ is an elliptic operator, one can show that it is actually essentially self-adjoint, i.e., it admits a unique self-adjoint extension in $L^2(\Pi)$. In the same way (of course, $2i\pi\mathcal{E}$ is not an elliptic operator: however, it is first-order, so that matters are elementary too, for other reasons), one can see that the operator $2i\pi\mathcal{E}$ in $L^2(\mathbb{R}^2)$, with initial domain $C_0^\infty(\mathbb{R}^2\backslash\{0\})$, is essentially self-adjoint too.

Both functions in (1.1.38) can be expressed in terms of the sole argument $\rho = \frac{|x-z\xi|^2}{\text{Im } z}$: for any such function, regarding the argument as a function of $z \in \Pi$ or of $(x, \xi) \in \mathbb{R}^2$, the remaining variables being treated as parameters, one can apply the operator Δ or the Euler operator to it. One can then prove the identity

$$\pi^2 \mathcal{E}^2 k(\rho) = - \left(\rho \frac{d}{d\rho} + \frac{1}{2} \right)^2 k(\rho)$$

$$= \left(\Delta - \frac{1}{4} \right) k(\rho) \tag{1.1.41}$$

for any C^2 function k on $[0, \infty[$. Indeed, since

$$W(\phi_{g.z}^j, \phi_{g.z}^j) = W(\phi_z^j, \phi_z^j) \circ g^{-1} \tag{1.1.42}$$

(the covariance formula again) and since the Euler operator is invariant under the action of $SL(2, \mathbb{R})$ on $\mathbb{R}^2$ by linear changes of coordinates, it suffices to prove the first line of (1.1.41) in the case when $z = i$, which provides some simplification; using the invariance of Δ under the action $(g, z) \mapsto g.z$ of $SL(2, \mathbb{R})$ in Π, it simplifies the computation of $\left(\Delta - \frac{1}{4} \right) k(\rho)$ to reduce it to the case when $x = 0$.

Remark 1.1.b. The map $h \mapsto (f_0, f_1)$ in Theorem 1.1.1 is *almost* continuous from $L^2_{\text{even}}(\mathbb{R}^2)$ to $L^2(\Pi) \times L^2(\Pi)$: comparing (1.1.43) and (1.2.15) below, one will notice that only the pole of $\Gamma(-\frac{i\lambda}{2})$ at $\lambda = 0$ prevents it from being continuous. At the same time, it is one-to-one, but far from having a continuous inverse. Actually, as mentioned and proved in [63, Theorem 2.1], one has

$$\|f_0\|_{L^2(\Pi)} = 2\|\Gamma(i\pi\mathcal{E})h\|_{L^2(\mathbb{R}^2)} \tag{1.1.43}$$

if $h \in L^2(\mathbb{R}^2)$ lies in the image of $L^2(\mathbb{R}^2)$ under $2i\pi\mathcal{E}$ and is the symbol of an even-even operator (*i.e*, if h is even and $\mathcal{G}$-invariant), while

$$\|f_1\|_{L^2(\Pi)} = 4\|\Gamma(1 + i\pi\mathcal{E})h\|_{L^2(\mathbb{R}^2)} \tag{1.1.44}$$

if $h \in L^2(\mathbb{R}^2)$ is the symbol of an odd-odd operator. These two equations will also be a corollary of some facts proved in Section 2.1.

The following theorem already shows a possible way to combine pseudo-differential analysis and modular form theory. Note that we often denote distributions in $\mathbb{R}^2$ which, by their very nature, must be singular, by letters such as $\mathfrak{S}, \dots$ instead of h.

Theorem 1.1.3. *Let $\Gamma = SL(2, \mathbb{Z})$. Define an automorphic distribution to be any distribution $\mathfrak{S} \in \mathcal{S}'(\mathbb{R}^2)$ such that $\mathfrak{S} \circ g^{-1} = \mathfrak{S}$ for every $g \in \Gamma$, where the distribution on the left-hand side is defined by the formula $\langle \mathfrak{S} \circ g^{-1}, h \rangle = \langle \mathfrak{S}, h \circ g \rangle$ for every $g \in SL(2, \mathbb{R})$ and $h \in \mathcal{S}(\mathbb{R}^2)$. Given any automorphic distribution $\mathfrak{S}$, the functions f_0, f_1 defined in the upper half-plane by the equations*

$$f_0(z) = (\phi_z^0 | \mathrm{Op}(\mathfrak{S}) \phi_z^0), \quad f_1(z) = (\phi_z^1 | \mathrm{Op}(\mathfrak{S}) \phi_z^1), \tag{1.1.45}$$

i.e., $\begin{pmatrix} f_0 \\ f_1 \end{pmatrix} = \Theta \mathfrak{S}$, are automorphic. If an automorphic distribution $\mathfrak{S}$ is moreover homogeneous of degree $-1 - i\lambda$ for some $\lambda \in \mathbb{R}$, in which case we shall call it a modular distribution, the functions f_0 and f_1 are generalized eigenfunctions of Δ for the eigenvalue $\frac{1+\lambda^2}{4}$, hence are non-holomorphic modular forms.

Proof. That f_0 and f_1 are automorphic if $\mathfrak{S}$ is a consequence of (1.1.42). The second part of the theorem follows from (1.1.41). $\square$

Remarks 1.1.c. (i) It is generally easier to decompose a distribution in $\mathbb{R}^2$ into its homogeneous components than to decompose a function in Π according to the spectral theory of Δ: this is an advantage of using the plane rather than the half-plane.

(ii) Most orbits of the linear action of Γ in $\mathbb{R}^2$ are dense in $\mathbb{R}^2$: this is in striking contrast with the action of this group in Π, and with the existence of a related fundamental domain. It follows that no continuous function in $\mathbb{R}^2$ can be an automorphic distribution, unless it is a constant.

(iii) It is of course for simplicity that we specialize in the arithmetic group $SL(2, \mathbb{Z})$ in our present investigations: it would certainly involve much work, though possibly not so many new ideas, relying on Iwaniec's treatment [21] of automorphic forms for congruence groups, to extend the results of [61] to more general such groups.

(iv) In ([61], p. 11), the factor 2^n (with $n = 1$ there) was absent from the right-hand side of the equation similar to (1.1.17) here. As already explained (and apologized for) elsewhere, all calculations in [61] were done, starting from the correct formula (including the factor 2), with the exception of Section 18. In that section (devoted to odd automorphic distributions and Maass forms of nonzero weight), the results of all calculations based on the use of Wigner functions ought to be multiplied by 2.

1.2   Two composition formulas

In this section, we discuss composition formulas in a representation-theoretic way, a point of view introduced in [61, Sec. 19]. Whenever two operators $Op(h_1)$ and $Op(h_2)$ can be composed so as to give a linear operator from $\mathcal{S}(\mathbb{R}^n)$ to $\mathcal{S}'(\mathbb{R}^n)$ as a result, the sharp product $h_1 \# h_2$ is that which makes the formula

$$Op(h_1)Op(h_2) = Op(h_1 \# h_2) \tag{1.2.1}$$

valid. There are two species of composition formulas, according to whether one puts the emphasis on the covariance formula (1.1.18) or (1.1.20): in each case, the trick is to decompose the space of symbols under consideration, say $L^2(\mathbb{R}^{2n})$, according to the group-action on this space involved in the covariance formula under consideration.

In the first formula, it is the additive group $\mathbb{R}^{2n}$ that acts on $L^2(\mathbb{R}^{2n})$ by translations. The differential operators on $L^2(\mathbb{R}^{2n})$ commuting with this action are exactly all differential operators with constant coefficients, and the joint generalized eigenfunctions of these are just the exponentials $(x, \xi) \mapsto e^{2i\pi(\langle x,\eta\rangle - \langle y,\xi\rangle)}$ with $(y, \eta) \in \mathbb{R}^{2n}$. The corresponding composition formula calls for a decomposition of the sharp product of two such exponentials as a superposition (in this case, just one term will do), again, of exponentials. Now, what has been said just after the definition (1.1.14) of the Weyl calculus can be rephrased by saying that the operator $\tau_{y,\eta} = \exp(2i\pi(\langle \eta, Q\rangle - \langle y, P\rangle))$ coincides with the operator with symbol $(x, \xi) \mapsto e^{2i\pi(\langle x,\eta\rangle - \langle y,\xi\rangle)}$. The sought-after composition formula is thus none other than Weyl's formula (1.1.7). Since the (symplectic) Fourier transformation provides the decomposition of an arbitrary symbol, say in $\mathcal{S}(\mathbb{R}^{2n})$, into such exponentials, a combination of (1.1.12) and (1.1.7) gives the general formula

$$(h_1 \# h_2)(X) = 2^{2n} \int_{\mathbb{R}^{2n} \times \mathbb{R}^{2n}} h_1(Y)h_2(Z)e^{-4i\pi[Y-X,Z-X]}dY\,dZ, \tag{1.2.2}$$

in which we have set $X = (x, \xi), Y = (y, \eta), Z = (z, \zeta)$: this can also be written (as seen with the help of a Fourier transformation) as

$$(h_1 \# h_2)(X) = [\exp(i\pi L)(h_1 \otimes h_2)](X, X) \tag{1.2.3}$$

with

$$i\pi L = \frac{1}{4i\pi} \sum_{j=1}^{n} \left(-\frac{\partial^2}{\partial y_j \partial \zeta_j} + \frac{\partial^2}{\partial z_j \partial \eta_j} \right). \tag{1.2.4}$$

Expanding the exponential into a series, one obtains the so-called Moyal formula

$$(h_1 \# h_2)(x, \xi) \tag{1.2.5}$$

$$= \sum \frac{(-1)^{|\alpha|}}{\alpha!\beta!} \left(\frac{1}{4i\pi} \right)^{|\alpha|+|\beta|} \left(\frac{\partial}{\partial x} \right)^{\alpha} \left(\frac{\partial}{\partial \xi} \right)^{\beta} h_1(x, \xi) \left(\frac{\partial}{\partial x} \right)^{\beta} \left(\frac{\partial}{\partial \xi} \right)^{\alpha} h_2(x, \xi),$$

valid at least when the series terminates at some point. This is the case when one of the two symbols is a polynomial in (x, ξ): then, it could also be derived by purely algebraic means from the special cases

$$x_j \# h = x_j h - \frac{1}{4i\pi} \frac{\partial h}{\partial \xi_j}, \quad \xi_j \# h = \xi_j h + \frac{1}{4i\pi} \frac{\partial h}{\partial x_j},$$

$$h \# x_j = x_j h + \frac{1}{4i\pi} \frac{\partial h}{\partial \xi_j}, \quad h \# \xi_j = \xi_j h - \frac{1}{4i\pi} \frac{\partial h}{\partial x_j}. \tag{1.2.6}$$

These four equations are also a simple consequence of the defining equation (1.1.14). From (1.2.6), it is immediate that the symbols of (linear) differential operators, say with coefficients in $\mathcal{S}'(\mathbb{R})$, are exactly those which are polynomial with respect to the variables ξ, and that such symbols can be explicitly computed, starting from the fact, immediate from (1.1.14), that the symbol of the operator of multiplication by some function (or distribution) a on $\mathbb{R}^n$ is just the function $h(x, \xi) = a(x)$.

As an application of equations (1.2.6), let us prove Mehler's formula (1.1.29). The sharp composition, on the left-hand side, by the symbol of L, is the operator

$$\pi \left[\left(x - \frac{1}{4i\pi} \frac{\partial}{\partial \xi} \right)^2 + \left(\xi + \frac{1}{4i\pi} \frac{\partial}{\partial x} \right)^2 \right] = \pi \left[x^2 + \xi^2 - \frac{1}{16\pi^2} \left(\frac{\partial^2}{\partial x^2} + \frac{\partial^2}{\partial \xi^2} \right) \right], \tag{1.2.7}$$

which reduces, on functions of $\rho = x^2 + \xi^2$, to

$$-\frac{1}{4\pi} \left(\rho \frac{d^2}{d\rho^2} + \frac{d}{d\rho} \right) + \pi \rho. \tag{1.2.8}$$

It follows, solving an elementary differential equation, that, when $|t| < \pi$, the symbol of the operator e^{-itL} is the function

$$h_t(x, \xi) = \frac{1}{\cos \frac{t}{2}} \exp \left(-2i\pi(x^2 + \xi^2) \tan \frac{t}{2} \right) : \tag{1.2.9}$$

the link (1.1.13) between the symbol and the integral kernel of the same operator yields Mehler's formula.

Equation (1.2.5) is the only generally known composition formula: in applications of pseudo-differential analysis to partial differential equations, most classes of symbols under use are defined so that this formula, without being an exact one, should still be valid in some asymptotic (do not confuse the adjective with "formal") sense.

On the other hand, this formula is absolutely inapplicable in a variety of situations, starting with the *automorphic* pseudo-differential analysis. In this case, still using the Weyl calculus, automorphic distributions are used as symbols. Now, not one term from the would-be Moyal expansion of the sharp product of two automorphic symbols could be meaningful: automorphic distributions are simply too singular for that. Calling for a composition formula of possible interest in arithmetic, a subject developed at length in [61], and which will be considered again in Section 3.4, will take us to the second species of composition formula (relative to the Weyl calculus) we have in mind. In the remainder of the present section, we consider only the sharp product of two symbols in $\mathcal{S}(\mathbb{R})$, so that this will be far from covering the case of automorphic symbols, but the harmonic analysis (as opposed to the arithmetic) is the same.

We follow the same path as previously, replacing however the action by translations of the space $\mathbb{R}^{2n}$ on itself by the linear action of the symplectic group. The only differential operator commuting with this action is the Euler operator $2i\pi\mathcal{E}$, to which one may add the (non-local) operator on symbols associated to the change of coordinates $(x, \xi) \mapsto (-x, -\xi)$. The decomposition of $L^2(\mathbb{R}^{2n})$ into irreducible components of this action is obtained from the decomposition of symbols as integral superpositions of symbols of given parities, homogeneous of degrees $-1 - i\lambda, \lambda \in \mathbb{R}$. We here make it explicit for $n = 1$, though there is no significant difference in the higher-dimensional case: only, in general, the L^2-theory demands that one should consider homogeneous functions of degrees having $-n$ as a real part.

Given $s \in \mathbb{R}\backslash\{0\}$ and $\alpha \in \mathbb{C}$, set

$$|s|_\delta^\alpha = \begin{cases} |s|^\alpha & \text{if } \delta = 0, \\ |s|^\alpha \text{sign } s & \text{if } \delta = 1. \end{cases} \tag{1.2.10}$$

Every function $h \in \mathcal{S}(\mathbb{R}^2)$ can be decomposed according to the equation

$$h = \sum_{\delta=0,1} \int_{-\infty}^\infty h_{i\lambda,\delta} d\lambda, \tag{1.2.11}$$

where

$$h_{i\lambda,\delta}(x, \xi) = \frac{1}{2\pi} \int_0^\infty t^{i\lambda} h_\delta(tx, t\xi) dt, \quad |x| + |\xi| \neq 0, \tag{1.2.12}$$

and h_δ denotes the even, or odd, part of h, according to whether $\delta = 0$ or 1. The function $h_{i\lambda,\delta}$ occurring in this decomposition is homogeneous of degree $-1 - i\lambda$ and has the parity specified by the index δ: it is characterized by the function $h_{i\lambda,\delta}^\flat$ on the line defined as

$$h_{i\lambda,\delta}^\flat(s) = h_{i\lambda,\delta}(s, 1), \tag{1.2.13}$$

since

$$h_{i\lambda,\delta}(x, \xi) = |\xi|_\delta^{-1-i\lambda} h_{i\lambda,\delta}^\flat\left(\frac{x}{\xi}\right). \tag{1.2.14}$$

All that precedes is perfectly elementary, and so is the equation

$$\|h\|^2_{L^2(\mathbb{R}^2)} = 4\pi \int_{-\infty}^{\infty} \|h^{\flat}_{i\lambda,\delta}\|^2_{L^2(\mathbb{R})}d\lambda, \tag{1.2.15}$$

valid for $h \in \mathcal{S}(\mathbb{R}^2)$ with the parity associated to δ. This equation makes it possible to extend the preceding set of formulas to the case when $h \in L^2(\mathbb{R}^2)$, in which $h^{\flat}_{i\lambda,\delta}$ is well-defined as an element of $L^2(\mathbb{R})$ for almost every λ.

Note that, in the case when $h \in \mathcal{S}(\mathbb{R}^2)$, one can define

$$h_{\nu,\delta}(x,\xi) = \frac{1}{2\pi} \int_0^{\infty} t^{\nu} h_{\delta}(tx,t\xi)dt, \quad |x| + |\xi| \neq 0 \tag{1.2.16}$$

for $\nu \in \mathbb{C}, \operatorname{Re}\nu > -1$, and perform a related change of contour in the integral on the right-hand side of (1.2.11).

Given $g = \left(\begin{smallmatrix} a & b \\ c & d \end{smallmatrix}\right)$, one notes after a straightforward computation the identity

$$(h \circ g^{-1})^{\flat}_{i\lambda,\delta}(s) = |-cs + a|_{\delta}^{-1-i\lambda} h^{\flat}_{i\lambda,\delta}\left(\frac{ds - b}{-cs + a}\right), \tag{1.2.17}$$

which indicates the way the representation $(g, h) \mapsto h \circ g^{-1}$ restricts to functions of a given parity and degree of homogeneity. It is customary to set

$$(\pi_{i\lambda,\delta}(g)u)(s) = |-cs + a|_{\delta}^{-1-i\lambda} u\left(\frac{ds - b}{-cs + a}\right), \tag{1.2.18}$$

defining in this way the two principal series (with $\delta = 0$ or 1) of unitary representations of $SL(2,\mathbb{R})$. Only the one with $\delta = 0$ is of so-called class one, i.e., for every real λ, there is a function in $L^2(\mathbb{R})$ invariant under all transformations $\pi_{i\lambda,0}(g)$ with $g \in SO(2)$. Note that the identity $(\mathcal{F}^{\text{symp}}h)_{-i\lambda,\delta} = \mathcal{F}^{\text{symp}}h_{i\lambda,\delta}$, immediate from (1.2.11) and the fact that the (symplectic) Fourier transform of a function homogeneous of degree $-1 - i\lambda$ is homogeneous of degree $-1 + i\lambda$, explains the well-known fact that the representations $\pi_{i\lambda,\delta}$ and $\pi_{-i\lambda,\delta}$ are unitarily equivalent.

It is then clear that a composition formula absolutely distinct from the Moyal expansion (1.2.5) or its variants exists: it is the one that should express the components of given parities and degrees of homogeneity of a composition $h = h_1 \# h_2$ explicit according to a formula such as

$$h^{\flat}_{i\lambda,\delta}(s) = \frac{1}{4\pi} \sum_{\delta_1+\delta_2\equiv\delta} \int_{-\infty}^{\infty}\int_{-\infty}^{\infty} d\lambda_1 d\lambda_2$$

$$\times \int_{\mathbb{R}^2} K^{\delta_1,\delta_2;\delta}_{i\lambda_1,i\lambda_2;i\lambda}(s_1, s_2; s)(h_1)^{\flat}_{i\lambda_1,\delta_1}(s_1)(h_2)^{\flat}_{i\lambda_2,\delta_2}(s_2)ds_1 ds_2 : \tag{1.2.19}$$

the condition $\delta_1 + \delta_2 \equiv \delta \bmod 2$ imposes itself since an indicator such as δ just makes it clear whether the operator $\operatorname{Op}(h_{\delta})$ preserves or reverses the parity of functions it is applied to. If $g = \begin{pmatrix} a & b \\ c & d \end{pmatrix}$, the operator in (1.2.19) also gives the expression of $\pi_{i\lambda,\delta}(g^{-1})h_{i\lambda,\delta}^{\flat}$ in terms of the family of pairs $(\pi_{i\lambda_1,\delta_1}(g^{-1})(h_1)_{i\lambda_1,\delta_1}^{\flat}$, $\pi_{i\lambda_2,\delta_2}(g^{-1})(h_2)_{i\lambda_2,\delta_2}^{\flat})$: this is a consequence of (1.2.17) and of the covariance property (1.1.20) of the Weyl calculus. Evaluating in this way the function $\pi_{i\lambda,\delta}(g^{-1})h_{i\lambda,\delta}^{\flat}$ at the point $\frac{ds-b}{-cs+a}$ and performing on the right-hand side of the identity just alluded to the change of variables $(s_1, s_2) \mapsto \left(\frac{ds_1-b}{-cs_1+a}, \frac{ds_2-b}{-cs_2+a} \right)$, one sees with the help of (1.2.18) that the integral kernel $K_{i\lambda_1,i\lambda_2;i\lambda}^{\delta_1,\delta_2;\delta}$ must satisfy the identity

$$K_{i\lambda_1,i\lambda_2;i\lambda}^{\delta_1,\delta_2;\delta} \left(\frac{ds_1-b}{-cs_1+a}, \frac{ds_2-b}{-cs_2+a}; \frac{ds-b}{-cs+a} \right)$$
$$= |-cs_1+a|_{\delta_1}^{1-i\lambda_1}|-cs_2+a|_{\delta_2}^{1-i\lambda_2}|-cs+a|_{\delta}^{1+i\lambda} K_{i\lambda_1,i\lambda_2;i\lambda}^{\delta_1,\delta_2;\delta}(s_1, s_2; s). \quad (1.2.20)$$

Now, given $\delta_1, \delta_2, \delta$ and $\lambda_1, \lambda_2, \lambda$, there are only two linearly independent functions satisfying this required covariance identity, to wit the functions

$$\chi_{i\lambda_1,i\lambda_2;i\lambda}^{\varepsilon_1,\varepsilon_2;\varepsilon}(s_1, s_2; s) = |s_1-s_2|_{\varepsilon}^{\frac{-1+i(\lambda_1+\lambda_2+\lambda)}{2}} |s_2-s|_{\varepsilon_1}^{\frac{-1+i(-\lambda_1+\lambda_2-\lambda)}{2}} |s-s_1|_{\varepsilon_2}^{\frac{-1+i(\lambda_1-\lambda_2-\lambda)}{2}}$$

$$(1.2.21)$$

with

$$\varepsilon_1 \equiv j + \delta_1, \quad \varepsilon_2 \equiv j + \delta_2, \quad \varepsilon \equiv j + \delta \mod 2 \quad (1.2.22)$$

for $j = 0$ or 1. This is a consequence of the fact that $SL(2,\mathbb{R})$ acts in a transitive way (by means of fractional-linear transformations) on the set of triples (s_1, s_2, s) for which the sign of $\frac{s_1-s_2}{(s-s_1)(s_2-s)}$ is fixed.

The considerations which precede, though essentially heuristic since we have deliberately overlooked the fact that fractional-linear transformations do not act on the real line, only on its projective completion, nevertheless make the following theorem (the proof of which follows) an unsurprising one, except for the determination of the coefficients $C_{i\lambda_1,i\lambda_2;i\lambda}^{\varepsilon_1,\varepsilon_2;\varepsilon}$.

A shorthand will be handy in the statement of the theorem.

Definition 1.2.1. Given $\varepsilon = 0$ or 1, we set

$$B_\varepsilon(\mu) = (-i)^\varepsilon \pi^{\mu-\frac{1}{2}} \frac{\Gamma(\frac{1-\mu+\varepsilon}{2})}{\Gamma(\frac{\mu+\varepsilon}{2})}, \quad (1.2.23)$$

so that one has the identities (with $0 < \operatorname{Re}\mu < 1$ in the first one)

$$(\mathcal{F}(|s|_\varepsilon^{-\mu}))(\sigma) = B_\varepsilon(\mu)|\sigma|_\varepsilon^{\mu-1}, \quad B_\varepsilon(\mu)B_\varepsilon(1-\mu) = (-1)^\varepsilon. \quad (1.2.24)$$

Theorem 1.2.2. *Recall the definition* (1.2.21) *of the functions* $\chi_{i\lambda_1,i\lambda_2;i\lambda}^{\varepsilon_1,\varepsilon_2;\varepsilon}$. *Let* h_1 *and* h_2 *be two symbols in* $\mathcal{S}(\mathbb{R}^2)$. *Then,* $h_1 \# h_2$ *lies in* $\mathcal{S}(\mathbb{R}^2)$ *too and one has for* *every pair (λ, δ) the equation, in the weak sense in $L^2(\mathbb{R})$,*

$$h^b_{i\lambda,\delta}(s) = \frac{1}{4\pi} \sum_{\substack{\delta_1+\delta_2 \equiv \delta \\ \bmod 2}} \sum_{j=0,1} \int_{-\infty}^{\infty} \int_{-\infty}^{\infty} d\lambda_1 d\lambda_2 \tag{1.2.25}$$

$$\times \int_{\mathbb{R}^2} \left[K^{\delta_1,\delta_2;\delta}_{i\lambda_1,i\lambda_2;i\lambda} \right]_j (s_1, s_2; s)(h_1)^b_{i\lambda_1,\delta_1}(s_1)(h_2)^b_{i\lambda_2,\delta_2}(s_2) ds_1 ds_2,$$

where the integral kernel $\left[K^{\delta_1,\delta_2;\delta}_{i\lambda_1,i\lambda_2;i\lambda} \right]_j$ is given as

$$\left[K^{\delta_1,\delta_2;\delta}_{i\lambda_1,i\lambda_2;i\lambda} \right]_j (s_1, s_2; s) = C^{\varepsilon_1,\varepsilon_2;\varepsilon}_{i\lambda_1,i\lambda_2;i\lambda} X^{\varepsilon_1,\varepsilon_2;\varepsilon}_{i\lambda_1,i\lambda_2;i\lambda}(s_1, s_2; s) \tag{1.2.26}$$

with $\varepsilon_1, \varepsilon_2, \varepsilon = 0$ or 1,

$$\varepsilon \equiv j + \delta, \quad \varepsilon_1 \equiv j + \delta_1, \quad \varepsilon_2 \equiv j + \delta_2 \mod 2, \tag{1.2.27}$$

and

$$C^{\varepsilon_1,\varepsilon_2;\varepsilon}_{i\lambda_1,i\lambda_2;i\lambda} = (-1)^{\varepsilon_1+\varepsilon_2+\varepsilon} 2^{\frac{-1+i(\lambda_1+\lambda_2-\lambda)}{2}} B_{\varepsilon_1} \left(\frac{1+i(-\lambda_1+\lambda_2-\lambda)}{2} \right)$$

$$\times B_{\varepsilon_2} \left(\frac{1+i(\lambda_1-\lambda_2-\lambda)}{2} \right) B_\varepsilon \left(\frac{1+i(\lambda_1+\lambda_2+\lambda)}{2} \right). \tag{1.2.28}$$

Proof. That the sharp product $h_1 \# h_2$ makes sense and lies in $\mathcal{S}(\mathbb{R}^2)$ is a consequence of the fact that functions in this space are exactly the symbols of continuous linear operators from $\mathcal{S}'(\mathbb{R})$ to $\mathcal{S}(\mathbb{R})$. That, given λ_1, λ_2, the $ds_1 ds_2$ - integral on the right-hand side of (1.2.25) becomes convergent when tested against any function $u = u(s) \in L^2(\mathbb{R})$ is an elementary matter in view of estimates such as

$$\left| h^b_{i\lambda,\delta}(s) \right| \le C(1+s^2)^{-\frac{1}{2}}, \tag{1.2.29}$$

a consequence of (1.2.13) and (1.2.12) in the case when $h \in \mathcal{S}(\mathbb{R}^2)$: details can be found in [61, Lemma 17.4] if desired. The $d\lambda_1 d\lambda_2$ - integration creates no supplementary problem since arbitrary powers of λ_1, λ_2 can be saved with the help of relations such as $(2i\pi \mathcal{E}h_1)_{i\lambda_1,\delta_1} = -i\lambda_1(h_1)_{i\lambda_1,\delta_1}$.

We now break the symmetry between h_1 and h_2, using a $\mathcal{G}$-transformation and polar coordinates in the plane to write h_1 as an integral superposition

$$h_1(x_1, \xi_1) = \int_0^{2\pi} h^\theta_1(x_1 \sin\theta - \xi_1 \cos\theta) d\theta \tag{1.2.30}$$

with

$$h^\theta_1(x_1) = 2 \int_0^\infty (\mathcal{G}h_1)(r\cos\theta, r\sin\theta) e^{4i\pi r x_1} r \, dr, \tag{1.2.31}$$

and we note the immediate fact that the functions $h_1^\theta(x_1)$ and $x_1 h_1^\theta(x_1)$ are uniformly bounded when $h_1 \in \mathcal{S}(\mathbb{R}^2)$. It suffices to establish the main formula (1.2.25) in the case when the function $h_1^\theta(x_1 \sin\theta - \xi_1 \cos\theta)$, for any fixed θ, has been substituted for h_1. Next, using the covariance of both sides of (1.2.25), it is no loss of generality to consider only the case when $\theta = \pi$, i.e., assume that $h_1(x_1, \xi_1) = h_1^o(\xi_1)$, a function any derivative of which is majorized by $C(1 + |\xi_1|)^{-1}$. Much simplification occurs already since, now, the function $(h_1)_{i\lambda_1, \delta_1}^\flat(s_1)$ is actually independent of s_1. Next, we decompose the function h_1^o into components of given degrees of homogeneity and parities, performing however a change of contour to write, with some a_1 such that $-1 < a_1 < 0$, the equation

$$h_1^o = \sum_{\delta_1 = 0,1} \frac{1}{i} \int_{\mathrm{Re}\,\nu_1 = a_1} (h_1^o)_{\nu_1, \delta_1}\, d\nu_1 \tag{1.2.32}$$

with

$$(h_1^o)_{\nu_1, \delta_1}(\xi_1) = \frac{1}{2\pi} \int_0^\infty t^{\nu_1} (h_1^o)_{\delta_1}(t\xi_1)\, dt, \quad \xi_1 \neq 0. \tag{1.2.33}$$

Finally, we are left with the problem of proving the main formula (1.2.25) in the case when $h_1(x_1, \xi_1) = |\xi_1|_{\delta_1}^{-1-\nu_1}$ for some ν_1 such that $-1 < \mathrm{Re}\,\nu_1 < 0$. In this case, h_1 coincides with $(h_1)_{\nu_1, \delta_1}$, so that $(h_1)_{\nu_1, \delta_1}^\flat(s_1) = 1$, and our task is to prove the equation, in which $\delta_2 \equiv \delta + \delta_1 \bmod 2$,

$$(|\xi|_{\delta_1}^{-1-\nu_1} \# h_2)_{i\lambda, \delta}^\flat(s)$$

$$= \frac{1}{4\pi} \sum_{j=0,1} \int_{-\infty}^\infty d\lambda_2 \int_{\mathbb{R}^2} \left[K_{\nu_1, i\lambda_2; i\lambda}^{\delta_1, \delta_2; \delta} \right]_j (s_1, s_2; s)(h_2)_{i\lambda_2, \delta_2}^\flat(s_2)\, ds_1 ds_2, \tag{1.2.34}$$

after we have noted that for generic values of the argument $(s_1, s_2; s)$, the function $\left[K_{\nu_1, i\lambda_2; i\lambda}^{\delta_1, \delta_2; \delta} \right]_j (s_1, s_2; s)$ extends as an entire function of $(\lambda_1, \lambda_2; \lambda)$.

We start from the equation

$$\xi \# h = \left(\xi + \frac{1}{4i\pi} \frac{\partial}{\partial x} \right) h. \tag{1.2.35}$$

Our problem involves the consideration of powers of the operator $\xi + \frac{1}{4i\pi}\frac{\partial}{\partial x}$, together with decompositions into (generalized) eigenfunctions of the Euler operator $2i\pi\mathcal{E}$. Now, these two operators do not commute, but they both fit within a 3-dimensional Lie algebra in view of the commutation relation

$$\left[2i\pi\mathcal{E}, \xi \pm \frac{1}{4i\pi}\frac{\partial}{\partial x} \right] = \xi \mp \frac{1}{4i\pi}\frac{\partial}{\partial x}. \tag{1.2.36}$$

Given $(a, b, c) \in \mathbb{R}^3$ with $a > 0$ and a function $h = h(x, \xi)$, set

$$(\pi(a, b, c)h)(x, \xi) = ah\left(a(x + \frac{b}{2}), a\xi \right) e^{2i\pi c\xi}. \tag{1.2.37}$$

As shown by a straightforward calculation, one has the identity

$$\pi(a_1, b_1, c_1)\pi(a, b, c) = \pi(aa_1, a_1^{-1}b + b_1, ca_1 + c_1); \tag{1.2.38}$$

on the other hand, one has the set of relations

$$\frac{1}{2i\pi} \frac{\partial}{\partial a}\Big|_{a=1} \pi(a, 0, 0) = \mathcal{E}, \quad \frac{1}{2i\pi} \frac{\partial}{\partial b}\Big|_{b=0} \pi(1, b, 0) = \frac{1}{4i\pi} \frac{\partial}{\partial x},$$

$$\frac{1}{2i\pi} \frac{\partial}{\partial c}\Big|_{c=0} \pi(1, 0, c) = (\xi), \tag{1.2.39}$$

where (ξ) denotes the operator of multiplication by ξ.

The set of points (a, b, c) with $a > 0$, given together with the composition rule $(a_1, b_1, c_1)(a, b, c) = (aa_1, a_1^{-1}b + b_1, ca_1 + c_1)$, is an unusual realization of the Poincaré group in $(1+1)$-dimensional spacetime (a fact mentioned only to satisfy a possible curiosity: it is more immediately apparent that is just the full covariance group of the so-called one-dimensional Fuchs calculus [59]), and each of the three operators involved in (1.2.36) is an infinitesimal operator of the representation π: we may then take advantage of the decomposition of the representation π into irreducible components.

This requires that a function $h \in \mathcal{S}(\mathbb{R}^2)$ should be decomposed in the open set where $\xi \neq 0$ according to the rule

$$h = \int_{-\infty}^{\infty} h^{[\beta]} d\beta \tag{1.2.40}$$

with

$$h^{[\beta]}(x, \xi) = f^{[\beta]}(\xi) \exp\left(2i\pi\beta \frac{x}{\xi}\right),$$

$$f^{[\beta]}(\xi) = \int_{-\infty}^{\infty} h(t\xi, \xi)e^{-2i\pi\beta t} dt. \tag{1.2.41}$$

The function h is thus characterized by the family of functions $(f^{[\beta]})_{\beta \neq 0}$ of one variable: we prefer using the rescaled version such that

$$(Mh)(\beta, s) = f^{[\beta]}(\gamma s) = \int_{-\infty}^{\infty} h(\gamma ts, \gamma s)e^{-2i\pi\beta t} dt, \tag{1.2.42}$$

where $\gamma > 0$ and $\kappa = \pm 1$ are defined by the equations

$$\gamma = \left|\frac{\beta}{2}\right|^{\frac{1}{2}}, \quad \kappa = \text{sign } \beta. \tag{1.2.43}$$

Starting from the equation

$$\int_{-\infty}^{\infty} |(Mh)(\beta, s)|^2 ds = \gamma^{-1} \int_{-\infty}^{\infty} \left|\int_{-\infty}^{\infty} h(ts, s)e^{-2i\pi\beta t} dt\right|^2 ds, \tag{1.2.44}$$

one obtains the isometry identity

$$\int_{\mathbb{R}^2} \left| \frac{\beta}{2} \right|^{\frac{1}{2}} |(Mh)(\beta, s)|^2 d\beta ds = \int_{\mathbb{R}^2} |h(x, \xi)|^2 \frac{dx d\xi}{|\xi|}.$$

(1.2.45)

Under the map M, the operators

$$(\xi), \quad \frac{1}{4i\pi} \frac{\partial}{\partial x}, \quad \mathcal{E}$$

(1.2.46)

transfer to the operators

$$(\gamma s), \quad \left(\frac{\kappa \gamma}{s} \right), \quad s\frac{d}{ds} + 1.$$

(1.2.47)

For every polynomial $P(\xi)$, one has for every symbol h the identity $P(\xi)\#h = P\left(\xi + \frac{1}{4i\pi}\frac{\partial}{\partial x}\right) h$. Using first the continuous functional calculus of self-adjoint operators in the two Hilbert spaces involved in (1.2.45), next analytic continuation, one obtains that, under M, the sharp composition by the symbol $|\xi|_{\delta_1}^{-1-\nu_1}$ transfers to the operator $\left| \gamma s + \frac{\kappa \gamma}{s} \right|_{\delta_1}^{-1-\nu_1}$.

We remark now that the M-transform of h is related to the Fourier transform of the function $h_{i\lambda,\delta}^\flat$ by the pair of reciprocal equations

$$(Mh)(\beta, s) = \sum_{\delta=0,1} \int_{-\infty}^{\infty} |\gamma s|_\delta^{-1-i\lambda} \widehat{h_{i\lambda,\delta}^\flat}(\beta) d\lambda,$$

$$\widehat{h_{i\lambda,\delta}^\flat}(\beta) = \frac{1}{4\pi} \gamma^{1+i\lambda} \int_{-\infty}^{\infty} |s|_\delta^{i\lambda} (Mh)(\beta, s) ds.$$

(1.2.48)

The equation $h = |\xi|_{\delta_1}^{-1-\nu_1} \# h_2$ thus implies the equation

$$\widehat{h_{i\lambda,\delta}^\flat}(\beta) = \frac{1}{4\pi} \gamma^{1+i\lambda} \int_{-\infty}^{\infty} |s|_\delta^{i\lambda} (Mh)(\beta, s) |\gamma(s + \kappa s^{-1})|_{\delta_1}^{-1-\nu_1} ds$$

(1.2.49)

$$= \frac{1}{4\pi} \gamma^{1+i\lambda} \int_{-\infty}^{\infty} |s|_\delta^{i\lambda} |\gamma(s + \kappa s^{-1})|_{\delta_1}^{-1-\nu_1} ds \int_{-\infty}^{\infty} |\gamma s|_{\delta_2}^{-1-i\lambda_2} \widehat{(h_2)_{i\lambda_2,\delta_2}^\flat}(\beta) d\lambda_2,$$

where we recall from (1.2.19) that the value of $\delta_2 = 0$ or 1 is determined by the condition $\delta \equiv \delta_1 + \delta_2 \mod 2$. Hence,

$$\widehat{h_{i\lambda,\delta}^\flat}(\beta) = \frac{1}{4\pi} \int_{-\infty}^{\infty} \left| \frac{\beta}{2} \right|^{\frac{-1-\nu_1+i(\lambda-\lambda_2)}{2}} I_\kappa(i(\lambda - \lambda_2)) \widehat{(h_2)_{i\lambda_2,\delta_2}^\flat}(\beta) d\lambda_2$$

(1.2.50)

with

$$I_\kappa(\mu) = \int_{-\infty}^{\infty} |s|_{\delta_1}^{-1+\mu} |s + \kappa s^{-1}|_{\delta_1}^{-1-\nu_1} ds$$

(1.2.51)

(recall that (δ_1, ν_1) is fixed in our present computation).

Lemma 1.2.3. *Defining $\varepsilon_1 = 0$ or 1 by the condition $j + \delta_1 \equiv \varepsilon_1$ mod 2, one has, recalling that $\kappa = \pm 1$,*

$$I_\kappa(\mu) = \frac{1}{2}B_{\delta_1}(1 + \nu_1) \sum_{j=0,1} \kappa^j B_{\varepsilon_1}\left(\frac{1 - \nu_1 - \mu}{2}\right) B_j\left(\frac{1 - \nu_1 + \mu}{2}\right). \tag{1.2.52}$$

Proof. The proof of the lemma degenerates to high-school trigonometry and repeated applications of the formula of complements and of the duplication formula of the Gamma function. On one hand,

$$I_1(\mu) = 2\int_0^\infty s^{\nu_1 + \mu}(1 + s^2)^{-1-\nu_1} ds$$

$$= \frac{\Gamma\left(\frac{1+\nu_1+\mu}{2}\right)\Gamma\left(\frac{1+\nu_1-\mu}{2}\right)}{\Gamma(1 + \nu_1)}$$

$$= \frac{\Gamma(-\nu_1)\Gamma\left(\frac{1+\nu_1+\mu}{2}\right)\Gamma\left(\frac{1+\nu_1-\mu}{2}\right)}{\pi} \times (-\sin \pi\nu_1). \tag{1.2.53}$$

In the same way, splitting the integral at $s = 1$, one obtains

$$I_{-1}(\mu) = \frac{\Gamma(-\nu_1)\Gamma\left(\frac{1+\nu_1-\mu}{2}\right)}{\Gamma\left(\frac{1-\nu_1-\mu}{2}\right)} + (-1)^{\delta_1}\frac{\Gamma(-\nu_1)\Gamma\left(\frac{1+\nu_1+\mu}{2}\right)}{\Gamma\left(\frac{1-\nu_1+\mu}{2}\right)}. \tag{1.2.54}$$

One has

$$I_{-1}(\mu) = \pi\frac{\Gamma(-\nu_1)}{\Gamma\left(\frac{1-\nu_1-\mu}{2}\right)\Gamma\left(\frac{1-\nu_1+\mu}{2}\right)}\left[\frac{1}{\cos \pi\frac{\nu_1-\mu}{2}} + (-1)^{\delta_1}\frac{1}{\cos \pi\frac{\nu_1+\mu}{2}}\right] \tag{1.2.55}$$

$$= \frac{\Gamma(-\nu_1)\Gamma\left(\frac{1+\nu_1+\mu}{2}\right)\Gamma\left(\frac{1+\nu_1-\mu}{2}\right)}{\pi} \times \begin{cases} 2\cos\frac{\pi\nu_1}{2}\cos\frac{\pi\mu}{2} & \text{if } \delta_1 = 0, \\ -2\sin\frac{\pi\nu_1}{2}\sin\frac{\pi\mu}{2} & \text{if } \delta_1 = 1, \end{cases}$$

an expression which compares easily to (1.2.53). It is now convenient to write

$$I_\kappa(\mu) = \sum_{j=0,1} \kappa^j K_j(\mu), \tag{1.2.56}$$

with

$$K_j(\mu) = \frac{1}{2}\left[I_1(\mu) + (-1)^j I_{-1}(\mu)\right]$$

$$= \frac{1}{2}I_1(\mu) \times \begin{cases} \dfrac{\sin\frac{\pi\nu_1}{2} + (-1)^{j+1}\cos\frac{\pi\mu}{2}}{\sin\frac{\pi\nu_1}{2}} & \text{if } \delta_1 = 0, \\[3mm] \dfrac{\cos\frac{\pi\nu_1}{2} + (-1)^j \sin\frac{\pi\mu}{2}}{\cos\frac{\pi\nu_1}{2}} & \text{if } \delta_1 = 1. \end{cases} \tag{1.2.57}$$

We use the relations

$$\sin \frac{\pi \nu_1}{2} - \cos \frac{\pi \mu}{2} = -2 \cos \frac{\pi(1 + \nu_1 - \mu)}{4} \cos \frac{\pi(1 + \nu_1 + \mu)}{4},$$

$$\cos \frac{\pi \nu_1}{2} + \sin \frac{\pi \mu}{2} = 2 \cos \frac{\pi(1 + \nu_1 - \mu)}{4} \sin \frac{\pi(1 + \nu_1 + \mu)}{4},$$

$$\sin \frac{\pi \nu_1}{2} + \cos \frac{\pi \mu}{2} = 2 \sin \frac{\pi(1 + \nu_1 - \mu)}{4} \sin \frac{\pi(1 + \nu_1 + \mu)}{4},$$

$$\cos \frac{\pi \nu_1}{2} - \sin \frac{\pi \mu}{2} = 2 \sin \frac{\pi(1 + \nu_1 - \mu)}{4} \cos \frac{\pi(1 + \nu_1 + \mu)}{4}, \qquad (1.2.58)$$

finding as a result

$$I_\kappa(\mu) = \frac{2}{\pi} \sum_{j=0,1} \kappa^j \Gamma(-\nu_1) \Gamma\left(\frac{1 + \nu_1 + \mu}{2}\right) \Gamma\left(\frac{1 + \nu_1 - \mu}{2}\right)$$

$$\times \begin{cases} \cos \frac{\pi \nu_1}{2} \cos \frac{\pi(1+\nu_1-\mu)}{4} \cos \frac{\pi(1+\nu_1+\mu)}{4} & \text{if } j = 0, \delta_1 = 0, \\ -\sin \frac{\pi \nu_1}{2} \cos \frac{\pi(1+\nu_1-\mu)}{4} \sin \frac{\pi(1+\nu_1+\mu)}{4} & \text{if } j = 0, \delta_1 = 1, \\ -\cos \frac{\pi \nu_1}{2} \sin \frac{\pi(1+\nu_1-\mu)}{4} \sin \frac{\pi(1+\nu_1+\mu)}{4} & \text{if } j = 1, \delta_1 = 0, \\ -\sin \frac{\pi \nu_1}{2} \sin \frac{\pi(1+\nu_1-\mu)}{4} \cos \frac{\pi(1+\nu_1+\mu)}{4} & \text{if } j = 1, \delta_1 = 1. \end{cases} \qquad (1.2.59)$$

Using repeatedly the pair of identities

$$\Gamma(\tau) \cos \frac{\pi \tau}{2} = \frac{1}{2} (2\pi)^\tau B_0(1 - \tau),$$

$$\Gamma(\tau) \sin \frac{\pi \tau}{2} = \frac{i}{2} (2\pi)^\tau B_1(1 - \tau), \qquad (1.2.60)$$

one obtains the lemma.   □

End of proof of Theorem 1.2.2. Let us compute the right-hand side of (1.2.34), an expression temporarily denoted as $(h^{\text{right}})^b_{i\lambda,\delta}(s)$: what we have to do is to show that it coincides with $h^b_{i\lambda,\delta}(s)$. We start from the computation of the (convergent) integral

$$\int_{-\infty}^{\infty} \chi^{\varepsilon_1,\varepsilon_2;\varepsilon}_{\nu_1,i\lambda_2;i\lambda}(s_1, s_2; s)ds_1 \qquad (1.2.61)$$

$$= |s_2 - s|_{\varepsilon_1}^{\frac{-1-\nu_1-i(\lambda-\lambda_2)}{2}} \left(|t|_{\varepsilon_2}^{\frac{-1+\nu_1-i(\lambda+\lambda_2)}{2}} * |t|_{\varepsilon}^{\frac{-1+\nu_1+i(\lambda+\lambda_2)}{2}} \right) (t = s - s_2) :$$

with the help of (1.2.24), this can be written as

$$|s_2 - s|_{\varepsilon_1}^{\frac{-1-\nu_1-i(\lambda-\lambda_2)}{2}} B_{\varepsilon_2}\left(\frac{1-\nu_1+i(\lambda+\lambda_2)}{2}\right) B_\varepsilon\left(\frac{1-\nu_1-i(\lambda+\lambda_2)}{2}\right) B_{\delta_1}(-\nu_1) |s - s_2|_{\delta_1}^{\nu_1}$$

$$= (-1)^{\delta_1} B_{\delta_1}(1 + \nu_1) B_{\varepsilon_2}\left(\frac{1-\nu_1+i(\lambda+\lambda_2)}{2}\right) B_\varepsilon\left(\frac{1-\nu_1-i(\lambda+\lambda_2)}{2}\right) |s_2 - s|_j^{\frac{-1+\nu_1-i(\lambda-\lambda_2)}{2}}.$$

$$\qquad (1.2.62)$$

Then,

$$(h^{\mathrm{right}})^\flat_{i\lambda,\delta}(s) = \frac{(-1)^{\delta_1}}{4\pi} \sum_{j=0,1} B_{\delta_1}(1+\nu_1) \int_{-\infty}^{\infty} C^{\varepsilon_1,\varepsilon_2;\varepsilon}_{\nu_1,i\lambda_2;i\lambda} B_{\varepsilon_2}\left(\frac{1-\nu_1+i(\lambda+\lambda_2)}{2}\right)$$

$$\times B_\varepsilon\left(\frac{1-\nu_1-i(\lambda+\lambda_2)}{2}\right) d\lambda_2 \int_{-\infty}^{\infty} |s_2-s|_j^{\frac{-1+\nu_1-i(\lambda-\lambda_2)}{2}} (h_2)^\flat_{i\lambda_2,\delta_2}(s_2) ds_2.$$

$$(1.2.63)$$

Again, the integral on the right of the right-hand side defines a function of s which is a convolution, and the Fourier transform of this function is the function

$$\beta \mapsto (-1)^j B_j\left(\frac{1-\nu_1+i(\lambda-\lambda_2)}{2}\right) |\beta|_j^{\frac{-1+\nu_1-i(\lambda-\lambda_2)}{2}} \widehat{(h_2)^\flat_{i\lambda_2,\delta_2}}(\beta). \qquad (1.2.64)$$

Finally,

$$\widehat{(h^{\mathrm{right}})^\flat_{i\lambda,\delta}}(\beta) = \frac{1}{4\pi} \sum_{j=0,1} (-1)^{\varepsilon_1} B_{\delta_1}(1+\nu_1) \int_{-\infty}^{\infty} B_j\left(\tfrac{1-\nu_1+i(\lambda-\lambda_2)}{2}\right) |\beta|_j^{\frac{-1-\nu_1+i(\lambda-\lambda_2)}{2}}$$

$$\times C^{\varepsilon_1,\varepsilon_2;\varepsilon}_{\nu_1,i\lambda_2;i\lambda} B_{\varepsilon_2}\left(\frac{1-\nu_1+i(\lambda+\lambda_2)}{2}\right) B_\varepsilon\left(\frac{1-\nu_1-i(\lambda+\lambda_2)}{2}\right) \widehat{(h_2)^\flat_{i\lambda_2,\delta_2}}(\beta) d\lambda_2.$$

$$(1.2.65)$$

This must be shown to be identical to the expression (1.2.50), rewritten with the help of Lemma 1.2.3 as

$$\widehat{h^\flat_{i\lambda,\delta}}(\beta) = \frac{1}{8\pi} B_{\delta_1}(1+\nu_1) \sum_{j=0,1} (\operatorname{sign}\beta)^j \int_{-\infty}^{\infty} \left|\frac{\beta}{2}\right|^{\frac{-1-\nu_1+i(\lambda-\lambda_2)}{2}}$$

$$\times B_{\varepsilon_1}\left(\frac{1-\nu_1-i(\lambda-\lambda_2)}{2}\right) B_j\left(\frac{1-\nu_1+i(\lambda-\lambda_2)}{2}\right) \widehat{(h_2)^\flat_{i\lambda_2,\delta_2}}(\beta) d\lambda_2. \quad (1.2.66)$$

It is indeed the case if

$$C^{\varepsilon_1,\varepsilon_2;\varepsilon}_{\nu_1,i\lambda_2;i\lambda} = (-1)^{\varepsilon_1} 2^{\frac{1+\nu_1-i(\lambda-\lambda_2)}{2}} \frac{B_{\varepsilon_1}\left(\frac{1-\nu_1-i(\lambda-\lambda_2)}{2}\right)}{B_{\varepsilon_2}\left(\frac{1-\nu_1+i(\lambda+\lambda_2)}{2}\right) B_\varepsilon\left(\frac{1-\nu_1-i(\lambda+\lambda_2)}{2}\right)},$$

$$(1.2.67)$$

an expression identical to the one obtained from (1.2.28) in view of (1.2.24). $\square$

Remarks 1.2.a (i) A detailed proof of Theorem 1.2.2 appeared in [61, Section 17]: our present proof is a lot shorter. We had started there from a decomposition of the special sharp product $|x|_{\delta_1}^{-1-\nu_1} \# |\xi|_{\delta_2}^{-1-\nu_2}$ as an explicit integral superposition of functions

$$(x,\xi) \mapsto |x|_\varepsilon^{\frac{-1-\nu_1+\nu_2-i\lambda}{2}} |\xi|_{\varepsilon'}^{\frac{-1+\nu_1-\nu_2-i\lambda}{2}} \qquad (1.2.68)$$

and from the decomposition, provided by (1.2.30), (1.2.31), of h_1 or h_2 as a super-position of symbols, each one depending only on some special linear combination of x and ξ. Now, two linear forms on $\mathbb{R}^2$ are generically transversal: if this condition is satisfied, there is a linear transformation in $SL(2, \mathbb{R})$ which takes these two linear forms to x and ξ respectively. There was, however, one serious difficulty (dealt with in [61, p. 204-209]), originating from the fact that "bad factors" occur when the two linear forms are close to being proportional. This difficulty has disappeared from the present proof, in which the decomposition (1.2.30) has been used in connection with the first symbol h_1 only.

(ii) Theorem 1.2.2 and either proof of it extend to the case of the Weyl calculus in n-dimensional space without significant differences. Only, symbols must in this case be decomposed as integral superpositions of functions homogeneous of degrees $-n - i\lambda$ rather than $-1 - i\lambda$ and a factor such as $B_\varepsilon \left(\frac{1+i(\lambda_1+\lambda_2+\lambda)}{2} \right)$ on the right-hand side of (1.2.28) must be replaced by $B_\varepsilon \left(\frac{2-n+i(\lambda_1+\lambda_2+\lambda)}{2} \right)$: details were given in [27]. Please note the unfortunate typo in the main formula (6.1) and in (6.17) there: the denominator 2 in the arguments of all Gamma factors ought to be replaced by 4 (just as in the one-dimensional case), as indeed it follows from (6.16) and the fact that the expression $c(\rho, \delta)$ there was the same as our present $B_\delta(-\rho)$.

(iii) Theorem 1.2.2 also has an extension to the case of pseudo-differential analysis of operators acting on *complex-valued* functions defined on local fields, such as $\mathbb{Q}_p$. This generalization was made by Bechata [1]. It is important to note that, in the p-adic case, the Moyal expansion (1.2.5) does not exist, since there are no differential operators (or polynomials) in such a setting.

(iv) There has been much recent interest, in connection with arithmetic, in triple products, i.e., trilinear forms defined in terms of the integral kernels $(s_1, s_2; s) \mapsto \chi_{i\lambda_1, i\lambda_2; i\lambda}^{\varepsilon_1, \varepsilon_2; \varepsilon}(s_1, s_2; s)$, or in related triple products of L-functions [37, 20] or [65] (a reference taken from [24, p. 722], unavailable to us). Propositions 2.1.3 and 2.1.4 will connect, by means of a Radon transformation, operations of the first species to the pointwise product and the Poisson bracket of functions in the hyperbolic half-plane. In [61] (and [60] for a preparation), these triple products were given a role analogous to the one presented in this chapter, but in the automorphic situation: up to some point, they were shown to be associated to a corresponding notion of triple product of L-functions, in Section 16 of [61]. Some of the main points of the automorphic Weyl calculus will be very briefly presented in Section 3.4.

1.3   The totally radial Weyl calculus

We make the tacit assumption that $n \geq 2$ in all this section (see, however, Remark 1.3.a.(ii) at the end of the chapter): we also set $\tau = \frac{n-2}{2}$ as we shall sometimes replace this half-integer by any real number > -1. A radial function $u = u(x)$ on $\mathbb{R}^n$ is a function $U(|x|)$ of the canonical length of x, in other words a function on $\mathbb{R}^n$ invariant under the linear changes of coordinates associated to matrices in $SO(n)$. Such a function can be identified with the function Ru on the half-line such that

$$(Ru)(t) = \omega_n (2t)^{\frac{n-2}{2}} U(\sqrt{2t}), \qquad (1.3.1)$$

where $\omega_n = \frac{2\pi^{\frac{n}{2}}}{\Gamma(\frac{n}{2})}$ is the superficial measure of the unit sphere of $\mathbb{R}^n$. Note the relation

$$\|u\|_{L^2(\mathbb{R}^n)}^2 = \omega_n \int_0^\infty s^{n-1} |U(s)|^2 ds$$

$$= \frac{\Gamma(\frac{n}{2})}{(2\pi)^{\frac{n}{2}}} \int_0^\infty t^{\frac{2-n}{2}} |(Ru)(t)|^2 dt. \qquad (1.3.2)$$

The n-dimensional metaplectic representation in $L^2(\mathbb{R}^n)$ of (the twofold cover of) the symplectic group $Sp(n, \mathbb{R})$ as made explicit (on generators, up to phase factors ± 1) in Section 1.1, does not act within the subspace of radial functions. However, if one replaces $Sp(n, \mathbb{R})$ by the image of $SL(2, \mathbb{R}) = Sp(1, \mathbb{R})$ under the map $\left(\begin{smallmatrix} a & b \\ c & d \end{smallmatrix}\right) \mapsto \left(\begin{smallmatrix} aI & bI \\ cI & dI \end{smallmatrix}\right)$ from $SL(2, \mathbb{R})$ to $Sp(n, \mathbb{R})$, one obtains a representation $\mathrm{Met}^{(n)}$ in the space $L^2_{\mathrm{rad}}(\mathbb{R}^n)$. This representation is a genuine representation of $SL(2, \mathbb{R})$ in the case when n is even, defined on generators of $SL(2, \mathbb{R})$ by the following table:

$$\left(\mathrm{Met}^{(n)}\left(\left(\begin{smallmatrix} 1 & 0 \\ c & 1 \end{smallmatrix}\right)\right) u\right)(x) = u(x) e^{i\pi c |x|^2},$$

$$\mathrm{Met}^{(n)}\left(\left(\begin{smallmatrix} 0 & 1 \\ -1 & 0 \end{smallmatrix}\right)\right) u = e^{-\frac{i\pi n}{4}} \mathcal{F}^{\mathrm{euc}} u,$$

$$\left(\mathrm{Met}^{(n)}\left(\left(\begin{smallmatrix} a & 0 \\ 0 & a^{-1} \end{smallmatrix}\right)\right) u\right)(x) = a^{-\frac{n}{2}} u(a^{-1} x), \quad a > 0, \qquad (1.3.3)$$

where $\mathcal{F}^{\mathrm{euc}}$ is the Euclidean (usual) Fourier transformation on $\mathbb{R}^n$. When n is odd, the representation $\mathrm{Met}^{(n)}$ is, as in the one-dimensional case, a representation of the twofold cover of $SL(2, \mathbb{R})$; if preferred, one may consider it as a projective representation of $SL(2, \mathbb{R})$, in this case a representation up to indeterminacy factors ± 1.

Operators from the n-dimensional Weyl calculus commuting with linear changes of coordinates associated to matrices in $SO(n)$ can be characterized by the following property of their symbols $\mathfrak{S}$: these should remain invariant under the changes $(x, \xi) \mapsto (\omega x, \omega \xi)$ for every $\omega \in SO(n)$. Note that, even though such an operator indeed preserves the space of radial functions, it cannot be identified, in general, with its restriction to radial functions: we shall come back to this question later. Meanwhile, let us just note the covariance identity

$$\mathrm{Met}^{(n)}\left(\left(\begin{smallmatrix} a & b \\ c & d \end{smallmatrix}\right)\right) \mathrm{Op}(\mathfrak{S}) \mathrm{Met}^{(n)}\left(\left(\begin{smallmatrix} a & b \\ c & d \end{smallmatrix}\right)\right)^{-1} = \mathrm{Op}\left(\mathfrak{S} \circ \left(\begin{smallmatrix} aI & bI \\ cI & dI \end{smallmatrix}\right)^{-1}\right), \qquad (1.3.4)$$

involving operators on radial functions only.

From Pukanszky's list [40] of unitary representations of the universal covering group of $SL(2,\mathbb{R})$, let us extract, for $\tau > 0$, the representation $\pi_{\tau+1}$ characterized up to indeterminacy factors in the group $\exp(2i\pi\tau\mathbb{Z})$ by the equation

$$\left(\pi_{\tau+1}\left(\left(\begin{smallmatrix} a & b \\ c & d \end{smallmatrix}\right)\right) f\right)(z) = (-cz + a)^{-\tau-1} f\left(\frac{dz - b}{-cz + a}\right). \tag{1.3.5}$$

This is a unitary representation in the Hilbert space $\widetilde{H}_{\tau+1}$ of holomorphic functions f in Π such that

$$\|f\|^2 = \int_\Pi |f(z)|^2 (\mathrm{Im}\, z)^{\tau+1} dm(z) < \infty. \tag{1.3.6}$$

When $\tau = 1, 2, \ldots$, one gets a genuine representation of $SL(2,\mathbb{R})$, and the corresponding sequence is called the holomorphic discrete series of representations of this group [26].

Consider the map $v \mapsto f$ from the space $H_{\tau+1} = L^2((0,\infty); t^{-\tau} dt)$ (compare to the right-hand side of (1.3.2) with $\tau = \frac{n-2}{2}$) to $\widetilde{H}_{\tau+1}$, essentially a Laplace transformation, defined as follows:

$$f(z) = \frac{(4\pi)^{\frac{\tau}{2}}}{(\Gamma(\tau))^{\frac{1}{2}}} z^{-\tau-1} \int_0^\infty v(t) e^{-2i\pi tz^{-1}} dt. \tag{1.3.7}$$

It is an isometry, and it intertwines the representation $\pi_{\tau+1}$ with the representation $\mathcal{D}_{\tau+1}$ in $H_{\tau+1}$ characterized, again up to indeterminacy factors in the group $\exp(2i\pi\tau\mathbb{Z})$, by the following two facts:

(i) if $b = 0$, $a > 0$, $(\mathcal{D}_{\tau+1}(g)v)(t) = a^{\tau-1} v(a^{-2}t) e^{2i\pi \frac{c}{a} t}$;

(ii) if $b > 0$ and $v \in C_0^\infty(]0, \infty[)$,

$$(\mathcal{D}_{\tau+1}(g)v)(s) = e^{-i\pi\frac{\tau+1}{2}} \frac{2\pi}{b} \int_0^\infty v(t) \left(\frac{s}{t}\right)^{\frac{\tau}{2}} \exp(2i\pi\frac{ds + at}{b}) J_\tau\left(\frac{4\pi}{b}\sqrt{st}\right) dt. \tag{1.3.8}$$

The verification of (1.3.8), taken from [58], p. 91, goes as follows. Up to an unspecified factor in the group $\exp(2i\pi\tau\mathbb{Z})$, one has

$$(\pi_{\tau+1}(g)f)\left(-\frac{1}{z}\right) = \left(-\frac{1}{z}\right)^{-\tau-1} (bz + d)^{-\tau-1} \left(\frac{bz + d}{-az - c}\right)^{\tau+1} f\left(\frac{bz + d}{-az - c}\right). \tag{1.3.9}$$

If f is linked to v by (1.3.7) and $b > 0$, one thus has

$$(\pi_{\tau+1}(g)f)\left(-\frac{1}{z}\right) = \frac{(4\pi)^{\frac{\tau}{2}}}{(\Gamma(\tau))^{\frac{1}{2}}} \left(-\frac{1}{z}\right)^{-\tau-1} (bz + d)^{-\tau-1} \int_0^\infty v(t) e^{2i\pi t \frac{az+c}{bz+d}} dt \tag{1.3.10}$$

and, in view of the inversion formula

$$v(s) = \frac{(\Gamma(\tau))^{\frac{1}{2}}}{(4\pi)^{\frac{\tau}{2}}} \int_{i\varepsilon-\infty}^{i\varepsilon+\infty} (-\frac{1}{z})^{\tau+1} f(-\frac{1}{z}) e^{-2i\pi sz} dz, \tag{1.3.11}$$

one obtains

$$(\mathcal{D}_{\tau+1}(g)v)(s) = \int_{i\varepsilon-\infty}^{i\varepsilon+\infty} e^{-2i\pi sz}(bz+d)^{-\tau-1}dz \int_0^\infty v(t)e^{2i\pi t\frac{az+c}{bz+d}}dt. \tag{1.3.12}$$

After changing $bz + d$ to z, one sees that the integral kernel of the transformation $\mathcal{D}_{\tau+1}(g)$ is

$$k(s,t) = \frac{1}{b} e^{\frac{2i\pi}{b}(ds+at)} \int_{i\varepsilon-\infty}^{i\varepsilon+\infty} z^{-\tau-1} e^{-\frac{2i\pi}{b}(sz+\frac{t}{z})} dz. \tag{1.3.13}$$

Performing the change of variable $z \mapsto \frac{ib}{2\pi s} z$ in the integral and using the equation

$$J_\tau(2\rho) = \frac{1}{2i\pi} \rho^\tau \int_{\varepsilon-i\infty}^{\varepsilon+i\infty} e^{z-\frac{\rho^2}{z}} z^{-\tau-1} dz, \tag{1.3.14}$$

a consequence of [36, p. 83], one obtains the expression of the operator $\mathcal{D}_{\tau+1}(g)$ indicated in (1.3.8).

One of the advantages of the realization $\mathcal{D}_{\tau+1}$, as opposed to $\pi_{\tau+1}$, is that it makes it possible to replace the condition $\tau > 0$ by the condition $\tau > -1$ without changing the definition of the space $H_{\tau+1}$: on the contrary, when $\tau \leq 0$, the space $\widetilde{H}_{\tau+1}$ can no longer be defined by the condition (1.3.6). Note that, even though the representation obtained when $\tau \leq 0$ is still unitary, it ceases to be square-integrable. Also, the representation $\mathrm{Met}^{(n)}$ is immediately seen (with the help of the intertwining operator R in (1.3.1) from radial functions in $\mathbb{R}^n$ to functions on the half-line) to be equivalent to the representation $\mathcal{D}_{\tau+1}$ with $\tau = \frac{n-2}{2}$, in view of Hecke's (or Bochner's) formula $[\mathcal{F}^{\mathrm{euc}}(y \mapsto U(|y|))](x) = V(|x|)$, with

$$V(s) = 2\pi s^{\frac{2-n}{2}} \int_0^\infty U(t) t^{\frac{n}{2}} J_{\frac{n-2}{2}}(2\pi st) dt. \tag{1.3.15}$$

Number theorists are of course quite familiar with this since holomorphic modular forms of weight $\frac{n}{2}$ (a half-integral number only when n is odd), in other words — up to equivalence — objects transforming in specific ways under transformations $\pi_{\frac{n}{2}}(g)$ with $g \in SL(2,\mathbb{Z})$, can often be realized as theta series in several variables: this is especially useful in questions centering around the representation of numbers as sums of squares.

Still assuming $n \geq 2$, we wish to consider totally radial operators in $\mathbb{R}^n$, by which we mean operators on functions in $\mathbb{R}^n$ which commute with rotations and change every function into a radial one: in other words, such operators are not changed if multiplied on either side by the operator of averaging with respect to the action of rotations. For instance, this is the case with the operator of orthogonal projection on the ground state of the harmonic oscillator: its Weyl symbol is the function $(x, \xi) \mapsto 2\exp(-2\pi(|x|^2+|\xi|^2))$. Even though such operators do have $(2n)$-dimensional Weyl symbols, it is much more efficient, as will be seen, to represent symbols of operators commuting with rotations by functions of 3 variables only, and symbols of totally radial operators by functions of 3 variables satisfying some special differential equation.

Recall that the modified Bessel function $K_\nu(t)$, which is for $t > 0$ an entire function of ν, can be defined [36, p. 66] as the linear combination

$$K_\nu(t) = \frac{1}{2}\Gamma(\nu)\Gamma(1-\nu)\left[I_{-\nu}(t) - I_\nu(t)\right], \tag{1.3.16}$$

with

$$I_\nu(t) = \sum_{m=0}^{\infty} \frac{1}{m!\,\Gamma(\nu+m+1)}\left(\frac{t}{2}\right)^{\nu+2m}. \tag{1.3.17}$$

Among the solutions of the equation

$$u''(t) + \frac{1}{t}u'(t) - \left(1 + \frac{\nu^2}{t^2}\right)u(t) = 0, \quad t > 0, \tag{1.3.18}$$

it is characterized, up to multiplication by a constant, by the property of being a rapidly decreasing function of t as $t \to \infty$.

Theorem 1.3.1. *Given $x, \xi \in \mathbb{R}^n$, define the number $|x \wedge \xi| \geq 0$ by the equation*

$$|x \wedge \xi|^2 = \sum_{j<k}(x_j\xi_k - x_k\xi_j)^2 = |x|^2|\xi|^2 - (\langle x, \xi\rangle)^2. \tag{1.3.19}$$

To each function $f \in L^2(\Pi)$, associate the function $\theta f = (\theta f)(s, z)$ on $\mathbb{R}_+^\times \times \Pi$ defined by the spectral-theoretic equation

$$(\theta f)(s, z) = s^{-\frac{1}{2}}\left[K_{i\sqrt{\Delta-\frac{1}{4}}}(4\pi s)f\right](z), \tag{1.3.20}$$

where Δ is the hyperbolic Laplacian on Π. Then, the function Λf on $\mathbb{R}^n \times \mathbb{R}^n$ defined by the equation

$$(\Lambda f)(x, \xi) = (\theta f)\left(|x \wedge \xi|, -\frac{|x|^2}{\langle x, \xi\rangle + i|x \wedge \xi|}\right)$$

$$= (\theta f)\left(|x \wedge \xi|, \frac{-\langle x, \xi\rangle + i|x \wedge \xi|}{|\xi|^2}\right) \tag{1.3.21}$$

is the Weyl symbol of a totally radial operator. Moreover, the operator in question commutes with the harmonic oscillator if and only if, for $z \in \Pi$, $f(z)$ depends only on the hyperbolic distance from i to z.

On the other hand, the symbolic calculus $f \mapsto \mathrm{Op}(\Lambda f)$ of totally radial oper-ators in $L^2(\mathbb{R}^n)$ with symbols in Π is covariant under the representation $\mathrm{Met}^{(n)}$, in the sense that

$$\mathrm{Met}^{(n)}(g)\mathrm{Op}(\Lambda f)\mathrm{Met}^{(n)}(g)^{-1} = \mathrm{Op}(\Lambda(f \circ g^{-1})), \quad g \in SL(2, \mathbb{R}). \tag{1.3.22}$$

Proof. An operator $\mathrm{Op}(h)$ is totally radial if and only if, on one hand, it commutes with rotations, on the other hand, it satisfies the identities $(x_j D_k - x_k D_j)\mathrm{Op}(h) = 0$, with $D_k = \mathrm{Op}(\xi_k)$. Taking benefit from the covariance (1.1.20) of the n-dimensional Weyl calculus, one can express the first condition by the fact that the symbol h is invariant under the linear changes of coordinates associated to block-matrices $\left(\begin{smallmatrix} \omega & 0 \\ 0 & \omega'^{-1} \end{smallmatrix}\right) = \left(\begin{smallmatrix} \omega & 0 \\ 0 & \omega \end{smallmatrix}\right)$ with $\omega \in SO(n)$: in other words, one must have

$$h(x, \xi) = F\left(\frac{|x|^2 + |\xi|^2}{2}, \langle x, \xi \rangle, \frac{|x|^2 - |\xi|^2}{2}\right) \tag{1.3.23}$$

for some function $F = F(p, q, r)$ in the cone $\{(p, q, r): p > 0, p^2 - q^2 - r^2 > 0\}$.

Using the equations (1.2.6), one obtains that the second condition is equiva-lent to the fact that the symbol

$$x_j \xi_k h + \frac{1}{4i\pi}\left(x_j \frac{\partial h}{\partial x_k} - \xi_k \frac{\partial h}{\partial \xi_j}\right) + \frac{1}{16\pi^2}\frac{\partial^2 h}{\partial \xi_j \partial x_k} \tag{1.3.24}$$

is invariant under the change $(j, k) \mapsto (k, j)$. Expressing this in terms of the coordinates p, q, r, one obtains after a straightforward calculation, deleting terms satisfying already the symmetry condition demanded, that the symbol

$$x_j \xi_k \left[F + \frac{1}{16\pi^2}\frac{\partial^2 F}{\partial q^2}\right] + \frac{1}{16\pi^2}\xi_j x_k \left(\frac{\partial^2 F}{\partial p^2} - \frac{\partial^2 F}{\partial r^2}\right) \tag{1.3.25}$$

must be a symmetric function of (j, k). This is the case if and only if the Klein-Gordon equation

$$\frac{\partial^2 F}{\partial p^2} - \frac{\partial^2 F}{\partial q^2} - \frac{\partial^2 F}{\partial r^2} = 16\pi^2 F \tag{1.3.26}$$

is satisfied. Let us also note that the operator $\mathrm{Op}(h)$ will commute with the har-monic oscillator (the operator with symbol $\pi(|x|^2 + |\xi|^2)$) if and only if the Poisson bracket $\{|x|^2 + |\xi|^2, h\}$ is zero: such a simple characterization of a commutation condition is true whenever one at least of the two operators involved is an infinites-imal operator of the metaplectic representation, as it follows from the covariance property. In terms of the coordinates p, q, r, the condition reduces to

$$r\frac{\partial F}{\partial q} - q\frac{\partial F}{\partial r} = 0: \tag{1.3.27}$$

it expresses the fact that $F(p, q, r)$ is actually a function of $(p, q^2 + r^2)$.

No longer assuming this latter condition, we forget the original coordinates $x, \xi \in \mathbb{R}^n$ and introduce a point $z = x + iy \in \Pi$ together with a variable $s > 0$. The cone where (p, q, r) varies can be parametrized by the pair $(s, z) \in \mathbb{R}_+^\times \times \Pi$, by means of the reciprocal sets of equations

$$p = s\frac{1 + |z|^2}{2y}, \qquad q = s\frac{x}{y}, \qquad r = s\frac{|z|^2 - 1}{2y},$$

$$z = \frac{q + is}{p - r} = \frac{p + r}{q - is}. \tag{1.3.28}$$

As observed in [34, p. 10], one has

$$dp^2 - dq^2 - dr^2 = ds^2 - s^2\frac{dx^2 + dy^2}{y^2} \tag{1.3.29}$$

so that, under the new coordinates, the operator $\frac{\partial^2}{\partial p^2} - \frac{\partial^2}{\partial q^2} - \frac{\partial^2}{\partial r^2}$ becomes

$$\frac{\partial^2}{\partial s^2} + \frac{2}{s}\frac{\partial}{\partial s} - \frac{y^2}{s^2}\left(\frac{\partial^2}{\partial x^2} + \frac{\partial^2}{\partial y^2}\right) = \frac{\partial^2}{\partial s^2} + \frac{2}{s}\frac{\partial}{\partial s} + s^{-2}\Delta \tag{1.3.30}$$

if one introduces the hyperbolic Laplacian Δ on Π, an operator which certainly commutes with the operator of multiplication by the last coordinate $s = (p^2 - q^2 - r^2)^{\frac{1}{2}}$: note that, in terms of the coordinates x, ξ, one has $(p^2 - q^2 - r^2)^{\frac{1}{2}} = |x \wedge \xi|$.

The equation (1.3.26) has been transformed to

$$\frac{\partial^2 F}{\partial s^2} + \frac{2}{s}\frac{\partial F}{\partial s} + s^{-2}\Delta F = 16\pi^2 F. \tag{1.3.31}$$

To solve this equation, we separate the variables and use the spectral decomposition in terms of generalized eigenfunctions of the Laplacian: with $F(p, q, r) = G(s, z)$, the equation reduces in the case when $G(s, z) = u(s)f(z)$ for some generalized eigenfunction f of Δ for the eigenvalue $\frac{1 + \lambda^2}{4}$ to the equation

$$u''(s) + \frac{2}{s}u'(s) + s^{-2}\frac{1 + \lambda^2}{4}u = 16\pi^2 u, \tag{1.3.32}$$

a solution of which is the function

$$u(s) = s^{-\frac{1}{2}}K_{\frac{i\lambda}{2}}(4\pi s). \tag{1.3.33}$$

This function does not generate linearly all solutions of (1.3.32), but the other solutions are wildly increasing as $s \to \infty$.

This concludes the proof of the first part of Theorem 1.3.1. Let us recall that the fact that the operator with symbol h commutes with the harmonic oscillator has already been transformed to (1.3.27): now, when $s = (p^2 - q^2 - r^2)^{\frac{1}{2}}$ has been fixed, a function of the pair $(p, q^2 + r^2)$ is just the same as a function of p alone, or of $\frac{1+|z|^2}{2y} = \cosh d(i, z)$.

We finally address the question of proving the covariance relation (1.3.22). Since the totally radial Weyl calculus is covariant (1.3.4) under $\text{Met}^{(n)}$, it suffices to show that, when $(x, \xi) \in \mathbb{R}^n \times \mathbb{R}^n$ is changed to $(dx - b\xi, -cx + a\xi)$, $s = |x \wedge \xi|$ is unchanged and that z, as given in (1.3.28), is changed to $\frac{dz-b}{-cz+a}$. The first point follows from the fact that the linear transformation with matrix $\left(\begin{smallmatrix} aI & bI \\ cI & dI \end{smallmatrix} \right)$ is symplectic. Using the intermediary provided by the coordinates p, q, r, all three of which are linear combinations of $|x|^2, \langle x, \xi \rangle, |\xi|^2$, one sees, using the equation $s = \sqrt{p^2 - q^2 - r^2}$, that under the change of coordinates indicated,

$$p - r \longmapsto (p - r)a^2 - 2qac + (p + r)c^2$$
$$= (p - r)\left(a - \frac{q + is}{p - r}c \right)\left(a - \frac{q - is}{p - r}c \right) = (p - r)| - cz + a|^2. \quad (1.3.34)$$

On the other hand,

$$q + is \longmapsto -cd(p + r) + (ad + bc)q - ab(p - r) + is. \quad (1.3.35)$$

We now compute the expression

$$(p - r)(-c\bar{z} + a)(dz - b) = (p - r)\left[-cd|z|^2 + (ad + bc)\text{Re } z - ab + i\text{Im } z \right]$$
$$= (p - r)\left[-cd\frac{q^2 + s^2}{(p - r)^2} + (ad + bc)\frac{q}{p - r} - ab + i\frac{s}{p - r} \right]$$
$$= -cd(p + r) + (ad + bc)q - ab(p - r) + is : \quad (1.3.36)$$

comparing this to (1.3.35), and using (1.3.34) as well, we verify that, indeed, under the change of coordinates under examination, $z = \frac{q+is}{p-r}$ transforms to

$$\frac{(p - r)(-c\bar{z} + a)(dz - b)}{(p - r)| - cz + a|^2} = \frac{dz - b}{-cz + a}. \quad (1.3.37)$$

This concludes the proof of Theorem 1.3.1. □

Remarks 1.3.a. (i) Operators on $L^2(\mathbb{R}^n)$ which commute with the action of $SO(n)$ by linear changes of coordinates are generally not totally radial, and are not characterized by their restrictions to radial functions. Rather, they may be regarded as a collection of operators, parametrized by a linear basis (P_α) of solid spherical harmonics (homogeneous polynomials in the nullspace of the operator $\sum \frac{\partial^2}{\partial x_j^2}$), the operator associated with P_α being a linear endomorphism of a space of functions given as the products of P_α by radial functions. Now, Hecke's (or Bochner's) formula (1.3.15) extends to this case: if P_α is a solid spherical harmonic of degree k, the Euclidean Fourier transform of the product $U(|x|)P_\alpha(x)$ is (see for instance [50]) the product $V(|x|)P_\alpha(x)$, with

$$V(s) = 2\pi i^{-k} s^{\frac{2-n}{2}-k} \int_0^\infty U(t) t^{\frac{n}{2}+k} J_{\frac{n-2}{2}+k}(2\pi st)dt. \tag{1.3.38}$$

On the representation-theoretic level, the only difference lies in the shift $\tau = \frac{n-2}{2} \mapsto \tau + k$. No novelty could thus be expected from the consideration of functions of the kind $x \mapsto U(|x|)P_\alpha(x)$ in place of radial functions, or from the corresponding extension of the totally radial calculus.

(ii) Even though we discarded it at the beginning of this section, the one-dimensional case is not to be excluded: but, when $\tau = -\frac{1}{2}$, only the representation $\mathcal{D}_{\tau+1}$, not the representation $\pi_{\tau+1}$, is meaningful. The pair $\{1, x\}$ constitutes a basis of spherical harmonics on the real line, and it is a special case of what precedes that the even (*resp.* odd) part of the one-dimensional metaplectic representation is unitarily equivalent with the representation $\mathcal{D}_{\frac{1}{2}}$ (*resp.* $\mathcal{D}_{\frac{3}{2}}$).

Chapter 2

The Radon transformation and applications

The Radon transformation and its dual, or their associates, of which each component of the transformation Θ in Theorem 1.1.1 is a special case, connect the analysis of functions in the plane and in the hyperbolic half-plane: when enriched with an automorphy condition, the dual Radon transformation will also set up a correspondence, in Chapter 3, from automorphic distribution theory (in the plane) to automorphic function theory (in the half-plane).

After having recalled the Iwasawa decomposition $G = NAK$ of the group $G - SL(2, \mathbb{R})$, we consider the Radon transformation V from the homogeneous space G/K to the space G/MN, with $M = \{\pm I\}$, and the dual Radon transformation, which acts in the reverse direction. The space G/MN can be regarded as the quotient of $\mathbb{R}^2 \backslash \{0\}$ by the equivalence that identifies (x, ξ) with $(-x, -\xi)$, while the space G/K is just the hyperbolic half-plane Π: consequently, the dual Radon transformation may be considered as a map from even functions in $\mathbb{R}^2$ to functions in Π. Besides, the maps V and V^* have associates, obtained by multiplying them, on the appropriate side, by arbitrary functions, in the spectral-theoretic sense, of the Euler operator in $\mathbb{R}^2$. All norm computations involving $SL(2, \mathbb{R})$-covariant maps from even functions in the plane to functions in Π rely on the results of calculations involving the Radon transformation and its associates. This is in particular the case for the map $h \mapsto f_0$ introduced in Theorem 1.1.1; we shall rely on these again to complete, in Section 2.2, our study of the totally radial Weyl calculus, as initiated in Section 1.3. The rest of the chapter is concerned with a family of bihomogeneous functions $\hom_{\rho,\nu}$ in the plane, the dual Radon transforms of which will play a basic role in our construction, in Chapter 4, of a new class of non-holomorphic modular forms. Splitting such transforms into two terms, we shall obtain a two-parameter family of functions $z \mapsto (\operatorname{Im} z)^{\frac{\rho-1}{2}} \chi_{\rho,\nu} \left(\frac{\operatorname{Re} z}{\operatorname{Im} z}\right)$ in the hyperbolic half-plane: these functions will constitute the starting points of the

Poincaré series to be introduced there. The functions $\chi_{\rho,\nu}$ are studied with much care in Section 2.3 and the functions in Π just mentioned are expressed in a natural way involving the resolvent of the Laplace operator Δ on Π in Section 2.4.

2.1   The Radon transformation

Consider the transformation $\Theta = (\Theta_0, \Theta_1): h \mapsto (f_0, f_1)$ introduced in Theorem 1.1.1 or, using (1.1.38) and starting, more generally, from a distribution,

$$(\Theta_0 \mathfrak{S})(z) = \langle \mathfrak{S}, (x, \xi) \mapsto 2 \exp\left(-2\pi \frac{|x - z\xi|^2}{\mathrm{Im}\, z}\right)\rangle,$$

$$(\Theta_1 \mathfrak{S})(z) = \langle \mathfrak{S}, (x, \xi) \mapsto 2 \left[\frac{4\pi}{\mathrm{Im}\, z}|x - z\xi|^2 - 1\right] \exp\left(-2\pi \frac{|x - z\xi|^2}{\mathrm{Im}\, z}\right)\rangle. \quad (2.1.1)$$

The two functions just introduced are linked by the equation

$$\Theta_1 \mathfrak{S} = \Theta_0(2i\pi \mathcal{E} \mathfrak{S}), \quad (2.1.2)$$

as it follows immediately from the fact that the transpose of the operator $2i\pi\mathcal{E}$ is $-2i\pi\mathcal{E}$. As a consequence, identities involving Θ_0 will always have analogues involving Θ_1, which we shall dispense with making explicit unless clarity demands it. We shall also use, consistently and without reference, the fact that the conjugate of the operator $2i\pi\mathcal{E}$ under the symplectic Fourier transformation, or under $\mathcal{G}$, is $-2i\pi\mathcal{E}$.

The map Θ connects even distributions in the plane to pairs of functions in the hyperbolic half-plane Π, and it has many nice properties; only, do not confuse x, the first of the pair of variables (x, ξ) in the plane (the standard notation in pseudo-differential analysis) with the real part of z. First, recall that Θ is covariant under the pair of actions of $SL(2, \mathbb{R})$ on $\mathbb{R}^2$ and on Π, which means that one always has

$$(\Theta(\mathfrak{S} \circ g^{-1}))(z) = (\Theta\mathfrak{S})(g^{-1}.z) \quad (2.1.3)$$

if, given $g = \left(\begin{smallmatrix} a & b \\ c & d \end{smallmatrix}\right) \in SL(2, \mathbb{R})$, one sets $g(x, \xi) = (ax + b\xi, cx + d\xi)$ and $g.z = \frac{az+b}{cz+d}$. Also, Θ kills all odd functions on $\mathbb{R}^2$, so we may as well restrict it to the space $\mathcal{S}'_{\mathrm{even}}(\mathbb{R}^2)$: another symmetry expresses itself in terms of the transformation $\mathcal{G}$ in (1.1.24), as the pair of identities

$$\Theta_0(\mathcal{G}\mathfrak{S}) = \Theta_0\mathfrak{S}, \quad \Theta_1(\mathcal{G}\mathfrak{S}) = -\Theta_1\mathfrak{S}. \quad (2.1.4)$$

The first one, say, can be seen by remarking that the (even) function $(x, \xi) \mapsto 2 \exp\left(-2\pi \frac{|x - z\xi|^2}{\mathrm{Im}\, z}\right)$ is $\mathcal{G}$-invariant: to see this, it suffices, taking benefit from the covariance property, to verify that the function $2 \exp(-2\pi(x^2 + \xi^2))$ is $\mathcal{G}$-invariant, which is immediate. Another proof consists (*cf.* what follows (1.1.24)) in remarking, a consequence of (1.1.34), that this function is the symbol of an even-even operator. From the fact that the set $(\phi_z^0)_{z\in\Pi}$ is total in $L^2_{\text{even}}(\mathbb{R})$, one can then see that, when restricted to even $\mathcal{G}$-invariant tempered distributions, Θ_0 becomes one-to-one.

In view of (1.1.41), one always has the identity

$$\Theta(\pi^2 \mathcal{E}^2 \mathfrak{S}) = \left(\Delta - \frac{1}{4}\right)\Theta\mathfrak{S}. \tag{2.1.5}$$

In other words, if $\mathfrak{S}$ is homogeneous of degree $-1-\nu$ or $-1+\nu$, the function $\Theta\mathfrak{S}$ is a pair of (generalized) eigenfunctions of Δ for the eigenvalue $\frac{1-\nu^2}{4}$.

The covariance property of Θ_0 (or Θ_1), as well as the way it exchanges the operators $\pi^2\mathcal{E}^2$ and $\Delta - \frac{1}{4}$, are shared by a family of transformations, linked to the so-called Radon transformation, which we need to recall in the case of the group $SL(2,\mathbb{R})$: the Radon transformation has been studied by Helgason [17, 18] in a considerable generality. We here follow with a few changes the exposition, in the case of $SL(2,\mathbb{R})$, made in [60, Sec.4], which is more immediately adapted to our needs related to pseudo-differential analysis, besides being of necessity simpler since it deals only with a rank-one case.

We parametrize the generic elements of the subgroups N, A, K entering the Iwasawa decomposition of $G = SL(2,\mathbb{R}) = NAK$ as

$$n = \begin{pmatrix} 1 & b \\ 0 & 1 \end{pmatrix}, \quad a = \begin{pmatrix} e^{\frac{r}{2}} & 0 \\ 0 & e^{-\frac{r}{2}} \end{pmatrix}, \quad k = \begin{pmatrix} \cos\frac{\theta}{2} & \sin\frac{\theta}{2} \\ -\sin\frac{\theta}{2} & \cos\frac{\theta}{2} \end{pmatrix}, \tag{2.1.6}$$

where $b \in \mathbb{R}, r \in \mathbb{R}, 0 \leq \theta < 4\pi$. Following the normalizations in ([18], ch.II,3), we set $dn = \pi^{-1}db, dk = (4\pi)^{-1}d\theta$. The homogeneous space G/K is identified with the hyperbolic half-plane Π in the usual way, sending gK to $z = g.i$. On the other hand, the space $\Xi = G/MN$, with $M = \{\pm I\}$, is identified with the quotient of $\mathbb{R}^2\backslash\{0\}$ by the equivalence $\left(\begin{smallmatrix} x \\ \xi \end{smallmatrix}\right) \sim \left(\begin{smallmatrix} -x \\ -\xi \end{smallmatrix}\right)$, under the map $\left(\begin{smallmatrix} a & b \\ c & d \end{smallmatrix}\right)MN \mapsto \pm\left(\begin{smallmatrix} a \\ c \end{smallmatrix}\right)$: one must be careful, again, not to use in the same formula x to denote the first coordinate of $\left(\begin{smallmatrix} x \\ \xi \end{smallmatrix}\right)$ (or (x,ξ)) in $\mathbb{R}^2$ and the real part of $z = x + iy \in \Pi$. On Π, we use the invariant measure $dm(z) = \frac{dx\,dy}{y^2}$ and, identifying functions on Ξ with even functions on $\mathbb{R}^2$, we use there the standard Lebesgue measure on the full plane. Let us also recall that the hyperbolic distance d on Π associated to the (squared) line element $ds^2 = \frac{dx^2+dy^2}{y^2}$ is G-invariant, i.e., that $d(g.z, g.z')$ is independent of g, and characterized as such by the equation $\cosh d(i, x+iy) = \frac{1+x^2+y^2}{2y}$. The Radon transformation V from functions f on Π to even functions on $\mathbb{R}^2$ is defined by the equation

$$(Vf)(g.\left(\begin{smallmatrix} 1 \\ 0 \end{smallmatrix}\right)) = \int_N f((gn).i)dn : \tag{2.1.7}$$

the integral is convergent, yielding a continuous function Vf if, say, $|f(z)| \leq C(\cosh d(i,z))^{-\frac{1}{2}-\varepsilon}$ for some $\varepsilon > 0$. Explicitly, completing if $x \neq 0$ the column

$(\frac{x}{\xi})$ into the matrix $\begin{pmatrix} x & 0 \\ \xi & x^{-1} \end{pmatrix}$,

$$(Vf)(\pm(\tfrac{x}{\xi})) = \frac{1}{\pi} \int_{-\infty}^{\infty} f\left(\frac{x^2(i+b)}{x\xi(i+b)+1}\right) db, \quad x \neq 0; \tag{2.1.8}$$

the dual Radon transform V^*, the formal adjoint of V, is defined by

$$(V^*h)(g.i) = \int_K h((gk).(\tfrac{1}{0}))dk \tag{2.1.9}$$

or, in coordinates,

$$(V^*h)(x+iy) = \frac{1}{2\pi} \int_0^{2\pi} h\left(\pm\left(\begin{matrix} y^{\frac{1}{2}}\cos\frac{\theta}{2} - xy^{-\frac{1}{2}}\sin\frac{\theta}{2} \\ -y^{-\frac{1}{2}}\sin\frac{\theta}{2} \end{matrix}\right)\right) d\theta. \tag{2.1.10}$$

We abbreviate the representation $\pi_{i\lambda,0}$, as defined in (1.2.18), as $\pi_{i\lambda}$ — it lies in the principal series of $SL(2,\mathbb{R})$, whereas the representation $\pi_{\tau+1}$ in (1.3.5) lies in the extended projective discrete series of this group — and abbreviate $h_{i\lambda,0}$ (resp. $h_{i\lambda,0}^b$) as $h_{i\lambda}$ (resp. $h_{i\lambda}^b$): in the present section, we only interest ourselves in even functions in the plane. Through the dual Radon transformation, the representation $\pi_{i\lambda}$ can be realized in some Hilbert space of functions in Π: we need to make this explicit.

We have already defined the Euler operator $2i\pi\mathcal{E} = x\frac{\partial}{\partial x} + \xi\frac{\partial}{\partial \xi} + 1$. It is essentially self-adjoint on $L^2(\mathbb{R}^2)$ (i.e., it admits a unique self-adjoint extension) if given the initial domain $C_0^\infty(\mathbb{R}^2\backslash\{0\})$. This makes it possible to define, in the spectral-theoretic sense, functions of $\mathcal{E}$. We shall need in particular the operator (a scalar when restricted to even functions of a given degree of homogeneity)

$$T = \left(\frac{\pi}{2}\right)^{\frac{1}{2}} \frac{\Gamma(\frac{1}{2} - i\pi\mathcal{E})}{\Gamma(-i\pi\mathcal{E})} = \pi^{-\frac{1}{2}}(-i\pi\mathcal{E}) \int_0^{\infty} t^{-\frac{1}{2}}(1+t)^{-1+i\pi\mathcal{E}} dt : \tag{2.1.11}$$

also, observe that $(t^{2i\pi\mathcal{E}}h)(x,\xi) = th(tx,t\xi)$ for $t > 0$.

We now give useful expressions of the transformation TV and its formal adjoint V^*T^*, with the help of the following special case of (1.2.14):

$$h_{i\lambda}(x,\xi) = |\xi|^{-1-i\lambda} h_{i\lambda}^b(\frac{x}{\xi}). \tag{2.1.12}$$

As a consequence of (2.1.11), T acts on $(Vf)_{i\lambda}$ as the scalar $(\frac{\pi}{2})^{\frac{1}{2}} \frac{\Gamma(\frac{1+i\lambda}{2})}{\Gamma(\frac{i\lambda}{2})}$, and it then follows from (2.1.12) and (2.1.8) that, assuming that, say, $f \in C_0^\infty(\Pi)$, one has for almost all λ the equation

$$(TVf)_{i\lambda}^b(s) = (2\pi)^{-\frac{3}{2}} \frac{\Gamma(\frac{1+i\lambda}{2})}{\Gamma(\frac{i\lambda}{2})} \int_0^{\infty} t^{i\lambda-2} dt \int_{-\infty}^{\infty} f\left(\frac{s^2(i+b)}{s(i+b)+t^2}\right) db : \tag{2.1.13}$$

performing the change of variable such that

$$z = \frac{s^2(i+b)}{s(i+b)+t^2}, \quad dm(z) = \frac{2dtdb}{t}, \tag{2.1.14}$$

so that $t^2 = \frac{|z-s|^2}{\operatorname{Im} z}$, one gets

$$(TVf)^{\flat}_{i\lambda}(s) = \frac{1}{2}(2\pi)^{-\frac{3}{2}} \frac{\Gamma(\frac{1+i\lambda}{2})}{\Gamma(\frac{i\lambda}{2})} \int_{\Pi} \left(\frac{|z-s|^2}{\operatorname{Im} z}\right)^{-\frac{1}{2}-\frac{i\lambda}{2}} f(z)dm(z). \tag{2.1.15}$$

In the reverse direction, we use the second equation (2.1.12) and (2.1.10), obtaining (after one has set $s = -y\cot an\frac{\theta}{2} + x$ in the latter formula) that

$$(V^*T^*h_{i\lambda})(z) = (2\pi)^{-\frac{1}{2}} \frac{\Gamma(\frac{1-i\lambda}{2})}{\Gamma(-\frac{i\lambda}{2})} \int_{-\infty}^{\infty} h^{\flat}_{i\lambda}(s) \left(\frac{|z-s|^2}{\operatorname{Im} z}\right)^{-\frac{1}{2}+\frac{i\lambda}{2}} ds: \tag{2.1.16}$$

note that the integral on the right-hand side is bounded if $h^{\flat}_{i\lambda} \in L^2(\mathbb{R})$.

From its very definition (2.1.7), the Radon transformation (as well as its dual) is obviously covariant under the two actions of G, on functions defined on Π and on $\mathbb{R}^2\backslash\{0\}$, through the fractional-linear change of complex coordinate and the linear change of real coordinates associated to the same matrix g. On the other hand, all functions, in the spectral-theoretic sense, of the Euler operator commute with the second action. Consequently, the transformations V and V^* preserve their covariance if multiplied on the left (*resp.* on the right) by an "arbitrary" function of $2i\pi\mathcal{E}$. Operators obtained as products of the Radon (*resp.* dual Radon) transformation by a function of the Euler operator on the left (*resp.* right) side will be called *associates* of the Radon or dual Radon transformation. A subclass consists of operators obtained in a comparable way, only replacing the function of the Euler operator by a function of the hyperbolic Laplacian on the other side: as a consequence of the last assertion in Theorem 2.1.2 below, even functions of $2i\pi\mathcal{E}$ can be replaced by appropriate functions of Δ with no change. We now show that the map Θ_0 introduced in (2.1.1) is an associate of the dual Radon transformation: of course, the same will then be true of the map Θ_1 in view of (2.1.2).

Proposition 2.1.1. *One has*

$$\Theta_0 = V^*(2\pi)^{\frac{1}{2}-i\pi\mathcal{E}}\Gamma\left(\frac{1}{2}+i\pi\mathcal{E}\right). \tag{2.1.17}$$

Proof. Starting from the decomposition (1.2.11) and applying the definition (2.1.1) of Θ, we obtain

$$(\Theta_0 h_{i\lambda})(z) = \frac{1}{2\pi} \int_0^{\infty} t^{i\lambda} \Theta_0((x,\xi) \mapsto h(tx,t\xi))dt$$

$$= \frac{1}{\pi} \int_0^\infty t^{i\lambda} dt \int_{\mathbb{R}^2} h(tx, t\xi) \exp\left(-2\pi \frac{|x - z\xi|^2}{\operatorname{Im} z}\right) dx d\xi$$

$$= \frac{1}{\pi} \int_{\mathbb{R}^2} h(x, \xi) dx d\xi \int_0^\infty t^{i\lambda-2} \exp\left(-2\pi \frac{|x - z\xi|^2}{t^2 \operatorname{Im} z}\right) dt. \qquad (2.1.18)$$

The integral is easily computed, which leads to the equation

$$(\Theta_0 h_{i\lambda})(z) = (2\pi)^{\frac{i\lambda-3}{2}} \Gamma(\frac{1 - i\lambda}{2}) \int_{\mathbb{R}^2} h(x, \xi) \left(\frac{|x - z\xi|^2}{\operatorname{Im} z}\right)^{\frac{i\lambda-1}{2}} dx d\xi. \qquad (2.1.19)$$

On the other hand, using (1.2.11) again and (2.1.16), we have

$$(V^*T^* h_{i\lambda})(z) = (2\pi)^{-\frac{3}{2}} \frac{\Gamma(\frac{1-i\lambda}{2})}{\Gamma(\frac{-i\lambda}{2})} \int_{-\infty}^\infty \left(\frac{|z - s|^2}{\operatorname{Im} z}\right)^{-\frac{1}{2}+\frac{i\lambda}{2}} ds \int_0^\infty t^{i\lambda} h(ts, t) dt :$$

$$(2.1.20)$$

we make the change of variable

$$t = \xi, \quad s = \frac{x}{\xi}, \quad ds dt = \xi^{-1} dx d\xi, \qquad (2.1.21)$$

and take advantage of the fact that h is assumed to be even to change the domain $\{(x, \xi) : \xi > 0\}$ to $\mathbb{R}^2$, ending up with the equation

$$(V^*T^* h_{i\lambda})(z) = \frac{1}{2}(2\pi)^{-\frac{3}{2}} \frac{\Gamma(\frac{1-i\lambda}{2})}{\Gamma(\frac{-i\lambda}{2})} \int_{\mathbb{R}^2} h(x, \xi) \left(\frac{|x - z\xi|^2}{\operatorname{Im} z}\right)^{\frac{i\lambda-1}{2}} dx d\xi. \qquad (2.1.22)$$

Comparing it with (2.1.19), we obtain

$$\Theta_0 h_{i\lambda} = 2(2\pi)^{\frac{i\lambda}{2}} \Gamma(-\frac{i\lambda}{2}) V^* T^* h_{i\lambda} \qquad (2.1.23)$$

or, since $2i\pi\mathcal{E} h_{i\lambda} = -i\lambda h_{i\lambda}$,

$$\Theta_0 = V^* T^* 2(2\pi)^{-i\pi\mathcal{E}} \Gamma(i\pi\mathcal{E}) : \qquad (2.1.24)$$

as $T^* = (\frac{\pi}{2})^{\frac{1}{2}} \frac{\Gamma(\frac{1}{2}+i\pi\mathcal{E})}{\Gamma(i\pi\mathcal{E})}$, this leads to Proposition 2.1.1.   □

This proposition explains several facts. First, since, according to (2.1.5), the operator $\pi^2 \mathcal{E}^2$ on $\mathbb{R}^2$ on Π transfers under Θ_0 to the operator $\Delta - \frac{1}{4}$, the same is true for the Radon transformation or its dual, whether it has been multiplied on the appropriate side with a function of the Euler operator or not. Next, consider the formal adjoint of Θ_0, defined by the equation

$$(\Theta_0^* f)(x, \xi) = 2 \int_\Pi f(z) \exp\left(-2\pi \frac{|x - z\xi|^2}{\operatorname{Im} z}\right) dm(z), \qquad (2.1.25)$$

or

$$\Theta_0^* = 2(2\pi)^{i\pi\mathcal{E}}\Gamma(-i\pi\mathcal{E})TV. \tag{2.1.26}$$

As already noticed, the range (the image) of Θ_0^* is $\mathcal{G}$-invariant: also, $\mathcal{G}(i\pi\mathcal{E}) = (-i\pi\mathcal{E})\mathcal{G}$. As a consequence,

$$\text{Ran}(TV) \quad \text{is invariant under the involution}(2\pi)^{-2i\pi\mathcal{E}}\frac{\Gamma(i\pi\mathcal{E})}{\Gamma(-i\pi\mathcal{E})}\mathcal{G}. \tag{2.1.27}$$

We now recall (with a better proof) a theorem given in [60, p. 27].

Theorem 2.1.2. *The transformation TV, initially defined on the space of continuous functions on Π with a compact support, extends as an isometry from $L^2(\Pi)$ onto the subspace $\text{Ran}(TV)$ of $L^2_{\text{even}}(\mathbb{R}^2)$ consisting of all functions invariant under the unitary involution $(2\pi)^{-2i\pi\mathcal{E}}\frac{\Gamma(i\pi\mathcal{E})}{\Gamma(-i\pi\mathcal{E})}\mathcal{G}$. The operator V^*T^* extends on $\text{Ran}(TV)$ as the inverse of TV, and is zero on the subspace $(\text{Ran}(TV))^\perp$ of $L^2_{\text{even}}(\mathbb{R}^2)$ consisting of all functions changing to their negatives under the same involution. Moreover, the isometry TV intertwines the two actions of G on $L^2(\Pi)$ and $L^2_{\text{even}}(\mathbb{R}^2)$ respectively, and transforms the operator $\Delta - \frac{1}{4}$ on $L^2(\Pi)$ into the operator $\pi^2\mathcal{E}^2$ on $L^2_{\text{even}}(\mathbb{R}^2)$.*

Proof. The isometry property is a very special case of ([18], ch.II,3), but sorting out notation is not that easy. An alternative proof is as follows. From (2.1.15) and (2.1.11), one has

$$\|(TVf)^\flat_{i\lambda}\|^2_{L^2(\mathbb{R})} = \frac{1}{32\pi^3}\left(\frac{\lambda}{2}\tanh\frac{\pi\lambda}{2}\right) \tag{2.1.28}$$

$$\times \int_{\Pi\times\Pi} f(z)\overline{f}(w)dm(z)dm(w)\int_{-\infty}^{\infty}\left(\frac{|z-s|^2}{\text{Im }z}\right)^{-\frac{1}{2}-\frac{i\lambda}{2}}\left(\frac{|w-s|^2}{\text{Im }z}\right)^{-\frac{1}{2}+\frac{i\lambda}{2}}ds.$$

Now, one has

$$\frac{1}{\pi}\int_{-\infty}^{\infty}\left(\frac{|z-s|^2}{\text{Im }z}\right)^{-\frac{1}{2}-\frac{i\lambda}{2}}\left(\frac{|w-s|^2}{\text{Im }z}\right)^{-\frac{1}{2}+\frac{i\lambda}{2}}ds = \mathfrak{P}_{-\frac{1}{2}+\frac{i\lambda}{2}}(\cosh d(z,w)), \tag{2.1.29}$$

a consequence of Plancherel's formula together with [36, p, 401]

$$\pi^{-\frac{1}{2}}\int_{-\infty}^{\infty}\left(\frac{|z-s|^2}{\text{Im }z}\right)^{-\frac{1}{2}-\frac{i\lambda}{2}}e^{-2i\pi s\sigma}ds = y^{\frac{1}{2}}e^{-2i\pi\sigma x}\frac{2\pi^{\frac{i\lambda}{2}}}{\Gamma(\frac{1+i\lambda}{2})}|\sigma|^{\frac{i\lambda}{2}}K_{\frac{i\lambda}{2}}(2\pi|\sigma|y) \tag{2.1.30}$$

and [36, p. 413]

$$\int_0^{\infty}K_{\frac{i\lambda}{2}}(2\pi\sigma\text{Im }z)K_{\frac{i\lambda}{2}}(2\pi\sigma\text{Im }w)\cos(2\pi\sigma\text{Re }(z-w)))d\sigma$$

$$= \frac{1}{8}(\text{Im }z\text{Im }w)^{-\frac{1}{2}}\Gamma(\frac{1+i\lambda}{2})\Gamma(\frac{1-i\lambda}{2})\mathfrak{P}_{-\frac{1}{2}+\frac{i\lambda}{2}}(\cosh d(z,w)). \tag{2.1.31}$$

The isometry property is then a consequence of (1.2.15) and of Mehler's decomposition [36, p. 398] of functions $f \in C_0^\infty(\Pi)$ provided by the pair of formulas

$$f(z) = \int_0^\infty f_\lambda(z) \left(\frac{\pi\lambda}{2} \tanh \frac{\pi\lambda}{2} \right) d\lambda,$$

$$f_\lambda(z) = \frac{1}{4\pi^2} \int_\Pi f(w) \mathfrak{P}_{-\frac{1}{2}+\frac{i\lambda}{2}} (\cosh d(z,w)) dm(w). \qquad (2.1.32)$$

The factor

$$\frac{\pi\lambda}{2} \tanh \frac{\pi\lambda}{2} = \pi \frac{\Gamma\left(\frac{1+i\lambda}{2}\right) \Gamma\left(\frac{1-i\lambda}{2}\right)}{\Gamma\left(\frac{i\lambda}{2}\right) \Gamma\left(\frac{-i\lambda}{2}\right)} \qquad (2.1.33)$$

appears repeatedly in connection with Mehler's transformation. That the range of TV is invariant under the involution under consideration has been established before the statement of the theorem; that it is the full subspace of $L^2_{\text{even}}(\mathbb{R}^2)$ characterized by this invariance or, what amounts to the same, that the image is dense, can be obtained by linking this to a property of Θ_0, with the help of Proposition 2.1.1. $\qquad \square$

Note that if $\mathcal{H}_{i\lambda}$ denotes the completion of the space of all f_λ ($f \in C_0^\infty(\Pi)$) under the norm such that

$$\|f_\lambda\|^2_{\mathcal{H}_{i\lambda}} = (4\pi^2)^{-2} \int_{\Pi \times \Pi} f(z)\bar{f}(w) \mathfrak{P}_{-\frac{1}{2}+\frac{i\lambda}{2}} (\cosh d(z,w)) dm(z)dm(w)$$

$$= (4\pi^2)^{-1} (f_\lambda | f)_{L^2(\Pi)}, \qquad (2.1.34)$$

one has the identity

$$\|f\|^2_{L^2(\Pi)} = 4\pi^2 \int_0^\infty \|f_\lambda\|^2_{\mathcal{H}_{i\lambda}} \left(\frac{\pi\lambda}{2} \tanh \frac{\pi\lambda}{2} \right) d\lambda. \qquad (2.1.35)$$

The following consequence of (2.1.24) and Theorem 2.1.2 was announced in (1.1.43): if a $\mathcal{G}$-invariant function $h \in L^2_{\text{even}}(\mathbb{R}^2)$ is the image under $2i\pi\mathcal{E}$ of some function in $L^2_{\text{even}}(\mathbb{R}^2)$, so that $\Gamma(i\pi\mathcal{E})h \in L^2_{\text{even}}(\mathbb{R}^2)$ too, one has

$$\|\Theta_0 h\|_{L^2(\Pi)} = 2\|\Gamma(i\pi\mathcal{E})h\|_{L^2(\mathbb{R}^2)}. \qquad (2.1.36)$$

Equation (1.1.44) follows from the preceding one and from (2.1.2).

Restricting the dual Radon transform to K-invariant functions, and using analytic continuation to replace $i\lambda$ by a more general complex number ν, one observes from (2.1.10) that if $h(x,\xi) = (x^2 + \xi^2)^{\frac{-1-\nu}{2}}$, one has

$$(V^* h)(iy) = \frac{1}{2\pi} \int_0^{2\pi} (y \cos^2 \frac{\theta}{2} + y^{-1} \sin^2 \frac{\theta}{2})^{\frac{-1-\nu}{2}} d\theta \qquad (2.1.37)$$

$$= \frac{1}{2\pi} \int_0^{2\pi} \left[\frac{y+y^{-1}}{2} + \frac{y-y^{-1}}{2} \cos\theta \right]^{\frac{-1-\nu}{2}} d\theta = \mathfrak{P}_{\frac{-1-\nu}{2}} \left(\frac{y+y^{-1}}{2} \right)$$

[36, p. 184] or, more generally, using covariance,

$$(V^*h)(z) = \mathfrak{P}_{\frac{-1-\nu}{2}}(\cosh d(i, z)), \qquad (2.1.38)$$

where d is the hyperbolic distance in Π.

The study of the restriction of the Radon, or dual Radon, transform and their associates to K-invariant functions is very classical: even more so, it is usually a preparation for the more general theory. All this fits within the so-called theory of Gelfand pairs and spherical function theory [18, 8]. In Section 2.3, we shall consider the way these transformations can be restricted to MA-invariant functions.

Consider the part of the Weyl calculus concerned with operators preserving the parity of functions, in other words the one defined from the consideration of even symbols only. It would be perfectly possible, if hardly advisable in general, to define a variant of this calculus in which symbols would be pairs of functions in Π, the images of the "true" symbol under the map Θ in (2.1.1). We here mention this possibility since, in the automorphic case, such a transfer will make it possible to bypass some technical difficulties inherent in the automorphic Weyl calculus, the source of which will be described in Section 3.4. One of our main interests in pseudo-differential analysis lies in the composition formulas: in view of Theorem 1.2.2, all we have to do is transferring under any associate of the dual Radon transformation the operations obtained from the integral kernels $\chi^{\varepsilon_1,\varepsilon_2;\varepsilon}_{i\lambda_1,i\lambda_2;i\lambda}(s_1, s_2; s)$. Actually, since we are only dealing with even symbols, one must take $\delta = \delta_1 = \delta_2 = 0$ with the notation from the theorem just referred to, so that, from (1.2.27), only the two cases in which $\varepsilon_1 = \varepsilon_2 = \varepsilon = 0$ or 1 must be considered.

As will be seen presently, when dealing with homogeneous symbols of given degrees of homogeneity, the operator with integral kernel $\chi^{0,0,0}_{i\lambda_1,i\lambda_2;i\lambda}$ (*respectively* $\chi^{1,1,1}_{i\lambda_1,i\lambda_2;i\lambda}$) will appear, up to scalar factors, as the transfer under any associate of the Radon transformation of the operator of pointwise multiplication (*respectively.* the Poisson bracket) on functions on Π. The simplicity of the result should not lead one to believe that a non-computational proof should exist as well: for, when restricted to pairs of (generalized) eigenfunctions of Δ for specific eigenvalues, a bilinear operator as simple as the pointwise product of functions may have a variety of quite complicated disguises. Given $h \in \mathcal{S}_{\text{even}}(\mathbb{R}^2)$, let us not confuse, in what follows, the function $h_{i\lambda}$ (a function on $\mathbb{R}^2\backslash\{0\}$, homogeneous of degree $-1 - i\lambda$) and the function $h^\flat_{i\lambda}$ on the line (to be precise, on the projective completion of the line).

Proposition 2.1.3. *Let $\lambda_1, \lambda_2, \lambda$ be real numbers, and let h_1, h_2 be two even functions in $\mathcal{S}(\mathbb{R}^2)$. One has the identity*

$$(TV((V^*T^*(h_1)_{i\lambda_1}).(V^*T^*(h_2)_{i\lambda_2})))^\flat_{i\lambda}(s) = 2^{-\frac{9}{2}}\pi^{-2} \tag{2.1.39}$$

$$\times \frac{\Gamma\left(\frac{1-i(\lambda+\lambda_1+\lambda_2)}{4}\right)\Gamma\left(\frac{1+i(\lambda-\lambda_1+\lambda_2)}{4}\right)\Gamma\left(\frac{1+i(\lambda+\lambda_1-\lambda_2)}{4}\right)\Gamma\left(\frac{1+i(\lambda-\lambda_1-\lambda_2)}{4}\right)}{\Gamma\left(-\frac{i\lambda_1}{2}\right)\Gamma\left(-\frac{i\lambda_2}{2}\right)\Gamma\left(\frac{i\lambda}{2}\right)}$$

$$\times \int_{\mathbb{R}^2} \chi^{0,0,0}_{i\lambda_1,i\lambda_2;i\lambda}(s_1,s_2;s)(h_1)^\flat_{\lambda_1}(s_1)(h_2)^\flat_{\lambda_2}(s_2)ds_1ds_2.$$

Proof. As already noted in (1.2.29), one has the estimate $|(h_1)^\flat_{\lambda_1}(s_1)| \leq C(1+s_1^2)^{-\frac{1}{2}}$ and a similar one relative to $(h_2)^\flat_{\lambda_2}$. Using (2.1.15) and (2.1.16), one can write the left-hand side of the identity to be proved as

$$\frac{1}{2}(2\pi)^{-\frac{5}{2}}\frac{\Gamma\left(\frac{1-i\lambda_1}{2}\right)\Gamma\left(\frac{1-i\lambda_2}{2}\right)\Gamma\left(\frac{1+i\lambda}{2}\right)}{\Gamma\left(-\frac{i\lambda_1}{2}\right)\Gamma\left(-\frac{i\lambda_2}{2}\right)\Gamma\left(\frac{i\lambda}{2}\right)}$$

$$\times \int_{\mathbb{R}^2} A_{i\lambda_1,i\lambda_2;i\lambda}(s_1,s_2;s)(h_1)^\flat_{\lambda_1}(s_1)(h_2)^\flat_{\lambda_2}(s_2)ds_1ds_2 \tag{2.1.40}$$

with

$$A_{i\lambda_1,i\lambda_2;i\lambda}(s_1,s_2;s)$$

$$= \int_\Pi \left(\frac{|z-s_1|^2}{\mathrm{Im}\,z}\right)^{-\frac{1}{2}+\frac{i\lambda_1}{2}}\left(\frac{|z-s_2|^2}{\mathrm{Im}\,z}\right)^{-\frac{1}{2}+\frac{i\lambda_2}{2}}\left(\frac{|z-s|^2}{\mathrm{Im}\,z}\right)^{-\frac{1}{2}-\frac{i\lambda}{2}}dm(z). \tag{2.1.41}$$

Using the identity

$$A_{i\lambda_1,i\lambda_2;i\lambda}\left(\frac{as_1+b}{cs_1+d},\frac{as_2+b}{cs_2+d};\frac{as+b}{cs+d}\right)$$

$$= |cs_1+d_1|^{1-i\lambda_1}|cs_2+d_1|^{1-i\lambda_2}|cs+d|^{1+i\lambda}A_{i\lambda_1,i\lambda_2;i\lambda}(s_1,s_2;s), \tag{2.1.42}$$

a consequence of

$$\frac{|z-\frac{as+b}{cs+d}|^2}{\mathrm{Im}\,z} = (cs+d)^{-2}\frac{|g^{-1}.z-s|^2}{\mathrm{Im}\,(g^{-1}.z)}, \quad g = \begin{pmatrix} a & b \\ c & d \end{pmatrix}, \tag{2.1.43}$$

and noting that if s_1,s_2,s are the images of $0,1,\infty$ under the fractional-linear transformation associated to the matrix $\begin{pmatrix} a & b \\ c & d \end{pmatrix}$, then

$$\chi^{0,0,0}_{i\lambda_1,i\lambda_2;i\lambda}(s_1,s_2;s) = |d|^{1-i\lambda_1}|c+d|^{1-i\lambda_2}|c|^{1+i\lambda}, \tag{2.1.44}$$

one gets

$$A_{i\lambda_1,i\lambda_2;i\lambda}(s_1,s_2;s) = I(i\lambda_1,i\lambda_2;i\lambda)\chi^{0,0,0}_{i\lambda_1,i\lambda_2;i\lambda}(s_1,s_2;s) \tag{2.1.45}$$

with

$$I(i\lambda_1,i\lambda_2;i\lambda) = \int_\Pi \left(\frac{|z|^2}{\mathrm{Im}\,z}\right)^{-\frac{1}{2}+\frac{i\lambda_1}{2}}\left(\frac{|z-1|^2}{\mathrm{Im}\,z}\right)^{-\frac{1}{2}+\frac{i\lambda_2}{2}}(\mathrm{Im}\,z)^{\frac{1}{2}+\frac{i\lambda}{2}}dm(z),$$

$$\tag{2.1.46}$$

a convergent integral. The justification of all that precedes is based on (1.2.29) and on the easily proved estimate

$$\int_{\mathbb{R}^3} |(s_1 - s_2)(s_2 - s)(s - s_1)|^{-\frac{1}{2}} ((1 + s_1^2)(1 + s_2^2)(1 + s^2))^{-\frac{1}{2}} ds_1 ds_2 ds < \infty. \quad (2.1.47)$$

Using (2.1.30) and the Plancherel formula for the dx-integration, we obtain

$$I(i\lambda_1, i\lambda_2; i\lambda) = \frac{8\pi^{1 - \frac{i(\lambda_1 + \lambda_2)}{2}}}{\Gamma(\frac{1 - i\lambda_1}{2})\Gamma(\frac{1 - i\lambda_2}{2})} \int_0^\infty y^{-\frac{1}{2} + \frac{i\lambda}{2}} dy \int_0^\infty \sigma^{-\frac{i(\lambda_1 + \lambda_2)}{2}} \cos(2\pi\sigma)$$
$$\times K_{\frac{i\lambda_1}{2}}(2\pi\sigma y) K_{\frac{i\lambda_2}{2}}(2\pi\sigma y) d\sigma, \quad (2.1.48)$$

where the $d\sigma$-integration has to be carried first. Integrating instead with respect to dy first so as to take advantage of [36, p. 101], one would formally obtain

$$I(i\lambda_1, i\lambda_2; i\lambda) = \frac{\pi^{\frac{1}{2}}}{2\Gamma(\frac{1 - i\lambda_1}{2})\Gamma(\frac{1 - i\lambda_2}{2})\Gamma(\frac{1 + i\lambda}{2})} \times \Gamma\left(\frac{1 - i(\lambda + \lambda_1 + \lambda_2)}{4}\right) \quad (2.1.49)$$
$$\times \Gamma\left(\frac{1 + i(\lambda - \lambda_1 + \lambda_2)}{4}\right) \Gamma\left(\frac{1 + i(\lambda + \lambda_1 - \lambda_2)}{4}\right) \Gamma\left(\frac{1 + i(\lambda - \lambda_1 - \lambda_2)}{4}\right),$$

and the process can be justified if one first inserts under the right-hand side of (2.1.48) the factor $h(\varepsilon\sigma)$ for some $h \in \mathcal{S}(\mathbb{R})$ with $h(0) = 1$, letting at the end ε go to zero. $\square$

Even though we shall not need this result in our main applications in Chapter 4, let us mention the following analogue of Proposition 2.1.3, in which the Poisson bracket of two smooth functions in Π is defined as

$$\{f_1, f_2\} = y^2 \left(-\frac{\partial f_1}{\partial y}\frac{\partial f_2}{\partial x} + \frac{\partial f_1}{\partial x}\frac{\partial f_2}{\partial y}\right). \quad (2.1.50)$$

Proposition 2.1.4. *Under the assumptions of Proposition 2.1.3, one has*

$$(TV(\{V^*T^*(h_1)_{i\lambda_1}) \cdot (V^*T^*(h_2)_{i\lambda_2}\}))^{\flat}_{i\lambda}(s) = 2^{-\frac{7}{2}}\pi^{-2} \quad (2.1.51)$$
$$\times \frac{\Gamma\left(\frac{3 - i(\lambda + \lambda_1 + \lambda_2)}{4}\right)\Gamma\left(\frac{3 + i(\lambda - \lambda_1 + \lambda_2)}{4}\right)\Gamma\left(\frac{3 + i(\lambda + \lambda_1 - \lambda_2)}{4}\right)\Gamma\left(\frac{3 + i(\lambda - \lambda_1 - \lambda_2)}{4}\right)}{\Gamma\left(-\frac{i\lambda_1}{2}\right)\Gamma\left(-\frac{i\lambda_2}{2}\right)\Gamma\left(\frac{i\lambda}{2}\right)}$$
$$\times \int_{\mathbb{R}^2} \chi^{1,1,1}_{i\lambda_1, i\lambda_2; i\lambda}(s_1, s_2; s)(h_1)^{\flat}_{\lambda_1}(s_1)(h_2)^{\flat}_{\lambda_2}(s_2) ds_1 ds_2.$$

The proof of this proposition, fully similar to that of Proposition 2.1.3, can be found if desired in [60, p. 73].

2.2   Back to the totally radial Weyl calculus

In this section, we examine the exact way in which the map Λ in Theorem 1.3.1 differs from an isometry, and we connect the totally radial calculus, with symbols living on Π, to the so-called Berezin calculus [2]. This ought to please people interested in quantization theory, by which we here mean the development of analogous (covariant) pseudo-differential analyses in which symbols are functions on rather general homogeneous spaces, in particular hermitian symmetric spaces. Even so, this is not yet the "good" symbolic calculus of totally radial operators: as will be seen in Chapter 6, calculations of an arithmetic character demand that symbols should live on the plane rather than the half-plane.

Lemma 2.2.1. *For every function $F(p, q, r)$ on the cone*

$$C = \{(p, q, r) \colon p > (q^2 + r^2)^{\frac{1}{2}}\},$$

one has, assuming summability, and recalling that $\omega_n = \frac{2\pi^{\frac{n}{2}}}{\Gamma(\frac{n}{2})}$,

$$
\begin{aligned}
I : &= \int_{\mathbb{R}^n \times \mathbb{R}^n} F\left(\frac{|x|^2 + |\xi|^2}{2}, \langle x, \xi \rangle, \frac{|x|^2 - |\xi|^2}{2}\right) dx d\xi \\
&= \frac{\omega_n \omega_{n-1}}{2} \int_C F(p, q, r)[p^2 - q^2 - r^2]^{\frac{n-3}{2}} dp dq dr.
\end{aligned}
\tag{2.2.1}
$$

Proof. Set

$$F(p, q, r) = H(p + r, q, p - r) = H(a, b, c), \tag{2.2.2}$$

so that

$$I = \int_{\mathbb{R}^n \times \mathbb{R}^n} H(|x|^2, \langle x, \xi \rangle, |\xi|^2) dx d\xi. \tag{2.2.3}$$

Given x, there is an x-dependent rotation in ξ-space which transforms $\langle x, \xi \rangle$ to $|x| \xi_1$. Hence, with $\xi = (\xi_1, \xi_*)$,

$$
\begin{aligned}
I &= \int_{\mathbb{R}^n \times \mathbb{R}^n} H(|x|^2, |x| \xi_1, |\xi|^2) dx d\xi \\
&= \omega_{n-1} \int_{\mathbb{R}^n} dx \int_{-\infty}^{\infty} d\xi_1 \int_0^{\infty} t^{n-2} H(|x|^2, |x| \xi_1, \xi_1^2 + t^2) dt \\
&= \omega_n \omega_{n-1} \int_{-\infty}^{\infty} d\xi_1 \int_0^{\infty} \int_0^{\infty} s^{n-1} t^{n-2} H(s^2, s\xi_1, \xi_1^2 + t^2) ds dt \\
&= \frac{1}{4} \omega_n \omega_{n-1} \int_{a>0, c>0, |b|<\sqrt{ac}} (ac - b^2)^{\frac{n-3}{2}} H(a, b, c) da db dc,
\end{aligned}
\tag{2.2.4}
$$

which leads to the expression indicated. $\qquad\square$

Theorem 2.2.2. *Assuming* $n \geq 2$, *let* $f \in L^2(\Pi)$. *One has the identity (in which* Λ *is the map* Λ *in Theorem 1.3.1)*

$$\|\Lambda f\|^2_{L^2(\mathbb{R}^{2n})} = \frac{2^{1-2n}\pi}{(\Gamma(\frac{n}{2}))^2}\|\Gamma\left(\frac{n-1}{2}+i\sqrt{\Delta-\frac{1}{4}}\right)f\|^2_{L^2(\Pi)}. \tag{2.2.5}$$

Proof. Since $(p^2 - q^2 - r^2)^{\frac{1}{2}} = s$, one obtains after a straightforward computation of the jacobian

$$\left|\frac{D(p,q,r)}{D(s,x,y)}\right| = \left|\frac{D(p,q,p+r)}{D(s,x,y)}\right| = \left|\frac{D(s\frac{1+|z|^2}{2y},\frac{sx}{y},\frac{s}{y})}{D(s,x,y)}\right| = \frac{s^2}{y^2} \tag{2.2.6}$$

the expression

$$[p^2 - q^2 - r^2]^{\frac{n-3}{2}}dpdqdr = s^{n-1}ds\frac{dxdy}{y^2} \tag{2.2.7}$$

in terms of the coordinates $(s,z) = (s, x+iy)$ linked to (p,q,r) by (1.3.28).

If f satisfies the first identity (2.1.32), (1.3.20) can be rewritten as

$$(\theta f)(s,z) = s^{-\frac{1}{2}}\int_0^\infty K_{\frac{i\lambda}{2}}(4\pi s)f_\lambda(z)\left(\frac{\pi\lambda}{2}\tanh\frac{\pi\lambda}{2}\right)d\lambda. \tag{2.2.8}$$

Then, using Lemma 2.2.1, (2.2.7) and (2.1.35), one obtains

$$\|\Lambda f\|^2_{L^2(\mathbb{R}^{2n})}$$

$$= 4\pi^2\frac{\omega_n\omega_{n-1}}{2}\int_0^\infty s^{n-2}ds\int_0^\infty |K_{\frac{i\lambda}{2}}(4\pi s)|^2\|f_\lambda\|^2_{i\lambda}\left(\frac{\pi\lambda}{2}\tanh\frac{\pi\lambda}{2}\right)d\lambda. \tag{2.2.9}$$

Now, according to [36, p. 101], and using the duplication formula for the Gamma function,

$$\int_0^\infty s^{n-2}\left[K_{\frac{i\lambda}{2}}(4\pi s)\right]^2 ds = 2^{-2n}\pi^{-n+\frac{3}{2}}\frac{\Gamma(\frac{n-1}{2})}{\Gamma(\frac{n}{2})}\Gamma\left(\frac{n-1+i\lambda}{2}\right)\Gamma\left(\frac{n-1-i\lambda}{2}\right). \tag{2.2.10}$$

The theorem follows. $\square$

Theorem 2.2.3. *For every* $w \in \Pi$, *define on* $\mathbb{R}^n$ *the radial function*

$$\phi_w(x) = \left(2\text{Im}\left(-\frac{1}{w}\right)\right)^{\frac{n}{4}}\exp\left(\frac{i\pi}{\bar{w}}|x|^2\right), \tag{2.2.11}$$

generalizing (1.1.32). For every function $f \in L^2(\Pi)$, *one has*

$$(\phi_w|\text{Op}(\Lambda f)\phi_w)_{L^2(\mathbb{R}^n)} = 2^{-n-\frac{1}{2}}\int_\Pi f(z)|(\phi_w|\phi_z)|^2 dm(z)$$

$$= 2^{-\frac{n}{2}-\frac{1}{2}}\int_\Pi f(z)(1 + \cosh d(z,w))^{-\frac{n}{2}}dm(z). \tag{2.2.12}$$

Proof. Using the covariance property, it is no loss of generality to assume that $w = i$. One has

$$W(\phi_i, \phi_i)(x, \xi) = 2^n \exp(-2\pi(|x|^2 + |\xi|^2)) = 2^n e^{-4\pi p}$$

in terms of the coordinates p, q, r introduced just after (1.3.23). Using Lemma 2.2.1, then (1.3.28) to express p in terms of s, z, one can write

$$(\phi_i|\mathrm{Op}(\Lambda f)\phi_i) = 2^{n-1} \omega_n \omega_{n-1} \int_0^\infty s^{n-1} ds$$

$$\times \int_\Pi s^{-\frac{1}{2}} \left(K_{i\sqrt{\Delta-\frac{1}{4}}}(4\pi s) f \right)(z) \exp\left(-4\pi s \frac{1+|z|^2}{2\mathrm{Im}\, z} \right) dm(z): \quad (2.2.13)$$

we set

$$\delta = \cosh d(i, z) = \frac{1+|z|^2}{2\mathrm{Im}\, z}. \tag{2.2.14}$$

To continue the calculation, we must integrate by parts, letting the self-adjoint operator $K_{i\sqrt{\Delta-\frac{1}{4}}}(4\pi s)$ act on the function $z \mapsto \exp(-4\pi s \cosh d(i, z))$ rather than on f. On functions of $\delta = \cosh d(i, z)$, the operator Δ acts as the ordinary differential operator $(1 - \delta^2)\frac{d^2}{d\delta^2} - 2\delta \frac{d}{d\delta}$ and, on the interval $(1, \infty)$, Legendre functions provide generalized eigenfunctions, since

$$\left[(1 - \delta^2)\frac{d^2}{d\delta^2} - 2\delta \frac{d}{d\delta} \right] \mathfrak{P}_{-\frac{1}{2}+\frac{i\lambda}{2}}(\delta) = \frac{1+\lambda^2}{4} \mathfrak{P}_{-\frac{1}{2}+\frac{i\lambda}{2}}(\delta). \tag{2.2.15}$$

Mehler's inversion formula (2.1.32) then gives the integral decomposition

$$e^{-4\pi s\delta} = \int_0^\infty \psi(\lambda) \mathfrak{P}_{-\frac{1}{2}+\frac{i\lambda}{2}}(\delta) d\lambda \tag{2.2.16}$$

if

$$\psi(\lambda) = \frac{\lambda}{4} \tanh \frac{\pi\lambda}{2} \int_1^\infty e^{-4\pi s\delta} \mathfrak{P}_{-\frac{1}{2}+\frac{i\lambda}{2}}(\delta) d\delta$$

$$= 2^{-\frac{3}{2}} \pi^{-1} \frac{\Gamma(\frac{1+i\lambda}{2})\Gamma(\frac{1-i\lambda}{2})}{\Gamma(\frac{i\lambda}{2})\Gamma(\frac{-i\lambda}{2})} s^{-\frac{1}{2}} K_{\frac{i\lambda}{2}}(4\pi s), \tag{2.2.17}$$

where we have used on one hand (2.1.33), on the other hand [36, p. 194] to compute the last integral. Then, the image of the function $e^{-4\pi s \cosh d(i, z)}$ under the operator $K_{i\sqrt{\Delta-\frac{1}{4}}}(4\pi s)$ is

$$K_{i\sqrt{\Delta-\frac{1}{4}}}(4\pi s) \left(e^{-4\pi s\delta} \right)$$

$$= 2^{-\frac{3}{2}} \pi^{-1} \int_0^\infty \frac{\Gamma(\frac{1+i\lambda}{2})\Gamma(\frac{1-i\lambda}{2})}{\Gamma(\frac{i\lambda}{2})\Gamma(\frac{2-i\lambda}{2})} s^{-\frac{1}{2}} \left[K_{\frac{i\lambda}{2}}(4\pi s) \right]^2 \mathfrak{P}_{-\frac{1}{2}+\frac{i\lambda}{2}}(\delta) d\lambda \tag{2.2.18}$$

and, from (2.2.13),

$$(\phi_i|\mathrm{Op}(\Lambda f)\phi_i) = 2^{n-1}\omega_n\omega_{n-1}2^{-\frac{3}{2}}\pi^{-1}\int_0^\infty s^{n-2}ds$$

$$\times \int_\Pi f(z)dm(z)\int_0^\infty \frac{\Gamma(\frac{1+i\lambda}{2})\Gamma(\frac{1-i\lambda}{2})}{\Gamma(\frac{i\lambda}{2})\Gamma(\frac{-i\lambda}{2})}\left[K_{\frac{i\lambda}{2}}(4\pi s)\right]^2 \mathfrak{P}_{-\frac{1}{2}+\frac{i\lambda}{2}}(\delta)d\lambda. \quad (2.2.19)$$

Now [36, p. 101], one has (*cf.* (2.2.10))

$$\int_0^\infty s^{n-2}\left[K_{\frac{i\lambda}{2}}(4\pi s)\right]^2 ds = 2^{-2n}\pi^{\frac{3}{2}-n}\frac{\Gamma(\frac{n-1}{2})}{\Gamma(\frac{n}{2})}\Gamma(\frac{n-1+i\lambda}{2})\Gamma(\frac{n-1-i\lambda}{2}):$$

$$(2.2.20)$$

with the help of the last two equations, one obtains the equation

$$(\phi_w|\mathrm{Op}(\Lambda f)\phi_w) = \frac{2^{-n-\frac{1}{2}}}{(\Gamma(\frac{n}{2}))^2}\int_\Pi f(z)dm(z) \quad (2.2.21)$$

$$\times \int_0^\infty \frac{\Gamma(\frac{1+i\lambda}{2})\Gamma(\frac{1-i\lambda}{2})}{\Gamma(\frac{i\lambda}{2})\Gamma(\frac{-i\lambda}{2})}\Gamma(\frac{n-1+i\lambda}{2})\Gamma(\frac{n-1-i\lambda}{2})\mathfrak{P}_{-\frac{1}{2}+\frac{i\lambda}{2}}(\cosh d(z,w))d\lambda.$$

We transform now the right-hand side of (2.2.12), still under the assumption that $w = i$, by decomposing the function $z \mapsto (1+\cosh d(i,z))^{-\frac{n}{2}}$ into generalized eigenfunctions of Δ. Again, Mehler's inversion formula gives the answer. Using first the Gamma integral, next the integral already used in (2.2.17), we obtain

$$\int_1^\infty (1+\delta)^{-\frac{n}{2}}\mathfrak{P}_{-\frac{1}{2}+\frac{i\lambda}{2}}(\delta)d\delta = \frac{(4\pi)^{\frac{n}{2}}}{\Gamma(\frac{n}{2})}\int_0^\infty s^{\frac{n}{2}}e^{-4\pi s}ds\int_1^\infty e^{-4\pi s\delta}\mathfrak{P}_{\frac{1}{2}+\frac{i\lambda}{2}}(\delta)d\delta$$

$$= \frac{1}{2^{\frac{1}{2}}\pi}\frac{(4\pi)^{\frac{n}{2}}}{\Gamma(\frac{n}{2})}\int_0^\infty s^{\frac{n-3}{2}}e^{-4\pi s}K_{\frac{i\lambda}{2}}(4\pi s)ds$$

$$= 2^{-\frac{n}{2}+1}\frac{\Gamma\left(\frac{n-1+i\lambda}{2}\right)\Gamma\left(\frac{n-1-i\lambda}{2}\right)}{(\Gamma(\frac{n}{2}))^2}: \quad (2.2.22)$$

at the last point, we have used the integral given in [13], p. 98. Then,

$$(1+\delta)^{-\frac{n}{2}} = \int_0^\infty \psi(\lambda)\mathfrak{P}_{\frac{1}{2}+\frac{i\lambda}{2}}(\delta)d\lambda \quad (2.2.23)$$

with

$$\psi(\lambda) = 2^{-\frac{n}{2}}\frac{\Gamma(\frac{1+i\lambda}{2})\Gamma(\frac{1-i\lambda}{2})}{\Gamma(\frac{i\lambda}{2})\Gamma(\frac{-i\lambda}{2})}\frac{\Gamma\left(\frac{n-1+i\lambda}{2}\right)\Gamma\left(\frac{n-1-i\lambda}{2}\right)}{(\Gamma(\frac{n}{2}))^2}. \quad (2.2.24)$$

This proves the identity of the right-hand sides of (2.2.12) and (2.2.21). One also observes, since

$$\phi_{-z^{-1}}(x) = (2\mathrm{Im}\,z)^{\frac{n}{4}}e^{-i\pi\bar{z}|x|^2}, \quad (2.2.25)$$

that

$$|(\phi_{-z-1}|\phi_i)|^2 = 2^n (\mathrm{Im}\, z)^{\frac{n}{2}} [(\mathrm{Re}\, z)^2 + (1 + \mathrm{Im}\, z)^2]^{-\frac{n}{2}}$$

$$= \left(\frac{1 + \cosh d(i,z)}{2} \right)^{-\frac{n}{2}} = |(\phi_z|\phi_i)|^2. \qquad (2.2.26)$$

$\square$

Remark 2.2.a. The present short remark will not be used in all that follows, and addresses itself only to readers interested in quantization theory, in particular in the Berezin calculus [2]. The first equation (2.2.12) can be interpreted as the fact that the function $2^{-n-\frac{1}{2}} f$ coincides with the contravariant symbol of the operator (on functions defined on the half-line) $R\mathrm{Op}(\Lambda f)R^{-1}$, with R as defined in (1.3.1): more precisely, since Berezin considered only complex-type realizations of Hilbert spaces with reproducing kernels, one should consider the conjugate of the last operator under the Laplace transformation defined in (1.3.7).

2.3   The dual Radon transform of bihomogeneous distributions

N.B. This section, in which the function $\chi_{\rho,\nu}$, basic in Chapter 4, is analyzed, has no independent interest: we therefore suggest that the reader should be temporarily satisfied with a look at Proposition 2.3.2 and Proposition 2.3.5. Theorem 2.4.1, in the section to follow, will already give some explanation of our interest in the function $\chi_{\rho,\nu}$.

Theorem 1.1.3 has shown the relevance of homogeneous functions, or distributions on $\mathbb{R}^2$, to modular form theory. It is natural to refine the notion by the consideration of bihomogeneous symbols, considering the variables x, ξ separately. In other words, besides the Euler operator $\mathcal{E} = \frac{1}{2i\pi} \left(x \frac{\partial}{\partial x} + \xi \frac{\partial}{\partial \xi} + 1 \right)$, we wish to consider the operator $\mathcal{B} = \frac{1}{4i\pi} \left(x \frac{\partial}{\partial x} - \xi \frac{\partial}{\partial \xi} \right)$. Since the two operators commute, one may consider their joint spectral theory. Of course, the operator $\mathcal{B}$ does not commute with the action of $SL(2, \mathbb{R})$, or $SL(2, \mathbb{Z})$, and it will not be possible to consider (in Chapter 4) modular distributions which would be at the same time generalized eigenfunctions of it. But applying the Poincaré summation process, starting from functions on Π built from bihomogeneous symbols, will lead to a class of automorphic functions with interesting properties.

Here, we still concentrate on the non-arithmetic situation. Note the equation

$$(\mathcal{B}h)(x,\xi) = \frac{1}{2i\pi} \frac{d}{dr}\bigg|_{r=0} h(e^{\frac{r}{2}} x, e^{-\frac{r}{2}} \xi), \qquad (2.3.1)$$

which indicates that $\mathcal{B}$ is the infinitesimal operator of the action on symbols of the one-parameter group $A \subset SL(2, \mathbb{R})$ recalled in (2.1.6). In view of the covariance property (1.1.20), the operator $\mathcal{B}$ has an interpretation in the Weyl calculus, expressed by the commutation identity, in which h is an arbitrary symbol in $\mathcal{S}'(\mathbb{R}^2)$,

$$\text{Op}(\mathcal{B}h) = \frac{1}{2}\left[\frac{QP + PQ}{2}, \text{Op}(h)\right], \qquad (2.3.2)$$

involving the basic infinitesimal operators Q and P of Heisenberg's representation. By the way, the operator $\mathcal{E}$, too, has an interpretation in the symbolic calculus (not linked to covariance), to wit the general identity

$$\text{Op}(\mathcal{E}h) = P\text{Op}(h)Q - Q\text{Op}(h)P. \qquad (2.3.3)$$

Both formulas are easily obtained from (1.2.6).

In view of arithmetic applications, we consider only even functions of x, ξ in the plane, since the dual Radon transformation kills odd functions. As done in [61, Section 18], the consideration of odd functions of x, ξ is necessary if, besides (Maass) non-holomorphic modular forms of usual type, one interests oneself in so-called Maass forms of weight one [4, Section 2.1]. It is for simplicity that we shall consider here only functions separately even with respect to x and ξ. This will force us to restrict our interest, in Chapter 4, to non-holomorphic modular forms of even type under the symmetry $z \mapsto -\bar{z}$: this is not necessary, but it is sufficient for our main purpose there.

Then, joint generalized eigenfunctions of the pair $(\mathcal{E}, \mathcal{B})$, to wit separately even symbols satisfying the pair of equations

$$2i\pi\mathcal{E}h = \nu h, \quad 4i\pi\mathcal{B}h = (\rho - 1)h, \qquad (2.3.4)$$

are multiples of the function

$$\text{hom}_{\rho,\nu}(x, \xi) = |x|^{\frac{\rho+\nu-2}{2}}|\xi|^{\frac{\nu-\rho}{2}}. \qquad (2.3.5)$$

Theorem 1.2.2, more precisely (1.2.68), has shown how such symbols, with ν on the line $\text{Re}\,\nu = 0$ and ρ on the line $\text{Re}\,\rho = 1$ occur from the decomposition into homogeneous components of a sharp product such as $|x|^{-1-i\lambda_1}\#|\xi|^{-1-i\lambda_2}$.

Our task in the present section is the computation and analysis of the function on Π obtained from the function (2.3.5) by a dual Radon transformation. Transferring under such a transformation the operator $\mathcal{B}$, one will obtain an operator commuting with Δ. Starting from (2.1.8) and using the equation

$$\left(x\frac{\partial}{\partial x} - \xi\frac{\partial}{\partial \xi}\right)\frac{x^2(i + b)}{x\xi(i + b) + 1} = 2\frac{x^2(i + b)}{x\xi(i + b) + 1}, \qquad (2.3.6)$$

one obtains the general identity

$$\mathcal{B}Vf = V\left(\frac{1}{2i\pi}\left(z\frac{\partial}{\partial z} + \bar{z}\frac{\partial}{\partial \bar{z}}\right)f\right). \qquad (2.3.7)$$

Again, on Π, the Euler operator $z\frac{\partial}{\partial z} + \bar{z}\frac{\partial}{\partial \bar{z}}$ does not commute with the action of $SL(2,\mathbb{Z})$ by fractional-linear transformations, and this operator does not preserve automorphic functions: in Chapter 4, something will remain from it, however, in an automorphic situation.

From (2.1.10), we obtain

$$(V^*\hom_{\rho,\nu})(x+iy) = \frac{1}{2\pi}\int_0^{2\pi} |y^{-\frac{1}{2}}\sin\frac{\theta}{2}|^{\frac{\nu-1}{2}}|y^{\frac{1}{2}}\cos\frac{\theta}{2} - xy^{-\frac{1}{2}}\sin\frac{\theta}{2}|^{\frac{\rho+\nu-2}{2}}\,d\theta$$

$$= y^{\frac{\rho-1}{2}} \times \frac{1}{2\pi}\int_0^{2\pi} |\sin\frac{\theta}{2}|^{\frac{\nu-1}{2}}|\cos\frac{\theta}{2} - \frac{x}{y}\sin\frac{\theta}{2}|^{\frac{\rho+\nu-2}{2}}\,d\theta. \tag{2.3.8}$$

We must thus compute the integral obtained, a function of $\frac{x}{y}$ only. The simplest case is that for which $\rho = 1$, which corresponds to MA-invariant symbols. As will be seen, while simpler, it is often a singular case rather than a special case only: this will be even more apparent in Chapter 4. For the time being, the computation of the integral (2.3.8) is quite simple when $\rho = 1$. Indeed, setting $t = \frac{x}{y}$, we first write it as

$$(V^*\hom_{1,\nu})(x+iy) = \frac{1}{2\pi}2^{\frac{1-\nu}{2}}\int_0^{2\pi} |\sin\theta - t(1-\cos\theta)|^{\frac{\nu-1}{2}}\,d\theta : \tag{2.3.9}$$

after a t-dependent translation in the θ-variable, we can change $\sin\theta - t(1-\cos\theta)$ to $\sqrt{1+t^2}\cos\theta$, so that

$$(V^*\hom_{1,\nu})(x+iy) = 2^{\frac{-1-\nu}{2}}\pi^{-1}\int_0^{2\pi} |t - \sqrt{t^2+1}\cos\theta|^{\frac{\nu-1}{2}}\,d\theta. \tag{2.3.10}$$

Starting from the classical integral representation [36, p. 184] of Legendre functions

$$\mathfrak{P}_{\frac{\nu-1}{2}}(w) = \frac{1}{2\pi}\int_0^{2\pi} [w + \sqrt{w^2-1}\cos\theta]^{\frac{\nu-1}{2}}\,d\theta \tag{2.3.11}$$

and using the relation

$$e^{-\frac{i\pi(1-\nu)}{4}} + e^{\frac{i\pi(1-\nu)}{4}} = \frac{2\pi}{\Gamma(\frac{1+\nu}{4})\Gamma(\frac{3-\nu}{4})}, \tag{2.3.12}$$

one obtains, setting $t = \frac{x}{y}$ and assuming that $\operatorname{Re}\nu > -1$ for convergence,

$$(V^*\hom_{1,\nu})(x+iy) = 2^{\frac{-1-\nu}{2}}\pi^{-1}\Gamma(\frac{1+\nu}{4})\Gamma(\frac{3-\nu}{4})\left[\mathfrak{P}_{\frac{\nu-1}{2}}(it) + \mathfrak{P}_{\frac{\nu-1}{2}}(-it)\right] : \tag{2.3.13}$$

this is an analytic function of t on the whole real line, since one has [36, p. 153]

$$\mathfrak{P}_{\frac{\nu-1}{2}}(-it) = {}_2F_1\left(\frac{1-\nu}{2}, \frac{1+\nu}{2}; 1; \frac{1+it}{2}\right), \tag{2.3.14}$$

and the hypergeometric function is single-valued when a cut along the real line, from 1 to ∞, has been made in the plane.

We shall spend more time on the general case, in which ρ is arbitrary. From the second equation (2.3.4), the function $V^*\mathrm{hom}_{\rho,\nu}$ on Π must satisfy the transformation rule $V^*\mathrm{hom}_{\rho,\nu}(az) = a^{\frac{\rho-1}{2}} V^*\mathrm{hom}_{\rho,\nu}(z)$ for $a > 0$. From Theorem 2.1.2 and the fact that the operator T there commutes with $\mathcal{E}$, it must also satisfy the equation $\left(\Delta - \frac{1-\nu^2}{4}\right) V^*\mathrm{hom}_{\rho,\nu} = 0$; finally, it must be invariant under the map $z \mapsto -\bar{z}$. One must thus have

$$V^*\mathrm{hom}_{\rho,\nu}(z) = (\mathrm{Im}\, z)^{\frac{\rho-1}{2}} \chi\left(\frac{\mathrm{Re}\, z}{\mathrm{Im}\, z}\right) \tag{2.3.15}$$

for some even function $\chi = \chi(t)$ on the real line, chosen so that the right-hand side of this equation, as a function of z, should lie in the nullspace of $\Delta - \frac{1-\nu^2}{4}$. Temporarily forgetting the parity condition, it is a straightforward matter to verify that this is the case if and only if the function χ satisfies the ordinary differential equation

$$(1 + t^2)\chi''(t) + (3 - \rho)t\chi'(t) + \left[\frac{1 - \nu^2}{4} + \frac{(\rho - 1)(\rho - 3)}{4}\right]\chi(t) = 0. \tag{2.3.16}$$

We first solve this equation in each of the intervals $]-\infty, 0[$ and $]0, \infty[$. The WKB method shows that, as $t \to \pm\infty$, $\chi(t)$ must be equivalent to a constant times $|t|^{\frac{\mu+\rho-2}{2}}$, with $\mu = \pm\nu$: more precisely, it is so unless the real part of ν is zero. It is then natural to set

$$\chi(t) = \left(\frac{-1 - it}{2}\right)_+^{\frac{\rho+\nu-2}{2}} \psi(t), \tag{2.3.17}$$

where we now make our convention regarding powers of complex numbers with non-integral exponents explicit: we shall denote as z^α the complex power of a number z with $\mathrm{Im}\, z > 0$, when the argument is taken in $]0, \pi[$, and as z_+^α the corresponding complex power of z with $z \notin]-\infty, 0]$ when the argument is taken in $]-\pi, \pi[$. Then,

$$(-iz)_+^\alpha = e^{-\frac{i\pi\alpha}{2}} z^\alpha \quad \text{if } \mathrm{Im}\, z > 0. \tag{2.3.18}$$

Unless otherwise stated, the cut made to make the hypergeometric function a single-valued function will always be the interval $[1, \infty[$.

Lemma 2.3.1. *Given $\rho, \nu \in \mathbb{C}$ with $\nu \notin \mathbb{Z}, \rho + \nu \notin 2\mathbb{Z}$, the function*

$$\chi(t) = \left(\frac{-1 - it}{2}\right)_+^{\frac{\rho+\nu-2}{2}} {}_2F_1\left(\frac{1-\nu}{2}, \frac{2-\rho-\nu}{2} : 1 - \nu; \frac{2}{1+it}\right) \tag{2.3.19}$$

satisfies the equation (2.3.16) in $]-\infty, 0[\cup]0, \infty[$.

Proof. Set $\psi(t) = F(s)$ with $s = \frac{2}{1+it}$. The computations which follow are absolutely tedious but straightforward. If χ and ψ are linked by (2.3.17), one has

$$\frac{\chi'(t)}{\chi(t)} = \frac{\psi'(t)}{\psi(t)} + \frac{i(\rho + \nu - 2)}{4}s,$$

$$\frac{\chi''(t)}{\chi(t)} = \frac{\psi''(t)}{\psi(t)} + \frac{i(\rho + \nu - 2)}{2}s\frac{\psi'(t)}{\psi(t)} - \frac{(2 - \rho - \nu)(4 - \rho - \nu)}{16}s^2. \qquad (2.3.20)$$

Then, (2.3.16) reads

$$(1 + t^2)\psi''(t) + \left[\frac{i(\rho + \nu - 2)}{2}s(1 + t^2) + (3 - \rho)t\right]\psi'(t)$$

$$+ \left[-\frac{(2 - \rho - \nu)(4 - \rho - \nu)}{16}s^2(1 + t^2) \right.$$

$$\left. + \frac{i(3 - \rho)(\rho + \nu - 2)}{4}st + \frac{1 - \nu^2}{4} + \frac{(\rho - 1)(\rho - 3)}{4} \right]\psi(t) = 0. \qquad (2.3.21)$$

Now, one has

$$\psi'(t) = -\frac{i}{2}s^2 F'(s), \quad \psi''(t) = -\frac{s^3}{2}F'(s) - \frac{s^4}{4}F''(s). \qquad (2.3.22)$$

Also,

$$it = \frac{2}{s} - 1, \quad 1 + t^2 = \frac{4(s - 1)}{s^2}, \qquad (2.3.23)$$

and one obtains

$$4(1 - s)\left[\frac{s}{2}F' + \frac{s^2}{4}F''\right] + \left[\frac{i(\rho + \nu - 2)}{2}\frac{4(s - 1)}{s} - i(3 - \rho)(\frac{2}{s} - 1)\right]\left(-\frac{i}{2}s^2 F'\right)$$

$$+ \left[-\frac{(2 - \rho - \nu)(4 - \rho - \nu)}{4}(s - 1) + (3 - \rho)\frac{\rho + \nu - 2}{4}s(\frac{2}{s} - 1) \right.$$

$$\left. + \frac{1 - \nu^2}{4} + \frac{(\rho - 1)(\rho - 3)}{4} \right]F = 0. \qquad (2.3.24)$$

The coefficient of F here reduces to $\frac{(\rho + \nu - 2)(1 - \nu)}{4}s$. The coefficient of F'' is $s^2(1 - s)$, and the coefficient of F' is

$$(\rho + \nu - 2)s(s - 1) + \frac{\rho - 3}{2}(2s - s^2) - 2s(s - 1) = (\nu + \frac{\rho - 5}{2})s^2 + (1 - \nu)s. \qquad (2.3.25)$$

The equation for F equivalent to (2.3.21), hence to (2.3.16), is, after we have divided everything by s,

$$s(1 - s)F''(s) + [1 - \nu - \frac{5 - \rho - 2\nu}{2}s]F'(s) + \frac{(\rho + \nu - 2)(1 - \nu)}{4}F(s) = 0: \qquad (2.3.26)$$

a solution of it is the function $_2F_1\left(\frac{1 - \nu}{2}, \frac{2 - \rho - \nu}{2}; 1 - \nu; s\right)$. This proves the lemma.

$\square$

It is useful to make the way χ transforms under the symmetry $t \mapsto -t$ explicit. From [36, p. 47], one obtains if $\text{Re } z \neq 0$, with some care about determinations of power functions, the general identity

$$_2F_1(a, b; c; z) = e^{-i\pi b \text{sign}(\text{Im } z)}(-z^{-1})_+^b \left(\frac{1-z}{z}\right)_+^{-b} {}_2F_1\left(c - a, b; c; \frac{z}{z-1}\right):$$

(2.3.27)

if $z = \frac{2}{1+it}$ with $t \in \mathbb{R}, t \neq 0$, the signs of $\text{Im } z$ and of t are the negative of each other, and one has $\frac{z}{z-1} = \frac{2}{1-it}$ so that, starting from the hypergeometric function occurring in the definition of χ, one must read the product of power functions on the right-hand side of (2.3.27) as

$$\left(\frac{-1-it}{2}\right)_+^{\frac{2-\rho-\nu}{2}} \left(\frac{-1+it}{2}\right)_+^{\frac{-2+\rho+\nu}{2}}.$$

(2.3.28)

It follows that

$$\chi(t) = e^{\frac{i\pi(2-\nu-\rho)}{2}}\chi(-t) \quad \text{if } t > 0.$$

(2.3.29)

Remark 2.3.a. In the next proposition, we define the function $\chi_{\rho,\nu}$ as a certain multiple of the function χ in (2.3.19). The normalization is chosen so that one should have simply

$$\mathfrak{P}_{\frac{\nu-1}{2}}(-it) = \chi_{1,\nu}(t) + \chi_{1,-\nu}(t)$$

(2.3.30)

and, more important, that the quantities denoted as $C(\rho, \nu)$ and $I(\rho, \nu)$ in what follows should be odd functions of ν.

Proposition 2.3.2. *Assume that* $\nu \notin \mathbb{Z}$ *and* $\rho \pm \nu \notin 2\mathbb{Z}$*, and set*

$$\chi_{\rho,\nu}(t) = 2^{\nu-1}\pi^{-\frac{1}{2}}\frac{\Gamma(\frac{\nu}{2})}{\Gamma(\frac{2-\rho+\nu}{2})}$$

$$\times \left(\frac{-1-it}{2}\right)_+^{\frac{\rho+\nu-2}{2}} {}_2F_1\left(\frac{1-\nu}{2}, \frac{2-\rho-\nu}{2} : 1 - \nu; \frac{2}{1+it}\right). \quad (2.3.31)$$

This function is analytic in $\mathbb{R}\backslash\{0\}$ *and one has for some constant* $C > 0$ *the inequality*

$$|\chi_{\rho,\nu}(t)| \leq C(1 + |t|)^{\frac{\text{Re}(\rho+\nu)-2}{2}}, \quad t \neq 0.$$

(2.3.32)

It extends as a C^∞ *function to each of the two closed intervals* $]-\infty, 0]$ *and* $[0, \infty[$*. The negative of the jump at 0 of the first-order derivative is*

$$C(\rho, \nu) = 2^{2-\rho}\pi^{\frac{1}{2}}\frac{\Gamma(\frac{\nu}{2})\Gamma(\frac{2-\nu}{2})}{\Gamma(\frac{2-\rho+\nu}{2})\Gamma(\frac{2-\rho-\nu}{2})\Gamma(\frac{\rho+\nu}{4})\Gamma(\frac{\rho-\nu}{4})}.$$

(2.3.33)

For $\text{Re}(\rho + \nu) < 0$*, one has*

$$I(\rho, \nu) := \int_{-\infty}^{\infty} \chi_{\rho,\nu}(t)dt = \frac{4C(\rho, \nu)}{\nu^2 - \rho^2}:$$

(2.3.34)

we still denote as $I(\rho, \nu)$ the analytic continuation of this function.

Proof. Let us temporarily denote as $\chi^o_{\rho,\nu}$ the function in (2.3.19), so as not to have to carry the extra coefficient in the first line of (2.3.31) all the time. Similarly, we denote as $C^o(\rho, \nu)$ and $I^o(\rho, \nu)$ the quantities defined in the same way as $C(\rho, \nu)$ and $I(\rho, \nu)$, only with $\chi_{\rho,\nu}$ replaced by $\chi^o_{\rho,\nu}$.

We first consider the case when $\rho - 1 \notin 2\mathbb{Z}$. We need to analyze the function $\chi_{\rho,\nu}(t)$ as $t \to 0^+$ or 0^-, so that the argument $\frac{2}{1+it}$ of the hypergeometric function goes to 2: to avoid arguments close to the half-line $[1, \infty[$, we use [36, p. 48],

$$(-z)^b_+ {}_2F_1(a, b; c; z) = \frac{\Gamma(c)\Gamma(b-a)}{\Gamma(b)\Gamma(c-a)}(-z)^{b-a}_+ {}_2F_1(a, a-c+1; a-b+1; \frac{1}{z})$$
$$+ \frac{\Gamma(c)\Gamma(a-b)}{\Gamma(a)\Gamma(c-b)} {}_2F_1(b, b-c+1; b-a+1; \frac{1}{z}). \quad (2.3.35)$$

This equation, applied with $z = \frac{2}{1+it}$, shows, since in our case $b - a = \frac{1-\rho}{2}$ is assumed to lie outside $\mathbb{Z}$, that the function $\chi_{\rho,\nu}$, while continuous on each of the two intervals $]-\infty, 0]$ and $[0, \infty[$, has a discontinuity at 0. It is an easy, but unnecessary matter to compute the jump there of this function: actually, we shall kill this discontinuity later by considering only the even part of $\chi_{\rho,\nu}$.

It is clear, since the cut along $[1, \infty[$ made to define the hypergeometric function could be moved slightly, that, on each of the two closed intervals under consideration, the function $\chi_{\rho,\nu}$ is actually C^∞. We need to compute the jump of its first derivative at 0. From [36, p. 41], we pick the relations, for $z \notin [0, \infty[$,

$$z^2 \frac{d}{dz}\left[(-z)^{\frac{2-\rho-\nu}{2}}_+ {}_2F_1\left(\frac{1-\nu}{2}, \frac{2-\rho-\nu}{2}; 1-\nu; z\right)\right]$$
$$= \frac{\rho+\nu-2}{2}(-z)^{\frac{4-\rho-\nu}{2}}_+ {}_2F_1\left(\frac{1-\nu}{2}, \frac{4-\rho-\nu}{2}; 1-\nu; z\right) \quad (2.3.36)$$

and

$$z^2 \frac{d}{dz}\left[(-z)^{\frac{-\rho-\nu}{2}}_+ {}_2F_1\left(\frac{1-\nu}{2}, \frac{-\rho-\nu}{2}; 1-\nu; z\right)\right]$$
$$= \frac{\rho+\nu}{2}(-z)^{\frac{2-\rho-\nu}{2}}_+ {}_2F_1\left(\frac{1-\nu}{2}, \frac{2-\rho-\nu}{2}; 1-\nu; z\right). \quad (2.3.37)$$

With $z = \frac{2}{1+it}$, so that $\frac{dz}{dt} = -\frac{i}{2}z^2$, one obtains from these equations the relations

$$\frac{d}{dt}\chi^o_{\rho,\nu} = \frac{i(2-\rho-\nu)}{4}\chi^o_{\rho-2,\nu}, \quad \frac{d}{dt}\chi^o_{\rho+2,\nu} = -\frac{i(\rho+\nu)}{4}\chi^o_{\rho,\nu}(t). \quad (2.3.38)$$

We then apply the general identity (2.3.35) to the new hypergeometric function. When $z = \frac{2}{1+it}$, only the first term on the right-hand side (the one accompanied by a power of $-z$) has discontinuities at $t = 0$: one has

$$\left(\frac{-1-i0^+}{2}\right)_+^{a-b} - \left(\frac{-1-i0^-}{2}\right)_+^{a-b} = 2^{b-a}[e^{-i\pi(a-b)} - e^{i\pi(a-b)}]$$

$$= 2^{1+b-a}\frac{i\pi}{\Gamma(b-a)\Gamma(1+a-b)}. \qquad (2.3.39)$$

It follows on one hand that

$$C^o(\rho,\nu) = \frac{i(\rho+\nu-2)}{4}[\chi^o_{\rho-2,\nu}(0^+) - \chi^o_{\rho-2,\nu}(0^-)] = \frac{2-\rho-\nu}{4}2^{\frac{5-\rho}{2}}$$

$$\times \frac{\pi}{\Gamma(\frac{3-\rho}{2})\Gamma(\frac{\rho-1}{2})}\frac{\Gamma(1-\nu)\Gamma(\frac{3-\rho}{2})}{\Gamma(\frac{4-\rho-\nu}{2})\Gamma(\frac{1-\nu}{2})}{}_2F_1\left(\frac{1-\nu}{2},\frac{1+\nu}{2};\frac{\rho-1}{2};\frac{1}{2}\right), \qquad (2.3.40)$$

on the other hand that

$$I^o(\rho,\nu) = -\frac{4i}{\rho+\nu}[\chi^o_{\rho+2,\nu}(0^+) - \chi^o_{\rho+2,\nu}(0^-)] \qquad (2.3.41)$$

$$= \frac{4}{\rho+\nu}2^{\frac{1-\rho}{2}}\frac{\pi}{\Gamma(\frac{-1-\rho}{2})\Gamma(\frac{3+\rho}{2})}\frac{\Gamma(1-\nu)\Gamma(\frac{-1-\rho}{2})}{\Gamma(\frac{-\rho-\nu}{2})\Gamma(\frac{1-\nu}{2})}{}_2F_1\left(\frac{1-\nu}{2},\frac{1+\nu}{2};\frac{\rho+3}{2};\frac{1}{2}\right).$$

Now, one has [36, p. 41]

$${}_2F_1\left(\frac{1-\nu}{2},\frac{1+\nu}{2};\gamma;\frac{1}{2}\right) = 2^{1-\gamma}\pi^{\frac{1}{2}}\frac{\Gamma(\gamma)}{\Gamma(\frac{1+2\gamma-\nu}{4})\Gamma(\frac{1+2\gamma+\nu}{4})}. \qquad (2.3.42)$$

Obtaining the ratio $\frac{I(\rho,\nu)}{C(\rho,\nu)} = \frac{I^o(\rho,\nu)}{C^o(\rho,\nu)}$ is just a matter of applying the last three formulas, and simplifying a few factors by means of the functional equation of the function Gamma. To obtain $C(\rho,\nu)$, we must also apply the duplication formula, which leads to the equation

$$C^o(\rho,\nu) = 2^{3-\rho-\nu}\pi\frac{\Gamma(\frac{2-\nu}{2})}{\Gamma(\frac{2-\rho-\nu}{2})\Gamma(\frac{\rho+\nu}{4})\Gamma(\frac{\rho-\nu}{4})}, \qquad (2.3.43)$$

and finally to (2.3.33).

This completes the proof of Proposition 2.3.2 under the extra assumption that $\rho-1 \notin 2\mathbb{Z}$. The general case follows by a continuity argument: however, since the case when $\rho = 1$ will be very important in the sequel, let us just indicate the differences in a direct proof in this case. Equation (2.3.35) does not apply any more: instead, one has

$$\chi_{1,\nu}(t) = \frac{\Gamma(\nu)\Gamma(1-\nu)}{(\Gamma(\frac{1+\nu}{2})\Gamma(\frac{1-\nu}{2}))^2}\sum_{n\geq0}\frac{1}{(n!)^2}\frac{\Gamma(\frac{1-\nu}{2}+n)}{\Gamma(\frac{1-\nu}{2})}\frac{\Gamma(\frac{1+\nu}{2}+n)}{\Gamma(\frac{1+\nu}{2})} \qquad (2.3.44)$$

$$\times \left(\frac{1+it}{2}\right)^n\left[\log\left(-\frac{2}{1+it}\right) + 2\frac{\Gamma'(n+1)}{\Gamma(n+1)} - \frac{\Gamma'(\frac{1-\nu}{2}+n)}{\Gamma(\frac{1-\nu}{2}+n)} - \frac{\Gamma'(\frac{1-\nu}{2}-n)}{\Gamma(\frac{1-\nu}{2}-n)}\right];$$

it is understood, there, that the argument $-\frac{2}{1+it}$ of the logarithm is to be taken in the interval $]-\pi, \pi[$. Then, one can write

$$-C(1,\nu): = \left(\frac{d}{dt}\chi_{1,\nu}\right)(0^+) - \left(\frac{d}{dt}\chi_{1,\nu}\right)(0^-)$$

$$= 2i\pi \frac{\Gamma(\nu)\Gamma(1-\nu)}{(\Gamma(\frac{1+\nu}{2})\Gamma(\frac{1-\nu}{2}))^2} \sum_{n\geq 1} \frac{1}{(n)^2} \frac{\Gamma(\frac{1-\nu}{2}+n)}{\Gamma(\frac{1-\nu}{2})} \frac{\Gamma(\frac{1+\nu}{2}+n)}{\Gamma(\frac{1+\nu}{2})} \frac{in}{2^n}$$

$$= -2\pi \frac{\Gamma(\nu)\Gamma(1-\nu)}{(\Gamma(\frac{1+\nu}{2})\Gamma(\frac{1-\nu}{2}))^3} \sum_{n\geq 1} \frac{\Gamma(\frac{1-\nu}{2}+n)\Gamma(\frac{1+\nu}{2}+n)}{\Gamma(n)} \frac{2^{-n}}{n!}$$

$$= -\pi \frac{1-\nu^2}{4} \frac{\Gamma(\nu)\Gamma(1-\nu)}{(\Gamma(\frac{1+\nu}{2})\Gamma(\frac{1-\nu}{2}))^2} {}_2F_1\left(\frac{3-\nu}{2}, \frac{3+\nu}{2}; 2; \frac{1}{2}\right). \qquad (2.3.45)$$

Again, the special value of the hypergeometric function is to be found in [36, p. 40]: it is $\frac{\pi^{\frac{1}{2}}}{\Gamma(\frac{5+\nu}{4})\Gamma(\frac{5-\nu}{4})}$: using this, one obtains the case $\rho = 1$ of (2.3.33).

This was the only place where a special argument was needed when $\rho = 1$. This case will be important in Chapter 4, where its singularity will originate from the fact that the Eisenstein series E_s is undefined for $s = 1$. □

We can now make $V^*\text{hom}_{\rho,\nu}$ explicit: in particular, in view of (2.3.13), it will confirm (2.3.30).

Proposition 2.3.3. *Under the assumptions of Proposition 2.3.2, to be completed by* $\text{Re}\,\nu > \max(\text{Re}\,\rho - 2, -\text{Re}\,\rho)$, *one has*

$$(V^*\text{hom}_{\rho,\nu})(z) = (\text{Im}\,z)^{\frac{\rho-1}{2}} \qquad (2.3.46)$$

$$\times 2^{\frac{\rho-\nu}{2}} \pi^{-1} \frac{\Gamma(\frac{2-\rho+\nu}{2})\Gamma(\frac{\rho+\nu}{4})\Gamma(\frac{4-\rho-\nu}{4})}{\Gamma(\frac{\nu+1}{2})} \left[\chi_{\rho,\nu}^{\text{even}}\left(\frac{\text{Re}\,z}{\text{Im}\,z}\right) + \chi_{\rho,-\nu}^{\text{even}}\left(\frac{\text{Re}\,z}{\text{Im}\,z}\right)\right],$$

with

$$\chi_{\rho,\nu}^{\text{even}}(t) = \frac{1}{2}\left[\chi_{\rho,\nu}(t) + \chi_{\rho,\nu}(-t)\right]. \qquad (2.3.47)$$

Proof. The proof of Proposition 2.3.2 shows that the functions $\chi_{\rho,\pm\nu}^{\text{even}}$ are continuous on the real line, even at 0: they have a discontinuity of the first-order derivative there, expressed by the pair of equations

$$(\chi_{\rho,\nu}^{\text{even}})'(0^+) - (\chi_{\rho,\nu}^{\text{even}})'(0^-) = -C(\rho,\nu),$$

$$(\chi_{\rho,-\nu}^{\text{even}})'(0^+) - (\chi_{\rho,-\nu}^{\text{even}})'(0^-) = -C(\rho,-\nu): \qquad (2.3.48)$$

since the coefficient $C(\rho,\nu)$ is an odd function of ν, the sum $\chi_{\rho,\nu}^{\text{even}} + \chi_{\rho,-\nu}^{\text{even}}$ is a C^1 function on the line, actually a C^∞ function in view of the differential equation it satisfies on each of the two closed intervals $]-\infty, 0]$ and $[0, \infty[$. The function

$y^{\frac{1-\rho}{2}}(V^*\mathrm{hom}_{\rho,\nu})(x+iy)$ must coincide with a multiple of this function. The coefficient is obtained by considering an equivalent as $t = \frac{x}{y} \to \infty$: to obtain this
equivalent, we further assume, which is not a loss of generality because of the
possibility of analytic continuation of the formula obtained, that $\mathrm{Re}\,\nu > 0$. In this
case, it is immediate, from (2.3.8), that

$$y^{\frac{1-\rho}{2}}(V^*\mathrm{hom}_{\rho,\nu})(x+iy) \sim \left|\frac{x}{y}\right|^{\frac{\rho+\nu-2}{2}} \times \frac{1}{2\pi} \int_0^{2\pi} \left|\sin\frac{\theta}{2}\right|^{\nu-1} d\theta$$

$$= \pi^{-\frac{1}{2}} \frac{\Gamma(\frac{\nu}{2})}{\Gamma(\frac{\nu+1}{2})} \left|\frac{x}{y}\right|^{\frac{\rho+\nu-2}{2}}. \tag{2.3.49}$$

On the other hand, (2.3.31), (2.3.47) and the equation

$$e^{\frac{i\pi}{4}(\rho+\nu-2)} + e^{-\frac{i\pi}{4}(\rho+\nu-2)} = \frac{2\pi}{\Gamma(\frac{\rho+\nu}{4})\Gamma(\frac{4-\rho-\nu}{4})} \tag{2.3.50}$$

yield the equivalent, as $|t| \to \infty$,

$$\chi_{\rho,\nu}^{\mathrm{even}}(t) \sim 2^{\frac{\nu-\rho}{2}} \pi^{\frac{1}{2}} \frac{\Gamma(\frac{\nu}{2})}{\Gamma(\frac{2-\rho+\nu}{2})\Gamma(\frac{\rho+\nu}{4})\Gamma(\frac{4-\rho-\nu}{4})} |t|^{\frac{\rho+\nu-2}{2}}. \tag{2.3.51}$$

The proposition follows. $\square$

We need another lemma.

Lemma 2.3.4. *Under the assumptions that $\nu \notin \mathbb{Z}$, $\rho - 1 \notin 2\mathbb{Z}$ and $\rho \pm \nu \notin 2\mathbb{Z}$, one
has*

$$(1+t^2)^{\frac{1-\rho}{2}}\chi_{\rho,\nu}(t) = \frac{\Gamma(\frac{2+\rho-\nu}{4})\Gamma(\frac{2+\rho+\nu}{4})}{\Gamma(\frac{4-\rho-\nu}{4})\Gamma(\frac{4-\rho+\nu}{4})}\chi_{2-\rho,\nu}(t). \tag{2.3.52}$$

Proof. We start from [36, p. 47], writing when $\mathrm{Re}\,z \neq 0$ the identity

$$_2F_1\left(\frac{1-\nu}{2},\frac{\rho-\nu}{2};1-\nu;z\right) = \exp\left(i\pi\frac{1-\rho}{2}\mathrm{sign}(\mathrm{Im}\,z)\right)$$

$$\times (-z)_+^{\frac{1-\rho}{2}}\left(\frac{1-z}{z}\right)_+^{\frac{1-\rho}{2}} {}_2F_1\left(\frac{1-\nu}{2},\frac{2-\rho-\nu}{2};1-\nu;z\right). \tag{2.3.53}$$

With $z = \frac{2}{1+it}$, one has $(-z)^{-1} = \frac{-1-it}{2}$, $\frac{1-z}{z} = \frac{-1+it}{2}$, and one must read the
product of power functions on the right-hand side of (2.3.53) as

$$e^{i\pi\frac{\rho-1}{2}\mathrm{sign}\,t}\left(\frac{-1-it}{2}\right)_+^{\frac{\rho-1}{2}}\left(\frac{-1+it}{2}\right)_+^{\frac{1-\rho}{2}}. \tag{2.3.54}$$

One can then write

$$\chi^o_{2-\rho,\nu}(t) = \left(\frac{-1-it}{2}\right)_+^{\frac{\nu-\rho}{2}} {}_2F_1\left(\frac{1-\nu}{2},\frac{\rho-\nu}{2};1-\nu;\frac{2}{1+it}\right)$$

$$= e^{i\pi\frac{\rho-1}{2}\operatorname{sign}t}\left(\frac{-1-it}{2}\right)_+^{\frac{\nu-1}{2}}\left(\frac{-1+it}{2}\right)^{\frac{1-\rho}{2}}_+ {}_2F_1\left(\frac{1-\nu}{2},\frac{2-\rho-\nu}{2};1-\nu;\frac{2}{1+it}\right)$$

$$= e^{i\pi\frac{\rho-1}{2}\operatorname{sign}t}\left(\frac{1+t^2}{4}\right)^{\frac{1-\rho}{2}}\left(\frac{-1-it}{2}\right)_+^{\frac{\nu+\rho-2}{2}} {}_2F_1\left(\frac{1-\nu}{2},\frac{2-\rho-\nu}{2};1-\nu;\frac{2}{1+it}\right)$$

$$= e^{i\pi\frac{\rho-1}{2}\operatorname{sign}t}\left(\frac{1+t^2}{4}\right)^{\frac{1-\rho}{2}}\chi^o_{\rho,\nu}(t). \tag{2.3.55}$$

Hence,

$$\left(\frac{1+t^2}{4}\right)^{\frac{1-\rho}{2}}\left(\chi^o_{\rho,\nu}\right)^{\text{even}}(t) = \frac{1}{2}\left[e^{i\pi\frac{1-\rho}{2}}\chi^o_{2-\rho,\nu}(t)+e^{i\pi\frac{\rho-1}{2}}\chi^o_{2-\rho,\nu}(-t)\right] \tag{2.3.56}$$

or, using (2.3.29), one has for $t > 0$

$$\left(\frac{1+t^2}{4}\right)^{\frac{1-\rho}{2}}\left(\chi^o_{\rho,\nu}\right)^{\text{even}}(t) = \frac{1}{2}\left[e^{i\pi\frac{1-\rho}{2}}+e^{i\pi\frac{\rho-1}{2}}e^{i\pi\frac{\nu-\rho}{2}}\right]\chi^o_{2-\rho,\nu}(t)$$

$$= \frac{\pi e^{i\pi\frac{\nu-\rho}{4}}}{\Gamma(\frac{\rho+\nu}{4})\Gamma(\frac{4-\rho-\nu}{4})}\chi^o_{2-\rho,\nu}(t): \tag{2.3.57}$$

using (2.3.29) again,

$$\left(\chi^o_{2-\rho,\nu}\right)^{\text{even}}(t) = \frac{\pi e^{i\pi\frac{\nu-\rho}{4}}}{\Gamma(\frac{2-\rho+\nu}{4})\Gamma(\frac{2+\rho-\nu}{4})}\chi^o_{2-\rho,\nu}(t). \tag{2.3.58}$$

Hence,

$$\left(\frac{1+t^2}{4}\right)^{\frac{1-\rho}{2}}\left(\chi^o_{\rho,\nu}\right)^{\text{even}}(t) = \frac{\Gamma(\frac{2-\rho+\nu}{4})\Gamma(\frac{2+\rho-\nu}{4})}{\Gamma(\frac{\rho+\nu}{4})\Gamma(\frac{4-\rho-\nu}{4})}\left(\chi^o_{2-\rho,\nu}\right)^{\text{even}}(t). \tag{2.3.59}$$

Finally, using the extra coefficient from $\chi^o_{\rho,\nu}$ to $\chi_{\rho,\nu}$ given in (2.3.31), one obtains

$$\frac{\left(\frac{1+t^2}{4}\right)^{\frac{1-\rho}{2}}\left(\chi^o_{\rho,\nu}\right)^{\text{even}}(t)}{\left(\chi^o_{2-\rho,\nu}\right)^{\text{even}}(t)} = 2^{1-\rho}\frac{\Gamma(\frac{2-\rho+\nu}{4})\Gamma(\frac{2+\rho-\nu}{4})}{\Gamma(\frac{\rho+\nu}{4})\Gamma(\frac{4-\rho-\nu}{4})}\frac{\Gamma(\frac{\rho+\nu}{2})}{\Gamma(\frac{2-\rho+\nu}{2})}, \tag{2.3.60}$$

which simplifies to (2.3.52) by an application of the duplication formula.

From Proposition 2.3.2, one may note that the product $\Gamma(\frac{4-\rho+\nu}{4})\Gamma(\frac{4-\rho-\nu}{4}) \times C(\rho, \nu)$ is invariant under the map $\rho \mapsto 2 - \rho$. It follows from this, and Lemma 2.3.4, that

$$\int_{-\infty}^{\infty} (1+t^2)^{\frac{1-\rho}{2}} \chi_{\rho,\nu}^{\text{even}}(t) dt = \frac{(\nu - \rho)(\nu + \rho)}{(\nu - 2 + \rho)(\nu + 2 - \rho)} I(\rho, \nu). \tag{2.3.61}$$

$\square$

To kill the discontinuity of $\chi_{\rho,\nu}$ at the origin, we replace it by its symmetrized version $\chi_{\rho,\nu}^{\text{even}}$ as defined in (2.3.47): note that this does not change the jump of the first-order derivative at 0.

Proposition 2.3.5. *Assume that $\nu \notin \mathbb{Z}$ and that $\rho \pm \nu \notin 2\mathbb{Z}$. One has in the distribution sense*

$$\left[-(1+t^2)\frac{d^2}{dt^2} + (\rho - 3)t\frac{d}{dt} - \frac{1 - \nu^2}{4} - \frac{(\rho - 1)(\rho - 3)}{4} \right] \chi_{\rho,\nu}^{\text{even}}(t) = C(\rho, \nu)\delta.$$

$$\tag{2.3.62}$$

On the other hand, one has in Π the equation

$$\left(\Delta - \frac{1 - \nu^2}{4} \right) \left[z \mapsto (\text{Im } z)^{\frac{\rho-1}{2}} \chi_{\rho,\nu}^{\text{even}} \left(\frac{\text{Re } z}{\text{Im } z} \right) \right] = C(\rho, \nu)(\text{Im } z)^{\frac{\rho-1}{2}} \delta_{(0, i\infty)},$$

$$\tag{2.3.63}$$

where $\delta_{(0,i\infty)}$ is the measure $\frac{dy}{y}$ on the hyperbolic line from 0 to $i\infty$.

Proof. Since the function $\chi_{\rho,\nu}^{\text{even}}$ is C^∞ in $[0, \infty[$ (up to the boundary), continuous on the line, and since it satisfies in $]0, \infty[$ the differential equation (2.3.16), its image, in the distribution sense, under the operator on the left-hand side of (2.3.62), depends only on the discontinuity at 0 of its first-order derivative. One can then, without modifying the result, replace the complete operator by $-\frac{d^2}{dt^2}$, and χ_ν^{even} by the function, linear on $] - \infty, 0]$ and on $[0, \infty[$, with the same half-derivatives at 0: this leads to (2.3.62).

The second part of the lemma is a corollary of the first. $\square$

2.4 The symmetries $\nu \mapsto -\nu$ and $\rho \mapsto 2 - \rho$

In the function $\text{hom}_{\rho,\nu}$ in the plane defined in (2.3.5), the parameters ρ and ν appear in a clear way: in particular, $-1 + \nu$ stands for the global degree of homogeneity, so that the functions $\text{hom}_{\rho,\nu}$ and $h_{\rho,-\nu}$, when $\text{Re }\nu < 0$, could be regarded as "ingoing" and "outgoing" in the sense of scattering theory [34]. However, the $\mathcal{G}$-transform of a function homogeneous of degree $-1 + \nu$ is homogeneous of degree $-1 - \nu$, and, under the dual Radon transform or any of its associates, a homogeneous distribution and its $\mathcal{G}$-transform have proportional images (*cf.* Theorem 2.1.2). Indeed, Proposition 2.3.3 confirms that, up to multiplication by a scalar, $V^*\text{hom}_{\rho,\nu}$ depends only on the pair (ρ, ν^2).

The function $x + iy \mapsto (C(\rho,\nu))^{-1} y^{\frac{\rho-1}{2}} \chi_{\rho,\nu}^{\text{even}}\left(\frac{x}{y}\right)$ is the starting point from which, with the help of a Poincaré series process, we shall build automorphic functions of a new style in Chapter 4. As a consequence of Lemma 2.3.4, it changes to a multiple when ρ is changed to $2-\rho$. On the other hand, it is essential, in view of our applications in Chapter 4, to make a clearcut distinction between the function $(C(\rho,\nu))^{-1} y^{\frac{\rho-1}{2}} \chi_{\rho,\nu}^{\text{even}}\left(\frac{x}{y}\right)$ and its transform under the symmetry $\nu \mapsto -\nu$, even though they satisfy the same differential equation (2.3.63). More precisely, in order to consider integral superpositions of these functions with a fixed ρ and $\operatorname{Re}\nu < 0$, we need to characterize, when $\operatorname{Re}\nu < 0$, the first of these functions within the pair under consideration.

The answer is provided by the resolvent of Δ: as is well-known (say, from spherical function theory, i.e., the reduction to K-invariant functions), the operator Δ is essentially self-adjoint in $L^2(\Pi) = L^2(\Pi, dm)$, where $dm(x+iy) = \frac{dx\,dy}{y^2}$, if, say, one takes $C_0^\infty(\Pi)$ as its initial domain; it has a purely continuous spectrum, coinciding with the interval $[\frac{1}{4}, \infty[$, so that the resolvent $\nu \mapsto \left(\Delta - \frac{1-\nu^2}{4}\right)^{-1}$ is well-defined for $\operatorname{Re}\nu \neq 0$. It is usually made explicit in terms of its integral kernel $k_{\frac{1-\nu}{2}}(z, z')$, a function of $d = d(z, z')$, according to the general (Gelfand's) theory of point-pair invariants: reducing the problem to its special case concerned with K-invariant theory, one obtains explicitly, assuming $\operatorname{Re}\nu < 0$,

$$k_{\frac{1-\nu}{2}}(z, z') = \frac{1}{4\pi} \frac{(\Gamma(\frac{1-\nu}{2}))^2}{\Gamma(1-\nu)} (\cosh \frac{d}{2})^{\nu-1} {}_2F_1\left(\frac{1-\nu}{2}, \frac{1-\nu}{2}; 1-\nu; \frac{1}{\cosh^2 \frac{d}{2}}\right).$$

(2.4.1)

This can be found in many places, including [32, 55], and could also be derived from (2.4.11) below.

This formula does not lead to tractable integrals when $\left(\Delta - \frac{1-\nu^2}{4}\right)^{-1}$ has to be applied to general (not K-invariant) functions: the proof of the theorem to follow will rely on an alternative construction [60, p. 205] of the resolvent, based of M.Riesz's theory [43] dealing with the solid convex cone in $\mathbb{R}^3$. We shall dispense with giving a priori arguments showing that the resolvent extends to spaces of distributions containing measures such as the one occurring in the next theorem: this will result, instead, from the explicit form (2.4.9) of $\left(\Delta - \frac{1-\nu^2}{4}\right)^{-1}$.

Theorem 2.4.1. *Assume that $0 < \operatorname{Re}\rho < 2$ and $\operatorname{Re}\nu < 0$. It is convenient to set*

$$\psi(z) = \frac{\operatorname{Re} z}{\operatorname{Im} z}. \tag{2.4.2}$$

Recalling Proposition 2.3.5, denote as $\delta_{(0,i\infty)}^{(\rho)}$ the measure in Π supported by the hyperbolic line from 0 to $i\infty$, coinciding with $y^{\frac{\rho-1}{2}} \frac{dy}{y}$ in terms of $y = \operatorname{Im} z$. One has

$$\left[\left(\Delta - \frac{1-\nu^2}{4}\right)^{-1} \delta_{(0,i\infty)}^{(\rho)}\right](z) = \frac{1}{C(\rho,\nu)} (\mathrm{Im}\, z)^{\frac{\rho-1}{2}} \left(\chi_{\rho,\nu}^{\mathrm{even}} \circ \psi\right)(z), \qquad (2.4.3)$$

with $C(\rho, \nu)$ as defined in (2.3.33).

Proof. Let C be the cone in $\mathbb{R}^3$ consisting of points $\eta = (\eta_0, \eta_1, \eta_2)$ with $\eta_0 > 0$ and $\eta_0^2 - \eta_1^2 - \eta_2^2 > 0$, and let $\mathcal{H}$ be the sheet of hyperboloid defined within C by the equation $\eta_0^2 - \eta_1^2 - \eta_2^2 = 1$. It is a very classical fact that $\mathcal{H}$, provided with the (Riemannian) metric which is the restriction to it of the indefinite metric $-d\eta_0^2 + d\eta_1^2 + d\eta_2^2$ in C, is another model of $\Pi = G/K$: the map κ from $\mathcal{H}$ to Π providing the required isometry is defined as $\kappa(\eta) = \frac{\eta_2 + i}{\eta_0 - \eta_1}$. With

$$\square = \frac{\partial^2}{\partial \eta_0^2} - \frac{\partial^2}{\partial \eta_1^2} - \frac{\partial^2}{\partial \eta_2^2} \qquad (2.4.4)$$

and

$$E = \eta_0 \frac{\partial}{\partial \eta_0} + \eta_1 \frac{\partial}{\partial \eta_1} + \eta_2 \frac{\partial}{\partial \eta_2}, \qquad (2.4.5)$$

it is easily verified (this is an extension to hyperboloids of the classical theory of spherical harmonics) that if Ψ is a function in C homogeneous of degree $k \in \mathbb{C}$ satisfying the equation $\square \Psi = 0$, its restriction to $\mathcal{H}$ satisfies the eigenvalue equation

$$\Delta_{\mathcal{H}} \left(\Psi|_{\mathcal{H}}\right) = -k(k+1)\Psi|_{\mathcal{H}}, \qquad (2.4.6)$$

if one denotes as $\Delta_{\mathcal{H}}$ the operator obtained by transferring under κ the hyperbolic Laplacian Δ on Π.

M.Riesz's theory [43, 44] gives a fundamental solution at 0 of the operator $\square$ as the convolution by the function (supported in the closure of C)

$$Z_2 = \frac{1}{2\pi} \left(\eta_0^2 - \eta_1^2 - \eta_2^2\right)_{\mathrm{pos}}^{-\frac{1}{2}}, \qquad (2.4.7)$$

where the subscript indicates that the whole function is to be multiplied by the characteristic function of C. Then, if Ψ is homogeneous of degree $\frac{-5-\nu}{2}$ in C, the function $Z_2 * \Psi$ lies in the nullspace of $\square$ and is homogeneous of degree $\frac{-1-\nu}{2}$ so that, as a consequence of the equation (2.4.6) taken with $k = \frac{-1-\nu}{2}$, one has

$$\left(\Delta_{\mathcal{H}} - \frac{1-\nu^2}{4}\right) \left((Z_2 * \Psi)|_{\mathcal{H}}\right) = \Psi|_{\mathcal{H}}. \qquad (2.4.8)$$

This provides the following recipe for computing the image of the resolvent $\left(\Delta - \frac{1-\nu^2}{4}\right)^{-1}$ on a function $f \in C_0^\infty(\Pi)$, under the assumption (to be justified presently) that $\mathrm{Re}\, \nu < 0$: transfer f to the function $f \circ \kappa$ on $\mathcal{H}$, extend the function obtained to a function Ψ on C homogeneous of degree $\frac{-5-\nu}{2}$, restrict the function

$Z_2 * \Psi$ to $\mathcal{H}$, finally compose this restriction with κ^{-1}. Since, for $r > 0$ and $\xi \in \mathcal{H}$, one has $d(r\xi) = r^2 dr \frac{d\xi_1 d\xi_2}{\xi_0}$, one obtains the formula we have been looking for:

$$\left[\left(\Delta - \frac{1 - \nu^2}{4}\right)^{-1} f\right](\kappa(\eta)) = \frac{1}{2\pi} \int_0^\infty r^{\frac{-1-\nu}{2}} dr$$

$$\times \int_{\mathcal{H}} [(\eta_0 - r\xi_0)^2 - (\eta_1 - r\xi_1)^2 - (\eta_2 - r\xi_2)^2]_{\text{pos}}^{-\frac{1}{2}} f(\kappa(\xi)) \frac{d\xi_1 d\xi_2}{\xi_0}. \quad (2.4.9)$$

One easily checks that the measure $\frac{d\xi_1 d\xi_2}{\xi_0}$ coincides with $dm(\kappa(\xi))$, the transfer under κ of the invariant measure dm on Π.

When $\kappa(\eta) = z$ and $\kappa(\xi) = z'$, one has

$$\eta_0 \xi_0 - \eta_1 \xi_1 - \eta_2 \xi_2 = \cosh d(z, z'), \quad (2.4.10)$$

so that the integral kernel of the operator obtained in (2.4.9) is the function

$$(z, z') \mapsto \frac{1}{2\pi} \int_0^{e^{-d(z,z')}} r^{\frac{\nu-1}{2}} (1 - 2r \cosh d(z, z') + r^2)^{-\frac{1}{2}} dr$$

$$= \frac{1}{2\pi^{\frac{1}{2}}} \frac{\Gamma(\frac{1-\nu}{2})}{\Gamma(\frac{2-\nu}{2})} e^{\frac{(\nu-1)d(z,z')}{2}} {}_2F_1\left(\frac{1}{2}, \frac{1-\nu}{2}; \frac{2-\nu}{2}; e^{-2d(z,z')}\right). \quad (2.4.11)$$

With the help of two transformations (one of which is quadratic) of the hypergeometric function, one can see [60, p. 206] that this is identical to the integral kernel $k(z, z')$ in (2.4.1): however, the expression (2.4.9) of the operator will be more convenient in the MA-invariant case. From any of the two expressions of the integral kernel in (2.4.11), one sees that it is bounded by a constant times $|\log d(z, z')|$ near the diagonal and by a constant times $(\cosh d(z, z'))^{\frac{\operatorname{Re}\nu - 1}{2}}$ when $d(z, z') \geq 1$. Since this kernel is K-biinvariant and, in polar geodesic coordinates ρ, θ around $i \in \Pi$ (with $\rho = d(i, z)$), the invariant measure expresses itself as $\sinh \rho d\rho d\theta$, it follows from the most popular criterion regarding L^2-continuity that, provided that $\operatorname{Re}\nu < -1$, the operator $\left(\Delta - \frac{1-\nu^2}{4}\right)^{-1}$ defined in (2.4.9) is indeed the resolvent. Now, if one only has $\operatorname{Re}\nu < 0$, the criterion just mentioned does not apply but it is immediate from the same estimates regarding the kernel that the operator under consideration is continuous from $L^2(\Pi)$ to the Banach space of bounded continuous functions: as such, it depends analytically on ν, and must still coincide with the resolvent (hence, be a continuous endomorphism of $L^2(\Pi)$).

We now substitute for f the measure $\delta_{(0,i\infty)}^{(\rho)}$. When $\xi_2 = 0$ and $y = \frac{1}{\xi_0 - \xi_1}$, one has $\frac{dy}{y} = \frac{d\xi_1}{\xi_0}$. We must thus, in the preceding integral, replace $f(\kappa(\xi)) \frac{d\xi_1 d\xi_2}{\xi_0}$ by $\frac{d\xi_1}{\sqrt{1+\xi_1^2}}$ and set the variable ξ_2 at the value 0, getting as a result (with $\xi_0 = \sqrt{1 + \xi_1^2}$)

$$\ell_{\rho,\nu}(\kappa(\eta)): = \left[\left(\Delta - \frac{1-\nu^2}{4}\right)^{-1} \delta^{(\rho)}_{(0,i\infty)}\right](\kappa(\eta)) \tag{2.4.12}$$

$$= \frac{1}{2\pi} \int_0^\infty r^{\frac{-1-\nu}{2}} dr \int_{-\infty}^\infty [(\eta_0 - r\xi_0)^2 - (\eta_1 - r\xi_1)^2 - \eta_2^2]_{\text{pos}}^{-\frac{1}{2}} (\xi_0 - \xi_1)^{\frac{1-\rho}{2}} \frac{d\xi_1}{\xi_0}.$$

Now,

$$(\eta_0 - r\xi_0)^2 - (\eta_1 - r\xi_1)^2 - \eta_2^2 = 1 - 2r(\eta_0\xi_0 - \eta_1\xi_1) + r^2 : \tag{2.4.13}$$

recalling that $1 + \eta_2^2 = \eta_0^2 - \eta_1^2$, consider the matrix $(1 + \eta_2)^{-\frac{1}{2}} \left(\begin{smallmatrix} \eta_0 & \eta_1 \\ \eta_1 & \eta_0 \end{smallmatrix}\right)$, which corresponds to a Lorentz transformation (in $(1+1)$-dimensional spacetime) in the variable (ξ_0, ξ_1), thus preserving the measure $\frac{d\xi_1}{\xi_0}$. Under this transformation, the expression (2.4.13) transforms into $1 - 2r\xi_0\sqrt{1 + \eta_2^2} + r^2$, while $\xi_0 - \xi_1$ transforms into $(1 + \eta_2^2)^{-\frac{1}{2}}(\eta_0 - \eta_1)(\xi_0 - \xi_1)$. Hence, setting $z = \kappa(\eta)$, so that $\eta_0 - \eta_1 = (\text{Im}\, z)^{-1}, \eta_2 = \frac{\text{Re}\, z}{\text{Im}\, z} = \psi(z)$, one obtains

$$\ell_{\rho,\nu}(z) = (\text{Im}\, z)^{\frac{\rho-1}{2}} \left(1 + (\psi(z))^2\right)^{\frac{\rho-1}{4}} \tag{2.4.14}$$

$$\times \frac{1}{2\pi} \int_0^\infty r^{\frac{-1-\nu}{2}} dr \int_{-\infty}^\infty [1 - 2r\xi_0\sqrt{1 + (\psi(z))^2} + r^2]_{\text{pos}}^{-\frac{1}{2}} (\xi_0 - \xi_1)^{\frac{1-\rho}{2}} \frac{d\xi_1}{\xi_0}.$$

This is an even function of $t = \frac{\text{Re}\, z}{\text{Im}\, z}$, which can be written, after one has performed the change of variable $r \mapsto [2\xi_0\sqrt{1 + t^2}]^{-1} r$, as

$$\ell_{\rho,\nu}(z) = \frac{2^{\frac{\nu-1}{2}}}{2\pi} (\text{Im}\, z)^{\frac{\rho-1}{2}} (1 + t^2)^{\frac{\rho+\nu-2}{4}} \text{Int}_{\rho,\nu}(t), \tag{2.4.15}$$

with

$$\text{Int}_{\rho,\nu}(t) = \int_0^\infty r^{\frac{-\nu-1}{2}} dr \int_{-\infty}^\infty \left[1 - r + \frac{r^2}{4(1 + t^2)(1 + \xi_1^2)}\right]_{\text{pos}}^{-\frac{1}{2}} (\xi_0 - \xi_1)^{\frac{1-\rho}{2}} \xi_0^{\frac{\nu-3}{2}} d\xi_1. \tag{2.4.16}$$

As $|t| \to \infty$, the integral goes to

$$\text{Int}_{\rho,\nu}(\infty) = \int_0^1 r^{\frac{-\nu-1}{2}} (1 - r)^{-\frac{1}{2}} dr \int_{-\infty}^\infty (1 + \xi_1^2)^{\frac{\nu-3}{4}} (\xi_0 - \xi_1)^{\frac{-1+\rho}{2}} d\xi_1$$

$$= \pi^{\frac{1}{2}} \frac{\Gamma(\frac{1-\nu}{2})}{\Gamma(\frac{2-\nu}{2})} \times \int_{-\infty}^\infty (\cosh t)^{\frac{\nu-1}{2}} e^{\frac{(1-\rho)t}{2}} dt$$

$$= 2^{\frac{-1-\nu}{2}} \pi^{\frac{1}{2}} \frac{\Gamma(\frac{\rho-\nu}{4})\Gamma(\frac{2-\rho-\nu}{4})}{\Gamma(\frac{2-\nu}{2})}, \tag{2.4.17}$$

where we have used [36, p. 432] at the end: note that this expression does not change under the symmetry $\rho \mapsto 2 - \rho$. One thus has the equivalent, as $\frac{\text{Re}\, z}{\text{Im}\, z} \to \infty$,

$$(\text{Im}\, z)^{\frac{1-\rho}{2}} \ell_{\rho,\nu}(z) \sim \frac{\pi^{-\frac{1}{2}}}{4} \frac{\Gamma(\frac{\rho-\nu}{4})\Gamma(\frac{2-\rho-\nu}{4})}{\Gamma(\frac{2-\nu}{2})} \left|\frac{\text{Re}\, z}{\text{Im}\, z}\right|^{\frac{\rho+\nu-2}{2}}. \tag{2.4.18}$$

Since the image under $\Delta - \frac{1-\nu^2}{4}$ of the function $\ell_{\rho,\nu}$ is zero in the complement of the hyperbolic line from 0 to $i\infty$, it follows from the structure of $\ell_{\rho,\nu}(z)$ as the product of $(\operatorname{Im} z)^{\frac{1-\rho}{2}}$ by an even function of $\frac{\operatorname{Re} z}{\operatorname{Im} z}$ and from (2.3.16) that the function in (2.4.18) must be a linear combination of the functions $\chi_{\rho,\nu}^{\mathrm{even}}(t)$ and $\chi_{\rho,-\nu}^{\mathrm{even}}(t)$: in view of the equivalent of $\chi_{\rho,\nu}(t)$ as $|t| \to \infty$ resulting from (2.3.31), it has to be a multiple of the function $\chi_{\rho,\nu}^{\mathrm{even}}(t)$ only (recall the assumption that $\operatorname{Re}\nu < 0$). The proof of Theorem 2.4.1 now reduces to proving that the functions $\ell_{\rho,\nu}(t)$ and $\frac{1}{C(\rho,\nu)}\chi_\nu^{\mathrm{even}}(t)$ are equivalent as $|t| \to \infty$.

Comparing (2.4.18) to the equivalent (2.3.51) of the second function, using the duplication formula

$$\Gamma(\frac{2-\rho-\nu}{4})\Gamma(\frac{4-\rho-\nu}{4}) = (2\pi)^{\frac{1}{2}}2^{\frac{-1+\rho+\nu}{2}}\Gamma(\frac{2-\rho-\nu}{2}) \tag{2.4.19}$$

and the expression (2.3.33) of $C(\rho,\nu)$, we obtain (2.4.3).   □

Remark 2.4.a. Even though one has $\left(\Delta - \frac{1-\nu^2}{4}\right)\left((\operatorname{Im} z)^{\frac{\rho-1}{2}}\chi_{\rho,\nu} \circ \psi\right) = 0$ in the complement of a one-dimensional set, this phenomenon leaves no trace after one has applied to the function $(\operatorname{Im} z)^{\frac{\rho-1}{2}}\chi_{\rho,\nu} \circ \psi$ a non-local operator such as the Radon transformation. It would be somewhat misleading to regard the function under consideration as "almost" an eigenfunction of Δ.

Our construction of Poincaré series (of a novel kind) in Chapter 4 is based on the use of the functions $(\operatorname{Im} z)^{\frac{\rho-1}{2}}\chi_{\rho,\nu} \circ \psi$, with $\operatorname{Re}\nu < 0$. These Poincaré series will take the place usually taken by Eisenstein series. In a way similar to that which leads, classically, to so-called incomplete Eisenstein series (this terminology, borrowed from [21, 23], sounds more appropriate than the traditional one of incomplete theta series), we may consider integral superpositions of the functions $\chi_{\rho,\nu}$ for a fixed ρ. What we obtain as a result is a space of images of the measure $\delta_{(0,i\infty)}^{(\rho)}$ (introduced in Theorem 2.4.1) under fairly general functions, in the spectral-theoretic sense, of the Laplacian: an integral transform will make these explicit.

In view of the spectrum of the hyperbolic Laplacian Δ, whether in the free half-plane or in the automorphic situation, a function of Δ is the same as an even function H of the operator $2\sqrt{\Delta - \frac{1}{4}}$ (the factor 2 is of course for convenience only), provided that, in the second case, one interests oneself only in automorphic functions orthogonal to constants (so that the square root should not create a difficulty). Experience, in particular with quantization theory [63, p. 57-59], shows that, as a function of one real variable, it is most of the time the function H, rather than the corresponding function of Δ, that appears simple, or interesting. This may be considered, in view of (2.1.5), as one more argument in favor of using the plane, rather than the half-plane, in the study of Δ.

Theorem 2.4.2. *Let* $H = H(\mu)$ *be an even holomorphic function in some strip* $|\operatorname{Im} \mu| < \beta_0$, *such that* $\int_{\operatorname{Im} \mu = \beta} |\mu|^2 |H(\mu)|^2 d\mu < \infty$ *for every* β *with* $|\beta| < \beta_0$: *set* $G(\sigma) = \int_{-\infty}^{\infty} H(\lambda) e^{2i\pi\lambda\sigma} d\lambda$. *Assuming* $0 < \operatorname{Re}\rho < 2$, *the image of the measure* $\delta_{(0,i\infty)}^{(\rho)}$ *under the operator* $H\left(2\sqrt{\Delta - \frac{1}{4}}\right)$ *is a function* ϕ, *which can be made explicit in terms of* $\sinh(4\pi\tau) = \frac{\operatorname{Re} z}{\operatorname{Im} z}$ *as*

$$\phi(z) = -\frac{1}{4\pi} (\operatorname{Im} z)^{\frac{\rho-1}{2}} (\cosh 4\pi\tau)^{\frac{\rho-2}{2}} \int_{|\tau|}^{\infty} G'(\sigma) \mathfrak{P}_{\frac{\rho-2}{2}}\left(\frac{\cosh 4\pi\sigma}{\cosh 4\pi\tau}\right) d\sigma. \quad (2.4.20)$$

Given any number $\beta \in]0, \beta_0[$, *one has the identity*

$$\phi(z) = -\frac{1}{4i\pi} \int_{\operatorname{Re}\nu = -\beta} \nu H(i\nu) \left[\left(\Delta - \frac{1 - \nu^2}{4}\right)^{-1} \delta_{(0,i\infty)}^{(\rho)}\right](z) d\nu$$

$$= -\frac{1}{4i\pi} \int_{\operatorname{Re}\nu = -\beta} \nu \frac{H(i\nu)}{C(\rho, \nu)} (\operatorname{Im} z)^{\frac{\rho-1}{2}} \chi_{\rho,\nu}^{\text{even}}\left(\frac{\operatorname{Re} z}{\operatorname{Im} z}\right) d\nu. \quad (2.4.21)$$

In the case when one has $H(\lambda) = K_{\frac{i\lambda}{2}}(\alpha)$ *for some* $\alpha > 0$, *so that* $G(\sigma) = 2\pi \exp(-\alpha \cosh(4\pi\sigma))$, *one has*

$$\phi(z) = \left(\frac{2\alpha}{\pi}\right)^{\frac{1}{2}} |z|^{\frac{\rho-1}{2}} K_{\frac{\rho-1}{2}}\left(\frac{\alpha|z|}{\operatorname{Im} z}\right). \quad (2.4.22)$$

Proof. Setting $F = \frac{1}{2i\pi} G'$, one has $\widehat{F}(\mu) = \mu H(\mu)$: it follows from the assumption made about H that $\int_{-\infty}^{\infty} |G(\sigma)|^2 e^{4\pi|\beta|\sigma} d\sigma < \infty$ whenever $|\beta| < \beta_0$, and that the same holds with F substituted for G.

Recall from the proof of Theorem 2.4.1 that the function $\ell_{\rho,\nu}$ is the image of the measure $\delta_{(0,i\infty)}^{(\rho)}$ under $\left(\Delta - \frac{1-\nu^2}{4}\right)^{-1}$ and that, from (2.4.14),

$$\ell_{\rho,\nu}(z) = 2^{-\frac{3}{2}} \pi^{-1} (\operatorname{Im} z)^{\frac{\rho-1}{2}}$$
$$\times (1 + (\psi(z))^2)^{\frac{\rho-2}{4}} \int_0^{\infty} r^{\frac{-\nu-2}{2}} \operatorname{Int}_\rho\left(\frac{1+r^2}{2r}(1 + (\psi(z))^2)^{-\frac{1}{2}}\right) dr, \quad (2.4.23)$$

with

$$\operatorname{Int}_\rho(c) = \int_{-\infty}^{\infty} (c - \xi_0)_{\text{pos}}^{-\frac{1}{2}} (\xi_0 - \xi_1)^{\frac{1-\rho}{2}} \frac{d\xi_1}{\xi_0}. \quad (2.4.24)$$

This integral is zero if $c \leq 1$, and we now compute it for $c > 1$. Setting $c = \cosh a$ with $a > 0$ and making the change of variable $\xi_1 = \sinh\eta$, one obtains, using [36, p. 407] at the end,

$$\operatorname{Int}_\rho(c) = \int_{-a}^{a} [\cosh a - \cosh\eta]^{-\frac{1}{2}} e^{\frac{(\rho-1)\eta}{2}} d\eta = 2^{\frac{1}{2}} \pi \mathfrak{P}_{\frac{\rho-2}{2}}(\cosh a), \quad (2.4.25)$$

an expression invariant when changing ρ to $2 - \rho$.

As told in the statement of the proposition, we set $\frac{\mathrm{Re}\,z}{\mathrm{Im}\,z} = \sinh 4\pi\tau$ for $z \in \Pi$, and we also make the change of variable $r = e^{-4\pi\sigma}$ in this integral. We obtain

$$\ell_{\rho,\nu}(z) = 2\pi (\mathrm{Im}\,z)^{\frac{\rho-1}{2}} (\cosh 4\pi\tau)^{\frac{\rho-2}{2}} \int_{|\tau|}^{\infty} e^{2\pi\nu\sigma} \mathfrak{P}_{\frac{\rho-2}{2}} \left(\frac{\cosh 4\pi\sigma}{\cosh 4\pi\tau} \right) d\sigma. \qquad (2.4.26)$$

Since

$$F(\sigma) = \int_{-\infty}^{\infty} \widehat{F}(\lambda) e^{2i\pi\lambda\sigma} d\lambda = \frac{1}{i} \int_{\mathrm{Re}\,\nu = -\beta} \widehat{F}(-i\nu) e^{2\pi\nu\sigma} d\nu, \qquad (2.4.27)$$

the function ϕ defined in (2.4.20) can be written as

$$\frac{1}{2i} (\mathrm{Im}\,z)^{\frac{\rho-1}{2}} (\cosh 4\pi\tau)^{\frac{\rho-2}{2}} \int_{|\tau|}^{\infty} \mathfrak{P}_{\frac{\rho-2}{2}} \left(\frac{\cosh 4\pi\sigma}{\cosh 4\pi\tau} \right) d\sigma . \frac{1}{i} \int_{\mathrm{Re}\,\nu = \beta} \widehat{F}(-i\nu) e^{2\pi\nu\sigma} d\nu$$

$$= -\frac{1}{4\pi} \int_{\mathrm{Re}\,\mu = -\beta} \widehat{F}(-i\nu) \ell_{\rho,\nu}(z) d\nu. \qquad (2.4.28)$$

This leads to the pair of equations (2.4.21), of which we now consider the first.

When ν moves along the straight line from $-\beta - i\infty$ to $-\beta + i\infty$, the variable $\mu = \frac{1-\nu^2}{4}$ describes a parabola $\mathcal{P}^-$ enclosing the spectrum $[\frac{1}{4}, \infty[$ of Δ in the clockwise sense: denoting as $\mathcal{P}^+$ the negative of the contour that precedes, one transforms the integral under consideration into the integral

$$\frac{1}{2i\pi} \int_{\mathcal{P}^+} H \left(2\sqrt{\mu - \frac{1}{4}} \right) (\mu - \Delta)^{-1} d\mu, \qquad (2.4.29)$$

which completes the main part of Theorem 2.4.2, in view of Dunfords's integral representation of the resolvent of an operator.

When $H(\lambda) = K_{\frac{i\lambda}{2}}(\alpha)$, that $G(\sigma) = 2\pi \exp\left(-\alpha \cosh(4\pi\sigma)\right)$ follows from [36, p. 408]. Then,

$$\phi(z) = 2\pi\alpha (\mathrm{Im}\,z)^{\frac{\rho-1}{2}} (\cosh 4\pi\tau)^{\frac{\rho-2}{2}} \int_{\tau}^{\infty} \sinh(4\pi\sigma) e^{-\alpha \cosh(4\pi\sigma)} \mathfrak{P}_{\frac{\rho-2}{2}} \left(\frac{\cosh 4\pi\sigma}{\cosh 4\pi\tau} \right) d\sigma$$

$$= \frac{\alpha}{2} (\mathrm{Im}\,z)^{\frac{\rho-1}{2}} (\cosh 4\pi\tau)^{\frac{\rho}{2}} \int_{1}^{\infty} e^{-\alpha t \cosh(4\pi\tau)} \mathfrak{P}_{\frac{\rho-2}{2}}(t) dt$$

$$= \left(\frac{2\alpha}{\pi} \right)^{\frac{1}{2}} (\mathrm{Im}\,z)^{\frac{\rho-1}{2}} (\cosh 4\pi\tau)^{\frac{\rho-1}{2}} K_{\frac{\rho-1}{2}}(\alpha \cosh(4\pi\tau)) \qquad (2.4.30)$$

according to [36, p. 194]: now, $\cosh(4\pi\tau) = \frac{|z|}{\mathrm{Im}\,z}$, which leads to (2.4.22). □

Chapter 3

Automorphic functions and automorphic distributions

In the first section of this chapter, we recall the basic facts and notions relative to the modular Laplacian (the operator Δ in $L^2(\Gamma\backslash\Pi)$ with $\Gamma = SL(2,\mathbb{Z})$) and fix notation; we spend some more time on matters related to Roelcke-Selberg expansions. Section 3.2 will provide a short "dictionary" from automorphic distribution theory (in the plane) to automorphic function theory (in Π): there is slightly more information in an automorphic distribution than in an automorphic function, so that pairs of automorphic functions have to be used.

Hilbert space facts are better discussed, as is usual, in the context of automorphic function theory but since, as has been noted several times, the operator $\pi^2\mathcal{E}^2$ transfers to $\Delta - \frac{1}{4}$ when moving from the plane to the half-plane, explicit spectral decompositions are more easily obtained in the context of automorphic distribution theory. Note that, in Chapter 5, we shall provide a large space of automorphic distributions with a Hilbert space structure in a direct way. It would probably be possible, then, to dispense with automorphic function theory (in Π) entirely, but we have not pushed this point of view to such an extremity.

In Section 3.3, we introduce a certain sequence of Dirichlet series $\zeta_k(s,t)$ in two variables, and relate them to series of Kloosterman sums. These have been much studied, following Selberg's paper [45], and classical results would lead immediately to the continuation of the function ζ_k in a certain domain of the (s,t)-variable. This is, however, insufficient for our main application in Chapter 4, and the continuation to the required domain will be treated in Section 3.6, as a corollary of the decomposition, according to the spectral theory of the modular Laplacian, of the pointwise product of two Eisenstein series. This latter problem, treated in detail in Section 3.5, was examined in [60] in preparation for the problem of obtaining the decomposition into homogeneous automorphic components of the sharp product of two Eisenstein distributions: so far as this last question is concerned, we shall satisfy ourselves with a brief overview (Section 3.4). The more classical results regarding sums of Kloosterman series will be needed again in Chapter 5.

3.1 Automorphic background

We do not necessarily assume that the reader has any knowledge of basic facts regarding non-holomorphic modular form theory for the full modular group $\Gamma = SL(2, \mathbb{Z})$, though it would help. Let us recommend the books by Bump [4] and by Iwaniec [21], or Chapter 15 from the book by Iwaniec-Kowalski [23], or Zagier's [68], or the book by Terras [55].

Automorphic functions are (in this $SL(2, \mathbb{Z})$-specialized context) functions in the upper half-plane Π invariant under all fractional-linear transformations $z \mapsto \frac{az+b}{cz+d}$ with $\left(\begin{smallmatrix} a & b \\ c & d \end{smallmatrix}\right) \in \Gamma$: if continuous, such a function is then characterized by its restriction to the fundamental domain $D = \{z = x + iy \in \mathbb{C}: -\frac{1}{2} < x < \frac{1}{2}, |z| > 1\}$. Non-holomorphic modular forms are automorphic functions f in Π which are moreover C^∞ and solutions of some eigenvalue equation $\Delta f = \frac{1-\nu^2}{4} f$, where Δ is the hyperbolic Laplacian $-y^2 \left(\frac{\partial^2}{\partial x^2} + \frac{\partial^2}{\partial y^2}\right)$: when restricted to Γ-invariant functions, Δ is called the modular Laplacian. Since automorphy implies $\mathbb{Z}$-periodicity with respect to x, automorphic functions with some regularity admit Fourier expansions with respect to this variable. If an automorphic function f is actually a non-holomorphic modular form, i.e., if it satisfies the eigenvalue equation above, a separation of variables shows that the coefficients of its Fourier expansion with respect to x have to satisfy, as functions of y, some easily recognizable (Bessel) ordinary differential equation, which leads if $f(z)$ is bounded by some power of y as z goes to infinity while staying in the fundamental domain to an expansion of the kind

$$f(x+iy) = a_+ y^{\frac{1-\nu}{2}} + a_- y^{\frac{1+\nu}{2}} + y^{\frac{1}{2}} \sum_{k \neq 0} b_k e^{2i\pi kx} K_{\frac{\nu}{2}}(2\pi|k|y): \qquad (3.1.1)$$

the numbers b_k are called the Fourier coefficients of f, and the sum $a_+ y^{\frac{1-\nu}{2}} + a_- y^{\frac{1+\nu}{2}}$ is called the "constant term" of f. The crude estimate demanded from $f(z)$ was made so as to avoid the necessity of considering, instead of the sole function $K_{\frac{\nu}{2}}$, the two functions $I_{\pm\frac{\nu}{2}}$, which are wildly increasing at infinity.

A great bulk of non-holomorphic modular form theory is made up by the so-called Eisenstein series. If $\operatorname{Re}\nu < -1$, the series

$$E_{\frac{1-\nu}{2}}(z) = \frac{1}{2} \sum_{\substack{m,n \in \mathbb{Z} \\ (m,n)=1}} \left(\frac{|mz-n|^2}{\operatorname{Im} z}\right)^{\frac{\nu-1}{2}} \qquad (3.1.2)$$

(here, and elsewhere, (m,n) denotes the g.c.d. of the pair m, n) is convergent, and its sum is a non-holomorphic modular form for the eigenvalue $\frac{1-\nu^2}{4}$.

Let us fix some notation before continuing. The Riemann zeta function is the function $\zeta(s)$ defined, when $\mathrm{Re}\, s > 1$, as the sum of the series $\sum_{n \geq 1} n^{-s}$. It extends as a meromorphic function in the whole complex plane, with a single (simple) pole at $s = 1$. We denote as ζ^* the function

$$\zeta^*(s) = \pi^{-\frac{s}{2}}\Gamma(\frac{s}{2})\zeta(s): \tag{3.1.3}$$

its advantage, in comparison to the function $\zeta(s)$, lies in the functional equation $\zeta^*(s) = \zeta^*(1 - s)$; the function $\zeta^*(s)$ has poles at $s = 0$ and $s = 1$. We also set

$$E^*_{\frac{1-\nu}{2}}(z) = \zeta^*(1 - \nu)E_{\frac{1-\nu}{2}}(z). \tag{3.1.4}$$

There are several ways (including the Fourier expansion below) to show that the Eisenstein series $E_{\frac{1-\nu}{2}}(z)$ extends as a meromorphic function of ν (note that it is nowhere holomorphic with respect to the other variable !), satisfying the functional equation $E^*_{\frac{1-\nu}{2}} = E^*_{\frac{1+\nu}{2}}$.

Given a measurable automorphic function f, set (with $dm(x+iy) = y^{-2}dxdy$)

$$\|f\|^2 = \int_D |f(z)|^2 dm(z): \tag{3.1.5}$$

the space, indifferently denoted as $L^2(D)$ or $L^2(\Gamma\backslash\Pi)$, of such functions f satisfying moreover the condition $\|f\| < \infty$ is a Hilbert space, and the associated scalar product is the Petersson scalar product there. No function $E^*_{\frac{1-\nu}{2}}$ can lie in $L^2(\Gamma\backslash\Pi)$, as shown by its Fourier series expansion

$$\zeta^*(1-\nu)E_{\frac{1-\nu}{2}}(z) \tag{3.1.6}$$
$$= \zeta^*(1-\nu)y^{\frac{1-\nu}{2}} + \zeta^*(1+\nu)y^{\frac{1+\nu}{2}} + 2y^{\frac{1}{2}}\sum_{k\neq 0}|k|^{-\frac{\nu}{2}}\sigma_\nu(|k|)K_{\frac{\nu}{2}}(2\pi|k|y)e^{2i\pi kx},$$

in which $\sigma_\nu(|k|) = \sum_{1 \leq d|k} d^\nu$. Indeed, the "constant term" (i.e., the sum of the two terms independent of x) is never square-integrable in D with respect to dm: for if, say, $\mathrm{Re}\,\nu \geq 0$, the function $z \mapsto y^{\frac{1+\nu}{2}}$ is not square-integrable, while the coefficient $\zeta(1+\nu)$ cannot vanish either (trivial if $\mathrm{Re}\,\nu > 0$, Hadamard's theorem if $\mathrm{Re}\,\nu = 0$). For the same reason, not only $E^*_{\frac{1-i\lambda}{2}}$, but also $E_{\frac{1-i\lambda}{2}}$ is well-defined for real λ.

There do exist, however, *cusp-forms*, which are non-holomorphic modular forms square-summable in D, distinct from constants. No example of such form is known explicitly (for the group Γ), however, and though it is easy to see that there must exist such forms f with the property that $f(-\bar{z}) = -f(z)$ for $z \in \Pi$ (such forms may be called of odd type in the present context), simply proving that there exist (infinitely many, as it turns out) cusp-forms of even type (such that $f(-\bar{z}) = f(z)$: we shall be more concerned with these) requires at present the use of Selberg's trace formula [5, 23]. Cusp-forms are characterized, among non-holomorphic modular forms, by the fact that their "constant term" is zero, in other words that $\int_{-\frac{1}{2}}^{\frac{1}{2}} f(x + iy)dx = 0$ for every $y > 0$. As a consequence, they are indeed square-integrable in D, and the "generalized" eigenvalue $\frac{1-\nu^2}{4}$ is, for them, a true eigenvalue, $> \frac{1}{4}$ as can easily be proved: hence, it can be written as $\frac{1+\lambda^2}{4}$ for some $\lambda \in \mathbb{R}$. The corresponding numbers λ are, again, not known: only, as a consequence of Selberg's trace formula (assuming $\lambda > 0$ since only λ^2 enters the eigenvalue), they constitute a sequence $(\lambda_m)_{m \geq 1}$ going to ∞. The same is true if one specializes in eigenvalues for which there exists at least one corresponding cusp-form of even (or of odd) type: we shall qualify such an eigenvalue as being itself of even, or odd, type (we know of no proof that it can never be of both types).

The space of cusp-forms corresponding to any given eigenvalue $\frac{1+\lambda_m^2}{4}$ is, as is easily seen by some compactness argument, finite-dimensional: it is not believed that this space can have dimension ≥ 1, but no proof of this is known. How to deal in general with multiplicity is answered theoretically by a well-known principle of spectral theory or mathematical Physics: besides the operator under consideration, in our case Δ, introduce a "complete" set of operators commuting with it. One such operator, to wit the "parity operator" that changes an automorphic function f to the function $z \mapsto f(-\bar{z})$, has already been implicitly considered when discussing cusp-forms of even and odd types: note that Eisenstein series are, in this sense, of even type (when coupled with the existence of Roelcke-Selberg expansions, this observation is the reason why the existence of cusp-forms of odd type, contrary to that of cusp-forms of even type, is a trivial matter). The Hecke operators T_N, $N \geq 1$, defined by the equation

$$(T_N f)(z) = N^{-\frac{1}{2}} \sum_{\substack{ad=N, d>0 \\ b \bmod d}} f\left(\frac{az + b}{d}\right), \tag{3.1.7}$$

make up the required family: they can be shown to commute pairwise, while commuting with Δ and with the parity operator too. We shall make use of the fundamental formal relations

$$\sum_{\ell \geq 1} p^{-\ell s} T_{p^\ell} = \left(1 - p^{-s} T_p + p^{-2s}\right)^{-1}, \quad p \text{ prime},$$

$$\sum_{N \geq 1} N^{-s} T_N = \prod_p \left(1 - p^{-s} T_p + p^{-2s}\right)^{-1}. \tag{3.1.8}$$

These can be found, for instance, in [22, p. 97] or [55]: if using the first reference, note that the factor N^{-1} appears in place of $N^{-\frac{1}{2}}$ in the definition of T_N.

Hecke eigenforms are cusp-forms of a definite parity which are at the same time joint eigenfunctions of the family of Hecke operators. A Hecke eigenform f is then characterized, up to multiplication by a scalar, by the corresponding set of eigenvalues: indeed, its Fourier coefficient b_1, as taken from (3.1.1), is never zero, and one has $T_N f = \frac{b_N}{b_1} f$ for every $N \geq 1$. The Hecke operators are self-adjoint with respect to the Petersson scalar product, from which it follows that the eigenvalues $\frac{b_N}{b_1}$ are real: note (this is trivial) that $b_k = b_{-k}$ for non-holomorphic modular forms of even type. In some cases, it is better to repeat eigenvalues $\frac{1+\lambda_m^2}{4}$ according to multiplicity (this is the case when needing the equivalent, provided by Selberg's trace formula, of λ_m as $m \to \infty$); in other cases, it is better not to do so. In the second case, assuming that the multiplicity of some eigenvalue $\frac{1+\lambda_p^2}{4}$ is κ, we choose an orthonormal basis $(\mathcal{M}_{p,j})_{j=1,\ldots,\kappa}$ of the corresponding space of cusp-forms. Since the T_N's commute pairwise, it follows from standard Hilbert space arguments that one can assume that every cusp-form $\mathcal{M}_{p,j}$ is a Hecke eigenform (these are indeed pairwise orthogonal). We shall have it completely normalized (up to permutation of the $\mathcal{M}_{p,j}$'s for a given p) by the assumption that the Fourier coefficient b_1 of each such Hecke eigenform is real and positive: then, all its Fourier coefficients are real. Actually, another possible normalization is just as important: the eigenform $\mathcal{N}_{p,j}$, proportional to $\mathcal{M}_{p,j}$, is the one for which the coefficient b_1 from (3.1.1) is 1, so that the Nth Fourier coefficient b_N of $\mathcal{N}_{p,j}$ yields directly the eigenvalue equation $T_N \mathcal{N}_{p,j} = b_N \mathcal{N}_{p,j}$.

We shall denote the kth coefficient b_k from the expansion (3.1.1) as $b_{k;p,j}$ in the case when $f = \mathcal{M}_{p,j}$, and as $c_{k;p,j}$ in the case when $f = \mathcal{N}_{p,j}$: hence,

$$c_{1;p,j} = 1 \quad \text{and} \quad c_{k;p,j} = \|\mathcal{N}_{p,j}\| b_{k;p,j}, \tag{3.1.9}$$

where the norm is taken in the space $L^2(\Gamma \backslash \Pi) = L^2(D)$.

The L-function associated to a cusp-form $\mathcal{M}$ with the Fourier expansion

$$\mathcal{M}(x + iy) = y^{\frac{1}{2}} \sum_{k \neq 0} b_k e^{2i\pi kx} K_{\frac{i\lambda}{2}}(2\pi |k| y) \tag{3.1.10}$$

(then, λ must be one of the λ_m's) is the Dirichlet series defined, for $\operatorname{Re} s$ large enough, as

$$L(s, \mathcal{M}) = \sum_{k \geq 1} b_k k^{-s} : \tag{3.1.11}$$

we also set

$$L^*(s, \mathcal{M}) = \begin{cases} \pi^{-s} \Gamma(\frac{s}{2} + \frac{i\lambda}{4}) \Gamma(\frac{s}{2} - \frac{i\lambda}{4}) L(s, \mathcal{M}) & \text{if } \mathcal{M} \text{ is of even type} \\ \pi^{-s} \Gamma(\frac{s+1}{2} + \frac{i\lambda}{4}) \Gamma(\frac{s+1}{2} - \frac{i\lambda}{4}) L(s, \mathcal{M}) & \text{if } \mathcal{M} \text{ is of odd type,} \end{cases} \tag{3.1.12}$$

getting as a result [4, p. 107] a function which extends as an entire function of s, satisfying the functional equation $L^*(s, \mathcal{M}) = \pm L^*(1 - s, \mathcal{M})$, according to type.

When initially defined on the space of C^∞ functions f in Π, automorphic (this tacitly means for us Γ-automorphic with $\Gamma = SL(2, \mathbb{Z})$) and such that f and Δf lie in $L^2(\Gamma \backslash \Pi)$, the modular Laplacian Δ is essentially self-adjoint in $L^2(\Gamma \backslash \Pi)$ i.e., it admits a unique self-adjoint extension. One may thus decompose arbitrary functions in this space according to the spectral resolution of this operator. It has both a discrete spectrum, for which a complete set of eigenfunctions is made up by cusp-forms, together with constants, and a continuous spectrum, for which the Eisenstein series $E_{\frac{1-i\lambda}{2}}$, $\lambda \in \mathbb{R}$, make up a complete set of generalized eigen-functions. This leads to the Roelcke-Selberg expansion of an arbitrary function $f \in L^2(\Gamma \backslash \Pi)$, convergent in the $L^2(D)$-sense ([23, Section 15] or [21, p. 112]), given as

$$f(z) = \Phi^0 + \frac{1}{8\pi} \int_{-\infty}^{\infty} \Phi(\lambda) E_{\frac{1-i\lambda}{2}}(z) d\lambda + \sum_{p \geq 1} \sum_j \Phi^{p,j} \mathcal{M}_{p,j}(z). \qquad (3.1.13)$$

Since the two functions $E_{\frac{1\pm i\lambda}{2}}$ are proportional, uniqueness of the expansion demands that $\Phi(\lambda)$ and $\Phi(-\lambda)$ should be related in some definite way: so that the product $\Phi(\lambda) E_{\frac{1-i\lambda}{2}}(z)$ should be an even function of λ, one must assume that the function $\lambda \mapsto \zeta^*(1 + i\lambda)\Phi(\lambda)$ is even. Then, one has

$$\|f\|^2_{L^2(\Gamma \backslash \Pi)} = \frac{\pi}{3}|\Phi^0|^2 + \frac{1}{8\pi} \int_{-\infty}^{\infty} |\Phi(\lambda)|^2 d\lambda + \sum_{p \geq 1} \sum_j |\Phi^{p,j}|^2. \qquad (3.1.14)$$

Since the functions $\mathcal{M}_{p,j}$ make up an orthonormal set, one has $\Phi^{p,j} = (\mathcal{M}_{p,j}|f)$ while, as the area of the fundamental domain D is $\frac{\pi}{3}$ (remember the formula giving the sum of angles of a triangle in a space of constant curvature), one has $\Phi^0 = \frac{3}{\pi}(1|f)$: just as we already did in the non-automorphic case, we make the convention that scalar products should be antilinear with respect to the variable on the left. If one assumes that $f(z)$ remains square-integrable in the fundamental domain D even after it has been multiplied by some power of $\operatorname{Im} z$ with a positive exponent, one can write

$$\Phi(\lambda) = (E_{\frac{1-i\lambda}{2}}|f)_{L^2(D)}. \qquad (3.1.15)$$

In the case when $f(z) = f(-\bar{z})$ identically, only Hecke eigenforms $\mathcal{M}_{p,j}$ of even type (together with Eisenstein series) will occur in the decomposition. When a series of Hecke eigenforms is introduced by the summation symbol $\sum_{p,j\,\text{even}}$, this will indicate that only Hecke eigenforms of even type are retained.

From the coefficients of the Fourier series decomposition of a function $f \in L^2(D)$ such that $\Delta f \in L^2(D)$, to wit

$$f(x + iy) = \sum_{k \in \mathbb{Z}} a_k(y) e^{2i\pi k x}, \qquad (3.1.16)$$

it is possible, in principle, to obtain the coefficients of the Roelcke-Selberg decomposition of f. In the case when the automorphic function f is C^∞ in Π and (sufficiently) rapidly decreasing at infinity in the fundamental domain, the result is obtained with the help of the so-called Rankin-Selberg unfolding method, an extension of which will be given in Section 6.2. However, it is necessary for applications to dispense as much as possible with the assumption of rapid decay, as done in [70]. The version we need here was given in [60, Section 7].

Lemma 3.1.1. *One has for some constant $C > 0$ and every $a \in]0, \frac{1}{2}]$ the estimate*

$$\int_0^\infty e^{\frac{\pi\rho}{2}} \left(K_{\frac{i\rho}{2}}(a)\right)^2 \frac{d\rho}{1+\rho^2} \le C|\log a|. \tag{3.1.17}$$

Proof. Set, for $0 < s < 1$,

$$\psi(s) = \int_0^\infty \left(K_{\frac{i\rho}{2}}(a)\right)^2 \cosh(\frac{\pi\rho}{2}(1-s))d\rho. \tag{3.1.18}$$

According to [36, p. 106],

$$\psi(s) = \pi K_0 \left(2a \sin \frac{\pi s}{2}\right). \tag{3.1.19}$$

The integral $\int_0^1 s\psi(s)ds$ can be computed in two ways, leading to the identity

$$\frac{4}{\pi^2} \int_0^\infty \left(K_{\frac{i\rho}{2}}(a)\right)^2 \left(\cosh \frac{\pi\rho}{2} - 1\right) \frac{d\rho}{\rho^2} = \pi \int_0^1 sK_0 \left(2a \sin \frac{\pi s}{2}\right) ds, \tag{3.1.20}$$

from which the required estimate follows. ☐

Lemma 3.1.2. *Let f be an automorphic function such that f and Δf are square-integrable in the fundamental domain, with the Fourier series expansion (3.1.16); assume moreover that f is orthogonal to constants. Then, for some positive constants $A(k)$ independent of y, one has the estimates*

$$|a_0(y)| \le A(0)y^{\frac{1}{2}}, \quad 0 < y < \infty \tag{3.1.21}$$

and, for $k \ne 0$,

$$|a_k(y)| \le A(k)y^{\frac{1}{2}}|\log y|^{\frac{1}{2}}, \quad 0 < y < 1. \tag{3.1.22}$$

Proof. It is convenient, in this proof, to repeat the eigenvalues according to their multiplicity, i.e., to replace the double sequence $(\mathcal{M}_{p,j})$ by a sequence $(\mathcal{M}_m)$: we then denote as $b_{k;m}$ the kth Fourier coefficient of $\mathcal{M}_m$. Since f and Δf lie in $L^2(D)$, the coefficients of the Roelcke-Selberg expansion (3.1.13) of f satisfy the estimates

$$\int_{-\infty}^\infty |\Phi(\rho)|^2 (1+\rho^2)^2 d\rho < \infty \quad \sum_{m\ge 1} (1+\lambda_m^2)^2 |\Phi^m|^2 < \infty. \tag{3.1.23}$$

From the first of these two estimates, the function Φ is integrable over $\mathbb{R}$. On the other hand, it is a consequence of Selberg's trace formula (*cf.* for instance [55, p. 290]) that λ_m is of the order of $m^{\frac{1}{2}}$ as $m \to \infty$: it thus follows from the second estimate that the series $\sum |\Phi^m|$ is convergent. Using the Roelcke-Selberg expansion of f, recalling the way, made explicit right after (3.1.13), $\Phi(\lambda)$ and $\Phi(-\lambda)$ are related, finally using the Fourier series expansion (3.1.2) of Eisenstein series, one may write

$$a_0(y) = \frac{1}{4\pi} \int_{-\infty}^{\infty} \Phi(\rho) y^{\frac{1-i\rho}{2}} d\rho, \qquad (3.1.24)$$

which proves (3.1.21). Also, for $k \neq 0$,

$$a_k(y) = (a_k)_{\text{cont}}(y) + (a_k)_{\text{disc}}(y) \qquad (3.1.25)$$

with

$$(a_k)_{\text{cont}}(y) = \frac{1}{4\pi} \int_{-\infty}^{\infty} \Phi(\rho) \frac{\pi^{\frac{1-i\rho}{2}}}{\Gamma(\frac{1-i\rho}{2})\zeta(1-i\rho)} |k|^{-\frac{i\rho}{2}} \sigma_{i\rho}(|k|) y^{\frac{1}{2}} K_{\frac{i\rho}{2}}(2\pi|k|y) d\rho \qquad (3.1.26)$$

and

$$(a_k)_{\text{disc}}(y) = y^{\frac{1}{2}} \sum_{m \geq 1} \Phi^m b_{k;m} K_{\frac{i\lambda_m}{2}}(2\pi|k|y). \qquad (3.1.27)$$

The proof of (3.1.22) makes it necessary to look separately at the contributions of the two parts. For the continuous part, one uses the Cauchy-Schwarz inequality and the first inequality (3.1.23), together with (3.1.17) and the estimate [36, p. 13]

$$\left| \Gamma(\frac{1-i\rho}{2}) \right|^{-2} \leq C \cosh \frac{\pi\rho}{2}, \qquad (3.1.28)$$

finally the fact ([41, p. 100] or [54, p. 161]) that $(\zeta(1-i\rho))^{-1}$ is majorized by some power of $\log|\rho|, |\rho| \to \infty$.

To estimate the contribution of the discrete part, one uses the second inequality (3.1.23), a result of Smith [49], valid for L^2-normalized cusp-forms, according to which one has

$$|b_{k;m}| \leq A(k) \lambda_m^{\frac{1}{2}} \exp \frac{\pi\lambda_m}{4}, \qquad (3.1.29)$$

and the elementary inequality (a consequence of a power series expansion with respect to a)

$$|K_{\frac{i\lambda_m}{2}}(a)| \leq C \lambda_m^{-\frac{1}{2}} \exp\left(-\frac{\pi\lambda_m}{4}\right), \qquad 0 < a \leq a_0. \qquad (3.1.30)$$

$\square$

Proposition 3.1.3. *Let f be an automorphic function, such that f and Δf are square-integrable in D. Let (3.1.13) be its Roelcke-Selberg expansion, and let (3.1.16) be its Fourier expansion. The function*

$$C_0^-(\mu) = \frac{1}{8\pi} \int_0^1 a_0(y) y^{-\frac{3}{2}} \frac{(\pi y)^{-\frac{\mu}{2}}}{\Gamma(-\frac{\mu}{2})} dy \qquad (3.1.31)$$

is holomorphic in the half-plane $\operatorname{Re}\mu < -1$ and extends as a meromorphic function in the half-plane $\operatorname{Re}\mu < 0$ with an only possible simple pole at $\mu = -1$. The residue of $C_0^-(\mu)$ at this point is $-\frac{\Phi^0}{4\pi}$, so that, in particular, f is orthogonal to the constants if and only if the function C_0^- is holomorphic throughout the half-plane $\operatorname{Re}\mu < 0$. The function

$$C_0^+(\mu) = -\frac{1}{8\pi} \int_1^\infty a_0(y) y^{-\frac{3}{2}} \frac{(\pi y)^{-\frac{\mu}{2}}}{\Gamma(-\frac{\mu}{2})} dy \qquad (3.1.32)$$

is holomorphic in the half-plane $\operatorname{Re}\mu > 0$. Finally, the function $\lambda \mapsto C_0^-(-\varepsilon + i\lambda) - C_0^+(\varepsilon + i\lambda)$ has, as $\varepsilon \to 0$, a limit in the space $L^2_{\mathrm{loc}}(\mathbb{R})$, which coincides with the function

$$\lambda \mapsto \frac{1}{8\pi} \frac{\pi^{-\frac{i\lambda}{2}}}{\Gamma(-\frac{i\lambda}{2})} \Phi(-\lambda). \qquad (3.1.33)$$

Proof. As a consequence of Lemma 3.1.2 and (3.1.24), one has

$$a_0(y) = \Phi^0 + \frac{1}{4\pi} \int_{-\infty}^\infty \Phi(\rho) y^{\frac{1-i\rho}{2}} d\rho, \qquad (3.1.34)$$

which implies, when $\operatorname{Re}\mu < -1$,

$$C_0^-(\mu) = -\frac{1}{4\pi}(1+\mu)^{-1} \frac{\pi^{-\frac{\mu}{2}}}{\Gamma(-\frac{\mu}{2})} \Phi^0 + \frac{i}{16\pi^2} \frac{\pi^{-\frac{\mu}{2}}}{\Gamma(-\frac{\mu}{2})} \int_{-\infty}^\infty \Phi(\rho) \frac{d\rho}{\rho - i\mu}. \qquad (3.1.35)$$

The same formula gives the value of $C_0^+(\mu)$ in the half-plane $\operatorname{Re}\mu > 0$. This implies Proposition 3.1.3, since on one hand

$$\int_{-\infty}^\infty \Phi(\rho) \left[\frac{1}{\rho + \lambda + i\varepsilon} - \frac{1}{\rho + \lambda - i\varepsilon} \right] d\rho = 2i \int_{-\infty}^\infty \frac{\varepsilon}{(\rho+\lambda)^2 + \varepsilon^2} \Phi(\rho) d\rho, \qquad (3.1.36)$$

as a function of λ, goes to $2i\pi\Phi(-\lambda)$, in the space $L^2(\mathbb{R})$, when $\varepsilon \to 0$. On the other hand, whether we put the coefficient $\frac{\pi^{-\frac{\mu}{2}}}{\Gamma(-\frac{\mu}{2})}$ inside $C_0^\pm(\mu)$ to the right of the sign $\lim_{\varepsilon \to 0}$ or to the left (having changed μ to $i\lambda$) does not change the limit: for the difference of values of this function at the points $\pm\varepsilon + i\lambda$ and $i\lambda$ is a $O(\varepsilon)$ while, in view of the equation $\int_{-\infty}^\infty \frac{d\rho}{|\rho+\lambda+i\varepsilon|^2} = \frac{\pi}{\varepsilon}$ and of the Cauchy-Schwarz inequality, an integral such as $\int_{-\infty}^\infty \frac{\Phi(\rho)d\rho}{\rho+\lambda+i\varepsilon}$ is a $O(\varepsilon^{-\frac{1}{2}})$. $\qquad\square$

Remarks 3.1.a. (i) In the case when $a_0(y)$ is small at infinity, there is no need to split the integrals defining the functions $C_0^-(\mu)$ and $C_0^+(\mu)$ for $\operatorname{Re}\mu > 0$, and the result of Theorem 3.1.3 then coincides with what is classically obtained from the Rankin-Selberg method (*cf.* (3) in [68, p. 268]).

(ii) It may not seem very useful that the factor $\frac{\pi^{-\frac{\mu}{2}}}{\Gamma(-\frac{\mu}{2})}$ should occur in the definition of $C_0^{\pm}(\mu)$, only to disappear at the end in view of (3.1.33). This factor occurred naturally in our first derivation [60, Sec. 6] of Proposition 3.1.3, based on a generalization of the theory of the Radon transformation to an automorphic environment; also, one may note that this factor is more obviously useful in Proposition 3.1.4 below. Our normalization of the automorphic Radon transformation was chosen so as to separate the Archimedean factor from the arithmetic ones: this is made possible by the use of automorphic distribution theory (as opposed to automorphic function theory), and an example will show in (5.2.7), where no Gamma factors are present.

In order to obtain the discrete part of the Roelcke-Selberg expansion of f, we shall assume that it is orthogonal to constants and that the continuous part of its Roelcke-Selberg expansion has been removed from it.

Proposition 3.1.4. *Let f be an automorphic function such that f and Δf are square-integrable in the fundamental domain. Assume that its Fourier expansion is given by (3.1.16) and that its Roelcke-Selberg expansion reduces to*

$$f(z) = \sum_{p\geq 1}\sum_{j} \Phi^{p,j} \mathcal{M}_{p,j}(z). \tag{3.1.37}$$

Given $k \neq 0$, define

$$c_k(\mu) = \frac{1}{8\pi}\int_0^\infty a_k(y)y^{-\frac{3}{2}}\frac{(\pi y)^{-\frac{\mu}{2}}}{\Gamma(-\frac{\mu}{2})}\,dy \tag{3.1.38}$$

when $-1 < \operatorname{Re}\mu < 0$. Then, $c_k(\mu)$ extends as a meromorphic function of μ in the half-plane $\operatorname{Re}\mu > -4$: it has no pole with $\operatorname{Re}\mu < 4$ except the pure imaginary points $\pm i\lambda_p, p\geq 1$; all its poles are simple. For every p, the projection of f on the eigenspace of Δ in $L^2(\Gamma\backslash\Pi)$ corresponding to the eigenvalue $\frac{1+\lambda_p^2}{4}$ is given as

$$\sum_{j}\Phi^{p,j}\mathcal{M}_{p,j}(z) = y^{\frac{1}{2}}\sum_{k\neq 0} d_k K_{\frac{i\lambda_p}{2}}(2\pi|k|y)e^{2i\pi nx} \tag{3.1.39}$$

with

$$d_k = -8\pi|k|^{-\frac{i\lambda_p}{2}} \times \operatorname{Res}_{\mu=i\lambda_p} c_k(\mu). \tag{3.1.40}$$

Proof. Again, we temporarily revert to the notation for which eigenvalues are repeated according to their multiplicities, writing instead of (3.1.37) the identity

$$f(z) = \sum_{m\geq 1}\Phi^m \mathcal{M}_m(z). \tag{3.1.41}$$

We first show that $c_k(\mu)$ makes sense for $\operatorname{Re}\mu < 0$, and that it can be obtained by performing a term-by-term integration in (3.1.38), starting from the identity

$$a_k(y) = y^{\frac{1}{2}} \sum_{m \geq 1} \Phi^m b_{k;m} K_{\frac{i\lambda_m}{2}}(2\pi|k|y). \tag{3.1.42}$$

This has to be integrated against $y^{-\frac{3}{2}-\frac{\mu}{2}} dy$ from 0 to ∞, assuming $\operatorname{Re}\mu < 0$: estimating the integral from 0 to 1 follows from the estimate (3.1.22) and the way it has been proved, so all we have to do is showing that

$$\sum_{m \geq 1} |\Phi^m| |b_{k;m}| \int_1^\infty |K_{\frac{i\lambda_m}{2}}(2\pi|k|y)| y^{-1-\frac{\operatorname{Re}\mu}{2}} dy < \infty. \tag{3.1.43}$$

From [36, p. 101, p. 13], one has when $-a < \operatorname{Re}\mu < 0$ for some $a > 0$ the inequality, in which the constant C depends only on k and a,

$$\int_0^\infty (K_{\frac{i\lambda_m}{2}}(2\pi|k|y))^2 y^{-\operatorname{Re}\mu} dy \leq C \left| \Gamma\left(\frac{1 - \operatorname{Re}\mu + i\lambda_m}{2}\right) \Gamma\left(\frac{1 - \operatorname{Re}\mu - i\lambda_m}{2}\right) \right|$$

$$\leq C\lambda_m^{-\operatorname{Re}\mu} e^{-\frac{\pi\lambda_m}{2}}, \tag{3.1.44}$$

so that, by the Cauchy-Schwarz inequality, the series on the left-hand side of (3.1.43) is majorized by

$$C \sum_{m \geq 1} |\Phi^m| |b_{k;m}| \lambda_m^{-\frac{\operatorname{Re}\mu}{2}} e^{-\frac{\pi\lambda_m}{4}}. \tag{3.1.45}$$

Using the estimate (3.1.29) and the second inequality (3.1.23), finally the Selberg equivalent according to which λ_m is of the order of $m^{\frac{1}{2}}$, one sees, with the help of the Cauchy-Schwarz inequality, that the series is convergent for $\operatorname{Re}\mu > -2$.

When $-2 < \operatorname{Re}\mu < 0$, one obtains as the result of a term-by-term integration, using [36, p. 91],

$$c_k(\mu) = \frac{1}{32\pi} \frac{|k|^{\frac{\mu}{2}}}{\Gamma(-\frac{\mu}{2})} \sum_{m \geq 1} \Phi^m b_{k;m} \Gamma\left(\frac{-\mu - i\lambda_m}{4}\right) \Gamma\left(\frac{-\mu + i\lambda_m}{4}\right). \tag{3.1.46}$$

Using the estimate

$$\left| \Gamma\left(\frac{-\mu - i\lambda_m}{4}\right) \Gamma\left(\frac{-\mu + i\lambda_m}{4}\right) \right| \leq C\lambda_m^{-\frac{\operatorname{Re}\mu}{2}-1} e^{-\frac{\pi\lambda_m}{4}}, \quad \lambda_m \to \infty, \tag{3.1.47}$$

one sees in the same way as that used to estimate (3.1.45) that $c_k(\mu)$ indeed extends as meromorphic function of μ in the half-plane $\operatorname{Re}\mu > -4$, with simple poles at the points $\pm i\lambda_m, \pm i\lambda_m + 4, \pm i\lambda_m + 8, \dots$. The contribution to the residue at $\mu = i\lambda_m$ coming from the mth term is

$$-\frac{1}{8\pi} |k|^{\frac{i\lambda_m}{2}} \Phi^m b_{k;m}, \tag{3.1.48}$$

and we only have to collect these residues, when corresponding to all Hecke eigenforms $\mathcal{M}_{p,j}$ such that λ_p coincides with some given value. $\qquad\square$

3.2 Automorphic distributions

We now turn to the subject of automorphic distributions in the plane $\mathbb{R}^2$, refreshing the reader's memory of Theorem 1.1.3. Recall definition (2.1.1) of the transform $\Theta = (\Theta_0, \Theta_1)$ from (say, tempered) even distributions in the plane to pairs of functions in Π, and the pseudo-differential interpretation of Θ as the map $h \mapsto (f_0, f_1)$ in Theorem 1.1.1. Also recall from (2.1.4) the pair of identities $\Theta_j(\mathcal{G}\mathfrak{S}) = (-1)^j \Theta_j \mathfrak{S}$ valid for every tempered distribution $\mathfrak{S}$: we have indicated right after (2.1.4) how to prove that the knowledge of $\Theta_0 \mathfrak{S}$ entails that of the sum $\mathfrak{S} + \mathcal{G}\mathfrak{S}$, and a similar argument shows that the knowledge of $\Theta_1 \mathfrak{S}$ entails that of the difference $\mathfrak{S} - \mathcal{G}\mathfrak{S}$. It is often convenient to use, instead of a distribution $\mathfrak{S}$, its rescaled version

$$\mathfrak{S}^{\text{resc}} = 2^{-\frac{1}{2} + i\pi\mathcal{E}} \mathfrak{S}, \tag{3.2.1}$$

or $\langle \mathfrak{S}^{\text{resc}}, h \rangle = \frac{1}{2} \langle \mathfrak{S}, (x, \xi) \mapsto h(2^{-\frac{1}{2}}x, 2^{-\frac{1}{2}}\xi) \rangle$, since the equation $\mathcal{F}^{\text{symp}}\mathfrak{S} = \pm\mathfrak{S}$ is then equivalent to $\mathcal{G}\mathfrak{S}^{\text{resc}} = \pm\mathfrak{S}^{\text{resc}}$.

The fundamental covariance property is expressed by the identity

$$(\Theta(\mathfrak{S} \circ g^{-1}))(z) = (\Theta\mathfrak{S})(g^{-1}.z). \tag{3.2.2}$$

It implies that distributions in the plane which are invariant under the action, by linear changes of coordinates associated to matrices of Γ — such distributions, of necessity invariant under the change $(x, \xi) \mapsto (-x, -\xi)$, will be called automorphic distributions — become, after a Θ-transformation, pairs of automorphic functions in Π. Let us recall from (2.1.5) the identity

$$\Theta(\pi^2 \mathcal{E}^2 \mathfrak{S}) = \left(\Delta - \frac{1}{4} \right) \mathfrak{S}. \tag{3.2.3}$$

It follows, if one combines it with the covariance property, that the Θ-transform of any automorphic distribution $\mathfrak{S}$ homogeneous of degree $-1 - \nu$ is a pair of non-holomorphic modular forms corresponding to the generalized eigenvalue $\frac{1-\nu^2}{4}$ of Δ. The homogeneity condition means (by duality) that

$$\langle \mathfrak{S}, (x, \xi) \mapsto t^{-1} h(t^{-1}x, t^{-1}\xi) \rangle = t^{-\nu} \langle \mathfrak{S}, h \rangle, \quad h \in \mathcal{S}(\mathbb{R}^2) : \tag{3.2.4}$$

again, let us warn against the risk of using x to denote simultaneously the first coordinate in $\mathbb{R}^2$ and the real part of $z \in \Pi$. This can be put to use in the following way: it often makes it possible to replace the search for the decomposition of a given automorphic function as a superposition of non-holomorphic modular forms (the Roelcke-Selberg decomposition) by the technically simpler problem of decomposing an automorphic distribution into its homogeneous components. Homogeneous automorphic distributions will, quite naturally, be called modular distributions.

One word about the use of tempered distributions is needed here. Since the maps Θ_j are defined as the result of testing a distribution against specific families of functions of a Gaussian style, such a condition about a distribution $\mathfrak{S}$ is considerably more restrictive than the one needed to make $\Theta_0\mathfrak{S}$ and $\Theta_1\mathfrak{S}$ meaningful. On the other hand, a space like $\mathcal{S}'(\mathbb{R}^2)$ is quite naturally associated to the additive structure of $\mathbb{R}^2$ — more precisely, that of $\mathbb{R}^2 \times \mathbb{R}^2$ — since the dual space $\mathcal{S}(\mathbb{R}^2)$ is just the space of C^∞ vectors of the Heisenberg representation in $L^2(\mathbb{R}^2)$ while, in the present context, we are more interested in the action of linear changes of coordinates associated to matrices in $SL(2, \mathbb{R})$. Hilbert spaces (or, more generally, topological vector spaces) of functions, or distributions, or more general analytic functionals, invariant under such an action, can only be defined with the help of functions of the Euler operator, in a way which already showed itself when dealing with associates of the Radon, or dual Radon, transform. In general analysis, singularities are very often dealt with by means of integrations by parts, which require that certain differential operators should be applicable to the objects under study: in automorphic distribution theory, one also has, sometimes, to ask for the applicability of such objects as the resolvent $(2i\pi\mathcal{E} - \mu)^{-1}$ for certain values of $\mu \in \mathbb{C}$. In other words, it will be necessary, on several occasions, to deal with functions, or distributions, in the image of $\mathcal{S}(\mathbb{R}^2)$, or $\mathcal{S}'(\mathbb{R}^2)$, under some (generally simple) polynomial in the Euler operator.

Still, as will be seen, Eisenstein series as well as cusp-forms are indeed the images of well-defined modular (tempered) distributions. Let us start with a study of Eisenstein distributions: since references will be needed later, let us note that we now denote as $\mathfrak{E}_\nu$ the modular distribution denoted as $\mathfrak{E}_\nu^\sharp$ in [60, 61]; the same musical shift by a semi-tone has occurred in the notation concerning cusp-distributions.

Proposition 3.2.1. *For* $\operatorname{Re}\nu < -1$, *and* $h \in \mathcal{S}_{\text{even}}(\mathbb{R}^2)$, *set*

$$\langle \mathfrak{E}_\nu, h \rangle = \sum_{|m|+|n|\neq 0} \int_0^\infty t^{-\nu} h(tn, tm) dt, \qquad (3.2.5)$$

defining in this way a tempered distribution $\mathfrak{E}_\nu$, *automorphic and homogeneous of degree* $-1 - \nu$, *hence a modular distribution. The map* $\nu \mapsto \mathfrak{E}_\nu$ *with values in* $\mathcal{S}'(\mathbb{R}^2)$ *extends as a holomorphic function for* $\nu \neq \pm 1$, *with simple poles there; the corresponding residues are given as*

$$\operatorname{Res}_{\nu=-1}\mathfrak{E}_\nu = -1, \quad \operatorname{Res}_{\nu=1}\mathfrak{E}_\nu = \delta, \qquad (3.2.6)$$

the unit mass at the origin of $\mathbb{R}^2$. *One has* $\mathcal{F}^{\text{symp}}\mathfrak{E}_\nu = \mathfrak{E}_{-\nu}$. *For* $\operatorname{Re}\nu < 0, \nu \neq -1$, *one has the identity*

$$\langle \mathfrak{E}_\nu, h \rangle = \zeta(-\nu) \int_{-\infty}^\infty |t|^{-\nu-1}(\mathcal{F}_1^{-1}h)(0, t) dt + \zeta(1 - \nu) \int_{-\infty}^\infty |t|^{-\nu} h(t, 0) dt$$

$$+ \sum_{k \neq 0} \sigma_\nu(|k|) \int_{-\infty}^\infty |t|^{-\nu-1}(\mathcal{F}_1^{-1}h)\left(\frac{k}{t}, t\right) dt, \quad (3.2.7)$$

denoting as $\mathcal{F}_1^{-1}h$ the inverse Fourier transform of h with respect to the first variable. The Θ-transform of the rescaled distribution (cf. (3.2.1)) $\mathfrak{E}_\nu^{\mathrm{resc}} = 2^{\frac{-1-\nu}{2}}\mathfrak{E}_\nu$ is given by the pair of equations

$$\Theta_0\left(\mathfrak{E}_\nu^{\mathrm{resc}}\right) = E^*_{\frac{1-\nu}{2}}, \qquad \Theta_1\left(\mathfrak{E}_\nu^{\mathrm{resc}}\right) = -\nu E^*_{\frac{1-\nu}{2}}. \tag{3.2.8}$$

Proof. That $\langle \mathfrak{E}_\nu, h \rangle$ is a convergent expression when $\mathrm{Re}\,\nu < -1$, and defines a modular distribution, homogeneous of degree $-1-\nu$, presents no difficulty. Next, denote as $(\mathfrak{E}_\nu)_{\mathrm{princ}}$ (*resp.* $(\mathfrak{E}_\nu)_{\mathrm{res}}$) the distribution defined in the same way as $\mathfrak{E}_\nu$, except for the fact that the integral from 0 to ∞ in (3.2.5) is replaced by the same integral taken from 0 to 1 (*resp.* from 1 to ∞), and observe that the distribution $(\mathfrak{E}_\nu)_{\mathrm{res}}$ extends as an entire function of ν. As a consequence of Poisson's formula, one has when $\mathrm{Re}\,\nu < -1$ the identity

$$\int_1^\infty t^\nu \sum_{(n,m)\in\mathbb{Z}^2} (\mathcal{F}^{\mathrm{symp}}h)\,(tn, tm)dt = \int_1^\infty t^\nu \sum_{(n,m)\in\mathbb{Z}^2} t^{-2}h(t^{-1}n, t^{-1}m)dt$$

$$= \int_0^1 t^{-\nu} \sum_{(n,m)\in\mathbb{Z}^2} h(tn, tm)dt, \tag{3.2.9}$$

from which one obtains that

$$\langle \mathcal{F}^{\mathrm{symp}}\left(\mathfrak{E}_{-\nu}\right)_{\mathrm{res}}, h \rangle = \langle (\mathfrak{E}_\nu)_{\mathrm{princ}}, h \rangle + \frac{h(0,0)}{1-\nu} + \frac{(\mathcal{F}^{\mathrm{symp}}h)(0,0)}{1+\nu}. \tag{3.2.10}$$

From this identity, one finds the meromorphic continuation of the function $\nu \mapsto \mathfrak{E}_\nu$, including the residues at the two poles, as well as the fact that $\mathfrak{E}_\nu$ and $\mathfrak{E}_{-\nu}$ are the images of each other under $\mathcal{F}^{\mathrm{symp}}$.

Next, we prove (3.2.8): in view of (2.1.2), it suffices to prove the first of this pair of equations, for which one loses no generality in assuming that $\mathrm{Re}\,\nu < -1$. Using the equation

$$\frac{1}{2}\int_{-\infty}^\infty |t|^{-\nu}.2\exp\left(-\frac{2\pi t^2}{\mathrm{Im}\,z}|x - z\xi|^2\right)dt = (2\pi)^{\frac{\nu-1}{2}}\Gamma(\frac{1-\nu}{2})\left(\frac{|x-z\xi|^2}{\mathrm{Im}\,z}\right)^{\frac{\nu-1}{2}}$$

$$\tag{3.2.11}$$

and the definition of the Eisenstein series $E_{\frac{1-\nu}{2}}(z)$ as well as (3.1.4), we are done.

Finally, the quickest way to prove (3.2.7) is, in view of the fact that a tempered distribution is characterized by its (Θ_0, Θ_1)-transform, to verify that this equation is indeed correct (*cf.* (2.1.1)) in the case when h coincides with the function $h_0(x, \xi) = 2\exp\left(-2\pi\frac{|x-z\xi|^2}{\mathrm{Im}\,z}\right)$ or with the function $h_1 = 2i\pi\mathcal{E}h_0$. Considering the first case only, one has

$$\left(\mathcal{F}_1^{-1}h_0\right)(s,t) = (2\mathrm{Im}\,z)^{\frac{1}{2}}\exp\left(-\frac{\pi}{2}[(4t^2 + s^2)\mathrm{Im}\,z - 4istRe\,z]\right), \tag{3.2.12}$$

and

$$\int_{-\infty}^{\infty} |t|^{-\nu-1}(\mathcal{F}_1^{-1}h)(0,t)dt = 2^{\frac{1+\nu}{2}}\pi^{\frac{\nu}{2}}\Gamma(-\frac{\nu}{2})(\operatorname{Im} z)^{\frac{1+\nu}{2}},$$

$$\int_{-\infty}^{\infty} |t|^{-\nu}h(t,0)dt = 2^{\frac{1+\nu}{2}}\pi^{\frac{\nu-1}{2}}\Gamma(\frac{1-\nu}{2})(\operatorname{Im} z)^{\frac{1-\nu}{2}}, \tag{3.2.13}$$

$$\int_{-\infty}^{\infty} |t|^{-\nu-1}(\mathcal{F}_1^{-1}h)\left(\frac{k}{t},t\right)dt = 2^{\frac{3+\nu}{2}}|k|^{-\frac{\nu}{2}}(\operatorname{Im} z)^{\frac{1}{2}} K_{\frac{\nu}{2}}(2\pi|k|\operatorname{Im} z)e^{2i\pi k\operatorname{Re} z}.$$

A comparison with (3.1.6), using also the already proved equation (3.2.8) and, when needed, the functional equation of the zeta function, proves the identity of the two sides of (3.2.7), when applied to h_0. □

Using only Eisenstein distributions, one can already decompose some useful automorphic distributions as integral superpositions of homogeneous ones.

Proposition 3.2.2. *Consider the (automorphic) Dirac comb*

$$\mathfrak{D}(x,\xi) = 2\pi \sum_{|m|+|n|\neq 0} \delta(x-n)\delta(\xi-m): \tag{3.2.14}$$

its decomposition as a superposition of homogeneous distributions is given as

$$\mathfrak{D} = 2\pi + \int_{-\infty}^{\infty} \mathcal{E}_{i\lambda}d\lambda. \tag{3.2.15}$$

Proof. Starting from (1.2.11), (1.2.12) and using a change of contour, one may write, for any $a > 1$ and any function $h \in \mathcal{S}_{\text{even}}(\mathbb{R}^2)$,

$$h = \frac{1}{i}\int_{\operatorname{Re}\nu=-a} h_{-\nu}d\nu \quad \text{with} \quad h_{-\nu}(x,\xi) = \frac{1}{2\pi}\int_0^{\infty} t^{-\nu}h(tx,t\xi)dt. \tag{3.2.16}$$

Then,

$$\langle \mathfrak{D}, h_{-\nu}\rangle = \sum_{|m|+|n|\neq 0} \int_0^{\infty} t^{-\nu}h(tn,tm)dt \tag{3.2.17}$$

and, using (3.2.5),

$$\langle \mathfrak{D}, h\rangle = \frac{1}{i}\int_{\operatorname{Re}\nu=-a} \langle \mathcal{E}_\nu, h\rangle d\nu, \tag{3.2.18}$$

after which it suffices to move the contour of integration to the line $\operatorname{Re}\nu = 0$, using the first equation (3.2.6). □

Incidentally, note that the invariance (Poisson's formula) of the distribution $\mathfrak{D} + 2\pi\delta$ under the symplectic Fourier transformation yields, when coupled with (3.2.8), an unusual proof of the functional equation $E^*_{\frac{1-\nu}{2}} = E^*_{\frac{1+\nu}{2}}$. The same proof, with $\mathbb{R}^n$ substituting for $\mathbb{R}^2$, would also give an unusual proof [62, p. 1158] of the functional equation of Epstein's zeta function.

We now lift cusp-forms to automorphic distributions, taking a generalization of (3.2.7) as a definition.

Proposition 3.2.3. *Let* $\mathcal{M}$ *be a cusp-form with the Fourier expansion*

$$\mathcal{M}(x+iy) = y^{\frac{1}{2}} \sum_{k \neq 0} b_k K_{\frac{i\lambda}{2}}(2\pi|k|y)e^{2i\pi kx} : \qquad (3.2.19)$$

we assume that $\lambda > 0$. *Given* $\epsilon = \pm 1$, *define a modular distribution* $\mathfrak{M}_{(\epsilon)}$, *homogeneous of degree* $-1 - i\varepsilon\lambda$, *by setting, for every function* $h \in \mathcal{S}(\mathbb{R}^2)$,

$$\langle \mathfrak{M}_{(\epsilon)}, h \rangle = \frac{1}{2} \sum_{k \neq 0} |k|^{\frac{i\epsilon\lambda}{2}} b_k \int_{-\infty}^{\infty} |t|^{-i\epsilon\lambda - 1} \left(\mathcal{F}_1^{-1} h \right) \left(\frac{k}{t}, t \right) dt. \qquad (3.2.20)$$

One has the identity $\mathcal{F}^{\mathrm{symp}}\mathfrak{M}_{(1)} = \mathfrak{M}_{(-1)}$. *The* Θ-*transform of the rescaled version* $\mathfrak{M}_{(\epsilon)}^{\mathrm{resc}}$ *is then given by the pair of equations*

$$\left(\Theta_0 \mathfrak{M}_{(\epsilon)}^{\mathrm{resc}} \right)(z) = \mathcal{M}(z), \quad \left(\Theta_1 \mathfrak{M}_{(\epsilon)}^{\mathrm{resc}} \right)(z) = -i\epsilon\lambda\mathcal{M}(z). \qquad (3.2.21)$$

Proof. The proof is quite similar to that of (3.2.7) in Proposition 3.2.1: only, the coefficients have changed.   □

To obtain in a sense a full set of automorphic distributions, one must complete the set of Eisenstein distributions by a full set of cusp-distributions. Recall that, taking advantage of Hecke's theory, we have denoted as $(\mathcal{M}_{p,j})$ a complete set of L^2-normalized Hecke eigenforms, with the following conventions: p runs through the set $\{1, 2, \dots\}$, the eigenvalue $\frac{1+\lambda_p^2}{4}$ increases with p, and λ_p is assumed to be positive; next, for every p, the set of Hecke eigenforms $\mathcal{M}_{p,j}$ obtained by letting j run to all possible values (to wit, $1 \leq j \leq \kappa_p$ for some κ_p) generates linearly the corresponding eigenspace of Δ. Then, with the notation (3.2.20) and $\mathcal{M} = \mathcal{M}_{p,j}$, we set

$$\mathfrak{M}_{p,j} = \mathfrak{M}_{(1)}, \quad \mathfrak{M}_{-p,j} = \mathfrak{M}_{(-1)} : \qquad (3.2.22)$$

note that $\mathcal{M}_{p,j}$ is only defined for $p \geq 1$, while $\mathfrak{M}_{p,j}$ is defined for $p \neq 0$. Of course, starting from $\mathcal{N}_{p,j}$ in place of $\mathcal{M}_{p,j}$, one can define in just as canonical a way two cusp-distributions naturally denoted as $\mathfrak{N}_{p,j}$ and $\mathfrak{N}_{-p,j}$.

We have just seen how to lift Eisenstein series and cusp-forms to Eisenstein distributions and cusp-distributions. Finally, above the constant 1, a function on Π, we must consider the automorphic distributions 1 and δ, or their rescaled versions 1 and $\frac{\delta}{2}$: the Θ-transform of 1 is the pair $(1, 1)$, that of $\frac{\delta}{2}$ is the pair $(1, -1)$. Operations on automorphic functions also correspond to operations on automorphic distributions. The fact that $\Delta - \frac{1}{4}$ corresponds to $\pi^2 \mathcal{E}^2$ has already been recognized as a fundamental one. We now consider the Hecke operators $T_N, N \geq 1$, to be completed by the operator T_{-1} which acts as ± 1 on automorphic functions of even or odd type, i.e., the operator defined by the equation $(T_{-1}f)(z) = f(-\bar{z})$. Given an automorphic distribution $\mathfrak{S}$, we set, for $N \geq 1$,

$$\langle T_N^{\mathrm{dist}}\mathfrak{S}, h\rangle = N^{-\frac{1}{2}} \sum_{\substack{ad-N,d>0 \\ b\,\mathrm{mod}\,d}} \langle \mathfrak{S}, (x, \xi) \mapsto h\left(\frac{dx - b\xi}{\sqrt{N}}, \frac{a\xi}{\sqrt{N}}\right)\rangle \qquad (3.2.23)$$

and

$$\langle T_{-1}^{\mathrm{dist}}\mathfrak{S}, h\rangle = \langle \mathfrak{S}, (x, \xi) \mapsto h(-x, \xi)\rangle. \qquad (3.2.24)$$

Then, under the Θ-transformation, whether $N \geq 1$ or $N = -1$, T_N^{dist} transfers to T_N. Also, for $N \geq 1$,

$$T_N^{\mathrm{dist}}\mathfrak{E}_\nu = N^{-\frac{\nu}{2}}\sigma_\nu(N)\mathfrak{E}_\nu, \quad T_N^{\mathrm{dist}}\mathfrak{M}_{p,j} = b_{N;|p|,j}\mathfrak{M}_{p,j}, \qquad (3.2.25)$$

where we recall the definition which just precedes (3.1.9). The operator T_{-1}^{dist} acts as the identity on Eisenstein distributions, as ± 1 on cusp-distributions, according to type.

One can define the generating series

$$\mathcal{L}(s) - \sum_{N \geq 1} N^{-s}T_N^{\mathrm{dist}} = \prod_{p\,\mathrm{prime}} \left(1 - p^{-s}T_p^{\mathrm{dist}} + p^{-2s}\right)^{-1} : \qquad (3.2.26)$$

then, if $\mathrm{Re}\,\nu < -1$ and $\mathrm{Re}\,s > 1 - \frac{\mathrm{Re}\,\nu}{2}$, one has (a consequence of (3.1.8))

$$\mathcal{L}(s)\mathfrak{E}_\nu = \zeta\left(s - \frac{\nu}{2}\right)\zeta\left(s + \frac{\nu}{2}\right)\mathfrak{E}_\nu, \quad \mathcal{L}(s)\mathfrak{M}_{p,j} = L(s, M_{|p|,j})\mathfrak{M}_{p,j}. \qquad (3.2.27)$$

These formulas make an analytic continuation with respect to s possible. We shall also need the linear combination

$$\mathcal{L}'(s) = \pi^{\frac{1}{2}-s}\frac{\Gamma(\frac{s+i\pi\mathcal{E}}{2})}{\Gamma(\frac{1-s-i\pi\mathcal{E}}{2})}\mathcal{L}_{\mathrm{even}}(s) + \pi^{\frac{1}{2}-s}\frac{\Gamma(\frac{s+1+i\pi\mathcal{E}}{2})}{\Gamma(\frac{2-s-i\pi\mathcal{E}}{2})}\mathcal{L}_{\mathrm{odd}}(s), \qquad (3.2.28)$$

where $\mathcal{L}_{\mathrm{even}}(s)$ (*resp.* $\mathcal{L}_{\mathrm{odd}}(s)$) coincides with $\mathcal{L}(s)$ on automorphic distributions of even (*resp.* odd) type and is zero on automorphic distributions of odd (*resp.* even) type. Then, $\mathcal{L}'(s)$ acts as $\pm \mathcal{L}'(1 - s)$, according to the type of the automorphic distributions it is applied to. The straightforward calculations involving the action of Hecke operators on the automorphic distribution level, or the operator $\mathcal{L}(s)$, are made in more detail in [61, Section 5] if needed.

Let us give another example of automorphic distribution, which number theorists will find familiar in another disguise. Start from the "elementary" distribution $\mathfrak{b}_1(x, \xi) = e^{2i\pi x}\delta(\xi - 1)$ and, more generally,

$$\mathfrak{b}_\ell(x, \xi) = e^{2i\pi\sqrt{\ell}x}\delta(\xi - \sqrt{\ell}) = \ell^{-\frac{1}{2}}\mathfrak{b}_1\left(\sqrt{\ell}x, \frac{\xi}{\sqrt{\ell}}\right), \qquad (3.2.29)$$

where $\ell = 1, 2, \ldots$: it is invariant under all linear transformations associated to matrices in the group $\Gamma^{0}_{\infty} = \{(\begin{smallmatrix} 1 & b \\ 0 & 1 \end{smallmatrix}) : b \in \mathbb{Z}\} = \Gamma \cap N$, so that, relying on a Poincaré series process, one can define, at least formally, the *Bezout distribution*

$$\mathfrak{B}_{\ell} = \frac{1}{2} \sum_{g \in \Gamma/\Gamma^{0}_{\infty}} (\mathfrak{b}_{\ell})_{\text{even}} \circ g^{-1}, \tag{3.2.30}$$

where $(\mathfrak{b}_{\ell})_{\text{even}}$ denotes the even part of $\mathfrak{b}_{\ell}$. The name (you may argue against it) stems from the formal expression $\mathfrak{B}_{1} = \sum_{(m,n)=1} I_{n,m}$, with

$$\langle I_{n,m}, h \rangle = \int_{-\infty}^{\infty} h(nx + n_1, mx + m_1) e^{2i\pi x} dx, \tag{3.2.31}$$

where (n_1, m_1) is any pair such that $m_1 n - n_1 m = 1$. Some care is required here, since the series on the right-hand side does not converge in the space of distributions. But, for $q \geq 1$, $\mathfrak{B}_{\ell}^{q} = (i\pi\mathcal{E})_{q}(-i\pi\mathcal{E})_{q}\mathfrak{B}_{\ell}$ (as defined by a termwise application of the differential operator in front of it) does make sense as a tempered distribution, as can be shown by means of integrations by parts: we use the Pochhammer notation $(X)_{q} = X(X+1)\ldots(X+q-1)$. Details can be found in [61, p. 24].

The elementary distribution $\mathfrak{b}_{\ell}$ is invariant under $\mathcal{F}^{\text{symp}}$, and so is the automorphic distribution $\mathfrak{B}_{\ell}^{q}$: since we are more interested in $\mathcal{G}$ than in $\mathcal{F}^{\text{symp}}$, we recall instead that the rescaled distribution

$$\mathfrak{C}_{\ell}^{q} = (\mathfrak{B}_{\ell}^{q})^{\text{resc}} = 2^{-\frac{1}{2}+i\pi\mathcal{E}}\mathfrak{B}_{\ell}^{q} \tag{3.2.32}$$

is $\mathcal{G}$-invariant. As a consequence, it is characterized by its Θ_0-transform

$$(\Theta_0\mathfrak{C}_{\ell}^{q})(z) = \langle \mathfrak{B}_{\ell}^{q}, (x, \xi) \mapsto \exp\left(-\pi\frac{|x - z\xi|^2}{\operatorname{Im} z}\right)\rangle. \tag{3.2.33}$$

It is immediate that

$$\langle \mathfrak{b}_{\ell}, (x, \xi) \mapsto \exp\left(-\pi\frac{|x - z\xi|^2}{\operatorname{Im} z}\right)\rangle = y^{\frac{1}{2}} e^{2i\pi\ell z} : \tag{3.2.34}$$

using the fact that the operator $\pi^2\mathcal{E}^2$ transfers to $\Delta - \frac{1}{4}$ under Θ and the equation

$$\left(\Delta - \frac{1}{4}\right)\left(\Delta + \frac{3}{4}\right)\ldots\left(\Delta - \frac{1}{4} + (q-1)^2\right)\left(y^{\frac{1}{2}} e^{2i\pi\ell z}\right)$$

$$= (4\pi\ell)^q \frac{\Gamma(\frac{1}{2} + q)}{\Gamma(\frac{1}{2})} y^{q+\frac{1}{2}} e^{2i\pi\ell z}. \tag{3.2.35}$$

as seen by induction, one obtains

$$(\Theta_0 \mathfrak{C}_\ell^q)(z) = (4\pi\ell)^q \frac{\Gamma(\frac{1}{2}+q)}{\Gamma(\frac{1}{2})}$$

$$\times \frac{1}{2} \sum_{\left(\begin{smallmatrix} n_1 & n_2 \\ m_1 & m_2 \end{smallmatrix}\right) \in \Gamma/\Gamma_\infty^0} \left(\frac{\operatorname{Im} z}{|-m_1 z + n_1|^2}\right)^{q+\frac{1}{2}} \exp\left(2i\pi\ell \frac{m_2 z - n_2}{-m_1 z + n_1}\right). \quad (3.2.36)$$

This is an automorphic function in Π, a special case of a class introduced by Selberg [45] and used by many authors in the field [7, 12]. It is rapidly decreasing at infinity in the fundamental domain, and the standard Rankin-Selberg unfolding method makes it possible to obtain the coefficients of its (Roelcke-Selberg) decomposition into (generalized) eigenfunctions of Δ. The result obtained is the identity (see for instance [7, p. 244, 246])

$$(\Theta_0 \mathfrak{C}_\ell^q)(z) = \frac{1}{4\pi} \int_{-\infty}^\infty \frac{\Gamma(q+\frac{i\lambda}{2})\Gamma(q-\frac{i\lambda}{2})}{\zeta^*(i\lambda)\zeta^*(-i\lambda)} \frac{\sigma_{i\lambda}(\ell)}{\ell^{\frac{i\lambda}{2}}} E_{\frac{1-i\lambda}{2}}^*(z) d\lambda$$

$$+ \sum_{\substack{p,j \\ p \geq 1}} \Gamma(q - \frac{i\lambda_p}{2})\Gamma(q + \frac{i\lambda_p}{2}) b_{\ell;p,j} \mathcal{M}_{p,j}. \quad (3.2.37)$$

Equation (3.2.37) is equivalent (since we already know that the distribution $\mathfrak{B}_\ell^q$ is invariant under $\mathcal{F}^{\mathrm{symp}}$) to its version involving automorphic distributions, to wit

$$\mathfrak{B}_\ell^q = \frac{1}{4\pi} \int_{-\infty}^\infty \frac{\Gamma(q-\frac{i\lambda}{2})\Gamma(q+\frac{i\lambda}{2})}{\zeta^*(i\lambda)\zeta^*(-i\lambda)} \frac{\sigma_{i\lambda}(\ell)}{\ell^{\frac{i\lambda}{2}}} \mathfrak{E}_{i\lambda} d\lambda$$

$$+ \frac{1}{2} \sum_{\substack{p,j \\ p \in \mathbb{Z}^\times}} \frac{\Gamma(q - \frac{i\lambda_p}{2})\Gamma(q + \frac{i\lambda_p}{2})}{\|\mathcal{N}_{|p|,j}\|} b_{\ell;p,j} \mathfrak{N}_{p,j}. \quad (3.2.38)$$

A Fourier decomposition of Selberg's function, together with its spectral analysis, leads to expressions of the Fourier coefficients of Hecke eigenforms as residues of series of Kloosterman sums. Another method towards the analysis of such series is based on the use of the automorphic Green kernel: it can be found in [21], for the case of general congruence subgroups. The Bezout distribution is also the basic object involved in the composition, in the sense of the Weyl calculus, of two Eisenstein distributions, as will be explained in Section 3.4.

3.3 The Kloosterman-related series $\zeta_k(s,t)$

Our main interest in this section lies in the Dirichlet series in two variables $\zeta_k(s,t)$ and $\zeta_k^-(s,t)$ to be introduced below. Their analytic continuation to a sufficient domain (Theorem 3.6.2) will be essential in Chapter 4, and will be a byproduct of the analysis, in Section 3.5, of the pointwise product of two Eisenstein series. In some domain smaller than needed, this continuation can also be derived from known results regarding series of Kloosterman sums. Let us hasten to say that these well-known results are not contained, either, in Theorem 3.6.2: we shall rely on these, in an essential way, in Chapter 5. As a preparation for our study of the function ζ_k, we need the continuation of a function related to the Hurwitz zeta function.

Definition 3.3.1. For $\varepsilon = 0, 1$ and $s \in \mathbb{C}$, we recall (1.2.23):

$$B_\varepsilon(s) = (-i)^\varepsilon \pi^{s - \frac{1}{2}} \frac{\Gamma(\frac{1-s+\varepsilon}{2})}{\Gamma(\frac{s+\varepsilon}{2})}, \tag{3.3.1}$$

and we note that the functional equation of the zeta function can be written

$$\zeta(s) = B_0(s)\zeta(1 - s). \tag{3.3.2}$$

Lemma 3.3.2. Given $\xi \notin \mathbb{Z}$ and $s \in \mathbb{C}$ with $\operatorname{Re} s > -1$, set

$$\zeta[s, \xi] = |\xi|^{-s} + 2\zeta(s) + \frac{1}{2} \sum_{n \neq 0} [|n + \xi|^{-s} + |n - \xi|^{-s} - 2|n|^{-s}], \tag{3.3.3}$$

a convergent series. One has

$$\zeta[s, \xi] = \sum_{n \in \mathbb{Z}} |\xi + n|^{-s} \quad \text{when } \operatorname{Re} s > 1 \tag{3.3.4}$$

and

$$\zeta[s, \xi] = B_0(s) \sum_{j \neq 0} e^{2i\pi j\xi} |j|^{s-1} \quad \text{when } -1 < \operatorname{Re} s < 0. \tag{3.3.5}$$

In particular, this function extends as a meromorphic function of s in the complex plane, with an only pole at $s = 1$.

Proof. The series defining $\zeta[s, \xi]$ converges for $\operatorname{Re} s > -1$ because the bracket is a $O(|n|^{-\operatorname{Re} s - 2})$ as $|n| \to \infty$. Equation (3.3.4) is immediate. When $-1 < \operatorname{Re} s < 0$, one may write

$$\zeta[s, \xi] = 2\zeta(s) + \frac{1}{2} \sum_{n \in \mathbb{Z}} [|n + \xi|^{-s} + |n - \xi|^{-s} - 2|n|^{-s}]$$

$$= 2\zeta(s) + \frac{\pi^{\frac{s}{2}}}{\Gamma(\frac{s}{2})} \int_0^\infty x^{s-1} [e^{-\pi(n+\xi)^2 x^2} + e^{-\pi(n-\xi)^2 x^2} - 2e^{-\pi n^2 x^2}] dx. \tag{3.3.6}$$

Using Poisson's formula,

$$\zeta[s, \xi] = 2\zeta(s) + \frac{\pi^{\frac{s}{2}}}{\Gamma(\frac{s}{2})} \int_0^\infty x^{s-2} dx \sum_{j \in \mathbb{Z}} e^{-\frac{\pi j^2}{x^2}} (e^{2i\pi j\xi} + e^{-2i\pi j\xi} - 2), \tag{3.3.7}$$

where the term with $j = 0$ can of course be dispensed with. Hence,

$$\zeta[s, \xi] = 2\zeta(s) + \frac{\pi^{\frac{s}{2}}}{\Gamma(\frac{s}{2})} \sum_{j\neq 0} \frac{1}{2} \frac{\Gamma(\frac{1-s}{2})}{\pi^{\frac{1-s}{2}}} |j|^{s-1} (e^{2i\pi j\xi} + e^{-2i\pi j\xi} - 2)$$

$$= 2\zeta(s) + \frac{1}{2} \pi^{s-\frac{1}{2}} \frac{\Gamma(\frac{1-s}{2})}{\Gamma(\frac{s}{2})} \sum_{j\neq 0} (e^{2i\pi j\xi} + e^{-2i\pi j\xi} - 2)|j|^{s-1} \qquad (3.3.8)$$

so that (3.3.5) is proved with the help of the functional equation of zeta. $\qquad \square$

Remark 3.3.a. The important point is to connect the two functions on the right-hand sides of (3.3.4) and (3.3.5): note, however, that the initial domains of convergence of these two series, to wit $\operatorname{Re} s > 1$ and $\operatorname{Re} s < 0$, do not intersect.

One has more generally, the case when $\varepsilon = 1$ requiring in the same way an intermediate function so as to connect two disjoint domains, the identity

$$\sum_{n\in\mathbb{Z}} |\xi + n|_\varepsilon^{-s} = B_\varepsilon(s) \sum_{j\neq 0} e^{2i\pi j\xi} |j|_\varepsilon^{s-1}, \qquad \varepsilon = 0, 1: \qquad (3.3.9)$$

the case when $\varepsilon = 1$ follows from the other one by taking a $\frac{d}{d\xi}$-derivative.

Definition 3.3.3. With $k \in \mathbb{Z}$, we set, when $\operatorname{Re} s > 1, \operatorname{Re} t > 1$,

$$\zeta_k(s,t) = \frac{1}{4} \sum_{\substack{m_1 m_2 \neq 0 \\ (m_1, m_2)=1}} |m_1|^{-s} |m_2|^{-t} \exp\left(2i\pi k \frac{\overline{m_2}}{m_1}\right), \qquad (3.3.10)$$

where $\overline{m_2}$ is the number mod m_1 such that $\overline{m_2} m_2 \equiv 1$. We define the function $\zeta_k^-(s,t)$ by the same formula, in which the extra factor $\operatorname{sign}(m_1 m_2)$ has been added on the right-hand side.

Remark 3.3.b. One may also write, with $\Gamma \cap \overline{N} = \{(\begin{smallmatrix} 1 & 0 \\ \ell & 1 \end{smallmatrix}), \ell \in \mathbb{Z}\}$,

$$\zeta_k(s,t) = \frac{1}{8} \sum_{\substack{(\begin{smallmatrix} a & b \\ c & d \end{smallmatrix}) \in \Gamma/(\Gamma\cap\overline{N}) \\ bd\neq 0}} |d|^{-s} |b|^{-t} \exp\left(-2i\pi k \frac{c}{d}\right). \qquad (3.3.11)$$

Next, we recall [21, p. 51] or [23, p. 19] the definition of Kloosterman sums, in which j, k, m are integers and $m \geq 1$:

$$S(j, k; m) = \sum_{\substack{\xi, \eta \bmod m \\ \xi\eta\equiv 1}} \exp\left(\frac{2i\pi}{m}(j\xi + k\eta)\right). \qquad (3.3.12)$$

In particular [23, p. 481], one has $S(j, 0; m) = \sum_{1\leq d|(j,m)} \operatorname{M\ddot{o}b}(\frac{m}{d})d$, where Möb is the Möbius indicator function (usually denoted as μ, too useful as a spectral parameter for us), from which it follows that, for $\operatorname{Re} v > 1$, one has

$$\sum_{m\geq 1} m^{-v} S(j, 0; m) = (\zeta(v))^{-1} \sum_{1\leq d|j} d^{1-v} = (\zeta(v))^{-1} \sigma_{1-v}(j) \qquad (3.3.13)$$

(*cf* (3.1.6) for a definition of the last factor), a formula useful later. Also, it is immediate that

$$\zeta_0(s,t) = \frac{\zeta(s)\zeta(t)}{\zeta(s+t)}. \tag{3.3.14}$$

Proposition 3.3.4. *For every* $k \in \mathbb{Z}$, *the function* $\zeta_k(s,t)$ *extends as a holomorphic function, first to the domain* $(\operatorname{Re} s > 1, \operatorname{Re} t > -1, \operatorname{Re}(s+t) > 2, t \neq 1)$, *next to the domain* $(\operatorname{Re}(s+t) > \frac{3}{2}, \operatorname{Re} t < 0)$. *In this latter domain, one has*

$$\zeta_k(s,t) = \frac{1}{2} B_0(t) \sum_{m \geq 1} m^{-s-t} \sum_{j \neq 0} S(j,k;m)|j|^{t-1}. \tag{3.3.15}$$

Proof. Consider first the case when $k = 0$. Whenever $\operatorname{Re} t < 0$ and $\operatorname{Re}(s+t) > 1$, the right-hand side of (3.3.15) can be written, with the help of (3.3.13), as

$$B_0(t)(\zeta(s+t))^{-1} \sum_{j \geq 1} j^{t-1}\sigma_{1-v}(j) = B_0(t)\frac{\zeta(s)\zeta(1-t)}{\zeta(s+t)} = \frac{\zeta(s)\zeta(t)}{\zeta(s+t)}. \tag{3.3.16}$$

Assume now that $k \neq 0$, and that, to begin with, $\operatorname{Re} s > 1, \operatorname{Re} t > 1$. Then,

$$\zeta_k(s,t) = \frac{1}{2} \sum_{m_1 \geq 1} m_1^{-s} \sum_{\substack{m_2 \neq 0 \\ (m_1,m_2)=1}} |m_2|^{-t} \exp\left(2i\pi k\frac{\overline{m_2}}{m_1}\right)$$

$$= \zeta(t) + \frac{1}{2} \sum_{m_1 \geq 2} m_1^{-s-t} \sum_{\substack{1 \leq \ell < m_1 \\ (\ell,m_1)=1}} \sum_{q \in \mathbb{Z}} \left|\frac{\ell}{m_1} + q\right|^{-t} \exp\left(2i\pi k\frac{\overline{\ell}}{m_1}\right)$$

$$= \zeta(t) + \frac{1}{2} \sum_{m_1 \geq 2} m_1^{-s-t} \sum_{\substack{\xi,\eta \bmod m_1 \\ \xi\eta \equiv 1}} \zeta\left[t, \frac{\xi}{m_1}\right] \exp\left(2i\pi k\frac{\eta}{m_1}\right). \tag{3.3.17}$$

When $\operatorname{Re} t > -1$ and $0 < \xi < 1$, one may use the integral expression [36, p. 23]

$$\sum_{n \geq 0} (\xi + n)^{-t} = \frac{1}{2}\xi^{-t} - \frac{\xi^{1-t}}{1-t} + \frac{1}{\Gamma(t)} \int_0^\infty \left[\frac{1}{e^x - 1} - \frac{1}{x} + \frac{1}{2}\right] e^{-\xi x} x^{t-1} dx, \tag{3.3.18}$$

in which the bracket is a $O(x)$ as $x \to 0$. It shows that the left-hand side is bounded by $C\xi^{-\max(0,\operatorname{Re} t)}$ for some $C > 0$ (there is an extra factor $|\log \xi|$ in the case when $t = 0$, but this will not change the end-result, and we may forget about it). This gives the function $\zeta[t,\xi]$ the bound $C[\xi^{-\max(0,\operatorname{Re} t)} + (1-\xi)^{-\max(0,\operatorname{Re} t)}]$. Comparing a sum to a Riemann sum, we then obtain the bound

$$\sum_{\xi \in (\mathbb{Z}/m_1\mathbb{Z})^\times} \left|\zeta\left[t, \frac{\xi}{m_1}\right]\right| \leq \begin{cases} Cm_1 & \text{if } -1 < \operatorname{Re} t \leq 1, \\ Cm_1^{\operatorname{Re} t} & \text{if } \operatorname{Re} t \geq 1. \end{cases} \tag{3.3.19}$$

The expression (3.3.17) of $\zeta_k(s,t)$ is thus convergent for (s,t) in any of the two sets defined by the inequalities $(-1 < \operatorname{Re} t \leq 1, t \neq 1, \operatorname{Re}(s+t) > 2)$ or $(\operatorname{Re} t \geq 1,$

$t \neq 1$, $\operatorname{Re} s > 1$), the union of which contains the domain ($\operatorname{Re} s > 1$, $\operatorname{Re} t > -1$, $\operatorname{Re}(s+t) > 2, t \neq 1$).

Using (3.3.5), we obtain when $\operatorname{Re}(s+t) > 2, -1 < \operatorname{Re} t < 0$, the identity

$$\zeta_k(s,t) = \zeta(t) + \frac{1}{2}B_0(t) \sum_{m_1 \geq 2} m_1^{-s-t}$$

$$\times \sum_{\substack{\xi,\eta \bmod m_1 \\ \xi\eta \equiv 1}} \exp\left(2i\pi k \frac{\eta}{m_1}\right) \sum_{j \neq 0} |j|^{t-1} \exp\left(2i\pi j \frac{\xi}{m_1}\right), \quad (3.3.20)$$

or

$$\zeta_k(s,t) = \zeta(t) + \frac{1}{2}B_0(t) \sum_{m_1 \geq 2} m_1^{-s-t} \sum_{j \neq 0} S(j,k;m_1)|j|^{t-1}$$

$$= \frac{1}{2}B_0(t) \sum_{m_1 \geq 1} m_1^{-s-t} \sum_{j \neq 0} S(j,k;m_1)|j|^{t-1}. \quad (3.3.21)$$

The Weil estimate [67, 12]

$$|S(j,k;m)| \leq \sigma_0(m)m^{\frac{1}{2}}(|j|,|k|;m)^{\frac{1}{2}}, \quad (3.3.22)$$

where $\sigma_0(m)$ is the number of divisors of m, a $O(m^\varepsilon)$ for every $\varepsilon > 0$, makes it possible to extend the meaning of the right-hand side of (3.3.21) to the domain ($\operatorname{Re}(s+t) > \frac{3}{2}, \operatorname{Re} t < 0$). $\qquad\square$

We shall not need the following easy corollary, quoted for the sake of completeness.

Proposition 3.3.5. *The function $\zeta_k^-(s,t)$ can be analytically continued in the same conditions as the ones in Proposition 3.3.4, leading in the last domain mentioned to the identity*

$$\zeta_k^-(s,t) = \frac{1}{2}B_1(t) \sum_{m \geq 1} m^{-s-t} \sum_{j \neq 0} S(j,k;m)|j|_1^{t-1}. \quad (3.3.23)$$

The Kloosterman-Selberg series are Dirichlet series in one variable, analogous to (3.3.15) or (3.3.23) but involving no summation with respect to j: they have been considered by Selberg [45], Bruggeman [3], Kuznetsov [31], Goldfeld-Sarnak [12], Deshouillers-Iwaniec [7], Iwaniec [21], Iwaniec-Kowalski [23]. The results there would lead to the meromorphic continuation of the series $\zeta_k(s,t)$ to the domain where $\operatorname{Re} t < 0$. However, we need, in view of an application in Chapter 4, to understand this continuation to the domain where $\operatorname{Re} s > 0, \operatorname{Re} t > 0, |\operatorname{Re}(s-t)| < 1$: how to do this will be reviewed in Section 3.6. This is not to mean that results in that section contain the ones already known: there is no inclusion in either direction.

Meanwhile, let us remark that, setting

$$(\zeta_k)_{\mathrm{sym}}(s,t) = \frac{1}{4} \sum_{\substack{m_1 m_2 \neq 0 \\ (m_1, m_2) = 1}} |m_1|_\delta^{-s} |m_2|_\delta^{-t} \exp\left(2i\pi k \left(\frac{\overline{m_2}}{m_1} - \frac{1}{2m_1 m_2}\right)\right) \quad (3.3.24)$$

and expanding $\exp\left(-\frac{i\pi k}{m_1 m_2}\right)$ into a series, one has

$$(\zeta_k)_{\mathrm{sym}}(s,t) = \sum_{q\,\mathrm{even} \geq 0} \frac{(i\pi k)^q}{q!} \zeta_k(s+q, t+q) + \sum_{q\,\mathrm{odd} \geq 1} \frac{(-i\pi k)^q}{q!} \zeta_k^-(s+q, t+q),$$

$$(3.3.25)$$

so that the difference $(\zeta_k)_{\mathrm{sym}}(s,t) - \zeta_k(s,t)$ can be continued as a holomorphic function to the domain $(\mathrm{Re}\, s > 0, \mathrm{Re}\, t > 0)$ we are interested in. In view of the observation that, if $\left(\begin{smallmatrix} n_1 & n_2 \\ m_1 & m_2 \end{smallmatrix}\right) \in \Gamma$, so does the matrix $\left(\begin{smallmatrix} -n_2 & -n_1 \\ m_2 & m_1 \end{smallmatrix}\right)$, one finds the identity $(\zeta_k)_{\mathrm{sym}}(s,t) = (\zeta_{-k})_{\mathrm{sym}}(t,s)$: but there is a more important reason to use this modified version, which will appear in Chapter 4. By the same trick, the difference $\zeta_k(s,t) - \zeta_k(t,s)$ extends analytically to the domain $(\mathrm{Re}\, s > 0, \mathrm{Re}\, t > 0)$ too.

3.4   About the sharp product of two Eisenstein distributions

This short section is purely expository, and the remainder of the book will not depend on it: however, we believe that this volume, devoted to pseudo-differential analysis for number theorists, could not be totally silent on the subject of auto-morphic pseudo-differential analysis. Also, Remark 3.4.b. will connect the present section to the following two ones: Chapter 4 will depend strongly on some results reviewed there.

We wish to consider the composition, in the sense of the Weyl calculus, of two automorphic distributions. Relying on the transfer, from automorphic func-tion theory (in Π) to automorphic distribution theory in the plane and on Roelcke-Selberg decompositions, it suffices to analyze the sharp product of two "elemen-tary" distributions, this vocable meaning Eisenstein distributions $\mathfrak{E}_{i\lambda}$ or cusp-distributions. The whole volume [61] was devoted to this question: we shall con-sider, here, only the sharp product of two Eisenstein distributions, limiting our-selves to explaining in which way it can be made meaningful and recalling from the given reference the main formula.

Consider two Eisenstein distributions $\mathfrak{E}_{\nu_1}$ and $\mathfrak{E}_{\nu_2}$, and their rescaled ver-sions (cf. (3.2.1)) $\mathfrak{E}_{\nu_1}^{\mathrm{resc}}$ and $\mathfrak{E}_{\nu_2}^{\mathrm{resc}}$. Assume that $|\mathrm{Re}\,(\nu_1 \pm \nu_2)| < 1$. Let $A_j = \mathrm{Op}\left(\mathfrak{E}_{\nu_j}^{\mathrm{resc}}\right)$, $j = 1, 2$. We do not aim at giving the product $A = A_1 A_2$ a genuine symbol $\mathfrak{S} = \mathfrak{E}_{\nu_1}^{\mathrm{resc}} \# \mathfrak{E}_{\nu_2}^{\mathrm{resc}}$. The reason is that the two operators cannot be com-posed in a traditional sense, not even so as to yield a linear operator A from the linear space generated by all functions ϕ_z^0, ϕ_z^1 with $z \in \Pi$ to the algebraic dual of that space. In other words (*cf.* Theorem 1.1.3), one is prevented by a (mild) divergence from trying to define, for instance, the scalar product $(\phi_z^0 | A_1 A_2 \phi_z^0)$ as $(A_1^* \phi_z^0 | A_2 \phi_z^0)$: note that, simply, $A_1^* = \text{Op}(\mathfrak{E}_{\nu_1}^{\text{resc}})$.

Instead of defining $\mathfrak{S}$ as a genuine distribution, we shall satisfy ourselves with defining it up to the addition of an arbitrary distribution homogeneous of degree -1, actually a multiple of $\mathfrak{E}_0$ since we are only dealing with automorphic distributions: this can be done by giving the image of $\mathfrak{S}$ under $2i\pi\mathcal{E}$ a definition, without defining $\mathfrak{S}$. How such a thing is possible was indicated in (2.3.3): defining $2i\pi\mathcal{E}\mathfrak{S}$ only requires giving the operator $PAQ - QAP$ a symbol. Or, using the Θ-transformation again, we only need to give the scalar product

$$(P\phi_z^0 | AQ\phi_z^0) - (Q\phi_z^0 | AP\phi_z^0) \tag{3.4.1}$$

and the similar one with ϕ_z^0 replaced by ϕ_z^1 a meaning. Now, computing $P\phi_z^0$ and $Q\phi_z^0$ with the help of (1.1.32), one is led (easy details are to be found in [61, p. 132]), in the case when a product $A_1 A_2$ has a genuine symbol $\mathfrak{S}$, to the equation

$$(\phi_z^0 | \text{Op}(2i\pi\mathcal{E}\mathfrak{S})\phi_z^0) = 2i\pi(P\phi_z^0 | AQ\phi_z^0) - (Q\phi_z^0 | AP\phi_z^0)$$
$$= (\phi_z^1 | \text{Op}(\mathfrak{S})\phi_z^1) = (A_1^* \phi_z^1 | A_2 \phi_z^1). \tag{3.4.2}$$

A slightly more complicated calculation leads to the equation

$$(\phi_z^1 | \text{Op}(2i\pi\mathcal{E}\mathfrak{S})\phi_z^1) = 3(A_1^* \phi_z^2 | A_2 \phi_z^2) - 3^{\frac{1}{2}}[(A_1^* \phi_z^0 | A_2 \phi_z^2) + (A_1^* \phi_z^2 | A_2 \phi_z^0)], \tag{3.4.3}$$

involving the new function

$$\phi_z^2(x) = 2^{\frac{5}{4}} \pi^{\frac{1}{2}} (\text{Im}\,(-z^{-1}))^{\frac{5}{4}} x^2 \exp \frac{i\pi x^2}{z}. \tag{3.4.4}$$

Of course, in the situation we are dealing with, we shall make from the right-hand sides of the last two equations a definition of the Θ-transform of $2i\pi\mathcal{E}\mathfrak{S}$. One can indeed verify that the right-hand side of (3.4.2) is a convergent integral, whereas the right-hand side of (3.4.3), considered as a whole, is convergent as an improper integral (an integration by parts is required).

After these partly unsatisfactory preparations (an explanation follows), one is led to the following, taken from [61, p. 162]:

Theorem 3.4.1. *Assuming* $|\text{Re}\,(\nu_1 \pm \nu_2)| < 1$, *one has in the sense just explained, up to the addition of an arbitrary multiple of* $\mathfrak{E}_0$, *the equation*

$$\mathfrak{E}_{\nu_1}^{\text{resc}} \# \mathfrak{E}_{\nu_2}^{\text{resc}}$$
$$= \sum_{\varepsilon = \pm 1} \left[\frac{\zeta(\varepsilon\nu_1)\zeta(\varepsilon\nu_2)}{\zeta(\varepsilon(\nu_1 + \nu_2) - 1)} \mathfrak{E}_{1-\varepsilon(\nu_1+\nu_2)}^{\text{resc}} + \frac{\zeta(\varepsilon\nu_1)\zeta(-\varepsilon\nu_2)}{\zeta(\varepsilon(\nu_1 - \nu_2) - 1)} \mathfrak{E}_{\varepsilon(\nu_1-\nu_2)-1}^{\text{resc}} \right]$$
$$+ \mathcal{L}'\left(\frac{1+\nu_1+\nu_2}{2}\right) \mathcal{GL}'\left(\frac{1+\nu_1-\nu_2}{2}\right) \mathfrak{B}_1^{\text{resc}}, \tag{3.4.5}$$

where the Bezout distribution $\mathfrak{B}_1$ was formerly defined in (3.2.30) and $\mathcal{L}'(s)$ is the generating series of Hecke operators introduced in (3.2.28).

The proof of this theorem is quite lenghty. The reader may remember that the distribution $\mathfrak{B}_1$ could not be defined by the series in (3.2.30), only its image under $\pi^2 \mathcal{E}^2$ could: but the right-hand side of (3.2.38), obtained as the result of a Roelcke-Selberg decomposition, is still meaningful for $q = 0$ and can be taken as a definition of $\mathfrak{B}_k = \mathfrak{B}_k^0$. This is the sense given to $\mathfrak{B}_1$ in Theorem 3.4.1, and $\mathcal{L}'(s)$ is then to be applied under the integral sign and summation sign in (3.2.38), relying on the equations (3.2.27) and (3.2.28).

Remark 3.4.a. It would clearly be much nicer if we could give the product of the two operators with symbols $\mathfrak{E}_{i\lambda_1}^{\mathrm{resc}}$ and $\mathfrak{E}_{i\lambda_2}^{\mathrm{resc}}$, each of which is of course an operator from $\mathcal{S}(\mathbb{R})$ to $\mathcal{S}'(\mathbb{R})$, a meaning as a composition in the ordinary sense. This is not possible if using the Weyl calculus. However, let us remark that the composition formula (3.4.5), while having to rely in an essential way upon the covariance rule (1.1.20) since Γ-invariance is the key element there, does not have anything to do with the covariance rule (1.1.18) under the Heisenberg representation. In other words, this formula, though considerably more difficult to establish than its non-arithmetic counterpart, is in some sense of the same nature as the second formula discussed in Chapter 1, in Theorem 1.2.2. Now, if one gives up the first covariance relation (1.1.18), the Weyl calculus does no longer appear as a unique possibility, rather as one (the most singular, actually) in a "series" (in the sense of harmonic analysis, i.e., truly a sequence) of symbolic calculi, all of which do satisfy the required covariance rule (1.1.20).

It demands of course much work to define these "higher-level Weyl calculi", as was done in [61, Chapter 2]. Let us just indicate the principle of the construction. Consider the series of representations $(\pi_{\tau+1})_{\tau>0}$ in (1.3.5), called the projective discrete series of $SL(2, \mathbb{R})$: one still gets a unitary representation when $-1 < \tau \leq 0$. Given $\tau > -1$, the two maps $\mathrm{Sq}_{\mathrm{even}}^{\tau+1}$ and $\mathrm{Sq}_{\mathrm{odd}}^{\tau+2}$ defined by the equations

$$\left(\mathrm{Sq}_{\mathrm{even}}^{\tau+1} v\right)(t) = 2^{\frac{\tau-1}{2}} |t|^{\frac{1}{2}-\tau} v\left(\frac{t^2}{2}\right),$$

$$\left(\mathrm{Sq}_{\mathrm{odd}}^{\tau+2} v\right)(t) = 2^{\frac{\tau}{2}} |t|_1^{-\frac{1}{2}-\tau} v\left(\frac{t^2}{2}\right) \tag{3.4.6}$$

define two isometries, the first one from the space $H_{\tau+1} = L^2((0, \infty); t^{-\tau} dt)$ of the representation $\pi_{\tau+1}$ onto $L_{\mathrm{even}}^2(\mathbb{R})$, the second from $H_{\tau+2}$ onto $L_{\mathrm{odd}}^2(\mathbb{R})$. Taking $\tau = -\frac{1}{2}$, it is an easy matter, using (1.3.8), to verify that the first map is actually an intertwining operator from $\pi_{\frac{1}{2}}$ to the even part of the one-dimensional metaplectic representation, and that the second map is an intertwining operator from $\pi_{\frac{3}{2}}$ to the odd part of the metaplectic representation: incidentally, the first part of this statement is also the particular case $n = 1$ of what precedes (1.3.15), even though we decided, there, to assume $n \geq 2$. In other words, up to unitary equivalence, one can understand the metaplectic representation as being the sum $\pi_{\frac{1}{2}} \oplus \pi_{\frac{3}{2}}$, a fact already mentioned in Remark 1.3.a. (ii).

Now, given an integer $p = 0, 1, \ldots$, consider instead the sum $\pi_{p+\frac{1}{2}} \oplus \pi_{p+\frac{3}{2}}$: reconsidering what precedes with $\tau = p - \frac{1}{2}$, one builds on $L^2(\mathbb{R})$ a unitary representation, to be called the metaplectic representation of level p. Next, we keep the usual operator Q of multiplication of functions of t by t; but we change the operator P to the one defined as the matrix

$$P_p = \frac{1}{2i\pi} \begin{pmatrix} 0 & \frac{d}{dt} + \frac{p}{t} \\ \frac{d}{dt} - \frac{p}{t} & 0 \end{pmatrix}, \tag{3.4.7}$$

relative to the decomposition of functions on the line into their even and odd parts (beginning with those with the parity of p). The operator P_p occurs quite naturally in connection with the analysis of the free (Dirac-Weyl) neutrino or other relativistic wave equations [61, Section 8]; alternatively, some readers may wish to regard it as a case of Dunkl operator. For $p \geq 1$, the operators Q and P_p do no longer generate a finite-dimensional Lie algebra, but one can still define a symbolic calculus by means of the equation (1.1.12) (with $n = 1$). One obtains a symbolic calculus which is just as nice as Weyl's in all aspects forgetting the Heisenberg representation, except for the fact that formulas become more complicated (we have obtained some time ago already, but not published, the explicit, more complicated, analogue of level p of Theorem 1.2.2: these investigations led to the much improved proof of that theorem given in the present book). On the other hand, all convergence problems become less severe as p increases: in particular, in the symbolic calculus of level $p \geq 2$, the question raised at the beginning of this remark has an affirmative answer [61, p. 102]. One can even compose any given number of operators the symbols of which are elementary automorphic distributions (Eisenstein distributions or cusp-distributions), provided p is chosen large enough. This being said, we have not worked out an explicit analogue of level p of Theorem 3.4.1.

Two last remarks in this direction: when τ is a half-integer as is the case here, the Bessel function which occurs in the integral kernel in (1.3.8) is an elementary (trigonometric) function. On the other hand, the function ϕ_z^2 in (3.4.4) plays, in the calculus of level 2, exactly the same role as the one played by ϕ_z^0 in the usual calculus (note that ϕ_i^2, contrary to ϕ_i^0 and ϕ_i^1, is not an eigenfunction of the harmonic oscillator: analysis at level p has its own "p-harmonic" oscillator $\pi(Q^2 + P_p^2)$).

Remark 3.4.b. In the non-arithmetic case, one can use very nice symbols, say in $\mathcal{S}(\mathbb{R}^2)$, and it was for such pairs of symbols that we proved Theorem 1.2.2. It showed that, after decomposition into homogeneous components (relative to both entries and to their sharp product), the structure of this composition formula depends entirely, up to numerical factors, on the operators with integral kernels $\chi_{i\lambda_1,i\lambda_2;i\lambda}^{\varepsilon_1,\varepsilon_2;\varepsilon}(s_1, s_2; s)$: when dealing with even symbols, only the cases when

$(\varepsilon_1, \varepsilon_2; \varepsilon) = (0,0;0)$ or $(1,1;1)$ have to be considered. Now, one may assume (we are not claiming that we have proved it) that the same still holds in the automorphic case, granted that the two operators under consideration are much more difficult to analyse in this case. How to do that is, however, suggested by Propositions 2.1.3 and 2.1.4: what these two propositions mean is that, when dealing with nice symbols, the operators under consideration transfer under associates of the dual Radon transformation to the pointwise product and the Poisson bracket of functions in Π.

This suggests that an alternative to the analysis of the sharp-composition of two automorphic distributions could be to analyse the pointwise product and the Poisson bracket of two automorphic functions in Π. Indeed, the methods used, in [61], in connection with the Weyl calculus of operators with automorphic symbols, did rely on a preliminary study, in [60], of the second class of problems just mentioned. In view of our application in Chapter 4, it will be sufficient to study the pointwise product (only) of two arbitrary Eisenstein series: in the next section, we shall review the main lines of the solution to this problem.

3.5   The pointwise product of two Eisenstein series

Our aim in this section is to obtain the decomposition into homogeneous components of a pointwise product such as $E^*_{\frac{1-\nu_1}{2}} E^*_{\frac{1-\nu_2}{2}}$. The results obtained here (towards which complete, if concise, proofs are given) will be applied, in Section 3.6, to the problem of finding the analytic continuation, in a sufficient domain, of the Dirichlet series $\zeta_k(s,t)$ introduced in (3.3.10): there, we shall give precise ideas of the proofs, only skipping some easy, if lengthy, technical details and giving a reference instead.

In view of the Fourier series expansion (3.1.6) of Eisenstein series, and taking into account the poles of zeta, we must assume from the start that ν_1 and ν_2 are distinct from $-1, 0, 1$. Then, we remark that, from the "constant terms" of the series (3.1.6) for the two factors, the four terms $y^{\frac{2 \pm \nu_1 \pm \nu_2}{2}}$, with two independent signs $\pm$, will appear in the expansion of the product. Now, exponents with a real part less than $\frac{1}{2}$ correspond to terms square-integrable in the fundamental domain. We may then, in general, make from the product under examination a function in $L^2(\Gamma \backslash \Pi)$, in the following way:

Proposition 3.5.1. *Let Ω be the subset of $\mathbb{C}^2$ characterized by the four conditions* $\mathrm{Re}\,(\nu_1 \pm \nu_2) \neq \pm 1$. *Given $(\nu_1, \nu_2) \in \Omega$, let Σ be the set of pairs $(\varepsilon_1 = \pm 1, \varepsilon_2 = \pm 1)$ such that $\varepsilon_1 \mathrm{Re}\,\nu_1 + \varepsilon_2 \mathrm{Re}\,\nu_2 < 1$: Σ depends only on which of the nine components of Ω the point (ν_1, ν_2) lies. Also, denote as $\Omega^\times$ the subset of Ω characterized by the additional conditions*

$$\nu_1 \neq -1, 0, 1, \quad \nu_2 \neq -1, 0, 1, \quad \pm\nu_1 \pm \nu_2 \neq 0, 2, \tag{3.5.1}$$

which do not disconnect Ω any further. If $(\nu_1, \nu_2) \in \Omega^\times$, the function

$$f_{\nu_1,\nu_2} = E^*_{\frac{1-\nu_1}{2}} E^*_{\frac{1-\nu_2}{2}} - \sum_{(\varepsilon_1,\varepsilon_2)\in\Sigma} \zeta^*(1-\varepsilon_1\nu_1)\zeta^*(1-\varepsilon_2\nu_2)E_{\frac{2-\varepsilon_1\nu_1-\varepsilon_2\nu_2}{2}} \quad (3.5.2)$$

lies in $L^2(\Gamma\backslash\Pi)$: so does the function $\Delta f_{\nu_1,\nu_2}$.

Proof. One just has to inspect the Fourier series expansion of the function so defined. Observe that, on the right-hand side, we are dealing with the unstarred function $E_{\frac{2-\varepsilon_1\nu_1-\varepsilon_2\nu_2}{2}}$: that it is meaningful is a consequence of the condition $\mathrm{Re}\,(\varepsilon_1\nu_1 + \varepsilon_2\nu_2) < 1$, which implies $\zeta^*(2 - \varepsilon_1\nu_1 - \varepsilon_2\nu_2) \neq 0$. $\qquad\square$

Theorem 3.5.2. *For $(\nu_1,\nu_2) \in \Omega^\times$, the Roelcke-Selberg decomposition (3.1.13) of the function f_{ν_1,ν_2} has no constant term. The density $\Phi(\lambda) = \Phi(\nu_1,\nu_2;\lambda)$ of the continuous part of this decomposition is given as*

$$\Phi(\nu_1,\nu_2;\lambda) = \frac{\zeta^*\left(\frac{1+i\lambda-\nu_1+\nu_2}{2}\right)\zeta^*\left(\frac{1+i\lambda+\nu_1-\nu_2}{2}\right)\zeta^*\left(\frac{1-i\lambda-\nu_1-\nu_2}{2}\right)\zeta^*\left(\frac{1+i\lambda-\nu_1-\nu_2}{2}\right)}{\zeta^*(-i\lambda)}.$$

$$(3.5.3)$$

The discrete part

$$f^{\mathrm{disc}}_{\nu_1,\nu_2}(z) = f_{\nu_1,\nu_2}(z) - \frac{1}{8i\pi}\int_{-\infty}^{\infty} \Phi(\nu_1,\nu_2;\lambda)E_{\frac{1-i\lambda}{2}}\,d\lambda \quad (3.5.4)$$

extends as a holomorphic function of (ν_1,ν_2) to the subset $\Omega^\sharp \supset \Omega^\times$ of $\mathbb{C}^2$ characterized by the conditions (3.5.1) together with the fact that at most one of the four conditions $\mathrm{Re}\,(\nu_1 \pm \nu_2) = \pm 1$ can be satisfied.

Proof. To compute the constant term and the continuous term from the Roelcke-Selberg decomposition of f_{ν_1,ν_2}, we shall apply Proposition 3.1.3. Expanding the two Eisenstein series we are dealing with into Fourier series, we observe that the "constant term" $a_0(y)$ of the Fourier series expansion of f_{ν_1,ν_2} is the sum of the function

$$(a_0)_{\mathrm{maj}}(y) = 4y\sum_{k\neq 0} |k|^{\frac{-\nu_1-\nu_2}{2}} \sigma_{\nu_1}(|k|)\sigma_{\nu_2}(|k|)K_{\frac{\nu_1}{2}}(2\pi|k|y)K_{\frac{\nu_2}{2}}(2\pi|k|y) \quad (3.5.5)$$

and of a linear combination, with well-defined coefficients depending on ν_1 and ν_2, of powers y^α, where $\alpha = \frac{2-\varepsilon_1\nu_1-\varepsilon_2\nu_2}{2}$ and $\varepsilon_1\mathrm{Re}\,\nu_1 + \varepsilon_2\mathrm{Re}\,\nu_2 > 1$, and of powers y^β with $\beta = \frac{\varepsilon_1\nu_1+\varepsilon_2\nu_2}{2}$ and $\varepsilon_1\mathrm{Re}\,\nu_1 + \varepsilon_2\mathrm{Re}\,\nu_2 < 1$. We now analyse the functions $C_0^\pm(\mu)$, as introduced in Proposition 3.1.3, by looking at the contributions of the various parts of a_0. Of course, each such part, contrary to their sum, does not lie in $L^2(\Gamma\backslash\Pi)$, but we may still compute the contribution to $C_0^-(\mu)$ of each part for $(-\mathrm{Re}\,\mu)$ large enough. A function such as y^α contributes to $C_0^-(\mu)$ (as defined in (3.1.31)), provided that $\mathrm{Re}\,\mu < \min(0, 2\mathrm{Re}\,\alpha - 1)$, the value

$$\frac{1}{8\pi}\frac{\pi^{-\frac{\mu}{2}}}{\Gamma\left(-\frac{\mu}{2}\right)} \times \frac{1}{\alpha - \frac{\mu+1}{2}}. \quad (3.5.6)$$

When continued analytically, this function has no pole at $\mu = -1$, as seen from the value of α and from (3.5.1); on the other hand, this continuation is regular on the pure imaginary line since $\operatorname{Re}\alpha < \frac{1}{2}$, and its restriction to this line agrees with what would be obtained from the continuation of $C_0^+(\mu)$ instead. Exactly the same goes so far as the terms y^β are concerned. This shows that, in applying Proposition 3.1.3, we may forget about these extra terms and consider only $(a_0)_{\mathrm{maj}}(y)$.

In some sense, this is simpler, since, for this contribution, the integral defining $C_0^+(\mu)$ is always convergent, so that the function in (3.1.33) is simply the value at $\mu = i\lambda$, computed from "far to the left" of the pure imaginary line, of the function

$$\frac{1}{8\pi}\int_0^\infty \frac{\pi^{-\frac{\mu}{2}}}{\Gamma\left(-\frac{\mu}{2}\right)}y^{-\frac{3}{2}-\frac{\mu}{2}}dy \times 8y\sum_{k\geq 1}k^{\frac{-\nu_1-\nu_2}{2}}\sigma_{\nu_1}(k)\sigma_{\nu_2}(k)K_{\frac{\nu_1}{2}}(2\pi ky)K_{\frac{\nu_2}{2}}(2\pi ky).$$

$$(3.5.7)$$

Using the Weber-Schafheitlin integral [36, p. 101]

$$\int_0^\infty y^{-\frac{1}{2}-\frac{\mu}{2}}K_{\frac{\nu_1}{2}}(2\pi ky)K_{\frac{\nu_2}{2}}(2\pi ky)dy = \frac{1}{8}(\pi y)^{\frac{\mu-1}{2}}\Gamma\left(\frac{1+\nu_1+\nu_2-\mu}{4}\right)$$

$$\times\Gamma\left(\frac{1+\nu_1-\nu_2-\mu}{4}\right)\Gamma\left(\frac{1-\nu_1+\nu_2-\mu}{4}\right)\Gamma\left(\frac{1-\nu_1-\nu_2-\mu}{4}\right)\quad (3.5.8)$$

and the Ramanujan formula ([22, p. 232] or [54, p. 163]

$$\sum_{k\geq 1}k^{\frac{-1+\mu-\nu_1-\nu_2}{2}}\sigma_{\nu_1}(k)\sigma_{\nu_2}(k) = (\zeta(1-\mu))^{-1}\zeta\left(\frac{1+\nu_1+\nu_2-\mu}{2}\right)$$

$$\times\zeta\left(\frac{1+\nu_1-\nu_2-\mu}{2}\right)\zeta\left(\frac{1-\nu_1+\nu_2-\mu}{2}\right)\zeta\left(\frac{1-\nu_1-\nu_2-\mu}{2}\right),\quad (3.5.9)$$

we obtain the spectral density of f_{ν_1,ν_2} as indicated in (3.5.3). We obtain also the fact that the constant term (a true constant, this time) from the Roelcke-Selberg expansion of this function is zero.

To complete the proof of Theorem 3.5.2, what remains to be done is proving that, for $(\nu_1,\nu_2)\in\Omega^\sharp$, the discrete part $f_{\nu_1,\nu_2}^{\mathrm{disc}}(z)$ depends on (ν_1,ν_2) in an analytic way. Observe that neither of the two terms from the difference on the right-hand side of (3.5.4) does: the first one, because of the discontinuities in the definition of f_{ν_1,ν_2} while crossing any of the four hyperplanes $\operatorname{Re}(\nu_1\pm\nu_2)=\pm 1$; the integral term, because the integrand $\lambda\mapsto\Phi(\nu_1,\nu_2;\lambda)$ ceases to be continuous on the real line when (ν_1,ν_2) lies on any of these hyperplanes. It is not difficult, however, to verify that the discontinuities experienced by the two terms in (3.5.4) when crossing one, and only one, of the hyperplanes under consideration exactly cancel out. Details are to be found in [60, p. 135] if desired, but it suffices to observe on one hand that, for real λ, the function

$$\Phi(\nu_1, \nu_2; \lambda) + \left[\frac{2i}{\lambda + i(1 + \nu_1 + \nu_2)} - \frac{2i}{\lambda - i(1 + \nu_1 + \nu_2)} \right]$$
$$\times \zeta^*(1 + \nu_1)\zeta^*(1 + \nu_2)E_{\frac{2+\nu_1+\nu_2}{2}}(z) \quad (3.5.10)$$

has no singularity as (ν_1, ν_2) crosses the hyperplane $\mathrm{Re}\,(\nu_1 + \nu_2) = -1$ while staying away from each of the three other hyperplanes into consideration; on the other hand, that one has

$$\frac{1}{8\pi} \int_{-\infty}^{\infty} \left[\frac{2i}{\lambda + i(1 + \nu_1 + \nu_2)} - \frac{2i}{\lambda - i(1 + \nu_1 + \nu_2)} \right] d\lambda$$
$$= \begin{cases} \frac{1}{2} & \text{if } \mathrm{Re}\,(\nu_1 + \nu_2) > -1 \\ -\frac{1}{2} & \text{if } \mathrm{Re}\,(\nu_1 + \nu_2) < -1. \end{cases} \quad (3.5.11)$$
$\square$

Theorem 3.5.3. *Recall the definition, just before (3.1.9), of $\mathcal{M}_{p,j}$, and $\mathcal{N}_{p,j}$, and recall the definition (3.1.12) of the function $L^*(s, \mathcal{M})$ in general. Under the assumptions of Theorem 3.5.2, the discrete part of the spectral decomposition of the function f_{ν_1, ν_2} is given as*

$$f_{\nu_1,\nu_2}^{\mathrm{disc}} = \sum_{\substack{p \geq 1 \\ p, j \, \mathrm{even}}} \Phi^{p,j} \mathcal{M}_{p,j}, \quad (3.5.12)$$

with

$$\Phi^{p,j} = \frac{1}{2} \|\mathcal{N}_{p,j}\|^{-1} L^* \left(\frac{1 - \nu_1 - \nu_2}{2}, \mathcal{N}_{p,j} \right) L^* \left(\frac{1 + \nu_1 - \nu_2}{2}, \mathcal{N}_{p,j} \right). \quad (3.5.13)$$

Recall that the subscript " p, j even" means that only the pairs (p, j) such that the Hecke eigenform $\mathcal{M}_{p,j}$ is of even type under the symmetry $z \mapsto -\bar{z}$ are retained.

Proof. Before giving the proof, and though this is not needed for our purposes, only for a good comprehension, let us indicate that Hecke eigenforms of odd type reappear in the analysis of Poisson brackets of two Eisenstein series (*cf.* Proposition 2.1.4). One has

$$\Phi^{p,j} = \int_D \overline{\mathcal{M}_{p,j}}(z) E^*_{\frac{1-\nu_1}{2}}(z) E^*_{\frac{1-\nu_2}{2}}(z) dm(z). \quad (3.5.14)$$

Using analytic continuation, it is no loss of generality to assume that $\nu_2 = i\lambda$ for some $\lambda \in \mathbb{R}$. Consider the automorphic function

$$f(z) = \overline{\mathcal{M}_{p,j}}(z) E^*_{\frac{1-\nu_1}{2}}(z), \quad (3.5.15)$$

which is rapidly decreasing as a function of y when z lies in the fundamental domain. As a consequence of (3.1.15) and of the equation $\nu_2 = i\lambda$, one has also

$\Phi^{p,j} = \zeta^*(1 - i\lambda)\Psi(-\lambda)$ if Ψ is the spectral density of the integral term from the Roelcke-Selberg resolution of f. Hence, we can again use Proposition 3.1.3 or, in this case, the simpler version following from the usual Rankin-Selberg method ($cf.$ Remark 3.1.a.) to evaluate $\Psi(-\lambda)$.

The "constant term" of the Fourier series expansion of f is the function

$$a_0(y) = 2y \sum_{k \neq 0} |k|^{-\frac{\nu_1}{2}} \sigma_{\nu_1}(|k|) b_{k;p,j} K_{\frac{i\lambda_p}{2}}(2\pi|k|y) K_{\frac{\nu_1}{2}}(2\pi|k|y), \qquad (3.5.16)$$

where $b_{k;p,j}$ is the kth Fourier coefficient of $\mathfrak{M}_{p,j}$ (3.1.9): it is immediate that Hecke eigenforms of odd type disappear from the picture. Applying Proposition 3.1.3, we obtain that, when (p, j) is of even type, $\Psi(-\lambda)$ is the value at $\mu = i\lambda$, as obtained by continuation from the left side of the pure imaginary line, of the integral

$$\int_0^\infty a_0(y) y^{-\frac{3}{2} - \mu} dy$$

$$= 4 \int_0^\infty y^{-\frac{1}{2} - \frac{\mu}{2}} \sum_{k \geq 1} k^{-\frac{\nu_1}{2}} \sigma_{\nu_1}(k) b_{k;p,j} K_{\frac{i\lambda_p}{2}}(2\pi|k|y) K_{\frac{\nu_1}{2}}(2\pi|k|y) dy. \qquad (3.5.17)$$

Using then the last part of Theorem 3.5.2 and, again, the Weber-Schafheitlin integral, one obtains that, for $\mathrm{Re}\,(-\nu_2)$ large enough, the coefficient $\Phi^{p,j}$ is given as

$$\Phi^{p,j} = \frac{1}{2} \pi^{\frac{-1+\nu_2}{2}} \left(\Gamma\left(\frac{1-\nu_2}{2}\right) \right)^{-1} \zeta^*(1 - \nu_2) \Gamma\left(\frac{1 + \nu_1 - \nu_2 + i\lambda_p}{4}\right)$$

$$\times \Gamma\left(\frac{1 + \nu_1 - \nu_2 - i\lambda_p}{4}\right) \Gamma\left(\frac{1 - \nu_1 - \nu_2 + i\lambda_p}{4}\right) \Gamma\left(\frac{1 - \nu_1 - \nu_2 - i\lambda_p}{4}\right)$$

$$\times \sum_{k \geq 1} k^{\frac{-1-\nu_1+\nu_2}{2}} \sigma_{\nu_1}(k) b_{k;p,j}. \qquad (3.5.18)$$

What remains to be done is computing the sum of the arithmetic series on the last line, for which we use the fact that $c_{k;p,j} = \|\mathcal{N}_{p,j}\| b_{k;p,j}$ enters the eigenvalue equation $T_k \mathcal{N}_{p,j} = c_{k;p,j} \mathcal{N}_{p,j}$, which yields when $\mathrm{Re}\,s$ is large

$$\sum_{k \geq 1} k^{-s} T_k \mathcal{N}_{p,j} = \|\mathcal{N}_{p,j}\| \sum_{k \geq 1} k^{-s} b_{k;p,j} \mathcal{N}_{p,j} = L(s, \mathcal{N}_{p,j}) \mathcal{N}_{p,j}. \qquad (3.5.19)$$

Though we shall only need this later, let us also note the comparable relation

$$\sum_{k \geq 1} k^{-s} T_k E_{\frac{1-\nu}{2}} = \zeta\left(s - \frac{\nu}{2}\right) \zeta\left(s + \frac{\nu}{2}\right) E_{\frac{1-\nu}{2}}, \qquad (3.5.20)$$

a consequence of $T_N E_{\frac{1-\nu}{2}} = N^{-\frac{\nu}{2}} \sigma_\nu(N) E_{\frac{1-\nu}{2}}$.

We compute then (recall that $\mathrm{Re}\,(-\nu_2)$ is large) the sum of the series

$$\sum_{k\geq 1} k^{\frac{-1-\nu_1+\nu_2}{2}} \sigma_{\nu_1}(k) T_k = \prod_{p\,\text{prime}} \sum_{\ell\geq 0} p^{\ell(\frac{-1-\nu_1+\nu_2}{2})}(1+p^{\nu_1}+\cdots+p^{\ell\nu_1}) T_{p^\ell} \quad (3.5.21)$$

(we have used the basic fact that $(m,n)=1$ implies $T_m T_n = T_{mn}$), which can be written as

$$\prod_p \sum_{\ell\geq 0} p^{\ell(\frac{-1-\nu_1+\nu_2}{2})} \frac{1-p^{(\ell+1)\nu_1}}{1-p^{\nu_1}} T_{p^\ell}$$

$$= \zeta(-\nu_1) \prod_p \sum_{\ell\geq 0} \left[p^{\ell(\frac{-1-\nu_1+\nu_2}{2})} - p^{\nu_1+\ell(\frac{-1+\nu_1+\nu_2}{2})} \right] T_{p^\ell}. \quad (3.5.22)$$

Using the first relation (3.1.8), this becomes

$$\zeta(-\nu_1) \prod_p \Big[\Big(1 - p^{\frac{-1-\nu_1+\nu_2}{2}} T_p + p^{-1-\nu_1+\nu_2} \Big)^{-1}$$

$$- p^{\nu_1} \Big(1 - p^{\frac{-1+\nu_1+\nu_2}{2}} T_p + p^{-1+\nu_1+\nu_2} \Big)^{-1} \Big], \quad (3.5.23)$$

or

$$\zeta(-\nu_1) \prod_p \Big(1 - p^{\frac{-1-\nu_1+\nu_2}{2}} T_p + p^{-1-\nu_1+\nu_2} \Big)^{-1} \quad (3.5.24)$$

$$\times \prod_p \Big(1 - p^{\frac{-1+\nu_1+\nu_2}{2}} T_p + p^{-1+\nu_1+\nu_2} \Big)^{-1} \times \prod_p \Big(1 + p^{-1+\nu_1+\nu_2} - p^{\nu_1} - p^{-1+\nu_2} \Big).$$

Using (3.5.19) and the second relation (3.1.8), one obtains

$$\|\mathcal{N}_{p,j}\| \sum_{k\geq 1} k^{\frac{-1-\nu_1+\nu_2}{2}} \sigma_{\nu_1}(k) b_{k;p,j}$$

$$= \zeta(-\nu_1) L\left(\frac{1+\nu_1-\nu_2}{2}, \mathcal{N}_{p,j} \right) L\left(\frac{1-\nu_1-\nu_2}{2}, \mathcal{N}_{p,j} \right) \prod_p (1-p^{\nu_1})(1-p^{-1+\nu_2})$$

$$= (\zeta(1-\nu_2))^{-1} L\left(\frac{1+\nu_1-\nu_2}{2}, \mathcal{N}_{p,j} \right) L\left(\frac{1-\nu_1-\nu_2}{2}, \mathcal{N}_{p,j} \right). \quad (3.5.25)$$

Theorem 3.5.3 follows. $\qquad\qquad\square$

Remark 3.5.a. In a related way, the pointwise product of an Eisenstein series and a Maass-Hecke eigenform was analyzed in [38]. In [61, Sec.16], besides computing the sharp product of two Eisenstein distributions, as indicated in Section 3.4, we pushed the sharp composition table further, considering also the case of an Eisenstein distribution together with a distribution of Maass-Hecke type, or that of two such distributions. Again, in view of Propositions 2.1.3 and 2.1.4, the problem was transformed to one involving the pointwise product or the Poisson bracket of two eigenfunctions of the modular Laplacian. We were prevented from completing our table by our lack of solution to the following problem: express, in terms of L-functions of type $L(s, f \times g)$ and of central values of triple L-functions, the integral, over the fundamental domain, of the product of three Maass-Hecke eigenforms, or that of the product of such a form by the Poisson bracket of two others. Even though, as alluded to in [24, p. 722], some recent work seems to go into a related direction, in the spirit of [9] (which dealt with Hecke forms of holomorphic type), it does not seem that anything answering the question just raised has appeared.

3.6 The continuation of ζ_k

N.B. Some lengthy if easy technical details have been omitted from the proof of Theorem 3.6.1 below: a reference is indicated.

Combining the analysis of the pointwise product of two Eisenstein series done in the last section with Proposition 3.1.4, we shall obtain here the analytic continuation of the function ζ_k in the domain exactly needed for our application in Chapter 4. We consider again the function f_{ν_1, ν_2} introduced in (3.5.2), replacing however the conditions (3.5.1) by

$$\operatorname{Re}\nu_1 < -1, \quad \operatorname{Re}\nu_2 < -1, \quad |\operatorname{Re}(\nu_1 - \nu_2)| < 1: \tag{3.6.1}$$

these would imply that $(\nu_1, \nu_2) \in \Omega^\times$ if it were not for the missing demand $\nu_1 \neq \nu_2$. However, this condition, used for simplicity in the last section, was just meant to prevent the consideration (when $\nu_1 = \nu_2$ and $\varepsilon_1 \neq \varepsilon_2$) of the non-existing Eisenstein series E_1: still, as will be observed in (4.5.1), there is an ersatz $E_1^\natural$ just as good for our purposes, so that the results of the last section are still applicable.

When $\operatorname{Re}\nu < -1$, one may rewrite (3.1.2) as

$$E^*_{\frac{1-\nu}{2}}(z) = \frac{1}{2} \sum_{|m|+|n|\neq 0} \pi^{\frac{\nu-1}{2}} \Gamma\left(\frac{1-\nu}{2}\right) \left(\frac{|mz-n|^2}{\operatorname{Im} z}\right)^{\frac{\nu-1}{2}}. \tag{3.6.2}$$

Under the present conditions, one may, with an obvious notation, derive from (3.6.2) an expression of the product $E^*_{\frac{1-\nu_1}{2}}(z) E^*_{\frac{1-\nu_2}{2}}(z)$ as the sum of a series over the set of all 4tuples (m_1, n_1, m_2, n_2) such that $|m_1|+|n_1| \neq 0$ and $|m_2|+|n_2| \neq 0$. The set of such 4tuples satisfying the extra condition $m_1 n_2 - n_1 m_2 = 0$ can be parametrized by the set of 4tuples (ℓ_1, ℓ_2, m, n) with $\ell_1 \geq 1, \ell_2 \neq 0$ and $(m, n) = 1$, by setting $\binom{m_1}{n_1} = \ell_1 \binom{m}{n}$ and $\binom{m_2}{n_2} = \ell_2 \binom{m}{n}$. It follows that the sum of all terms with $m_1 n_2 - n_1 m_2 = 0$ from the product of series under consideration is $\zeta^*(1-\nu_1)\zeta^*(1-\nu_2) E_{\frac{2-\nu_1-\nu_2}{2}}(z)$, which is one of the three terms one has in this case to subtract from the product $E^*_{\frac{1-\nu_1}{2}}(z) E^*_{\frac{1-\nu_2}{2}}(z)$ to obtain the function denoted as

$f_{\nu_1,\nu_2}(z)$ in (3.5.2). Hence, not forgetting in the case when $\nu_1 = \nu_2$ to make the slight modification indicated above,

$$f_{\nu_1,\nu_2}(z) = \frac{1}{2}\pi^{\frac{\nu_1+\nu_2-2}{2}}\Gamma\left(\frac{1-\nu_1}{2}\right)\Gamma\left(\frac{1-\nu_2}{2}\right)$$

$$\times \sum_{m_1n_2-n_1m_2\neq 0}\left(\frac{|m_1z-n_1|^2}{y}\right)^{\frac{\nu_1-1}{2}}\left(\frac{|m_2z-n_2|^2}{y}\right)^{\frac{\nu_2-1}{2}} \tag{3.6.3}$$

$$- \zeta^*(-\nu_1)\zeta^*(1-\nu_2)E_{\frac{2+\nu_1-\nu_2}{2}}(z) - \zeta^*(1-\nu_1)\zeta^*(-\nu_2)E_{\frac{2-\nu_1+\nu_2}{2}}(z).$$

This is not the expression of $f_{\nu_1,\nu_2}(z)$ we used in the last section, because it is only available when $\mathrm{Re}\,\nu_1 < -1$ and $\mathrm{Re}\,\nu_2 < -1$. However, it has the advantage of being easily related to a simpler function. Indeed, define

$$f^0_{\nu_1,\nu_2}(z) = \frac{1}{2}\pi^{\frac{\nu_1+\nu_2-2}{2}}\Gamma\left(\frac{1-\nu_1}{2}\right)\Gamma\left(\frac{1-\nu_2}{2}\right)$$

$$\times \sum_{m_1n_2-n_1m_2=1}\left(\frac{|m_1z-n_1|^2}{y}\right)^{\frac{\nu_1-1}{2}}\left(\frac{|m_2z-n_2|^2}{y}\right)^{\frac{\nu_2-1}{2}} \tag{3.6.4}$$

$$- \pi^{\frac{\nu_1+\nu_2-1}{2}}\Gamma(-\frac{\nu_1}{2})\Gamma(\frac{1-\nu_2}{2})E_{\frac{2+\nu_1-\nu_2}{2}}(z)$$

$$- \pi^{\frac{\nu_1+\nu_2-1}{2}}\Gamma(\frac{1-\nu_1}{2})\Gamma(-\frac{\nu_2}{2})E_{\frac{2-\nu_1+\nu_2}{2}}(z).$$

Using the definition of the Hecke operators and the fact that the set $M_N(\mathbb{Z})$ of integral matrices with determinant N can be written [55, p. 238] as

$$M_N(\mathbb{Z}) = \bigcup_{\substack{ad=N,d>0 \\ b\,\mathrm{mod}\,d}}\Gamma\begin{pmatrix} a & -b \\ 0 & d \end{pmatrix}, \tag{3.6.5}$$

also the relation (3.5.20), one can show the identity

$$f_{\nu_1,\nu_2} = \sum_{N\geq 1}N^{\frac{\nu_1+\nu_2-1}{2}}T_N f^0_{\nu_1,\nu_2}. \tag{3.6.6}$$

From the results of the last section, together with (3.5.20) and (3.5.19), one obtains first that the function $f^0_{\nu_1,\nu_2}$ is orthogonal to constants and that the density of the integral term from its Roelcke-Selberg decomposition is

$$\Phi_0(\nu_1,\nu_2;\lambda) = \pi^{\frac{\nu_1+\nu_2-1}{2}}\Gamma\left(\frac{1-i\lambda-\nu_1-\nu_2}{4}\right)\Gamma\left(\frac{1+i\lambda-\nu_1-\nu_2}{4}\right)$$

$$\times \frac{\zeta^*\left(\frac{1+i\lambda-\nu_1+\nu_2}{4}\right)\zeta^*\left(\frac{1+i\lambda+\nu_1-\nu_2}{4}\right)}{\zeta^*(-i\lambda)}. \tag{3.6.7}$$

Next, the discrete part of this decomposition is the sum of the series $\sum \Phi_0^{p,j} \mathcal{M}_{p,j}$, where only Hecke eigenfunctions of even type are considered, and

$$\Phi_0^{p,j} = \left(L\left(\frac{1-\nu_1-\nu_2}{2}, \mathcal{N}_{p,j}\right) \right)^{-1} \Phi^{p,j} = \frac{1}{2}\pi^{\frac{\nu_1+\nu_2-1}{2}} \|\mathcal{N}_{p,j}\|^{-1} \tag{3.6.8}$$

$$\times \, \Gamma\left(\frac{1-\nu_1-\nu_2+i\lambda}{4}\right) \Gamma\left(\frac{1-\nu_1-\nu_2-i\lambda}{4}\right) L^*\left(\frac{1+\nu_1-\nu_2}{2}, \mathcal{N}_{p,j}\right).$$

We shall now apply Proposition 3.1.4 to the automorphic function f_{ν_1,ν_2}^0, with the first observation that Proposition 3.1.3 was used in the last section in a quite different way: there, we used it so as to compute a spectral density. Here, we already know, from a quite independent method, an expression of the discrete part of the spectral decomposition of this function, and we are going to make use of Proposition 3.1.4 or, more precisely, of (3.1.40), in the reverse direction, so as to obtain information about the analytic continuation of the function c_k occurring there.

The first point to note is that the function f_{ν_1,ν_2}^0 does not satisfy all assumptions of Proposition 3.1.4, since the integral term of its spectral resolution is certainly not zero. However, one can define the function $\mu \mapsto c_k(\mu)$ by (3.1.38) in terms of the full function f_{ν_1,ν_2}^0 (i.e., without having first removed from it the integral term of its spectral resolution). Then, as this integral term is explicit, given in (3.6.7), one can analyze, as done in [60, Theor. 11.1], the possible singularities of the contribution to the function c_k due to this integral term. One finds that, in the domain $\mathrm{Re}\,\mu < 1 - |\mathrm{Re}\,(\nu_1 - \nu_2)|$, no singularities appear from this contribution except for simple poles located at non-trivial zeros of the zeta function. When this has been done, and these singularities have been registered, one can apply Proposition 3.1.4 as if the function f_{ν_1,ν_2}^0 did satisfy all assumptions of this proposition.

From the function f_{ν_1,ν_2}^0 as expressed in (3.6.4), we delete the two extra Eisenstein series, as well as the terms with $m_1 m_2 = 0$ from the main series: these, as is easily seen, will not contribute singularities to the function c_k considered in Theorem 3.6.1 below in the relevant domain. Then, the kth Fourier coefficient

$$a_k(y) = \int_{-\frac{1}{2}}^{\frac{1}{2}} e^{-2i\pi kx} f_{\nu_1,\nu_2}^0(x+iy)dx \tag{3.6.9}$$

can be replaced by

$$(a_k)_{\mathrm{main}}(y) = \frac{1}{2}\pi^{\frac{\nu_1+\nu_2-2}{2}} \Gamma\left(\frac{1-\nu_1}{2}\right) \Gamma\left(\frac{1-\nu_2}{2}\right) \tag{3.6.10}$$

$$\times \sum_{\substack{m_1 m_2 \neq 0 \\ m_1 n_2 - n_1 m_2 = 1}} \int_{-\frac{1}{2}}^{\frac{1}{2}} e^{-2i\pi kx} \left(\frac{|m_1 z - n_1|^2}{y}\right)^{\frac{\nu_1-1}{2}} \left(\frac{|m_2 z - n_2|^2}{y}\right)^{\frac{\nu_2-1}{2}} dx,$$

after which the function $c_k(\mu)$ in Proposition 3.1.4 becomes the simpler function

$$(c_k)_{\text{main}}(\mu) = \frac{1}{16} \frac{\pi^{\frac{\nu_1+\nu_2-4-\mu}{2}}}{\Gamma(-\frac{\mu}{2})} \Gamma\left(\frac{1-\nu_1}{2}\right) \Gamma\left(\frac{1-\nu_2}{2}\right)$$

$$\times \sum_{m_1 m_2 \neq 0} |m_1|^{\nu_1-1} |m_2|^{\nu_2-1} \int_0^\infty y^{-\frac{3}{2}-\frac{\mu}{2}} I_{m_1,m_2}(y) dy, \quad (3.6.11)$$

with

$$I_{m_1,m_2}(y) = y^{\frac{2-\nu_1-\nu_2}{2}} \qquad\qquad\qquad (3.6.12)$$

$$\times \sum_{\substack{n_1,n_2 \text{ such that} \\ m_1 n_2 - n_1 m_2 = 1}} \int_{-\frac{1}{2}}^{\frac{1}{2}} e^{-2i\pi kx} \left[\left(x-\frac{n_1}{m_1}\right)^2 + y^2\right]^{\frac{\nu_1-1}{2}} \left[\left(x-\frac{n_2}{m_2}\right)^2 + y^2\right]^{\frac{\nu_2-1}{2}} dx.$$

It is at this point that the factor $\exp\left(2i\pi k \frac{\overline{m_2}}{m_1}\right)$, characteristic of Kloosterman-related functions, appears. Indeed, fixing n_1^0 such that $n_1^0 m_2 \equiv -1 \mod m_1$, one can parametrize the set of pairs (n_1, n_2) with $m_1 n_2 - n_1 m_2 = 1$ by $\ell \in \mathbb{Z}$, setting $n_1 = n_1^0 + \ell m_1$, n_2 being then determined as $n_2 = \frac{1+n_1 m_2}{m_1}$. Performing in the last integral the change of variable $x \mapsto x + \frac{n_1^0}{m_1}$ and summing with respect to ℓ, one can write the sum of the last series of integrals as

$$e^{-2i\pi k \frac{n_1^0}{m_1}} \int_{-\infty}^\infty e^{-2i\pi kx} (x^2 + y^2)^{\frac{\nu_1-1}{2}} \left[\left(x-\frac{1}{m_1 m_2}\right)^2 + y^2\right]^{\frac{\nu_2-1}{2}} dx. \quad (3.6.13)$$

Transforming this integral into a convolution integral, one obtains, using [36, p. 401],

$$I_{m_1,m_2}(y) = \frac{4\pi^{\frac{2-\nu_1-\nu_2}{2}}}{\Gamma(\frac{1-\nu_1}{2})\Gamma(\frac{1-\nu_2}{2})} y e^{2i\pi k \frac{\overline{m_2}}{m_1}}$$

$$\times \int_{-\infty}^\infty |k-\sigma|^{-\frac{\nu_1}{2}} |\sigma|^{-\frac{\nu_2}{2}} K_{\frac{\nu_1}{2}}(2\pi|k-\sigma|y) K_{\frac{\nu_2}{2}}(2\pi|\sigma|y) e^{-2i\pi \frac{\sigma}{m_1 m_2}} d\sigma. \quad (3.6.14)$$

When computing $(c_k)_{\text{main}}(\mu)$ with the help of (3.6.11), (3.6.12), the integral with respect to dy will produce, again, a Weber-Schafheitlin integral [36, p. 101]: since the arguments of the two Bessel functions are no longer identical, the result is somewhat more complicated than the analogue from the proof of Theorem 3.5.2, and involves a hypergeometric function. We obtain

$$(c_k)_{\text{main}}(\mu)$$

$$= 2^{-5}\pi^{-\frac{3}{2}} \frac{\Gamma\left(\frac{1+\nu_1+\nu_2-\mu}{4}\right) \Gamma\left(\frac{1+\nu_1-\nu_2-\mu}{4}\right) \Gamma\left(\frac{1-\nu_1+\nu_2-\mu}{4}\right) \Gamma\left(\frac{1-\nu_1-\nu_2-\mu}{4}\right)}{\Gamma\left(-\frac{\mu}{2}\right) \Gamma\left(\frac{1-\mu}{2}\right)}$$

$$\times \sum_{\substack{m_1 m_2 \neq 0 \\ (m_1, m_2)=1}} |m_1|^{\nu_1 - 1} |m_2|^{\nu_2 - 1} e^{2i\pi k \frac{m_2}{m_1}} \mathcal{F}\left[\sigma \mapsto |\sigma|^{\frac{-1-\nu_1-\nu_2+\mu}{2}}\right. \tag{3.6.15}$$

$$\left. \times \,_2F_1\left(\frac{1-\nu_1+\nu_2-\mu}{4}, \frac{1+\nu_1+\nu_2-\mu}{4}; \frac{1-\mu}{2}; \frac{k(2\sigma-k)}{\sigma^2}\right)\right]\left(\frac{1}{m_1 m_2}\right).$$

Since

$$\left(\mathcal{F}(\sigma \mapsto |\sigma|^{\frac{-1-\nu_1-\nu_2+\mu}{2}})\right)\left(\frac{1}{m_1 m_2}\right) = \pi^{\frac{\nu_1+\nu_2-\mu}{2}} \frac{\Gamma\left(\frac{1-\nu_1-\nu_2+\mu}{4}\right)}{\Gamma\left(\frac{1+\nu_1+\nu_2-\mu}{4}\right)} |m_1 m_2|^{\frac{1-\nu_1-\nu_2+\mu}{2}},$$

$$\tag{3.6.16}$$

it is clear that the series defining $\zeta_k \left(\frac{1-\nu_1+\nu_2-\mu}{2}, \frac{1+\nu_1-\nu_2-\mu}{2}\right)$ would appear if the constant 1 were substituted for the hypergeometric function. This will make the following theorem [60, Theor.11.3] at least not surprising, as an application of Proposition 3.1.4. It is indeed true that the constant 1 from the Taylor expansion of the hypergeometric function (as a function of its last argument) can be shown to be its sole part with any influence for our purposes. However, some estimates are required, in view of which, after having made a partition of unity to separate the regions where the argument of the hypergeometric function is close to $0, 1$ or ∞, we must take advantage in each case of a well-chosen so-called linear transformation of the hypergeometric function [36, p. 47-48]. The somewhat lengthy technical details are to be found in the given reference.

Theorem 3.6.1. *Assume that* $\operatorname{Re}\nu_1 < -1, \operatorname{Re}\nu_2 < -1$ *and* $|\operatorname{Re}(\nu_1 - \nu_2)| < 1$. *Consider, for* $k \in \mathbb{Z}^\times$, *the meromorphic function defined when* $\operatorname{Re}\mu < -1 - |\operatorname{Re}(\nu_1 - \nu_2)|$ *by the equation*

$$b_k(\mu) = \frac{1}{8}\pi^{\frac{-3+\nu_1+\nu_2-\mu}{2}} \frac{\Gamma\left(\frac{1-\nu_1-\nu_2+\mu}{4}\right)\Gamma\left(\frac{1+\nu_1-\nu_2-\mu}{4}\right)\Gamma\left(\frac{1-\nu_1+\nu_2-\mu}{4}\right)\Gamma\left(\frac{1-\nu_1-\nu_2-\mu}{4}\right)}{\Gamma\left(-\frac{\mu}{2}\right)\Gamma\left(\frac{1-\mu}{2}\right)}$$

$$\times \zeta_k\left(\frac{1-\nu_1+\nu_2-\mu}{2}, \frac{1+\nu_1-\nu_2-\mu}{2}\right). \tag{3.6.17}$$

It extends as a meromorphic function in the half-plane $\operatorname{Re}\mu < 1 - |\operatorname{Re}(\nu_1 - \nu_2)|$. *Besides the actual poles* $-1 - \nu_1 + \nu_2$ *and* $-1 + \nu_1 - \nu_2$ *(which correspond to* $s = 1$ *and* $t = 1$ *when the pair of arguments of the function* ζ_k *is denoted as* (s, t)), *and the non-trivial zeros of the zeta function with a real part* $< 1 - |\operatorname{Re}(\nu_1 - \nu_2)|$, *its only possible poles in the strip* $-1 + \operatorname{Re}(\nu_1 + \nu_2) < \operatorname{Re}\mu < 1 - |\operatorname{Re}(\nu_1 - \nu_2)|$ *(note that this strip contains the closed strip* $-3 \leq \operatorname{Re}\mu \leq 0$) *are points* $i\lambda_p$ *with* $\frac{1+\lambda_p^2}{4}$ *in the discrete part of the spectrum of the hyperbolic Laplacian: they are simple. Moreover, the Fourier coefficients* d_k *of the orthogonal projection* $\operatorname{Pr}_{\frac{1+\lambda_p^2}{4}} f_{\nu_1,\nu_2}$ *of the function* f_{ν_1,ν_2}^0 *defined in (3.6.4) on the corresponding eigenspace of* Δ, *characterized by the identity*

$$\left(\operatorname{Pr}_{\frac{1+\lambda_p^2}{4}} f_{\nu_1,\nu_2}^0\right)(z) = y^{\frac{1}{2}} \sum_{k \neq 0} d_k K_{\frac{i\lambda_k}{2}}(2\pi|k|y)e^{2i\pi kx}, \tag{3.6.18}$$

are given by the formula

$$d_k = -8\pi |k|^{-\frac{i\lambda_p}{2}} \times \text{Res}_{\mu=i\lambda_p} b_k(\mu). \tag{3.6.19}$$

Taking advantage of the fact that we know an explicit form (Theorem 3.5.3) of the discrete part from the Roelcke-Selberg expansion of f_{ν_1,ν_2}, and trading the pair $\left(\frac{1-\nu_1+\nu_2-\mu}{2}, \frac{1+\nu_1-\nu_2-\mu}{2}\right)$ for a pair (s,t), one obtains the following:

Theorem 3.6.2. *For $k \neq 0$, the function $\zeta_k(s,t)$, initially defined for $\text{Re}\, s > 1, \text{Re}\, t > 1$, extends as a meromorphic function for $\text{Re}\, s > 0, \text{Re}\, t > 0, |\text{Re}\,(s-t)| < 1, s \neq 1, t \neq 1$, holomorphic outside the set of points (s,t) such that $s+t = 1-i\lambda_p$ with $\frac{1+\lambda_p^2}{4}$ in the even part (with respect to the symmetry $z \mapsto -\bar{z}$) of the discrete spectrum of Δ in $L^2(\Gamma\backslash\Pi)$, or $s+t = \omega$, a non-trivial zero of the zeta function. The polar parts at points of the first species are given by the formula*

$$\text{Res}_{\mu=i\lambda_p} \zeta_k \left(\frac{1-\nu-\mu}{2}, \frac{1+\nu-\mu}{2}\right) \tag{3.6.20}$$

$$= -2^{i\lambda_p} \pi^{\frac{1+i\lambda_p}{2}} \frac{\Gamma(-i\lambda_p)}{\Gamma(\frac{1+\nu-i\lambda_p}{4})\Gamma(\frac{1-\nu-i\lambda_p}{4})} |k|^{\frac{i\lambda_p}{2}} \sum_{p, j\text{even}} b_{k;p,j} L^* \left(\frac{1+\nu}{2}, \mathcal{M}_{p,j}\right),$$

with $b_{k;p,j}$ as defined just before (3.1.9).

Proof. This is Proposition 14.6 in [60]. However, it is just a corollary of Theorem 3.6.1 together with Theorem 3.5.3, noting, as a consequence of (3.6.6), that the coefficient of a Hecke eigenform $\mathcal{M}_{p,j}$ in the spectral decomposition of f_{ν_1,ν_2} is the product of the same coefficient relative to the function $f^0_{\nu_1,\nu_2}$ by $L\left(\frac{1-\nu_1-\nu_2}{2}, \mathcal{N}_{p,j}\right)$. There is a detail, to wit that the line $s+t = 1$ does not contribute singularities apart from the ones already mentioned, though an application of Theorem 3.6.1 would leave the question whether the point $\mu = 0$ is a pole unsettled (because of the denominator $\Gamma(-\frac{\mu}{2})$ on the right-hand side of (3.6.17)): special examination [60, Prop.15.8] takes care of this. $\qquad \square$

Chapter 4

A class of Poincaré series

We consider in this chapter the automorphic function $f_{\rho,\nu}$ built by the usual Poincaré-type series process, starting from the unusual function

$$z \mapsto (\operatorname{Im} z)^{\frac{\rho-1}{2}} \chi_{\rho,\nu}\left(\frac{\operatorname{Re} z}{\operatorname{Im} z}\right).$$

A procedure similar in one respect, but involving more arithmetic and less analysis, was used in [60, Sec.19-20] in connection with a real quadratic extension of the rationals: it will be more profitable to comment on this later (end of Section 4.7). One of the byproducts of this approach lies in a realization of cusp-forms as residues of appropriate series, with considerable resemblance to Eisenstein series: again, it connects to some previous work [60, Sec.15], but there is much improvement, as these series now appear as the main parts of explicit automorphic functions.

The classical approach to non-holomorphic modular form theory starts with the construction of Eisenstein series and of integral superpositions thereof, to wit incomplete Eisenstein series [21, 23, 55]. The space of automorphic functions so constructed "misses" exactly the constants and all cusp-forms, while it exhausts the set of generalized functions corresponding to the continuous part of the spectrum of the modular Laplacian Δ: the (classical) self-adjoint realization of Δ to be considered in all such matters was briefly recalled just after (3.1.12). We wish to put the present constructions on a parallel footing, starting from a simple modification $f_{\rho,\nu}^{\sharp}$ of $f_{\rho,\nu}$ turning this function into a function square-integrable in the fundamental domain when $\operatorname{Re} \nu < 0$. This time, the linear space generated by the functions $f_{\rho,\nu}^{\sharp}$ for any given ρ "reaches" all Hecke eigenforms $\mathcal{M}_{p,j}$ invariant under the map $z \mapsto -\bar{z}$ such that the L-function $L(s, \mathcal{M}_{p,j})$ does not vanish at $\frac{\rho}{2}$: by this, we mean that all such eigenforms are present in the spectral decomposition of every function $f_{\rho,\nu}^{\sharp}$. On the other hand, the closure S_ρ in some appropriate topology (not in the $L^2(\Gamma\backslash\Pi)$-norm, though: the continuity of the spectral density must be preserved) of the linear space generated by the functions $f_{\rho,\nu}^{\sharp}$ for some given ρ may fail to fill up the part of the subspace of $L^2(\Gamma\backslash\Pi)$ under consideration orthogonal to cusp-forms.

But it does so in an interesting way: assuming that $0 < \operatorname{Re}\rho < 2$, the space S_ρ is orthogonal to all Eisenstein series $E_{\frac{1+i\lambda}{2}}$ such that $\frac{\rho-i\lambda}{2}$ or $\frac{\rho+i\lambda}{2}$ is a non-trivial zero of the zeta function. To put this orthogonality property in a different way, the knowledge of the space S_ρ determines whether $\operatorname{Re}\rho = 1$ or not and, in the affirmative case, the value of ρ. On the other hand, the question whether these spaces, considered when $\operatorname{Re}\rho \neq 1$, are identical, is totally out of reach at present.

The developments in this chapter are rather lengthy. We have chosen to familiarize ourselves with part of the structure by considering one example, in the frame of automorphic distributions (on $\mathbb{R}^2$) rather than automorphic functions (on Π): the arithmetic is the same, and the analysis is simpler. The series initially used to define $f_{\rho,\nu}$ is convergent only when $\operatorname{Re}\nu < -1 - |\operatorname{Re}\rho - 1|$, and the analytic continuation with respect to ν of the function obtained demands important developments: it relies in particular on the analytic continuation of the Dirichlet series $\zeta_k(s,t)$ as obtained in Section 3.6. Then, establishing the asymptotics of $f_{\rho,\nu}(x + iy)$ as $y \to \infty$ is again a non-trivial matter. It will enable us to compute the Roelcke-Selberg expansion of $f_{\rho,\nu}$. We then discuss "incomplete ρ-series" (elements of the space S_ρ) and relate the functions $f_{\rho,\nu}$ to the automorphic Green's kernel. Finally, dropping the parameter ν and concentrating on ρ only, we establish the "Roelcke-Selberg" expansions of a class of one-dimensional automorphic measures supported in the union of lines congruent, under Γ, to the hyperbolic line form 0 to $i\infty$. A duality will manifest itself between this problem and that of decomposing the restriction to this line of nonholomorphic modular forms into homogeneous components.

4.1   An automorphic distribution of a Poincaré series type

Let us start from the even distribution

$$\mathfrak{s}(x,\xi) = e^{4i\pi x\xi} : \qquad\qquad (4.1.1)$$

it is invariant under $\mathcal{G}$. On the other hand, it is not invariant under the linear action of any subgroup of Γ larger than $\{\pm I\}$, and more work is needed to tackle the convergence problems related to the Poincaré series which has to be considered, to wit (in a formal sense, for the time being)

$$\mathfrak{S} = \sum_{g\in\{\pm I\}\backslash\Gamma} \mathfrak{s} \circ g. \qquad\qquad (4.1.2)$$

Before we do so, however, let us remark that this example, though undoubtedly a special one, will bring methods applicable to more general situations. Indeed, having chosen $e^{4i\pi x\xi}$ rather than $e^{4i\pi tx\xi}$ for a more general real number t only makes it possible not to carry an unnecessary parameter all the way: then, all MA-invariant functions, say in $\mathcal{S}(\mathbb{R}^2)$, are integral superpositions of the functions $e^{4i\pi tx\xi}$. Note that the MA-invariant case is that which corresponds to the choice $\rho = 1$ in (2.3.4). Write

$$\mathfrak{S}(x,\xi) = \sum_{g=\left(\begin{smallmatrix} a & b \\ c & d \end{smallmatrix}\right)\in\{\pm I\}\backslash\Gamma} e^{4i\pi(ax+b\xi)(cx+d\xi)}$$

$$= \frac{1}{2}e^{4i\pi x\xi} \sum_{g=\left(\begin{smallmatrix} a & b \\ c & d \end{smallmatrix}\right)\in\Gamma} e^{4i\pi(acx^2+2bcx\xi+bd\xi^2)}. \tag{4.1.3}$$

In all this chapter, we shall have to rely on an appropriate partition of Γ and on a good parametrization of its main part.

We split Γ as $\Gamma = \Gamma_0 \cup \Gamma_1 \cup \Gamma_2 \cup \Gamma_3$, where Γ_0 consists of the 4 special matrices $\pm\left(\begin{smallmatrix} 1 & 0 \\ 0 & 1 \end{smallmatrix}\right)$ and $\pm\left(\begin{smallmatrix} 0 & 1 \\ -1 & 0 \end{smallmatrix}\right)$, Γ_1 consists of the matrices $\pm\left(\begin{smallmatrix} 1 & 0 \\ c & 1 \end{smallmatrix}\right)$ and $\pm\left(\begin{smallmatrix} a & 1 \\ -1 & 0 \end{smallmatrix}\right)$ with $c \neq 0, a \neq 0$, Γ_2 consists of the matrices $\pm\left(\begin{smallmatrix} 1 & b \\ 0 & 1 \end{smallmatrix}\right)$ and $\pm\left(\begin{smallmatrix} 0 & 1 \\ -1 & d \end{smallmatrix}\right)$ with $b \neq 0, d \neq 0$: then, the remaining part Γ_3 of Γ is the set of matrices the 4 entries of which are non-zero.

Lemma 4.1.1. *Given $g - \left(\begin{smallmatrix} a & b \\ c & d \end{smallmatrix}\right) \in \Gamma_2 \cup \Gamma_3$, in other words assuming $bd \neq 0$, set $bc = n, bd = m$. The image of the map $g \mapsto n, m$ is the set of pairs n, m with $n \in \mathbb{Z}, m \in \mathbb{Z}^\times$ such that $m|n(n+1)$. Every such pair is the image of exactly two such matrices, one the negative of the other. The image of Γ_3 is characterized by the additional condition $n \neq 0, -1$.*

Proof. One has $ac = \frac{(bc)(ad)}{bd} = \frac{n(n+1)}{m}$ so we must assume that $m|n(n+1)$. Conversely, given a pair m, n of integers with $m \neq 0$ satisfying this condition, there is, up to a simultaneous sign change of all its entries, a unique matrix $\left(\begin{smallmatrix} a & b \\ c & d \end{smallmatrix}\right)$ such that $bc = n, bd = m$. Indeed, the condition $(c, d) = 1$ imposes the choice $b = \pm(m, n)$, from which one finds $c = \frac{n}{b}$ and $d = \frac{m}{b}$. It remains to check that a, obtained from the equation $a = \frac{n(n+1)}{mc} = \pm\frac{(n+1)(m,n)}{m}$, is indeed an integer: set $m = m_1 m_2$ with $m_1|n+1$ and $1 \leq m_2|n$; then, $\frac{(n+1)(m,n)}{m} = \frac{(n+1)m_2}{m} = \frac{n+1}{m_1}$. Within $\Gamma_2 \cup \Gamma_3$, Γ_2 is characterized by the additional condition $ac = 0$, or $n(n+1) = 0$. $\square$

Coming back to $\mathfrak{S}$, note that the two choices of sign for b lead to the same value of the exponential, so that one has

$$\frac{1}{2}e^{4i\pi x\xi} \sum_{g=\left(\begin{smallmatrix} a & b \\ c & d \end{smallmatrix}\right)\in\Gamma_2\cup\Gamma_3} e^{4i\pi(acx^2+2bcx\xi+bd\xi^2)}$$

$$= e^{4i\pi x\xi} \sum_{\substack{m\in\mathbb{Z}^\times, n\in\mathbb{Z} \\ m|n(n+1)}} \exp\left(\frac{4i\pi}{m}[n(n+1)x^2 + 2mnx\xi + m^2\xi^2]\right). \tag{4.1.4}$$

The terms with $bd = 0$ are obtained either for ($b = 0, a = d = \pm 1, c$ arbitrary) or ($d = 0, -b = c = \pm 1, a$ arbitrary) and contribute to $\mathfrak{S}(x, \xi)$ the sums

$$e^{4i\pi x\xi} \sum_{c \in \mathbb{Z}} e^{4i\pi c x^2} \quad \text{and} \quad e^{-4i\pi x\xi} \sum_{a \in \mathbb{Z}} e^{4i\pi a x^2}. \tag{4.1.5}$$

Finally,

$$\mathfrak{S}(x, \xi) = e^{4i\pi x\xi} \sum_{c \in \mathbb{Z}} e^{4i\pi c x^2} + e^{-4i\pi x\xi} \sum_{c \in \mathbb{Z}} e^{4i\pi c x^2}$$

$$+ \sum_{\substack{m \in \mathbb{Z}^\times, n \in \mathbb{Z} \\ m \mid n(n+1)}} \exp\left(\frac{4i\pi}{m}[n(n+1)x^2 + m(2n+1)x\xi + m^2\xi^2]\right). \tag{4.1.6}$$

Just like the series (3.2.30), the present one fails from being convergent in the space $\mathcal{S}'(\mathbb{R}^2)$: we shall show, however, that it is well-defined as a continuous linear form on the image of $\mathcal{S}(\mathbb{R}^2)$ under some Pochhammer-like polynomial in the Euler operator (or, equivalently, that the image of $\mathfrak{S}$ under the transpose of the operator just alluded to makes sense as a tempered distribution). We need a few lemmas in order to tackle the question of convergence.

Lemma 4.1.2. *Consider the Lie algebra $\mathfrak{g}$ linearly generated by E, M, N with the commutation relations*

$$[M, N] = 0, \qquad [E, M] = -N, \qquad [E, N] = -M. \tag{4.1.7}$$

Reduce elements of the enveloping algebra $U(\mathfrak{g})$ to polynomials in M, N by means of the commutation relations together with a repeated application of the rule that any element $X_1 \ldots X_{k-1} X_k$ of $U(\mathfrak{g})$ such that $X_k = E$ can be replaced by $X_1 \ldots X_{k-1} M$. Then, for any $\ell = 1, 2, \ldots$, the element $(E - \ell + 1)(E - \ell + 2) \ldots (E + \ell - 1)$ can be reduced to a polynomial in (the commuting variables) M, N without any term of total degree $< \ell$, and the same holds concerning the product of the polynomial in E just considered by E.

Proof. In other words, the reduction procedure is that associated to dividing off $U(\mathfrak{g})$, as a vector space, by the left ideal $\mathcal{J}$ consisting of products $Q(E - M)$ with $Q \in U(\mathfrak{g})$; incidentally, though this is not needed for the proof, $\mathfrak{g}$ is the Lie algebra of the Poincaré (*aka* inhomogeneous Lorentz) group in two-dimensional spacetime. That the reduction procedure makes it possible to replace any element of $U(\mathfrak{g})$ by some unique equivalent polynomial in M, N is obvious: the only problem has to do with degrees. This Lie algebra can be realized, in a faithful way, as a Lie algebra of differential operators in the half-plane $\{(a, t) : a > 0, t \in \mathbb{R}\}$, through the correspondence

$$E \mapsto \frac{d}{dt}, \qquad M \mapsto a \sinh t, \qquad N \mapsto -a \cosh t. \tag{4.1.8}$$

Then, the left-ideal $\mathcal{J}$ becomes the set of operators, in the algebra generated by the three operators just introduced, which kill the function $\phi(a,t) = e^{a\cosh t}$. Indeed, one has $E\phi = M\phi$: in the reverse direction, applying the reduction procedure to any such operator killing the function ϕ, one is left with a polynomial in M, N with the same property, which has to be zero identically since M, N are algebraically independent as functions of (a,t). Consequently, what we have to show is that the function

$$\psi = \left(\frac{d}{dt} - \ell + 1\right) \cdots \left(\frac{d}{dt} + \ell - 1\right) \phi \tag{4.1.9}$$

is identical to $P(a\cosh t, a\sinh t)\phi$ for some unique polynomial without term of total degree $< \ell$. In other words, we need to show that $\psi = \psi(a,t)$ satisfies the condition that $\left(\frac{\partial}{\partial a}\right)^j \psi(a,t)$ vanishes at $a = 0$ for $j = 0, \ldots, \ell - 1$. Now,

$$\left[\left(\frac{\partial}{\partial a}\right)^j \psi\right](0,t) = \left(\frac{d}{dt} - \ell + 1\right) \cdots \left(\frac{d}{dt} + \ell - 1\right) \left[\left(\frac{\partial}{\partial a}\right)^j \phi\right](0,t)$$

$$= \left(\frac{d}{dt} - \ell + 1\right) \cdots \left(\frac{d}{dt} + \ell - 1\right) [(\cosh t)^j] : \tag{4.1.10}$$

this is zero when $0 \leq j \leq \ell - 1$ since, on one hand, $(\cosh t)^j$ is a linear combination of terms e^{kt} with $-j \leq k \leq j$, on the other hand the operator $\left(\frac{d}{dt} - \ell + 1\right) \cdots \left(\frac{d}{dt} + \ell - 1\right)$ kills all exponentials e^{kt} with $-\ell + 1 \leq k \leq \ell - 1$. $\qquad\square$

Lemma 4.1.3. *Consider in the (x, ξ)-space the commuting operators*

$$A = \frac{1}{8i\pi}\frac{\partial}{\partial x} - \frac{\xi}{2}, \qquad B = \frac{1}{8i\pi}\frac{\partial}{\partial x} + \frac{\xi}{2}. \tag{4.1.11}$$

Given $\ell = 1, 2, \ldots$, there exist differential operators $D_\ell, \ldots, D_{2\ell}$ with the following properties: (i) for every j, D_j is a homogeneous polynomial in the operators A, B of total degree $2j$; (ii) given $c \neq 0$ and the quadratic form $Q(x, \xi) = x(cx + \xi)$, one has the identity

$$[(i\pi\mathcal{E}) \times (i\pi\mathcal{E} - \ell + 1) \ldots (i\pi\mathcal{E} + \ell - 1)] e^{4i\pi Q} = \left[c^{-2\ell}D_{2\ell} + \cdots + c^{-\ell}D_\ell\right] e^{4i\pi Q}. \tag{4.1.12}$$

Proof. Note the commutation relations

$$[\mathcal{E}, A] = -\frac{1}{2i\pi}B, \qquad [\mathcal{E}, B] = -\frac{1}{2i\pi}A, \tag{4.1.13}$$

from which it follows that

$$[\mathcal{E}, AB] = -\frac{1}{2i\pi}(A^2 + B^2), \qquad [\mathcal{E}, A^2 + B^2] = -\frac{2}{i\pi}AB. \tag{4.1.14}$$

Next, set

$$E = i\pi\mathcal{E}, \qquad M = \frac{4i\pi}{c}AB, \qquad N = \frac{2i\pi}{c}(A^2 + B^2), \tag{4.1.15}$$

rewriting the preceding line as

$$[E, M] = -N, \qquad [E, N] = -M \tag{4.1.16}$$

and observing at once that these are just the commutation relations (4.1.7). Starting from the equations

$$Ae^{4i\pi Q} = cxe^{4i\pi Q}, \quad Be^{4i\pi Q} = (cx + \xi)e^{4i\pi Q}, \quad \frac{1}{4}\mathcal{E}e^{4i\pi Q} = \left(Q + \frac{1}{8i\pi}\right)e^{4i\pi Q},$$
$$\tag{4.1.17}$$

and using

$$ABe^{4i\pi Q} = A(cx + \xi)e^{4i\pi Q} = cx(cx + \xi)e^{4i\pi Q} + \frac{c}{8i\pi}e^{4i\pi Q}$$

$$= c\left(Q + \frac{1}{8i\pi}\right)e^{4i\pi Q} = \frac{c}{4}\mathcal{E}e^{4i\pi Q}, \tag{4.1.18}$$

one finds the equation

$$i\pi\mathcal{E}e^{4i\pi Q} = Me^{4i\pi Q}. \tag{4.1.19}$$

To "compute" the image of $e^{4i\pi Q}$ under any polynomial in the operator $i\pi\mathcal{E}$, we may use the identity just obtained and the commutation relations (4.1.16): but this just means applying the reduction procedure from Lemma 4.1.2 so as to substitute for the polynomial in $i\pi\mathcal{E}$ under consideration a polynomial in M, N. Remembering that M, N are of degree 2 with respect to A, B and paying attention to degrees, one obtains Lemma 4.1.3 if one does not fail to remember that the factor c^{-1} is hidden in M, N.   □

The series to be analyzed in the sequel rely on an easy lemma, of which we give two versions: the second one is essentially stronger, but makes it necessary to set aside the matrices $\left(\begin{smallmatrix} a & b \\ c & d \end{smallmatrix}\right)$ such that $abcd = 0$, which is not always desirable.

Lemma 4.1.4. *One has*

$$\sum_{g=\left(\begin{smallmatrix} a & b \\ c & d \end{smallmatrix}\right)\in\Gamma} (1 + |ac| + |bd|)^{\alpha} < \infty \qquad \text{if } \alpha < -1,$$

$$\sum_{\substack{g=\left(\begin{smallmatrix} a & b \\ c & d \end{smallmatrix}\right)\in\Gamma \\ abcd\neq 0}} |abcd|^{\beta} < \infty \qquad \text{if } \beta < -\frac{1}{2}. \tag{4.1.20}$$

Proof. Using the parametrization (4.1.1) of $(\Gamma_2 \cup \Gamma_3)/\{\pm 1\}$, the pair of conditions $bd \neq 0, ac \neq 0$ is equivalent to $(m \neq 0, n \neq 0, -1)$; also, $m|n(n + 1)$, so the second line follows from the elementary fact that the number of divisors of $n(n + 1)$ is a $O(|n|^{\varepsilon})$ for every $\varepsilon > 0$. The first line is a consequence of the first, since on one hand $|ac| + |bd| \geq 2\sqrt{|abcd|}$, on the other hand, when one of the entries of the matrix is zero, only the one in the opposite corner is allowed to take values $\neq \pm 1$.   □

Lemma 4.1.5. *The series* (4.1.3) *converges in the space of continuous linear forms on the space which is the image of* $S(\mathbb{R}^2)$ *under the operator*

$$(i\pi\mathcal{E}) \times (i\pi\mathcal{E} - 1)_3 = (i\pi\mathcal{E})^2(i\pi\mathcal{E} - 1)(i\pi\mathcal{E} + 1). \tag{4.1.21}$$

Proof. Starting from the expression (4.1.6) of $\mathfrak{S}$, we examine the first term $\sum_{c\in\mathbb{Z}} e^{4i\pi x(cx+\xi)}$ of the series: that the series $\sum_{c\in\mathbb{Z}} \int_{\mathbb{R}^2} h(x,\xi)e^{4i\pi x(cx+\xi)} dx d\xi$ converges if $h \in (i\pi\mathcal{E})(i\pi\mathcal{E} - 1)_3 S(\mathbb{R}^2)$ is the result of an integration by parts, since an application of Lemma 4.1.3 makes it possible to bound the term associated to $c \neq 0$ by a constant times c^{-2}.

We now examine the main term, given by a double series. One may assume, without loss of generality, that $|\frac{n(n+1)}{m}| \geq |m|$. Indeed, as noted in the proof of Lemma 4.1.1, one has $\frac{n(n+1)}{m} = ac$ and $m = bd$: but assuming that $|ac| \geq |bd|$ is always possible, splitting the double series in two parts and making in one of them the change of coordinates $(x,\xi) \mapsto (\xi, x)$. In view of Lemma 4.1.4, it thus suffices to estimate the general term of the double series

$$\sum_{\substack{m\in\mathbb{Z}^\times, n\in\mathbb{Z} \\ m|n(n+1)}} \int_{\mathbb{R}^2} h(x,\xi) \exp\left(\frac{4i\pi}{m}[n(n+1)x^2 + m(2n+1)x\xi + m^2\xi^2]\right) dx d\xi$$

$$\tag{4.1.22}$$

by a constant times $|\frac{n(n+1)}{m}|^{-2}$, under the assumption that h lies in the subspace of $S(\mathbb{R}^2)$ indicated. Set

$$\begin{aligned}
Q_{m,n}(x,\xi): &= \frac{n(n+1)x^2 + m(2n+1)x\xi + m^2\xi^2}{m} \\
&= \frac{(nx + m\xi)[(n+1)x + m\xi]}{m} \\
&= \left(x + \frac{m}{n}\xi\right)\left(\frac{n(n+1)}{m}x + n\xi\right).
\end{aligned} \tag{4.1.23}$$

If we perform the change of variables

$$\begin{pmatrix} y \\ \eta \end{pmatrix} = \begin{pmatrix} x + \frac{m}{n}\xi \\ -\xi \end{pmatrix} \qquad \text{and set } c' = \frac{n(n+1)}{m} \in \mathbb{Z}, \tag{4.1.24}$$

we have

$$Q_{m,n}(x,\xi) = y(c'y + \eta) \qquad \text{and} \qquad \frac{1}{8i\pi}\frac{\partial}{\partial y} \pm \frac{\eta}{2} = \frac{1}{8i\pi}\frac{\partial}{\partial x} \mp \frac{\xi}{2}, \tag{4.1.25}$$

so we are back to a case where we can apply Lemma 4.1.3, leading to the desired estimate. □

Remark 4.1.a.. So far as questions of convergence are concerned, we might as well have used the operator $(i\pi\mathcal{E} - 1)_3$ in place of the product of this operator by $i\pi\mathcal{E}$.

However, the function $\mathfrak{s}$ introduced in (4.1.1), which is the starting building block for the construction of $\mathfrak{S}$, is $\mathcal{G}$-invariant, and so is its image under the operator $-(i\pi\mathcal{E})(i\pi\mathcal{E} - 1)_3$, while its image under the operator $(i\pi\mathcal{E} - 1)_3$ changes to its negative under $\mathcal{G}$.

The series $S_\ell(z)$ in the following theorem may be considered as being in the same spirit as those of Selberg [45], a particular case of which occurred in (3.2.36). However, as will be seen, the series is quite different.

Theorem 4.1.6. *Let $\mathfrak{S}$ be the linear form on the appropriate subspace of $\mathcal{S}(\mathbb{R}^2)$ introduced in (4.1.3), as justified in Lemma 4.1.5. For $\ell = 1, 2, \ldots$, set*

$$\Box_\ell = \pi^2 \mathcal{E}^2 (\pi^2 \mathcal{E}^2 + 1)(\pi^2 \mathcal{E}^2 + 4) \ldots (\pi^2 \mathcal{E}^2 + (\ell - 1)^2) \tag{4.1.26}$$

so that, as soon as $\ell \geq 2$, $\Box_\ell \mathfrak{S}$ is well defined as an automorphic distribution. The Θ_0-transform (cf. (2.1.1)) of this distribution is the automorphic function given by the equation

$$(\Theta_0 \Box_\ell \mathfrak{S})(z) = \frac{1}{4\pi} \left(\Gamma(\ell + \frac{1}{2}) \right)^2 S_\ell(z), \tag{4.1.27}$$

with

$$S_\ell(z) = \sum_{\left(\begin{smallmatrix} a & b \\ c & d \end{smallmatrix} \right) \in \Gamma} \left(\frac{(az + b)(c\bar{z} + d)}{\mathrm{Im}\, z} \right)^{-\frac{1}{2} - \ell}. \tag{4.1.28}$$

One has for some constant $C_\ell > 0$ the uniform estimate

$$|S_\ell(z)| \leq C_\ell (\mathrm{Im}\, z)^{-\ell} \qquad \text{for } \mathrm{Im}\, z \geq 1. \tag{4.1.29}$$

Proof. Since $\mathfrak{S}$ is "generated" by $\mathfrak{s}(x, \xi) = e^{4i\pi x\xi}$ (*cf.* (4.1.1)), we first compute the integral

$$2 \int_{\mathbb{R}^2} e^{4i\pi x\xi} \exp\left(-\frac{2\pi}{\mathrm{Im}\, z} |x - z\xi|^2 \right) dx d\xi = 2 \int \exp(-\pi R(x, \xi)) dx d\xi, \tag{4.1.30}$$

with $R(x, \xi) = \frac{2|x - z\xi|^2}{\mathrm{Im}\, z} - 4ix\xi$: since the (symmetric) matrix associated with R, to wit

$$R = \frac{2}{\mathrm{Im}\, z} \begin{pmatrix} 1 & -z \\ -z & |z|^2 \end{pmatrix}, \tag{4.1.31}$$

has determinant $-\frac{8iz}{\mathrm{Im}\, z}$, the value of the integral (4.1.30) is

$$2(\det R)^{-\frac{1}{2}}) = \left(-\frac{8iz}{\mathrm{Im}\, z} \right)_+^{-\frac{1}{2}} = \frac{1}{2} z^{-\frac{1}{2}} (z - \bar{z})^{\frac{1}{2}} : \tag{4.1.32}$$

let us recall our convention (2.3.18) regarding powers of complex numbers with non-integral exponents.

Next, we use the identity (2.1.5) to transfer, through Θ_0, the operator $\square_\ell$, acting in the (x, ξ) coordinates, to the operator $(\Delta - \frac{1}{4})(\Delta + \frac{3}{4}) \dots (\Delta - \frac{1}{4} + (\ell-1)^2)$ acting on functions of $z \in \Pi$. Writing $\Delta = (z - \bar{z})^2 \frac{\partial^2}{\partial z \partial \bar{z}}$, one obtains, by induction,

$$(\Delta - \frac{1}{4})(\Delta + \frac{3}{4}) \dots (\Delta - \frac{1}{4} + (\ell - 1)^2) \left[z^{-\frac{1}{2}}(z - \bar{z})^{\frac{1}{2}} \right]$$
$$= \pi^{-1}(\Gamma(\ell + \frac{1}{2}))^2 z^{-\frac{1}{2} - \ell}(z - \bar{z})^{\frac{1}{2} + \ell}. \quad (4.1.33)$$

More generally, the covariance of Θ_0 yields the identity

$$(\Theta_0(\mathfrak{s} \circ g))(z) = (\Theta_0 \mathfrak{s})(g.z) \quad (4.1.34)$$

and, since $\frac{z}{\operatorname{Im} z}$ transforms to $\frac{(az+b)(c\bar{z}+d)}{\operatorname{Im} z}$ under $\left(\begin{smallmatrix} a & b \\ c & d \end{smallmatrix} \right)$, the image of the linear form $\mathfrak{S}$ on the function $\square_\ell((x, \xi) \mapsto 2 \exp\left(-\frac{2\pi}{\operatorname{Im} z}|x - z\xi|^2\right))$, can be obtained if $\ell \geq 2$, as it follows from Lemma 4.1.5, by performing the required summation with respect to $g \in \Gamma$.

To obtain the estimate (4.1.29), one may employ Lemma 4.1.3. As soon as $\ell \geq 2$, the convergence of the series defining the value of the linear form $\mathfrak{S}$ on the image of the function $(x, \xi) \mapsto \exp\left(-\frac{2\pi}{\operatorname{Im} z}|x - z\xi|^2\right)$ is ensured. On top of that, one can have an estimate of the sum of the series by taking advantage of the fact that, for $\operatorname{Im} z \geq 1$, one has for some constants C_j the estimates (with the notation of Lemma 4.1.3, in which $j \geq \ell$)

$$\left| D_j \left((x, \xi) \mapsto \exp\left(-\frac{2\pi}{\operatorname{Im} z}|x - z\xi|^2\right) \right) \right| \leq C_j (\operatorname{Im} z)^{-j} : \quad (4.1.35)$$

indeed, looking at the expressions (4.1.11) of A and B, it suffices to remark, after one has written

$$q_z(x, \xi) := \frac{|x - z\xi|^2}{\operatorname{Im} z} = \frac{(x - \xi \operatorname{Re} z)^2}{\operatorname{Im} z} + (\operatorname{Im} z)\xi^2, \quad (4.1.36)$$

that

$$\left| \frac{\partial q_z}{\partial x} \right| \leq 2(\operatorname{Im} z)^{-\frac{1}{2}} \times \frac{|x - \xi \operatorname{Re} z|}{(\operatorname{Im} z)^{\frac{1}{2}}}, \qquad \left| \frac{\partial^2 q_z}{\partial x^2} \right| \leq 2(\operatorname{Im} z)^{-1},$$
$$|\xi| \leq (\operatorname{Im} z)^{-\frac{1}{2}} \times ((\operatorname{Im} z)^{\frac{1}{2}}|\xi|), \quad (4.1.37)$$

where the extra factors set aside on the right are bounded by 1 or by $(q_z(x, \xi))^{\frac{1}{2}}$, hence do no harm as multipliers of $e^{-2\pi q_z(x, \xi)}$. $\square$

We come now to the question of decomposing the automorphic distribution $\square_\ell \mathfrak{S}$ into its homogeneous components. We shall not push its solution to the end: for what we are really interested in, instead, is the decomposition of $\Theta_0(\square_\ell \mathfrak{S})$

into (generalized) eigenfunctions of Δ. We have seen in Section 3.2 why these two problems are essentially equivalent to each other.

Recall the decomposition (1.2.11) of an arbitrary even function in $L^2_{\text{even}}(\mathbb{R}^2)$ as an integral superposition of even functions homogeneous of degrees $-1-i\lambda, \lambda \in \mathbb{R}$. Exercising care about convergence, one can use this for more general functions, and we apply it to the function $\mathfrak{s}(x,\xi) = e^{4i\pi x\xi}$. According to this recipe, one has $\mathfrak{s} = \int_{-\infty}^{\infty} \mathfrak{s}_{i\lambda} d\lambda$, with

$$\mathfrak{s}_{i\lambda}(x,\xi) = \frac{1}{2\pi} \int_0^{\infty} t^{i\lambda} e^{4i\pi t^2 x\xi} dt = \frac{1}{4\pi} \int_0^{\infty} t^{\frac{i\lambda-1}{2}} e^{4i\pi t x\xi} dt : \tag{4.1.38}$$

this integral is only semi-convergent, and may be computed as the limit, as $\varepsilon \to 0$, of the integral obtained after one has substituted $-(\varepsilon - 4i\pi t x\xi)$ for the exponent: we obtain

$$\mathfrak{s}_{i\lambda}(x,\xi) = \frac{\Gamma(\frac{1+i\lambda}{2})}{4\pi}(0 - 4i\pi x\xi)^{\frac{-1-i\lambda}{2}}. \tag{4.1.39}$$

More generally, we may set, for $|\operatorname{Re}\nu| < 1$,

$$\mathfrak{s}_{\nu}(x,\xi) = \frac{1}{4\pi} \int_0^{\infty} t^{\frac{\nu-1}{2}} e^{4i\pi t x\xi} dt$$

$$= \frac{\Gamma(\frac{1+\nu}{2})}{4\pi}(0 - 4i\pi x\xi)^{\frac{-1-\nu}{2}}. \tag{4.1.40}$$

Actually, it is not the Θ_0-transform of $\mathfrak{s}_{\nu}$, as obtained from that of $\mathfrak{s}$ together with the first line of (4.1.40), we wish to obtain: rather, that of its image under the operator

$$\square_\ell = (-1)^\ell (i\pi\mathcal{E})^2 (i\pi\mathcal{E} - 1)(i\pi\mathcal{E} + 1)\ldots(i\pi\mathcal{E} - \ell + 1)(i\pi\mathcal{E} + \ell - 1). \tag{4.1.41}$$

Now, on functions of $tx\xi$, the operator $i\pi\mathcal{E}$ has the same effect as the operator $\frac{1}{2} + t\frac{d}{dt}$, so that, after an integration by parts, one obtains the identity (between semi-convergent integrals)

$$\int_0^{\infty} t^{\frac{\nu-1}{2}} \square_\ell \left(e^{4i\pi t x\xi} \right) dt = \int_0^{\infty} \left(t\frac{d}{dt} + \frac{1}{2} \right)^2 \tag{4.1.42}$$

$$\times \left[\left(t\frac{d}{dt} + \frac{1}{2} \right)^2 - 1 \right]\ldots\left[\left(t\frac{d}{dt} + \frac{1}{2} \right)^2 - (\ell-1)^2 \right] \left(t^{\frac{\nu-1}{2}} \right) e^{4i\pi t x\xi} dt$$

$$= \left(\frac{\nu}{2} \right)^2 \left[\left(\frac{\nu}{2} \right)^2 - 1 \right] \left[\left(\frac{\nu}{2} \right)^2 - 4 \right]\ldots\left[\left(\frac{\nu}{2} \right)^2 - (\ell-1)^2 \right] \int_0^{\infty} t^{\frac{\nu-1}{2}} e^{4i\pi t x\xi} dt.$$

Hence,

$$(\square_\ell \mathfrak{s})_{\nu} = \frac{\nu}{2}\frac{\Gamma(\frac{\nu}{2} + \ell)}{\Gamma(\frac{\nu}{2} - \ell + 1)}\mathfrak{s}_{\nu} : \tag{4.1.43}$$

even though applying $\square_\ell$ to $\mathfrak{S}$ was necessary to make it a distribution, this would bring no change to the problem, to which we turn now, of analyzing the homogeneous components of the object under consideration.

One has

$$(\Theta_0\mathfrak{s}_\nu)(z) = \frac{1}{4\pi}\int_0^\infty t^{\frac{\nu-1}{2}}\Theta_0((x,\xi) \mapsto e^{4i\pi t x\xi})(z)dt. \tag{4.1.44}$$

We thus compute

$$\Theta_0((x,\xi) \mapsto e^{4i\pi t x\xi})(z) = 2\int_{\mathbb{R}^2} e^{4i\pi t x\xi}e^{-2\pi q_z(x,\xi)}dxd\xi, \tag{4.1.45}$$

with q_z as defined in (4.1.36): since the symmetric matrix associated to the quadratic form $q_z(x,\xi) - 2itx\xi$, to wit

$$\begin{pmatrix} \frac{1}{\operatorname{Im} z} & -\frac{\operatorname{Re} z}{\operatorname{Im} z} - it \\ -\frac{\operatorname{Re} z}{\operatorname{Im} z} - it & \frac{|z|^2}{\operatorname{Im} z} \end{pmatrix}, \tag{4.1.46}$$

has determinant $1 - 2it\frac{\operatorname{Re} z}{\operatorname{Im} z} + t^2$, one obtains

$$(\Theta_0\mathfrak{s}_\nu)(z) = \frac{1}{4\pi}\int_0^\infty t^{\frac{\nu-1}{2}}\left(1 - 2it\frac{\operatorname{Re} z}{\operatorname{Im} z} + t^2\right)_+^{-\frac{1}{2}}dt. \tag{4.1.47}$$

This integral can be computed:

$$\int_0^\infty t^{\frac{\nu-1}{2}}[t^2 - 2it\frac{\operatorname{Re} z}{\operatorname{Im} z} + 1]_+^{-\frac{1}{2}}dt = 2^{\frac{1}{2}}\int_0^\infty t^{\frac{\nu-1}{2}}dt\int_0^\infty e^{-2\pi r(t^2 - 2i\frac{\operatorname{Re} z}{\operatorname{Im} z}t+1)}r^{-\frac{1}{2}}dr$$

$$= 2^{\frac{1}{2}}\int_0^\infty t^{\frac{\nu-2}{2}}dt\int_0^\infty e^{-2\pi r(t - 2i\frac{\operatorname{Re} z}{\operatorname{Im} z}+t^{-1})}r^{-\frac{1}{2}}dr$$

$$= 2^{\frac{1}{2}}\int_0^\infty e^{4i\pi r\frac{\operatorname{Re} z}{\operatorname{Im} z}}r^{-\frac{1}{2}}dr\int_0^\infty t^{\frac{\nu-2}{2}}e^{-2\pi r(t+t^{-1})}dt$$

$$= 2^{\frac{3}{2}}\int_0^\infty e^{4i\pi r\frac{\operatorname{Re} z}{\operatorname{Im} z}}r^{-\frac{1}{2}}K_{\frac{\nu}{2}}(4\pi r)dr: \tag{4.1.48}$$

we have used the usual integral expression [36, p. 85] of modified Bessel functions on the last line. According to (*loc.cit.*, p. 92), the final result is (still assuming that $|\operatorname{Re}\nu| < 1$)

$$(\Theta_0\mathfrak{s}_\nu)(z) = \frac{\Gamma(\frac{1-\nu}{2})\Gamma(\frac{1+\nu}{2})}{4\pi}\mathfrak{P}_{\frac{\nu-1}{2}}\left(-i\frac{\operatorname{Re} z}{\operatorname{Im} z}\right) \tag{4.1.49}$$

in terms of the Legendre function $\mathfrak{P}_{\frac{\nu-1}{2}}$: this is, according to (2.3.30), a multiple of the function $\chi_{1,\nu}\left(\frac{\operatorname{Re} z}{\operatorname{Im} z}\right) + \chi_{1,-\nu}\left(\frac{\operatorname{Re} z}{\operatorname{Im} z}\right)$.

If $g = \left(\begin{smallmatrix} a & b \\ c & d \end{smallmatrix}\right) \in \Gamma$, one has

$$(\Theta_0(\mathfrak{s}_\nu \circ g))(z) = (\Theta_0 \mathfrak{s}_\nu)(g.z). \tag{4.1.50}$$

The automorphic distribution $\square_\ell \mathfrak{S}$ has played its role. From now on, we shall stay in the hyperbolic half-plane Π, where our job will be to analyze what can be done with the series

$$\frac{1}{2} \sum_{g \in \Gamma} \mathfrak{P}_{\frac{\nu-1}{2}} \left(-i \frac{\operatorname{Re}(g.z)}{\operatorname{Im}(g.z)} \right). \tag{4.1.51}$$

Doing this, or concerning ourselves, rather, with the generalization of this problem involving the parameter ρ, will keep us busy for the rest of this chapter.

The analysis of this series — which converges for no value of ν — has some similarity with a study made in [60, Sections 19-20]. There, starting from an irrational number τ in some real quadratic extension $\mathbb{Q}^{(2)}$ of $\mathbb{Q}$, and from the group $\Gamma_\tau \subset \Gamma$ stabilizing τ, we dealt with the series

$$\sum_{g \in \Gamma/\Gamma_\tau} p_\nu \left(\frac{-z + g.\bar{\tau}}{z + g.\tau} \right), \tag{4.1.52}$$

in which $p_\nu(z) = \frac{1}{2} \left[\mathfrak{P}_{\frac{\nu-1}{2}} \left(i \frac{\operatorname{Re} z}{\operatorname{Im} z} \right) + \mathfrak{P}_{\frac{\nu-1}{2}} \left(-i \frac{\operatorname{Re} z}{\operatorname{Im} z} \right) \right]$, and $\bar{\tau}$ denotes the conjugate of τ within $\mathbb{Q}^{(2)}$. Here, in the case when $\rho = 1$ (we did not consider the parameter ρ there), we shall use the same function p_ν, but whereas the group Γ_τ can be seen to be infinite, with the help of continued fraction theory, the group that takes the place of Γ_τ in our present endeavours reduces to $\{\pm I\}$, which completely changes the nature of the problem, so far as convergence problems are concerned. We shall comment again about all this at the end of Section 4.7.

4.2   The automorphic function $f_{\rho,\nu}$

The series (4.1.51) diverges for every value of ν. To do something with it, we must first break it into two parts, according to the equation (2.3.30) in which the function $\mathfrak{P}_{\frac{\nu-1}{2}}(-it)$ is split into two terms. Considering the generalization depending on the extra parameter ρ, we now study the function $f_{\rho,\nu}$ defined as the sum of the series

$$f_{\rho,\nu}(z) = \frac{1}{2} \sum_{g \in \Gamma} (\operatorname{Im}(g.z))^{\frac{\rho-1}{2}} \chi_{\rho,\nu}^{\text{even}}(\psi(g.z))$$

$$= \frac{1}{2} \sum_{g = \left(\begin{smallmatrix} a & b \\ c & d \end{smallmatrix}\right) \in \Gamma} \left(\frac{\operatorname{Im} z}{|cz + d|^2} \right)^{\frac{\rho-1}{2}} \chi_{\rho,\nu}^{\text{even}} \left(\frac{\operatorname{Re}(g.z)}{\operatorname{Im}(g.z)} \right). \tag{4.2.1}$$

Our reason for using the even part only of the function $\chi_{\rho,\nu}$ is that we are primarily interested in non-holomorphic modular forms of even type with respect to the symmetry $z \mapsto -\bar{z}$: it would complicate things only slightly to consider modular forms of odd type too, of necessity cusp-forms, which was the choice made in [60, 61] in another context (that of establishing composition formulas for the automorphic Weyl calculus), and with other methods. On the other hand, one may remark that, in the case when $\rho = 1$, using the series (4.2.1) based on the function $\chi_{1,\nu}$ or on the even part only of this function would not change the result. This is so because changing $g = \left(\begin{smallmatrix} a & b \\ c & d \end{smallmatrix}\right)$ to $\left(\begin{smallmatrix} -c & -d \\ a & b \end{smallmatrix}\right)$ has the effect of changing $\psi(g.z)$ to its negative. When $\rho \neq 1$, the same argument, together with Lemma 2.3.4, makes it possible to prove the identity (once convergence has been established)

$$f_{\rho,\nu} = \frac{\Gamma\left(\frac{2+\rho-\nu}{4}\right)\Gamma\left(\frac{2+\rho+\nu}{4}\right)}{\Gamma\left(\frac{4-\rho-\nu}{4}\right)\Gamma\left(\frac{4-\rho+\nu}{4}\right)} f_{2-\rho,\nu}. \tag{4.2.2}$$

The function $f_{\rho,\nu}$ is invariant under the change $z \mapsto -\bar{z}$, a consequence of the fact that $g_1.(-\bar{z}) = -\overline{g.z}$ if $g = \left(\begin{smallmatrix} a & b \\ c & d \end{smallmatrix}\right)$ and $g_1 = \left(\begin{smallmatrix} a & -b \\ -c & d \end{smallmatrix}\right)$.

We shall first show, in the present section, that the series is convergent when $\operatorname{Re}\nu < -1 - |\operatorname{Re}\rho - 1|$. In the section to follow, we shall then show that it extends as a meromorphic function for $\operatorname{Re}\nu < 1 - |\operatorname{Re}\rho - 1|$. So that the domains of the parameter ν to which the functions $f_{\rho,\nu}$ and $f_{\rho,-\nu}$ can be continued should intersect, which will be important later, we assume that $0 < \operatorname{Re}\rho < 2$.

We begin with geometric estimates of the arguments $\frac{\operatorname{Re}(g.z)}{\operatorname{Im}(g.z)}$ and $\operatorname{Im}(g.z)$ entering the two factors of the general term of the series (4.2.1).

Lemma 4.2.1. *Recall that $\psi(z) = \frac{\operatorname{Re}z}{\operatorname{Im}z}$, so that*

$$\psi(g.z) = \frac{\operatorname{Re}((az+b)(cz+d))}{y}, \qquad g = \left(\begin{smallmatrix} a & b \\ c & d \end{smallmatrix}\right) \in SL(2,\mathbb{R}). \tag{4.2.3}$$

If $g = \left(\begin{smallmatrix} a & b \\ c & d \end{smallmatrix}\right) \in \Gamma$, one has the estimate

$$\max(1, |\psi(g.z)|) \geq \frac{1}{1+\sqrt{2}} \max(|ac|, |bd|) e^{-d(i,z)}, \tag{4.2.4}$$

denoting as d the hyperbolic distance in Π.

Proof. One has

$$\psi(g.z) = \frac{ac|z|^2 + (1+2bc)x + bd}{y}. \tag{4.2.5}$$

Changing the matrix $g = \left(\begin{smallmatrix} a & b \\ c & d \end{smallmatrix}\right)$ to $g_1 = \left(\begin{smallmatrix} d & c \\ b & a \end{smallmatrix}\right)$ amounts to exchanging ac and bd, while preserving bc: on the other hand, if $z = g.i$, one has $g_1.i = \bar{z}^{-1}$, and one has $d(i,z) = d(i,\bar{z}^{-1})$ in view of the formula from hyperbolic geometry $\cosh d(i,z) = \frac{1+|z|^2}{2\operatorname{Im}z}$: incidentally, note that $e^{d(i,z)} \geq \frac{1}{2}\max(y, y^{-1})$ as a consequence. It follows that for the first part, one may assume, without loss of generality, that $|ac| \geq |bd|$, a reduction of cases already used in the proof of Lemma 4.1.5. Let us get rid first of the exceptional cases in which $bd = 0$, so that $\left(\begin{smallmatrix} a & b \\ c & d \end{smallmatrix}\right) = \pm\left(\begin{smallmatrix} 1 & 0 \\ c & 1 \end{smallmatrix}\right)$ or $\pm\left(\begin{smallmatrix} a & 1 \\ -1 & 0 \end{smallmatrix}\right)$. In the first case,

$$\psi(g.z) = \psi\left(\frac{z}{cz+1}\right) = \frac{c|z|^2 + x}{y} :$$
(4.2.6)

if $|x| \leq \frac{1}{3}|c||z|^2$, one has $|\psi(g.z)| \geq \frac{2}{3}\frac{|c||z|^2}{y} \geq \frac{1}{3}|c|e^{-d(i,z)}$. If $|x| \geq \frac{1}{3}|c||z|^2$, one has

$$1 \geq \frac{1}{3}|c|\frac{|z|^2}{|x|y}y \geq \frac{2}{3}|c|y \geq \frac{1}{3}|c|e^{-d(i,z)} :$$
(4.2.7)

in both cases,

$$\max(1, |\psi(g.z)|) \geq \frac{1}{3}\max(|ac|, |bd|)e^{-d(i,z)};$$
(4.2.8)

the second exceptional case is just the same since, then, $\psi(g.z) = \frac{c|z|^2 + x}{y}$.

We now assume that $bd \neq 0$ and we use the notation $bc = n, bd = m, ac = \frac{n(n+1)}{m}$ introduced in Lemma 4.1.1. Under our assumption that $n(n+1) \geq m^2$ (one has $n(n+1) \geq 0$ when $n \in \mathbb{Z}$), note that $n \neq 0, -1$ since $m \neq 0$. One has

$$\psi(g.z) = \frac{n(n+1)|z|^2 + m(2n+1)x + m^2}{my}$$

$$= \frac{n(n+1)}{my}\left[\left(x + \frac{m(n+\frac{1}{2})}{n(n+1)}\right)^2 + y^2 - \frac{1}{4}\frac{m^2}{n^2(n+1)^2}\right].$$
(4.2.9)

Set $\alpha = 1 + \sqrt{2}$. When $y \geq \frac{\alpha}{2}\frac{|m|}{n(n+1)}$, one has

$$|\psi(g.z)| \geq \frac{n(n+1)}{|m|y}\left(1 - \frac{1}{\alpha^2}\right)y^2$$

$$\geq \frac{1}{2}\left(1 - \frac{1}{\alpha^2}\right)\frac{n(n+1)}{|m|}e^{-d(i,z)}$$

$$= \frac{1}{\alpha}\max(|ac|, |bd|)e^{-d(i,z)}.$$
(4.2.10)

When $y \leq \frac{\alpha}{2}\frac{|m|}{n(n+1)}$, one has

$$1 \geq \frac{2}{\alpha}\frac{n(n+1)}{|m|}y \geq \frac{1}{\alpha}\max(|ac|, |bd|)e^{-d(i,z)}.$$
(4.2.11)

$\square$

Corollary 4.2.2. *Given $C > 0$ and a compact subset K of Π, one can have $z \in K, g \in \Gamma$ and $|\psi(g.z)| \leq C$ only for a finite set of matrices g.*

Proof. It is a consequence of the lemma, together with the fact that, for a matrix $\left(\begin{smallmatrix} a & b \\ c & d \end{smallmatrix}\right) \in \Gamma$, knowing a bound for $\max(|ac|, |bd|)$ gives a bound for all entries of the matrix, in view of the identity

$$1 + (ac + bd)^2 = (ad - bc)^2 + (ac + bd)^2 = (a^2 + b^2)(c^2 + d^2). \qquad (4.2.12)$$

$\square$

Lemma 4.2.3. *With $\psi(z) = \frac{\operatorname{Re} z}{\operatorname{Im} z}$, one has*

$$\frac{1}{2} e^{-d(i,z)} \leq (\operatorname{Im}(g.z))^{-1} \leq 60 e^{3d(i,z)} \max(1, |\psi(g.z)|^2). \qquad (4.2.13)$$

Proof. Recall that $\cosh d(i, z) = \frac{1+x^2+y^2}{2y}$ and that $\max(y, y^{-1}) \leq 2e^{d(i,z)}$. If $g = \left(\begin{smallmatrix} a & b \\ c & d \end{smallmatrix}\right)$, one has

$$(\operatorname{Im}(g.z))^{-1} = \frac{|cz + d|^2}{y} \geq \min\left(c^2 y, \frac{d^2}{y}\right) \geq \min(y, y^{-1}). \qquad (4.2.14)$$

In the other direction,

$$\frac{|cz + d|^2}{y} = \frac{c^2(x^2 + y^2) + d^2 + 2cdx}{y} \leq \frac{1 + x^2 + y^2}{y}(c^2 + d^2)$$
$$= 2(c^2 + d^2) \cosh d(i, z) \leq 2(c^2 + d^2)e^{d(i,z)}. \qquad (4.2.15)$$

Next, using (4.2.12),

$$c^2 + d^2 \leq 1 + (ac + bd)^2 \leq 1 + 4(\max(|ac|, |bd|))^2 \leq 5(\max(|ac|, |bd|))^2 : \quad (4.2.16)$$

combining this inequality with the preceding one and with (4.2.4), we obtain

$$\frac{|cz + d|^2}{y} \leq 10(1 + \sqrt{2})^2 e^{3d(i,z)} \max(1, |\psi(g.z)|^2). \qquad (4.2.17)$$

$\square$

Theorem 4.2.4. *Let $\Sigma \subset \Pi$ be the union of the (locally finite) collection of Γ-transforms of the hyperbolic line from 0 to $i\infty$. Let ρ, ν be complex numbers such that $\nu \notin \mathbb{Z}$, $\rho \pm \nu \notin 2\mathbb{Z}$ and $\operatorname{Re}\nu < \min(-\operatorname{Re}\rho, \operatorname{Re}\rho - 2) = -1 - |\operatorname{Re}\rho - 1|$. The series (4.2.1) converges uniformly on every compact subset of Π, defining a continuous automorphic function of z invariant under the symmetry $z \mapsto -\bar{z}$, depending on ν in a holomorphic way. The function $f_{\rho,\nu}$ is C^∞ in the complement of Σ, with discontinuities of its normal derivative along Σ. One has, in the distribution sense,*

$$\left(\Delta - \frac{1 - \nu^2}{4}\right) f_{\rho,\nu} = 2C(\rho, \nu) ds_\Sigma^{(\rho)}, \qquad (4.2.18)$$

where the measure $ds_\Sigma^{(\rho)}$ is supported in Σ, Γ-invariant and, as such, characterized by the fact that, on the hyperbolic line from 0 to $i\infty$, it is given by the density $\frac{1}{2}\left(y^{\frac{\rho-1}{2}} + y^{\frac{1-\rho}{2}}\right)\frac{dy}{y}$: recall that $C(\rho, \nu)$ was defined in (2.3.33).

Proof. First observe that, defining the fundamental domain D of Γ, in the usual way, by the conditions $|z| > 1, |\operatorname{Re} z| < \frac{1}{2}$, one has

$$\Sigma \cap \overline{D} = \{iy : y \geq 1\} \tag{4.2.19}$$

since, in the equivalence class modulo Γ of any point $z \in \overline{D}$, there cannot exist any point $w \in D, w \neq z$; also, matrices in Γ which fix i, i.e., matrices in the group Γ_0 as introduced before Lemma 4.1.1, also fix the whole hyperbolic line from 0 to $i\infty$. As a consequence, a point on Σ lies on exactly one line congruent to the line just considered.

The proof of Theorem 4.2.4 is based on the estimate (2.3.32), which leads to

$$(\operatorname{Im}(g.z))^{\frac{\operatorname{Re}\rho-1}{2}} \left|\chi_{\rho,\nu}^{\text{even}}(\psi(g.z))\right| \leq C(\operatorname{Im}(g.z))^{\frac{\operatorname{Re}\rho-1}{2}}(1+|\psi(g.z)|)^{\frac{\operatorname{Re}(\rho+\nu)-2}{2}} \tag{4.2.20}$$

and on geometric lemmas already proven. It is a consequence of Lemma 4.2.1 and Lemma 4.1.4 that a series $\sum_{g \in \Gamma}(1 + |\psi(g.z)|)^{\alpha}$ converges, uniformly on every compact subset of Π, if $\operatorname{Re}\alpha < -1$. We also have to take the extra factor $(\operatorname{Im}(g.z))^{\frac{\operatorname{Re}\rho-1}{2}}$ into account. When $\operatorname{Re}\rho \geq 1$, it follows from Lemma 4.2.3 that the extra factor is majorized by $\left(2e^{d(i,z)}\right)^{\frac{\operatorname{Re}\rho-1}{2}}$, a harmless factor when only local estimates are required, so that convergence is ensured provided that $\operatorname{Re}\nu < -\operatorname{Re}\rho$. Assuming on the contrary that $\rho < 1$, we obtain with some new absolute constant $C_1 > 0$ the estimate

$$(\operatorname{Im}(g.z))^{\frac{\operatorname{Re}\rho-1}{2}}|\chi_{\rho,\nu}(\psi(g.z))| \tag{4.2.21}$$

$$\leq C_1 \exp\left(\frac{3}{2}(1-\operatorname{Re}\rho)d(i,z)\right)(1+|\psi(g.z)|)^{1-\operatorname{Re}\rho}(1+|\psi(g.z)|)^{\frac{\operatorname{Re}(\rho+\nu)-2}{2}}:$$

the total exponent of $1 + |\psi(g.z)|$ is $\frac{1}{2}(\operatorname{Re}(\nu - \rho))$, so that the condition $\operatorname{Re}\nu < \operatorname{Re}\rho - 2$ will do. That the function $f_{\rho,\nu}$ so defined is automorphic is obvious.

All terms of the series for $f_{\rho,\nu}$ are killed in $\Pi\backslash\Sigma$ by the differential operator $\Delta - \frac{1-\nu^2}{4}$: this is a consequence of Proposition 2.3.5 and of the invariance of Δ under the action of $SL(2, \mathbb{R})$. Since the operator Δ is elliptic, the series for $f_{\rho,\nu}$ can be differentiated, term-by-term, as often as wished, while preserving uniform convergence on compact subsets of $\Pi\backslash\Sigma$. Consequently, one has $\left(\Delta - \frac{1-\nu^2}{4}\right)f_{\rho,\nu} = 0$ in $\Pi\backslash\Sigma$.

To obtain the more precise result (4.2.18), it suffices, in view of the facts that $f_{\rho,\nu}$ is automorphic and that Δ is Γ-invariant, to analyze the left-hand side of this equation near points on the hyperbolic line from 0 to $i\infty$. In the series obtained from an application of the operator $\Delta - \frac{1}{4}$ to the right-hand side of (4.2.1), the only terms singular on the line under consideration correspond to the 4 matrices $\pm\left(\begin{smallmatrix} 1 & 0 \\ 0 & 1 \end{smallmatrix}\right)$ and $\pm\left(\begin{smallmatrix} 0 & 1 \\ -1 & 0 \end{smallmatrix}\right)$, since all other matrices will move points on the line from 0 to $i\infty$ away from it. Equation (4.2.18) is thus a consequence of Proposition 2.3.5.

Let us observe at this point that the measure $ds_\Sigma^{(\rho)}$ is invariant under the change $\rho \mapsto 2 - \rho$. This was not the case for the measure $\delta_{(0,i\infty)}^{(\rho)}$ introduced in Theorem 2.4.1. In the summation process over Γ leading from the second measure to the first, the distinction between ρ and $2 - \rho$ has disappeared because the transformation $z \mapsto -z^{-1}$ associated to the matrices $\pm \left(\begin{smallmatrix} 0 & 1 \\ -1 & 0 \end{smallmatrix} \right)$ changes $\delta_{(0,i\infty)}^{(\rho)}$ to $\delta_{(0,i\infty)}^{(2-\rho)}$. □

Remark 4.2.a. With $ds_\Sigma = ds_\Sigma^{(1)}$, equation (4.2.18) is equivalent (once it is already known that $f_{\rho,\nu}$ is continuous and that $\left(\Delta - \frac{1-\nu^2}{4} \right) f_{\rho,\nu} = 0$ outside Σ) to the fact that the jump along Σ of the normal derivative of $f_{\rho,\nu}$ is $-2C(\rho,\nu)$ times the (Radon-Nikodym) density of the measure $ds_\Sigma^{(\rho)}$ with respect to ds_Σ: however, this notion of jump is somewhat unusual from the point of view of differential geometry. Using the Riemannian length element $ds^2 = y^{-2}(dx^2 + dy^2)$, there is no problem in defining it up to sign, and in particular the correctly normalized normal derivative at a point $z = x + iy$ on the hyperbolic line $i\mathbb{R}_+^*$ from 0 to $i\infty$ is $\pm y \frac{\partial}{\partial x}$: but there is no canonical way to orient the hyperbolic lines Σ is made of. On the other hand, if $z^0 \in \Sigma$, there are exactly 4 matrices $g \in \Gamma$ such that $g^{-1}.z^0$ lies on $i\mathbb{R}_+^*$: indeed, if g is such a matrix, the other ones are $-g$ and $\pm g \left(\begin{smallmatrix} 0 & 1 \\ -1 & 0 \end{smallmatrix} \right)$. Consider generally the changes of complex coordinate $z = g.w_1$ and $z = \left(g \left(\begin{smallmatrix} 0 & 1 \\ -1 & 0 \end{smallmatrix} \right) \right).w_2$ so that $w_1 = -w_2^{-1}$. The region, near z^0, where $\operatorname{Re} w_1 > 0$ is the same as that where $\operatorname{Re} w_2 < 0$ so that the jump of normal derivative at z^0 along the line under consideration is changed to its negative when $\pm g$ is changed to $\pm g \left(\begin{smallmatrix} 0 & 1 \\ -1 & 0 \end{smallmatrix} \right)$. But one also has $\frac{dw_1}{w_1} = -\frac{dw_2}{w_2}$, so that it has an intrinsic meaning to regard the jump of normal derivative as a multiple of $\frac{dw}{w}$, where $z^0 = g.w^0$ for some arbitrary pair $(g, w^0) \in \Gamma \times i\mathbb{R}_+^*$: only note that this is, so to speak, an arithmetic-geometric, rather than differential-geometric, notion, since only transformations associated to matrices in Γ must, or can, be used as changes of coordinates.

4.3 The analytic continuation of $f_{\rho,\nu}$

We make the standing assumptions, used in Proposition 2.3.2 in view of establishing the basic properties of the function $\chi_{\rho,\nu}$, that $\nu \notin \mathbb{Z}, \rho \pm \nu \notin 2\mathbb{Z}$. We shall also consistently assume that $0 < \operatorname{Re} \rho < 2$, for a different reason, stated just before Lemma 4.2.1. We display these assumptions for reference:

$$\nu \notin \mathbb{Z}, \quad \rho \pm \nu \notin 2\mathbb{Z}, \quad 0 < \operatorname{Re} \rho < 2. \tag{4.3.1}$$

We wish to analyze the continuation of the function $\nu \mapsto f_{\rho,\nu}$, as yet only defined for $\operatorname{Re} \nu < -1 - |\operatorname{Re} \rho - 1|$, to the domain $\operatorname{Re} \nu < 1 - |\operatorname{Re} \rho - 1|$. We first approach $f_{\rho,\nu}$, which is a continuous automorphic function but not a generalized eigenfunction of the hyperbolic Laplacian, only almost one (it satisfies the eigenvalue equation in the complement of Σ), by a simpler-looking function, this time a true generalized eigenfunction of Δ for the eigenvalue $\frac{1-\nu^2}{4}$, but no longer an auto-morphic function: only almost one. With our present limited aim in view, a good approximation is one for which the error term can be continued as a holomorphic function for $\mathrm{Re}\,\nu < 1 - |\mathrm{Re}\,\rho - 1|$. No uniformity whatsoever with respect to the variable z is discussed here: we shall have to come back to this in Section 4.4, in which we discuss asymptotics of $f_{\rho,\nu}(x + iy)$ as $y \to \infty$, for $\mathrm{Re}\,\nu < 1 - |\mathrm{Re}\,\rho - 1|$. The present section constitutes only a rough approach to the much more precise results which will be required there.

Definition 4.3.1. Given two functions f and f_1 of (ρ, ν, z), defined, under the additional conditions (4.3.1), for $\mathrm{Re}\,\nu < -1 - |\mathrm{Re}\,\rho - 1|$ and $z \in \Pi$, also holomorphic with respect to (ρ, ν), we shall say that f and f_1 are equivalent, and write $f \sim f_1$ to express the fact that the difference $f(\rho, \nu, z) - f_1(\rho, \nu, z)$ extends as a holomorphic function of ν for $\mathrm{Re}\,\nu < 1 - |\mathrm{Re}\,\rho - 1|$, still assuming (4.3.1).

Proposition 4.3.2. *Set*

$$H_{\rho,\nu}(z) = \frac{1}{2} \sum_{\substack{m \in \mathbb{Z}^\times, n \in \mathbb{Z} \\ m|n(n+1)}} \left| \frac{m_1}{m_2} \right|^{\frac{1-\rho}{2}} \left(\frac{|(n + \frac{1}{2})z + m|^2}{|m|y} \right)^{\frac{\nu-1}{2}}, \qquad (4.3.2)$$

where the pair m_1, m_2 is defined in terms of the pair m, n according to the rule

$$m = m_1 m_2, \qquad 1 \le m_1|n + 1, m_2|n. \qquad (4.3.3)$$

This series converges (uniformly on compact subsets of Π) when $\mathrm{Re}\,\nu < -1 - |\mathrm{Re}\,\rho - 1|$. Moreover, one has the equivalence, in the sense of Definition 4.3.1,

$$f_{\rho,\nu}(z) \sim 2^{\frac{\nu-\rho+2}{2}} \pi^{\frac{1}{2}} \frac{\Gamma(\frac{\nu}{2})}{\Gamma(\frac{2-\rho+\nu}{2})\Gamma(\frac{4-\rho-\nu}{4})\Gamma(\frac{\rho+\nu}{4})} H_{\rho,\nu}(z). \qquad (4.3.4)$$

Proof. Let us split $f_{\rho,\nu}(z)$, as defined when $\mathrm{Re}\,\nu < -1 - |\mathrm{Re}\,\rho - 1|$, as the sum of four series $[f_{\rho,\nu}]_j(z)$, the parameter $j = 0, 1, 2, 3$ indicating that the corresponding summation takes place with g describing the set Γ_j, as defined in Lemma 4.1.1. We shall prove that $[f_{\rho,\nu}]_j(z) \sim 0$ when $j = 0, 1, 2$ and that $[f_{\rho,\nu}]_2(z) + [f_{\rho,\nu}]_3(z)$ is equivalent to the right-hand side of (4.3.4).

Since Γ_0 reduces to a set of 4 matrices, it is obvious that $[f_{\rho,\nu}]_0 \sim 0$ in the sense of Definition 4.3.1. Also, since $\Gamma_2 \cdot \left(\begin{smallmatrix} 0 & 1 \\ -1 & 0 \end{smallmatrix} \right) = \Gamma_1$, the contributions to the sum on the right-hand side of (4.2.1) of the sets of parameters Γ_1 and Γ_2 are related to each other under the change $z \mapsto -z^{-1}$ so that we may dispense with the study of $[f_{\rho,\nu}]_1$ as long as we only look for analytic continuation, not for z-dependent estimates. We may thus concentrate on the series $[f_{\rho,\nu}]_{23} = [f_{\rho,\nu}]_2 + [f_{\rho,\nu}]_3$: the analysis will show at the same time that $[f_{\rho,\nu}]_2 \sim 0$.

We use the parametrisation in Lemma 4.1.1. Recall that $n = bc, m = bd$ and that the pair m, n, in which $m \ne 0$, determines the pair c, d up to a sign change (the same on both entries), so that it determines $|cz + d|$. Indeed, since $m|n(n+1)$, there is a unique decomposition $m = m_1 m_2$ such that (4.3.3) holds: since $n+1 = ad$, the pair m_1, m_2 thus determined in terms of m, n has to be such that $m_1 = |d|$ and $m_2 = b \operatorname{sign} d$. Then,

$$|cz + d| = \left|\frac{nz + m}{b}\right| = \left|\frac{nz + m}{m_2}\right|, \tag{4.3.5}$$

and, a consequence of (4.2.9),

$$\psi(g.z) = \frac{|(n + \frac{1}{2})z + m|^2}{my} - \frac{1}{4m}\frac{|z|^2}{y}. \tag{4.3.6}$$

One has

$$[f_{\rho,\nu}]_{23}(z) = \sum_{\substack{m \in \mathbb{Z}^\times, n \in \mathbb{Z} \\ m|n(n+1)}} |m_2|^{\rho-1} \left(\frac{|nz + m|^2}{y}\right)^{\frac{1-\rho}{2}} \chi_{\rho,\nu}^{\text{even}}\left(\frac{|(n + \frac{1}{2})z + m|^2}{my} - \frac{1}{4m}\frac{|z|^2}{y}\right). \tag{4.3.7}$$

The proof of the proposition consists in showing the absolute convergence, for $\operatorname{Re}\nu < 1 - |\operatorname{Re}\rho - 1|$, of the series of error terms obtained in the following sequence of approximations: first, replace the argument $\frac{|nz+m|^2}{y}$ of the power function in the right-hand side of (4.3.7) by the expression $\frac{|(n+\frac{1}{2})z+m|^2}{y}$ which occurs in the argument of $\chi_{\rho,\nu}^{\text{even}}$; next, substitute for the function $\chi_{\rho,\nu}^{\text{even}}(t)$ the function $\tilde{\chi}_{\rho,\nu}^{\text{even}}(t)$ obtained from (2.3.31) when the hypergeometric series there has been reduced to its first term 1; finally, replace the factor $\left(\frac{-1-it}{2}\right)_+^{\frac{\rho+\nu-2}{2}}$ which occurs in (2.3.31) by $\left(\frac{-it}{2}\right)_+^{\frac{\rho+\nu-2}{2}}$. We shall bound the errors committed one at a time. As a final reduction, since $\left(\begin{smallmatrix} a & b \\ c & d \end{smallmatrix}\right)\left(\begin{smallmatrix} 0 & 1 \\ -1 & 0 \end{smallmatrix}\right) = \left(\begin{smallmatrix} -b & a \\ -d & c \end{smallmatrix}\right)$, we may retain here only the matrices $\left(\begin{smallmatrix} a & b \\ c & d \end{smallmatrix}\right)$ such that $|bd| \geq |ac|$ or, in terms of the pair m, n, such that $|m| > |n|$.

Set

$$\left(\frac{|nz + m|^2}{y}\right)^{\frac{1-\rho}{2}} = \left(\frac{|(n + \frac{1}{2})z + m|^2}{y}\right)^{\frac{1-\rho}{2}} (1 + w). \tag{4.3.8}$$

Since

$$w = \left(1 - \frac{(n + \frac{1}{4})|z|^2 + mx}{|(n + \frac{1}{2})z + m|^2}\right)^{\frac{1-\rho}{2}} - 1, \tag{4.3.9}$$

it is clear that, whenever z varies in a given compact subset of Π, one has for some constant $C > 0$ the estimate

$$|w| \leq C \frac{|m|}{|(n + \frac{1}{2})z + m|^2} : \tag{4.3.10}$$

note that it is at this point that we use the inequality $|n| \leq |m|$. It follows that

$$
\left| \left(\frac{|nz + m|^2}{y} \right)^{\frac{1-\rho}{2}} - \left(\frac{|(n + \frac{1}{2})z + m|^2}{y} \right)^{\frac{1-\rho}{2}} \right|
$$

$$
\leq C|m|^{\frac{1-\mathrm{Re}\,\rho}{2}} \left(\frac{|(n + \frac{1}{2})z + m|^2}{|m|y} \right)^{\frac{-1-\mathrm{Re}\,\rho}{2}} . \qquad (4.3.11)
$$

The difference between the series (4.3.7) and that obtained when replacing, in the first factor, the argument $\frac{|nz+m|^2}{y}$ by $\frac{|(n+\frac{1}{2})z+m|^2}{y}$, is thus majorized, for some constant $C > 0$ (use the estimate (2.3.32)), by the series

$$
\sum_{\substack{m \in \mathbb{Z}^\times, n \in \mathbb{Z} \\ m | n(n+1)}} |m_2|^{\mathrm{Re}\,\rho - 1} |m|^{\frac{1-\mathrm{Re}\,\rho}{2}} \left(\frac{|(n + \frac{1}{2})z + m|^2}{|m|y} \right)^{\frac{\mathrm{Re}\,\nu - 3}{2}} . \qquad (4.3.12)
$$

One has $|m_2|^{\mathrm{Re}\,\rho - 1} |m|^{\frac{1-\mathrm{Re}\,\rho}{2}} = \left| \frac{m_2}{m_1} \right|^{\frac{\mathrm{Re}\,\rho - 1}{2}} \leq |m|^{\frac{|\mathrm{Re}\,\rho - 1|}{2}}$. Since, for $n \neq 0, -1$, the number of integers m dividing $n(n+1)$ is a $O(|n|^\varepsilon)$ for every $\varepsilon > 0$, this main part of the series is (absolutely) convergent for $\mathrm{Re}\,\nu < 1$; the part of the series obtained for $n = 0, -1$ is majorized by a constant times $\sum_{m \in \mathbb{Z}^\times} |m|^{\frac{1-\mathrm{Re}\,\rho}{2}} |m|^{\frac{\mathrm{Re}\,\nu - 3}{2}}$, and converges when $\mathrm{Re}\,\nu < 1 - |\mathrm{Re}\,\rho - 1|$, which is just what we need.

The next approximation consists in replacing $\chi^{\mathrm{even}}_{\rho,\nu}(t)$ by the function, also continuous on the whole real line,

$$
\tilde{\chi}^{\mathrm{even}}_{\rho,\nu}(t) = \frac{1}{2} 2^{\nu-1} \pi^{-\frac{1}{2}} \frac{\Gamma(\frac{\nu}{2})}{\Gamma(\frac{2-\rho+\nu}{2})} \left[\left(\frac{-1 - it}{2} \right)^{\frac{\rho+\nu-2}{2}}_+ + \left(\frac{-1 + it}{2} \right)^{\frac{\rho+\nu-2}{2}}_+ \right] .
$$
$$
(4.3.13)
$$

Then,

$$
\left| \chi^{\mathrm{even}}_{\rho,\nu}(t) - \tilde{\chi}^{\mathrm{even}}_{\rho,\nu}(t) \right| \leq C(1 + |t|)^{\frac{\rho+\nu-4}{2}} : \qquad (4.3.14)
$$

it follows from the main argument of the proof of Theorem 4.2.4, comparing the present exponent $\frac{\rho+\nu-4}{2}$ to the exponent $\frac{\rho+\nu-2}{2}$ of $1 + |\psi(g.z)|$ in (4.2.20), that, substituting the function $\tilde{\chi}^{\mathrm{even}}_{\rho,\nu}$ for the function $\chi^{\mathrm{even}}_{\rho,\nu}$ in the series currently under examination, we obtain a series equivalent to $f_{\rho,\nu}(z)$ in the sense of Definition 4.3.1.

Summing up the present state of affairs, we have obtained that the function $[f_{\rho,\nu}]_{23}(z)$ is equivalent to the series

$$
\sum_{\substack{m \in \mathbb{Z}^\times, n \in \mathbb{Z} \\ m | n(n+1)}} |m_2|^{\rho - 1} \left(\frac{|(n + \frac{1}{2})z + m|^2}{y} \right)^{\frac{1-\rho}{2}} \tilde{\chi}^{\mathrm{even}}_{\rho,\nu} \left(\frac{|(n + \frac{1}{2})z + m|^2}{my} - \frac{1}{4m} \frac{|z|^2}{y} \right) .
$$
$$
(4.3.15)
$$

Next, given two complex numbers w and h with the property that $w + sh \notin]-\infty, 0]$ for $0 \le s \le 1$, one has

$$(w+h)_+^{\frac{\nu-1}{2}} - w_+^{\frac{\nu-1}{2}} = \frac{\nu-1}{2} h \int_0^1 (w+sh)_+^{\frac{\nu-3}{2}} ds. \qquad (4.3.16)$$

This makes it possible, in view of the continuation problem under dicussion, to replace the argument $\psi(g.z)$ in (4.3.15) by an approximation $\psi^\sharp(g.z)$, provided that, in a way uniform relative to $s \in [0,1]$ and to g in the set of matrices under consideration, but not necessarily so relative to z, one has for some $C > 0$ the estimate

$$|1 \pm i[(1-s)\psi(g.z) + s\psi^\sharp(g.z)]| \ge C^{-1}|1 \pm i[\psi(g.z)]|. \qquad (4.3.17)$$

One observes that the first term in the sum on the right-hand side of (4.3.6) is always $\ge \frac{9}{4|m|}y$: when z varies in a compact set, it is $\ge \frac{1}{2|m|}\frac{|z|^2}{y}$ except for a finite number of values of n at the most. Then, setting $\psi^\sharp(g.z) = \frac{|(n+\frac{1}{2})z+m|^2}{my}$, the condition relative to the pair $(\psi(g.z), \psi^\sharp(g.z))$ just explained is satisfied. This leaves us with the problem of examining, in place of $[f_{\rho,\nu}]_{23}(z)$, the series

$$\frac{1}{2}2^{\nu-1}\pi^{-\frac{1}{2}}\frac{\Gamma(\frac{\nu}{2})}{\Gamma(\frac{2-\rho+\nu}{2})} \sum_{\substack{m\in\mathbb{Z}^\times, n\in\mathbb{Z} \\ m|n(n+1)}} |m_2|^{\rho-1}$$

$$\times \left(\frac{|(n+\frac{1}{2})z+m|^2}{y}\right)^{\frac{1-\rho}{2}} \left[\left(\frac{-1-it}{2}\right)_+^{\frac{\rho+\nu-2}{2}} + \left(\frac{-1+it}{2}\right)_+^{\frac{\rho+\nu-2}{2}}\right], \qquad (4.3.18)$$

where $\frac{|(n+\frac{1}{2})z+m|^2}{my}$ is to be substituted for t.

Finally, it is a quite elementary matter that an equivalent series, in the sense of Definition 4.3.1, is obtained if one substitutes for the bracket on the right-hand side the expression

$$\left(\frac{-it}{2}\right)_+^{\frac{\rho+\nu-2}{2}} + \left(\frac{it}{2}\right)_+^{\frac{\rho+\nu-2}{2}} = \frac{2^{\frac{4-\rho-\nu}{2}}\pi}{\Gamma(\frac{\rho+\nu}{4})\Gamma(\frac{4-\rho-\nu}{4})}|t|^{\frac{\rho+\nu-2}{2}}. \qquad (4.3.19)$$

This completes the proof of Proposition 4.3.2. $\qquad\square$

We are now left with the task of analyzing the continuation in the domain where $\text{Re}\,\nu < 1-|\text{Re}\,\rho-1|$ of the (non-automorphic) function $H_{\rho,\nu}(z)$. The starting point consists in making the Fourier series expansion of the periodic function

$$z \mapsto H_{\rho,\nu}(-z^{-1}) = \frac{1}{2} \sum_{\substack{m\in\mathbb{Z}^\times, n\in\mathbb{Z} \\ m|n(n+1)}} \left|\frac{m_1}{m_2}\right|^{\frac{1-\rho}{2}} |m|^{\frac{1-\nu}{2}} \left(\frac{|n+\frac{1}{2}-mz|^2}{|m|y}\right)^{\frac{\nu-1}{2}}, \qquad (4.3.20)$$

for $\text{Re}\,\nu < -1-|\text{Re}\,\rho-1|$, explicit: recall that m_1 and m_2 have been defined in (4.3.3).

Theorem 4.3.3. *For* $\operatorname{Re}\nu < -1 - |\operatorname{Re}\rho - 1|$, *one has*

$$H_{\rho,\nu}(-z^{-1}) = \pi^{\frac{1}{2}} \frac{\Gamma(-\frac{\nu}{2})}{\Gamma(\frac{1-\nu}{2})} \frac{\zeta(\frac{\rho-\nu}{2})\zeta(\frac{2-\rho-\nu}{2})}{\zeta(1-\nu)} y^{\frac{1+\nu}{2}} \tag{4.3.21}$$

$$+ \frac{2\pi^{\frac{1-\nu}{2}}}{\Gamma(\frac{1-\nu}{2})} y^{\frac{1}{2}} \sum_{k\neq 0} (\zeta_k)_{\text{sym}} \left(\frac{\rho-\nu}{2}, \frac{2-\rho-\nu}{2}\right) e^{2i\pi kx} |k|^{-\frac{\nu}{2}} K_{\frac{\nu}{2}}(2\pi|k|y).$$

Proof. Setting $n = r + \ell m$ with $0 \le r < m$, $\ell \in \mathbb{Z}$ the condition $m|n(n+1)$ becomes $r(r+1) \equiv 0 \bmod m$, and one has

$$H_{\rho,\nu}(-z^{-1}) = \frac{1}{2} \sum_{m\neq 0} \left|\frac{m_1}{m_2}\right|^{\frac{1-\rho}{2}} |m|^{\frac{1-\nu}{2}}$$

$$\times \sum_{\substack{r \bmod m \\ r(r+1)\equiv 0}} \sum_{\ell \in \mathbb{Z}} \left[\frac{(m(\ell - x) + r + \frac{1}{2})^2 + m^2 y^2}{y}\right]^{\frac{\nu-1}{2}}. \tag{4.3.22}$$

We use Poisson's formula to compute the series (with respect to ℓ) on the right-hand side: one has [36, p. 401]

$$\left[\mathcal{F}\left((1+s^2)^{\frac{\nu-1}{2}}\right)\right](\sigma) = \frac{2\pi^{\frac{1-\nu}{2}}}{\Gamma(\frac{1-\nu}{2})} |\sigma|^{-\frac{\nu}{2}} K_{\frac{\nu}{2}}(2\pi|\sigma|). \tag{4.3.23}$$

Writing

$$\frac{(m(s - x) + r + \frac{1}{2})^2 + m^2 y^2}{y} = m^2 y \left[1 + \left(\frac{s - x + \frac{r}{m}}{y}\right)^2\right], \tag{4.3.24}$$

one obtains after an elementary change of variable

$$\int_{-\infty}^{\infty} e^{-2i\pi s\sigma} (m^2 y)^{\frac{\nu-1}{2}} \left[1 + \left(\frac{s - x + \frac{r}{m} + \frac{1}{2m}}{y}\right)^2\right]^{\frac{\nu-1}{2}} ds$$

$$= \frac{2\pi^{\frac{1-\nu}{2}}}{\Gamma(\frac{1-\nu}{2})} |m|^{\nu-1} y^{\frac{1}{2}} e^{2i\pi(x - \frac{r}{m} - \frac{1}{2m})\sigma} |\sigma|^{-\frac{\nu}{2}} K_{\frac{\nu}{2}}(2\pi y|\sigma|), \tag{4.3.25}$$

which, if one also uses the fact that the value at $\sigma = 0$ of the continuous function $|\sigma|^{-\frac{\nu}{2}} K_{\frac{\nu}{2}}(2\pi|\sigma|)$ is $\frac{1}{2}\pi^{\frac{\nu}{2}}\Gamma(-\frac{\nu}{2})$, leads to

$$H_{\rho,\nu}(-z^{-1}) = \frac{\pi^{\frac{1-\nu}{2}}}{\Gamma(\frac{1-\nu}{2})} \sum_{m\neq 0} |m|^{\frac{\nu-1}{2}} \left|\frac{m_1}{m_2}\right|^{\frac{1-\rho}{2}} \sum_{\substack{r \bmod m \\ r(r+1)\equiv 0}} \left[\frac{1}{2}\pi^{\frac{\nu}{2}}\Gamma(-\frac{\nu}{2})y^{\frac{1+\nu}{2}}\right.$$

$$+ y^{\frac{1}{2}} \sum_{k\neq 0} e^{2i\pi kx} \exp\left(-2i\pi k\frac{r + \frac{1}{2}}{m}\right) |k|^{-\frac{\nu}{2}} K_{\frac{\nu}{2}}(2\pi|k|y)\bigg]. \tag{4.3.26}$$

We have used several times the fact that given $r \in \mathbb{Z}$, if $m \neq 0$ divides a product $r(r+1)$, one can find a unique decomposition $m = m_1 m_2$ such that $1 \leq m_1 | r+1$ and $m_2 | r$: we now observe that conversely, given a pair (m_1, m_2) of relatively prime nonzero integers with $m_1 \geq 1$, there is a unique $r \bmod m_1 m_2$ such that $r \equiv -1 \bmod m_1$ and $r \equiv 0 \bmod m_2$, and $\frac{r}{m_2} \equiv -\overline{m}_2 \bmod m_1$. Hence,

$$
\frac{1}{2} \sum_{m \neq 0} |m|^{\frac{\nu-1}{2}} \left|\frac{m_1}{m_2}\right|^{\frac{1-\rho}{2}} \sum_{\substack{r \bmod m \\ r(r+1) \equiv 0}} \exp\left(-2i\pi n \frac{r+\frac{1}{2}}{m}\right)
$$

$$
= \frac{1}{2} \sum_{\substack{m_1 \geq 1, m_2 \neq 0 \\ (m_1, m_2)=1}} |m_1|^{\frac{-\rho+\nu}{2}} |m_2|^{\frac{\rho+\nu-2}{2}} \exp\left(2i\pi k\left(\frac{\overline{m}_2}{m_1} - \frac{1}{m_1 m_2}\right)\right)
$$

$$
= (\zeta_k)_{\text{sym}}\left(\frac{\rho-\nu}{2}, \frac{2-\rho-\nu}{2}\right), \qquad (4.3.27)
$$

in terms of the function $(\zeta_k)_{\text{sym}}$ introduced in (3.3.24). This is valid whether $k \neq 0$ or $k = 0$: also recall that

$$
(\zeta_0)_{\text{sym}}(s,t) = \zeta_0(s,t) = \frac{\zeta(s)\zeta(t)}{\zeta(s+t)}. \qquad (4.3.28)
$$

Equation (4.3.21) follows. $\qquad \square$

Note for future reference that one can write the coefficient of $y^{\frac{1+\nu}{2}}$ in the right-hand side of (4.3.21) as

$$
\pi^{\frac{1}{2}} \frac{\Gamma(-\frac{\nu}{2})}{\Gamma(\frac{1-\nu}{2})} \frac{\zeta(\frac{\rho-\nu}{2})\zeta(\frac{2-\rho-\nu}{2})}{\zeta(1-\nu)} = \pi^{\frac{1}{2}} \frac{\Gamma(-\frac{\nu}{2})}{\Gamma(\frac{\rho-\nu}{4})\Gamma(\frac{2-\rho-\nu}{4})} \frac{\zeta^*(\frac{\rho-\nu}{2})\zeta^*(\frac{\rho+\nu}{2})}{\zeta^*(\nu)} : \quad (4.3.29)
$$

this expression will facilitate tracking the effect of the change $\nu \mapsto -\nu$.

Observing that all estimates leading to the continuation of the function $(\zeta_k)_{\text{sym}}$, though possibly non-uniform with respect to k, only involve possible losses of powers of k quite compensated for given y by the factors $K_{\frac{\nu}{2}}(2\pi|k|y)$, we obtain from an application of Theorem 3.6.2, resumming a Fourier series, the following:

Theorem 4.3.4. *Assume* $|\operatorname{Re}\rho - 1| < 1$. *The function defined, for* $\operatorname{Re}\nu < -1 - |\operatorname{Re}\rho - 1|, \nu \notin \mathbb{Z}, \rho \pm \nu \notin 2\mathbb{Z}$, *by the convergent series*

$$
H_{\rho,\nu}(z) = \frac{1}{2} \sum_{\substack{cm \in \mathbb{Z}^\times, n \in \mathbb{Z} \\ m | n(n-1)}} \left|\frac{m_1}{m_2}\right|^{\frac{1-\rho}{2}} |m|^{\frac{1-\nu}{2}} \left(\frac{|(n+\frac{1}{2})z + m|^2}{\operatorname{Im} z}\right)^{\frac{\nu-1}{2}}, \qquad (4.3.30)
$$

extends as a holomorphic function for $\operatorname{Re}\nu < 1 - |\operatorname{Re}\rho - 1|, \nu \notin \mathbb{Z}, \rho \pm \nu \notin 2\mathbb{Z}$, *except for the following possible poles: the non-trivial zeros of the zeta function,* and the points $i\lambda_p$ with $\frac{1+\lambda_p^2}{4}$ in the even part of the discrete spectrum of Δ. A point $i\lambda_p$ of the second species can only be a simple pole: it is one if and only if one has $L(\frac{2-\rho}{2}, \mathcal{M}) \neq 0$ for at least one even cusp-form corresponding to the eigenvalue $\frac{1+\lambda_p^2}{4}$. If such is the case, one has

$$\mathrm{Res}_{\nu = i\lambda_p} H_{\rho,\nu}(z) = -\pi^{\frac{\rho-1}{2}} \frac{\Gamma(-\frac{i\lambda_p}{2})\Gamma(\frac{2-\rho+i\lambda_p}{4})}{\Gamma(\frac{\rho-i\lambda_p}{4})} \mathcal{M}_p^\rho(z), \qquad (4.3.31)$$

where

$$\mathcal{M}_p^\rho = \sum_j L(\frac{2-\rho}{2}, \mathcal{M}_{p,j})\mathcal{M}_{p,j}, \qquad (4.3.32)$$

with the understanding that only Hecke eigenforms of even type (with respect to the symmetry $z \mapsto -\bar{z}$) are retained in the sum. In the open subset of the half-plane $\mathrm{Re}\,\nu < 1 - |\mathrm{Re}\,\rho - 1|$, just made explicit, where the function $\nu \mapsto H_{\rho,\nu}(z)$ can be analytically continued, the function $f_{\rho,\nu}$ can be continued just as well.

Proof. The last sentence is a consequence of (4.3.4).   □

Our analysis of the continuation of the function $f_{\rho,\nu}(z)$ to the half-plane $\mathrm{Re}\,\nu < 1 - |\mathrm{Re}\,\rho - 1|$, from which an explicit discrete set of points of an arithmetic nature has been deleted, is thus over. Also note that, in view of the functional equation $L^*(s, \mathcal{M}_{p,j}) = L^*(1 - s, \mathcal{M}_{p,j})$, with the notation in (3.1.12), the conditions $L(\frac{2-\rho}{2}, \mathcal{M}_p^\rho) \neq 0$ and $L(\frac{\rho}{2}, \mathcal{M}_p^\rho) \neq 0$ are equivalent, and that the expression (4.3.31) is fully invariant under the change $\rho \mapsto 2 - \rho$.

The Fourier expansion in Theorem 4.3.3 extends to the admissible values of ν with $\mathrm{Re}\,\nu < 1 - |\mathrm{Re}\,\rho - 1|$, so that one has for every such value the estimate

$$H_{\rho,\nu}(-z^{-1}) = \pi^{\frac{1}{2}} \frac{\Gamma(-\frac{\nu}{2})}{\Gamma(\frac{1-\nu}{2})} \frac{\zeta(\frac{\rho-\nu}{2})\zeta(\frac{2-\rho-\nu}{2})}{\zeta(1-\nu)} y^{\frac{1+\nu}{2}} + O(y^{-\infty}), \qquad y \to \infty, \quad (4.3.33)$$

where the way the remainder is denoted means that, for every N, its product by y^N is bounded when $y \geq 1$: termwise differentiation is possible as well.

Remark 4.3.a. Consider the slightly different function

$$z \mapsto F_{\rho,\nu}(-z^{-1}) = \frac{1}{2} \sum_{\substack{m \in \mathbb{Z}^\times, n \in \mathbb{Z} \\ m | n(n+1)}} \left| \frac{m_1}{m_2} \right|^{\frac{1-\rho}{2}} |m|^{\frac{1-\nu}{2}} \left(\frac{|n - mz|^2}{\mathrm{Im}\,z} \right)^{\frac{\nu-1}{2}} \qquad (4.3.34)$$

for $\mathrm{Re}\,\nu < -1 - |\mathrm{Re}\,\rho - 1|$, with m_1, m_2 defined in (4.3.3), and immediately note that changing the condition $m | n(n + 1)$ to $m | n(n - 1)$ would not modify the function (change the pair m, n to $-m, -n$). This is just the function which would have been denoted as $F_{-i\nu,1-\rho}(z)$ in [60, p. 153]. Exactly the same proof as that of Theorem 4.3.3 shows the identity [60, Prop.15.5]

$$F_{\rho,\nu}(-z^{-1}) = \pi^{\frac{1}{2}} \frac{\Gamma(-\frac{\nu}{2})}{\Gamma(\frac{1-\nu}{2})} \frac{\zeta(\frac{\rho-\nu}{2})\zeta(\frac{2-\rho-\nu}{2})}{\zeta(1-\nu)} y^{\frac{1+\nu}{2}} \tag{4.3.35}$$
$$+ \frac{2\pi^{\frac{1-\nu}{2}}}{\Gamma(\frac{1-\nu}{2})} y^{\frac{1}{2}} \sum_{k \neq 0} \zeta_k \left(\frac{\rho-\nu}{2}, \frac{2-\rho-\nu}{2} \right) e^{2i\pi kx} |k|^{-\frac{\nu}{2}} K_{\frac{\nu}{2}}(2\pi|k|y).$$

As indicated right after (3.3.25), the difference $(\zeta_k)_{\mathrm{sym}}\left(\frac{\rho-\nu}{2}, \frac{2-\rho-\nu}{2}\right) - \zeta_k\left(\frac{\rho-\nu}{2}, \frac{2-\rho-\nu}{2}\right)$ remains holomorphic for $\mathrm{Re}\,\nu < \min(\mathrm{Re}\,\rho, 2-\mathrm{Re}\,\rho) = 1-|\mathrm{Re}\,\rho - 1|$. This is why Theorem 4.3.4 is completely equivalent to the analogous one in which the function $F_{\rho,\nu}$ would substitute for $H_{\rho,\nu}$: this version was given as [60, Theor.15.6].

4.4 Asymptotics of $f_{\rho,\nu}(x+iy), y \to \infty$

We introduced in (4.2.1) the function $f_{\rho,\nu}(z)$ and showed that the series defining it is convergent for $\mathrm{Re}\,\nu < -1 - |\mathrm{Re}\,\nu - 1|$; next, we transformed it (Prop. 4.3.2) into the product of $H_{\rho,\nu}(z)$ by a simple factor, up to an error term extending as a holomorphic function of ν for $\mathrm{Re}\,\nu < 1 - |\mathrm{Re}\,\rho - 1|$. Finally, we dealt with the continuation of the function $H_{\rho,\nu}$ (which implies that of $f_{\rho,\nu}$) for $\mathrm{Re}\,\nu < 1 - |\mathrm{Re}\,\rho - 1|$: this demanded removing a discrete set of poles with arithmetic significance. We now consider the asymptotics of $f_{\rho,\nu}(z)$ as $y = \mathrm{Im}\,z \to \infty$.

Since $f_{\rho,\nu}$ is automorphic, one has $f_{\rho,\nu}(z) = f_{\rho,\nu}(-z^{-1})$, and we shall actually look for the expansion as $y \to \infty$ of the terms involved (from the partition of Γ introduced just before Definition 4.1.1) in the decomposition of $f_{\rho,\nu}(-z^{-1})$. The reason is of course that we have obtained the Fourier series expansion of $H_{\rho,\nu}(-z^{-1})$, not that of $H_{\rho,\nu}(z)$, which is not periodic. We shall examine separately the two terms of the decomposition

$$f_{\rho,\nu}(z) = [f_{\rho,\nu}]_{01}(-z^{-1}) + [f_{\rho,\nu}]_{23}(-z^{-1}), \tag{4.4.1}$$

with

$$[f_{\rho,\nu}]_{01}(-z^{-1}) = \frac{1}{2} \sum_{g \in \Gamma_0 \cup \Gamma_1} (\mathrm{Im}\,(g.(-z^{-1})))^{\frac{\rho-1}{2}} \chi_{\rho,\nu}^{\mathrm{even}}(\psi(g.(-z^{-1}))),$$
$$[f_{\rho,\nu}]_{23}(-z^{-1}) = \frac{1}{2} \sum_{g \in \Gamma_2 \cup \Gamma_3} (\mathrm{Im}\,(g.(-z^{-1})))^{\frac{\rho-1}{2}} \chi_{\rho,\nu}^{\mathrm{even}}(\psi(g.(-z^{-1}))). \tag{4.4.2}$$

From the results of Section 4.3, we already know that the first of these two functions extends as an analytic function of ν for $\mathrm{Re}\,\nu < 1 - |\mathrm{Re}\,\rho - 1|$, assuming also (4.3.1). As seen from Theorem 4.3.4, the second of these functions extends to the

same half-plane after one has removed the following points: the non-trivial zeros
of the zeta function, and the points $i\lambda_p$ with $\frac{1+\lambda_p^2}{4}$ in the even part of the discrete
spectrum of Δ.

Asymptotics of the first series will be obtained with the help of the Euler-
Maclaurin formula, but some preparation is needed since we shall be dealing with
a function presenting discontinuities. Recall (*e.g.,* [54, p. 6] or [64, p. 215]) the
equation

$$\sum_{n\in\mathbb{Z}} f(n) = \int_{-\infty}^{\infty} f(t)dt + \frac{(-1)^k}{(k+1)!} \int_{-\infty}^{\infty} B_{k+1}(t)f^{(k+1)}(t)dt, \qquad (4.4.3)$$

valid for instance under the assumption that f should be of class C^{k+1} on the line
and that, for some $\delta > 1$ and $j = 0, \ldots, k+1$, the estimate $|f^{(j)}(t)| \le C(1+|t|)^{-\delta}$
should hold. The function B_{k+1} is periodic of period 1, and its restriction to $[0, 1[$ is
characterized by the sequence of equations $B_0(t) = 1, B_r' = rB_{r-1}, \int_0^1 B_r(t)dt = 0$
for $r \ge 1$.

We need to have a version applicable to the case when, for some $x \notin \mathbb{Z}$, one
has $f(t) = h(t)\mathrm{sign}(x+t)$ for some smooth function h. To do so, we approach the
sign function by the function $t \mapsto \tanh\frac{x+t}{\varepsilon}$, with $\varepsilon > 0$ going to 0. Then, (4.4.3)
can still be applied, provided we replace the integral on the right-hand side by the
limit as $\varepsilon \to 0$ of the integral $I_k(\varepsilon)$ in the next lemma.

Lemma 4.4.1. *Let $h \in C^\infty(\mathbb{R})$ have a compact support disjoint from $\mathbb{Z}$, let $x \notin \mathbb{Z}$,
and consider, for $\varepsilon > 0$, the integral*

$$I_k(\varepsilon) = \int_{-\infty}^{\infty} B_{k+1}(t) \left(\frac{d}{dt}\right)^{k+1} \left[h(t)\tanh\frac{x+t}{\varepsilon}\right] dt. \qquad (4.4.4)$$

One has

$$I_k(\varepsilon) \to \int_{-\infty}^{\infty} B_{k+1}(t)h^{(k+1)}(t)\mathrm{sign}(x+t)dt$$

$$+ 2(k+1)! \sum_{r=0}^{k} \frac{(-1)^{k-r}}{r+1} B_{r+1}(-x)h^{(r)}(-x), \qquad \varepsilon \to 0. \quad (4.4.5)$$

Proof. One has

$$\left(\frac{d}{dt}\right)^{k+1} \left[h(t)\tanh\frac{x+t}{\varepsilon}\right] = \sum_{j=0}^{k+1} \binom{k+1}{j} h^{(j)}(t) \left(\frac{d}{dt}\right)^{k+1-j} (\tanh\frac{x+t}{\varepsilon}):$$

$$(4.4.6)$$

the contribution to the integral defining $I_k(\varepsilon)$ of the term with $j = k+1$ goes to
the integral

$$\int_{-\infty}^{\infty} B_{k+1}(t)h^{(k+1)}(t)\mathrm{sign}(x+t)dt \qquad (4.4.7)$$

as $\varepsilon \to 0$. The contribution of a term with $j \le k$, to wit

$$\binom{k+1}{j} \int_{-\infty}^{\infty} B_{k+1}(t)h^{(j)}(t) \left(\frac{d}{dt}\right)^{k+1-j} (\tanh\frac{x+t}{\varepsilon})dt$$

$$= (-1)^{k+1-j} \binom{k+1}{j} \int_{-\infty}^{\infty} \left(\frac{d}{dt}\right)^{k+1-j} \left[B_{k+1}(t)h^{(j)}(t)\right] \tanh\frac{x+t}{\varepsilon}dt, \quad (4.4.8)$$

has the limit

$$2(-1)^{k-j} \binom{k+1}{j} \left(\frac{d}{dt}\right)^{k-j} \left[B_{k+1}(t)h^{(j)}(t)\right](t=-x). \quad (4.4.9)$$

What remains to be done is transforming the sum

$$\lim_{\varepsilon \to 0} R_k(\varepsilon): = 2\sum_{j=0}^{k} (-1)^{k-j} \binom{k+1}{j} \left(\frac{d}{dt}\right)^{k-j} \left[B_{k+1}(t)h^{(j)}(t)\right](t=-x).$$

$$(4.4.10)$$

One has

$$\lim_{\varepsilon \to 0} R_k(\varepsilon) = 2\sum_{j=0}^{k} (-1)^{k-j} \binom{k+1}{j} \sum_{\ell=0}^{k-j} \binom{k-j}{\ell} B_{k+1}^{(k-j-\ell)}(-x)h^{(j+\ell)}(-x)$$

$$= 2 \sum_{\substack{0 \le j \le k \\ 0 \le \ell \le k-j}} \frac{(-1)^{k-j}}{k+1-j} \frac{(k+1)!}{j!\ell!(k-j-\ell)!} B_{k+1}^{(k-j-\ell)}(-x)h^{(j+\ell)}(-x). \quad (4.4.11)$$

Setting $j+\ell = r$, this can be rewritten as

$$\lim_{\varepsilon \to 0} R_k(\varepsilon) = \sum_{r=0}^{k} \alpha_{k,r} B_{k+1}^{(k-r)}(-x)h^{(r)}(-x), \quad (4.4.12)$$

with

$$\alpha_{k,r} = 2\frac{(k+1)!}{(k-r)!} \sum_{j=0}^{r} \frac{(-1)^{k-j}}{(k+1-j)j!(r-j)!}$$

$$= 2\frac{(k+1)!}{(k-r)!} \int_{0}^{1} (-x)^{k-r}(1-x)^r dx = 2(-1)^{k-r}r!. \quad (4.4.13)$$

One finishes the proof with the help of the equation $B_{k+1}^{(k-r)}(-x) = \frac{(k+1)!}{(r+1)!}B_{r+1}(-x)$.
□

Theorem 4.4.2. *Together with the standing assumptions* (4.3.1), *assume that* $\mathrm{Re}\,\nu < 1 - |\mathrm{Re}\,\rho - 1|$. *One has the asymptotic expansion, uniform with respect to* x, *up to arbitrary powers of* y^{-1},

$$[f_{\rho,\nu}]_{01}(-z^{-1}) \sim I(\rho,\nu)y^{\frac{\rho+1}{2}} + \frac{\Gamma(\frac{2+\rho-\nu}{4})\Gamma(\frac{2+\rho+\nu}{4})}{\Gamma(\frac{4-\rho-\nu}{4})\Gamma(\frac{4-\rho+\nu}{4})}I(2-\rho,\nu)y^{\frac{3-\rho}{2}} \qquad (4.4.14)$$

$$+ \sum_{j\geq 0} \frac{2^{2j}}{(2j+2)!}B_{2j+2}(x)\left[C(\rho,\nu)\frac{\Gamma(j+1-\frac{\nu+\rho}{4})\Gamma(j+1+\frac{\nu-\rho}{4})}{\Gamma(1-\frac{\nu+\rho}{4})\Gamma(1+\frac{\nu-\rho}{4})}y^{\frac{\rho-1}{2}-2j-1}\right.$$

$$\left. + \frac{\Gamma(\frac{2+\rho-\nu}{4})\Gamma(\frac{2+\rho+\nu}{4})}{\Gamma(\frac{4-\rho-\nu}{4})\Gamma(\frac{4-\rho+\nu}{4})}C(2-\rho,\nu)\frac{\Gamma(j+1-\frac{\nu+2-\rho}{4})\Gamma(j+1+\frac{\nu-2+\rho}{4})}{\Gamma(1-\frac{\nu+2-\rho}{4})\Gamma(1+\frac{\nu-2+\rho}{4})}y^{\frac{1-\rho}{2}-2j-1}\right].$$

Also, this asymptotic expansion can be differentiated with respect to x or y, when $x \notin \mathbb{Z}$, as many times as needed.

Proof. There are substantial differences (compare (2.3.35) and (2.3.44)) between the special case $\rho = 1$ and the case when $\rho \notin 1+2\mathbb{Z}$ so far as the transformation rule relating the values of the hypergeometric functions of interest at two arguments the inverse of each other are concerned. For a change, let us discuss the case when $\rho = 1$ first.

Recall that the matrices in $\Gamma_0 \cup \Gamma_1$ are all matrices $\pm\left(\begin{smallmatrix}1&0\\c&1\end{smallmatrix}\right)$ and $\pm\left(\begin{smallmatrix}a&1\\-1&0\end{smallmatrix}\right)$ with $a, c \in \mathbb{Z}$. It is immediate that

$$\psi\left(\left(\begin{smallmatrix}1&0\\c&1\end{smallmatrix}\right).(-z^{-1})\right) = \frac{c-x}{y}, \qquad \psi\left(\left(\begin{smallmatrix}a&1\\-1&0\end{smallmatrix}\right).(-z^{-1})\right) = \frac{-a+x}{y}, \qquad (4.4.15)$$

so that

$$\frac{1}{2}\sum_{g\in\Gamma_0\cup\Gamma_1} \chi_\nu(\psi(g.(-z^{-1}))) = \sum_{n\in\mathbb{Z}}\chi_{1,\nu}\left(\frac{x+n}{y}\right) + \sum_{n\in\mathbb{Z}}\chi_{1,\nu}\left(\frac{-x+n}{y}\right). \qquad (4.4.16)$$

Since this is a periodic function of x, nothing is lost in assuming in what follows that $-\frac{1}{2} \leq x \leq \frac{1}{2}$. We also assume, but only temporarily, that $x \neq 0$ as we wish to apply Lemma 4.4.1.

Set

$$\phi_{1,\nu}(t) = \frac{1}{2\pi}\frac{\Gamma(\frac{\nu}{2})\Gamma(\frac{2-\nu}{2})}{\Gamma(\frac{1+\nu}{2})\Gamma(\frac{1-\nu}{2})}{}_2F_1\left(\frac{1-\nu}{2},\frac{1+\nu}{2};1;\frac{1+it}{2}\right). \qquad (4.4.17)$$

From (2.3.44), one has the decomposition

$$\chi_{1,\nu}(t) = i\pi\phi_{1,\nu}(t)\operatorname{sign}t + \psi_{1,\nu}(t), \qquad (4.4.18)$$

where the function $\psi_{1,\nu}$, just as $\phi_{1,\nu}$, is C^∞ in the interval $]-1,1[$. Choose a function $\alpha \in C^\infty(\mathbb{R})$ with a compact support contained in $]-1,1[$, and such that $\alpha(t) = 1$ for t close to 0. Finally, set

$$h(t) = \phi_{1,\nu}\left(\frac{x+t}{y}\right)\alpha\left(\frac{x+t}{y}\right). \qquad (4.4.19)$$

In the decomposition

$$\chi_{1,\nu}\left(\frac{x+t}{y}\right) = i\pi h(t)\mathrm{sign}(x+t)$$

$$+ \psi_{1,\nu}\left(\frac{x+t}{y}\right)\alpha\left(\frac{x+t}{y}\right) + \chi_{1,\nu}\left(\frac{x+t}{y}\right)\left[1 - \alpha\left(\frac{x+t}{y}\right)\right], \quad (4.4.20)$$

only the first term has discontinuities. Applying (4.4.3) and Lemma 4.4.1 and regrouping the integrals, one obtains

$$\sum_{n\in\mathbb{Z}} \chi_{1,\nu}\left(\frac{x+n}{y}\right) = \int_{-\infty}^{\infty} \chi_{1,\nu}\left(\frac{x+t}{y}\right)dt + 2i\pi \sum_{r=0}^{k} \frac{(-1)^r}{r+1}\phi_{1,\nu}^{(r)}(0)B_{r+1}(-x)y^{-r}$$

$$+ \frac{(-1)^k}{(k+1)!}\int_{-\infty}^{\infty} B_{k+1}(t)\left(\frac{d}{dt}\right)^{k+1}\chi_{1,\nu}\left(\frac{x+t}{y}\right)dt, \quad (4.4.21)$$

where the last integral must be taken in the usual (i.e., not distribution) sense: the function under the integral sign has only a mild discontinuity at $t = -x$, since the function $\chi_{1,\nu}$, as defined and C^∞ in $\mathbb{R}\backslash\{0\}$, extends as a C^∞ function on each of the two intervals $]-\infty, 0]$ and $[0, \infty[$.

Note that, even though we started with the assumption $\mathrm{Re}\,\nu < -1$ in order to apply (4.4.3), this identity provides (again: this task has already been done in Section 4.3) the analytic continuation with respect to ν of the sum under consideration. Since the integral remainder is obviously a $O(y^{-k})$ as $y \to \infty$ (it is even a $O(y^{-k-1})$ since, for every $j \in \mathbb{Z}$, one has $\int_j^{j+1} B_{k+1}(t)dt = 0$), one obtains the asymptotic expansion

$$[f_{1,\nu}]_{01}(-z^{-1}) \sim 2I(1,\nu)y + 2i\pi \sum_{r\geq 0} \frac{(-1)^r}{r+1}\phi_\nu^{(r)}(0)[B_{r+1}(x) + B_{r+1}(-x)]y^{-r},$$

$$(4.4.22)$$

in view of the definition (2.3.34) of $I(1,\nu)$.

Applying a $\frac{\partial}{\partial x}$ or $\frac{\partial}{\partial y}$-derivative (when $x \notin \mathbb{Z}$) is immediate, using a relation such as $\frac{\partial}{\partial y}\sum_n \chi_{1,\nu}\left(\frac{x+n}{y}\right) = -y^{-1}\sum_n f(n)$, with $f(t) = \frac{x+t}{y}\chi'_{1,\nu}\left(\frac{x+t}{y}\right)$, or $\frac{\partial}{\partial x}\sum_n \chi_{1,\nu}\left(\frac{x+n}{y}\right) = y^{-1}\sum_n \chi_{1,\nu}\left(\frac{x+n}{y}\right)$. The result obtained in this latter case, together with the fact that the function $[f_{1,\nu}]_{01}(-z^{-1})$ is continuous, even at points on the line from 0 to $i\infty$, makes it possible to extend the validity of the expansion (4.4.22) at points $z = x + iy$ with $x \in \mathbb{Z}$, showing at the same time that this expansion is uniform with respect to x.

What remains to be done is observing first that, since the function B_{r+1} has the parity of $r + 1$, this reduces to

$$[f_{1,\nu}]_{01}(-z^{-1}) \sim 2I(1,\nu)y - 4i\pi \sum_{j\geq 0} \frac{a_j}{2j+2}B_{2j+2}(x)y^{-2j-1} \quad (4.4.23)$$

with $a_j = \phi_\nu^{(2j+1)}(0)$, next computing the coefficients a_j. First, one has

$$\phi'_{1,\nu}(0) = \frac{i}{4\pi} \frac{\Gamma(\frac{\nu}{2})\Gamma(\frac{2-\nu}{2})}{\Gamma(\frac{1+\nu}{2})\Gamma(\frac{1-\nu}{2})} \frac{1-\nu^2}{4} \, {}_2F_1\left(\frac{3-\nu}{2}, \frac{3+\nu}{2}; 2; \frac{1}{2}\right):$$
(4.4.24)

we have already encountered this special value of a hypergeometric function just
after (2.3.45), which leads to the first value $2i\pi a_0 = -C(1,\nu)$. Next, using the
relation $B''_{2j+2} = (2j+1)(2j+2)B_{2j}$, one has

$$(\Delta + (2j+1)(2j+2))\left(B_{2j+2}(x)y^{-2j-1}\right) = -(2j+1)(2j+2)B_{2j}(x)y^{-2j+1},$$
(4.4.25)

so that

$$\left(\Delta - \frac{1-\nu^2}{4}\right)\left[\frac{a_j}{j+1}B_{2j+2}(x)y^{-2j-1}\right] = -2(2j+1)a_j B_{2j}(x)y^{-2j+1}$$
$$- \frac{(2j+\frac{3-\nu}{2})(2j+\frac{3+\nu}{2})}{j+1}a_j B_{2j+2}(x)y^{-2j-1}. \quad (4.4.26)$$

Now, the asymptotic expansion of the function $\left(\Delta - \frac{1-\nu^2}{4}\right)[f_\nu]_{01}(-z^{-1})$ must be
formally zero for $-1 < x < 1, x \neq 0$, which leads, for $j \geq 1$, to the relation

$$2(2j+1)a_j = \frac{(2j+\frac{-1-\nu}{2})(2j+\frac{-1+\nu}{2})}{j}a_{j-1}$$
(4.4.27)

and, finally, to the general formula

$$2i\pi a_j = -2^{2j}\frac{C(1,\nu)}{(2j+1)!}\frac{\Gamma(j+\frac{3-\nu}{4})\Gamma(j+\frac{3+\nu}{4})}{\Gamma(\frac{3-\nu}{4})\Gamma(\frac{3+\nu}{4})}.$$
(4.4.28)

We consider now the case when $\rho \neq 1$. We complete (4.4.15) by

$$\mathrm{Im}\left(\left(\begin{smallmatrix} a & 1 \\ -1 & 0 \end{smallmatrix}\right).(-z^{-1})\right) = y, \qquad \mathrm{Im}\left(\left(\begin{smallmatrix} 1 & 0 \\ c & 1 \end{smallmatrix}\right).(-z^{-1})\right) = \frac{y}{|c-z|^2},$$
(4.4.29)

obtaining

$$[f_{\rho,\nu}]_{01}(-z^{-1}) = \sum_{a\in\mathbb{Z}} y^{\frac{\rho-1}{2}}\chi_{\rho,\nu}^{\mathrm{even}}\left(\frac{-a+x}{y}\right) + \sum_{c\in\mathbb{Z}}\left(\frac{y}{|c-z|^2}\right)^{\frac{\rho-1}{2}}\chi_{\rho,\nu}^{\mathrm{even}}\left(\frac{c-x}{y}\right)$$
(4.4.30)

or, with the help of (2.3.52),

$$[f_{\rho,\nu}]_{01}(-z^{-1})$$
(4.4.31)
$$= y^{\frac{\rho-1}{2}}\sum_{n\in\mathbb{Z}}\chi_{\rho,\nu}^{\mathrm{even}}\left(\frac{x+n}{y}\right) + \frac{\Gamma(\frac{2+\rho-\nu}{4})\Gamma(\frac{2+\rho+\nu}{4})}{\Gamma(\frac{4-\rho-\nu}{4})\Gamma(\frac{4-\rho+\nu}{4})}y^{\frac{1-\rho}{2}}\sum_{n\in\mathbb{Z}}\chi_{2-\rho,\nu}^{\mathrm{even}}\left(\frac{x-n}{y}\right).$$

We need to find an analogue of (4.4.18), concentrating the singularities of $\chi_{\rho,\nu}(t)$ at $t = 0$ in a single sign function. Recall that the function $\chi^0_{\rho,\nu}$ is that defined in (2.3.19) and that (2.3.35) gives an expression of it as a sum of two terms, only the first of which is singular at 0: this term is

$$\frac{\Gamma(1-\nu)\Gamma(\frac{1-\rho}{2})}{\Gamma(\frac{2-\rho-\nu}{2})\Gamma(\frac{1-\nu}{2})}\left(\frac{-1-it}{2}\right)_+^{\frac{\rho-1}{2}} {}_2F_1\left(\frac{1-\nu}{2},\frac{1+\nu}{2};\frac{\rho+1}{2};\frac{1+it}{2}\right). \tag{4.4.32}$$

Since

$$\left(\frac{-1-it}{2}\right)_+^{\frac{\rho-1}{2}} = e^{i\pi\frac{1-\rho}{2}\operatorname{sign} t}\left(\frac{1+it}{2}\right)_+^{\frac{\rho-1}{2}}, \tag{4.4.33}$$

one obtains a decomposition

$$\chi_{\rho,\nu}(t) = e^{i\pi\frac{1-\rho}{2}\operatorname{sign} t}\xi_{\rho,\nu}(t) + \psi_{\rho,\nu}(t), \tag{4.4.34}$$

in which the functions $\xi_{\rho,\nu}$ and $\psi_{\rho,\nu}$ are C^∞ on the whole real line. Not forgetting the coefficient from $\chi^0_{\rho,\nu}$ to $\chi_{\rho,\nu}$, as given in (2.3.31), one finds that

$$\xi_{\rho,\nu}(t) = 2^{\nu-1}\pi^{-\frac{1}{2}}\frac{\Gamma(\frac{\nu}{2})\Gamma(1-\nu)}{\Gamma(\frac{1-\nu}{2})}\frac{\Gamma(\frac{1-\rho}{2})}{\Gamma(\frac{2-\rho+\nu}{2})\Gamma(\frac{2-\rho-\nu}{2})}$$

$$\times \left(\frac{1+it}{2}\right)_+^{\frac{\rho-1}{2}} {}_2F_1\left(\frac{1-\nu}{2},\frac{1+\nu}{2};\frac{\rho+1}{2};\frac{1+it}{2}\right). \tag{4.4.35}$$

The derivative of the second line, evaluated at $t = 0$, is [36, p. 41]

$$\frac{i}{2}\frac{d}{dz}\left(z^{\frac{\rho-1}{2}} {}_2F_1\left(\frac{1-\nu}{2},\frac{1+\nu}{2};\frac{\rho+1}{2};z\right)\right)(z = \frac{1}{2}) \tag{4.4.36}$$

$$= \frac{i(\rho-1)}{4}(\frac{1}{2})^{\frac{\rho-3}{2}} {}_2F_1\left(\frac{1-\nu}{2},\frac{1+\nu}{2};\frac{\rho-1}{2};\frac{1}{2}\right) = 2^{2-\rho}i\pi^{\frac{1}{2}}\frac{\Gamma(\frac{\rho+1}{2})}{\Gamma(\frac{\rho+\nu}{4})\Gamma(\frac{\rho-\nu}{4})}.$$

Hence,

$$\xi'_{\rho,\nu}(0) = 2^{\nu-\rho+1}i\frac{\Gamma(\frac{\nu}{2})\Gamma(1-\nu)}{\Gamma(\frac{1-\nu}{2})}\frac{\Gamma(\frac{1-\rho}{2})\Gamma(\frac{1+\rho}{2})}{\Gamma(\frac{2-\rho+\nu}{2})\Gamma(\frac{2-\rho-\nu}{2})\Gamma(\frac{\rho+\nu}{4})\Gamma(\frac{\rho-\nu}{4})}. \tag{4.4.37}$$

On the other hand,

$$\chi^{\operatorname{even}}_{\rho,\nu}(t) = \frac{i\pi}{\Gamma(\frac{1-\rho}{2})\Gamma(\frac{1+\rho}{2})}\xi^{\operatorname{odd}}_{\rho,\nu}(t)\operatorname{sign} t + \frac{\pi}{\Gamma(\frac{\rho}{2})\Gamma(\frac{2-\rho}{2})}\xi^{\operatorname{even}}_{\rho,\nu}(t) + \psi^{\operatorname{even}}_{\rho,\nu}(t). \tag{4.4.38}$$

The function accompanied by the factor $i\pi\operatorname{sign} t$ in this decomposition, which plays therefore the same role as the function denoted as $\phi_{1,\nu}(t)$ in (4.4.18), is the function

$$\phi_{\rho,\nu}(t) = \frac{1}{\Gamma(\frac{1-\rho}{2})\Gamma(\frac{1+\rho}{2})}\xi^{\operatorname{odd}}_{\rho,\nu}(t). \tag{4.4.39}$$

One has

$$\phi'_{\rho,\nu}(0) = 2^{\nu-\rho+1}i\frac{\Gamma(\frac{\nu}{2})\Gamma(1-\nu)}{\Gamma(\frac{1-\nu}{2})\Gamma(\frac{2-\rho+\nu}{2})\Gamma(\frac{2-\rho-\nu}{2})\Gamma(\frac{\rho+\nu}{4})\Gamma(\frac{\rho-\nu}{4})} \tag{4.4.40}$$

or, using (2.3.33),

$$2i\pi\phi'_{\rho,\nu}(0) = -C(\rho,\nu). \tag{4.4.41}$$

From this point on, the search for an asymptotic expansion of the function $[f_{\rho,\nu}]_{01}(-(x+iy)^{-1})$ as $y \to \infty$ is entirely similar to the proof of the case when $\rho = 1$, provided that we do not forget the powers of y which show in (4.4.31). The integral term is

$$y^{\frac{\rho+1}{2}}\int_{-\infty}^{\infty}\chi_{\rho,\nu}(t)dt + \frac{\Gamma(\frac{2+\rho-\nu}{4})\Gamma(\frac{2+\rho+\nu}{4})}{\Gamma(\frac{4-\rho-\nu}{4})\Gamma(\frac{4-\rho+\nu}{4})}y^{\frac{3-\rho}{2}}\int_{-\infty}^{\infty}\chi_{2-\rho,\nu}(t)dt$$

$$= I(\rho,\nu)y^{\frac{\rho+1}{2}} + \frac{\Gamma(\frac{2+\rho-\nu}{4})\Gamma(\frac{2+\rho+\nu}{4})}{\Gamma(\frac{4-\rho-\nu}{4})\Gamma(\frac{4-\rho+\nu}{4})}I(2-\rho,\nu)y^{\frac{3-\rho}{2}}. \tag{4.4.42}$$

Following the proof written in the case when $\rho = 1$, we obtain that, besides the integral term just made explicit, the complete asymptotic expansion of $[f_{\rho,\nu}]_{01}(-z^{-1})$ comprises an asymptotic series

$$-2i\pi\sum_{j\geq 0}\frac{B_{2j+2}(x)}{2j+2}\left[a_{\rho,j}y^{\frac{\rho-1}{2}-2j-1} + \frac{\Gamma(\frac{2+\rho-\nu}{4})\Gamma(\frac{2+\rho+\nu}{4})}{\Gamma(\frac{4-\rho-\nu}{4})\Gamma(\frac{4-\rho+\nu}{4})}a_{2-\rho,j}y^{\frac{1-\rho}{2}-2j-1}\right],$$

$$\tag{4.4.43}$$

where $2i\pi a_{\rho,0} = -C(\rho,\nu)$ and the other coefficients can be determined by the fact that the image of this asymptotic expansion under the operator $\Delta - \frac{1-\rho^2}{4}$ must be formally zero: note that, since $\rho - 1 \notin 2\mathbb{Z}$, the same must be true for each of the two terms in (4.4.43). As

$$\left(\Delta - \frac{1-\nu^2}{4}\right)\left[\frac{a_{\rho,j}}{j+1}B_{2j+2}(x)y^{\frac{\rho-1}{2}-2j-1}\right] = -2(2j+1)a_{\rho,j}B_{2j}(x)y^{\frac{\rho-1}{2}-2j+1}$$

$$- \frac{(2j+\frac{4-\rho-\nu}{2})(2j+\frac{4-\rho+\nu}{2})}{j+1}a_{\rho,j}B_{2j+2}(x)y^{\frac{\rho-1}{2}-2j-1}, \tag{4.4.44}$$

this leads to the relation, for $j \geq 1$,

$$2(2j+1)a_{\rho,j} = \frac{(2j-\frac{\nu+\rho}{2})(2j+\frac{\nu-\rho}{2})}{j}a_{\rho,j-1}, \tag{4.4.45}$$

and to the general formula

$$2i\pi a_{\rho,j} = -2^{2j}\frac{C(\rho,\nu)}{(2j+1)!}\frac{\Gamma(j+1-\frac{\nu+\rho}{4})\Gamma(j+1+\frac{\nu-\rho}{4})}{\Gamma(1-\frac{\nu+\rho}{4})\Gamma(1+\frac{\nu-\rho}{4})}. \tag{4.4.46}$$

Theorem 4.4.2 follows. $\square$

Remark 4.4.a. Explicitly, the expansion of $[f_{1,\nu}]_{01}$ and of its first-order derivatives start as follows: if $y \to \infty$ while $z = x + iy \in D$, one has

$$[f_{1,\nu}]_{01}(-z^{-1}) = 2I(1,\nu)y + C(1,\nu)(x^2 - |x| + \frac{1}{6})y^{-1} + O(y^{-3}),$$

$$\frac{\partial}{\partial x}\left([f_{1,\nu}]_{01}(-z^{-1})\right) = C(1,\nu)(2x - \operatorname{sign} x)y^{-1} + O(y^{-3}),$$

$$\frac{\partial}{\partial y}\left([f_{1,\nu}]_{01}(-z^{-1})\right) = 2I(1,\nu) - C(1,\nu)(x^2 - |x| + \frac{1}{6})y^{-2} + O(y^{-4}), \quad (4.4.47)$$

with the understanding that, when $x = 0$, only one-sided $\frac{\partial}{\partial x}$-derivatives are considered. When $\rho \neq 1$, extra powers of $y^{\frac{\rho-1}{2}}$ and $y^{\frac{1-\rho}{2}}$ enter the picture, while coefficients are of course more complicated.

We now turn to the study of the sum $[f_{\rho,\nu}]_{23}(-z^{-1})$, as defined in (4.4.2). Recall that the set $\Gamma_2 \cup \Gamma_3$ is precisely the one for which the parametrisation from Lemma 4.1.1 by pairs m,n with $m \neq 0$ is available. In the study of this term, it will not be necessary to set apart the case when $\rho = 1$.

When $\operatorname{Re}\nu < -1 - |\operatorname{Re}\rho - 1|$ and the standing assumptions (4.3.1) are satisfied, this sum is given by the convergent series

$$[f_{\rho,\nu}]_{23}(-z^{-1}) \tag{4.4.48}$$

$$= \sum_{\substack{m \in \mathbb{Z}^\times, n \in \mathbb{Z} \\ m|n(n+1)}} |m_2|^{\rho-1} \left(\frac{|mz-n|^2}{y}\right)^{\frac{1-\rho}{2}} \chi_{\rho,\nu}^{\mathrm{even}}\left(\frac{|n+\frac{1}{2}-mz|^2}{my} - \frac{1}{4m}\frac{|z|^2}{y}\right).$$

We shall analyze the behaviour, as $\operatorname{Im} z \to \infty$, of various error terms already known (from the last two sections) to admit analytic continuation for $\operatorname{Re}\nu < 1-|\operatorname{Re}\rho-1|$, thus reducing the problem to the analysis of the funtion $H_{\rho,\nu}(-z^{-1})$ in (4.3.21).

Note that here, contrary to what was the case in our discussion following (4.3.7), in which we only dealt with local estimates, it is impossible, in order to study the error term Err which will occur when substituting $\frac{|n+\frac{1}{2}-mz|^2}{y}$ for the argument $\frac{|mz-n|^2}{y}$ of the power function, to reduce the set of indices m,n by the further constraint $|n| < |m|$. On the other hand, there is the advantage that we may now assume that $z \in D$, in particular that $|x| \leq \frac{1}{2}$ and $y \leq |z| \leq \frac{2}{\sqrt{3}}y$. Finally, note that the error term Err is not present in the case when $\rho = 1$.

When $z \in D$ and $m \neq 0$, one has, for every $\delta > 0$,

$$\frac{|n+\frac{1}{2}-mz|^2}{|m|y} - \frac{1}{4m}\frac{|z|^2}{y} \geq \frac{(n+\frac{1}{2})^2 - |m(n+\frac{1}{2})| + m^2y^2}{|m|y} - \frac{y}{3|m|}$$

$$\geq \frac{(1-\delta^2)(n+\frac{1}{2})^2 + m^2(y^2 - \frac{1}{4\delta^2}) - \frac{y^2}{3}}{|m|y}$$

$$\geq (1 - \delta^2)\frac{(n + \frac{1}{2})^2}{2|m|y} + \frac{2}{3}(1 - \frac{1}{4\delta^2})|m|y : \qquad (4.4.49)$$

choosing $\delta \in]\frac{1}{2}, 1[$, one sees that the argument of the function $\chi^{even}_{\rho,\nu}$ in (4.4.48) dominates both $\frac{(n+\frac{1}{2})^2}{2|m|y}$ and $|m|y$, as well as $\frac{|n+\frac{1}{2}-mz|^2}{|m|y}$.

Since we are evaluating the error term Err defined above for the argument $-z^{-1}$, we now replace (4.3.8), (4.3.9) and (4.3.10) by

$$\left(\frac{|mz - n|^2}{y}\right)^{\frac{1-\rho}{2}} = \left(\frac{|n + \frac{1}{2} - mz|^2}{y}\right)^{\frac{1-\rho}{2}}(1 + w), \qquad (4.4.50)$$

and

$$w = \left(1 - \frac{n + \frac{1}{4} - mx}{|(n + \frac{1}{2})z + m|^2}\right)^{\frac{1-\rho}{2}} - 1, \qquad (4.4.51)$$

finally, with an absolute constant $C > 0$,

$$|w| \leq C\frac{|n| + |m|}{|n + \frac{1}{2} - mz|^2}. \qquad (4.4.52)$$

Hence,

$$\left|\left(\frac{|mz - n|^2}{y}\right)^{\frac{1-\rho}{2}} - \left(\frac{|n + \frac{1}{2} - mz|^2}{y}\right)^{\frac{1-\rho}{2}}\right|$$

$$\leq Cy^{-1}\left(\frac{|n + \frac{1}{2} - mz|^2}{|m|y}\right)^{\frac{-1-\mathrm{Re}\,\rho}{2}}\left[|m|^{\frac{1-\mathrm{Re}\,\rho}{2}} + |m|^{\frac{-1-\mathrm{Re}\,\rho}{2}}|n|\right], \qquad (4.4.53)$$

and, using (2.3.32),

$$|\mathrm{Err}| \leq C\sum_{\substack{m \in \mathbb{Z}^\times, n \in \mathbb{Z} \\ m|n(n+1)}}|\frac{m_1}{m_2}|^{\frac{1-\mathrm{Re}\,\rho}{2}}\left(\frac{|n + \frac{1}{2} - mz|^2}{|m|y}\right)^{\frac{\mathrm{Re}\,\nu-3}{2}}\left[y^{-1} + \frac{|n|}{|m|y}\right]. \qquad (4.4.54)$$

We need a lemma.

Lemma 4.4.3. *If $\mathrm{Re}\,\nu < 1 - |\mathrm{Re}\,\rho - 1|$ and $k \geq 1$, there is a constant $C > 0$ such that, for every $z \in \overline{D}$,*

$$S: = \sum_{\substack{m \in \mathbb{Z}^\times, n \in \mathbb{Z} \\ m|n(n+1)}}|\frac{m_1}{m_2}|^{\frac{1-\mathrm{Re}\,\rho}{2}}\left(\frac{|n + \frac{1}{2} - mz|^2}{my} - \frac{1}{4m}\frac{|z|^2}{y}\right)^{\frac{\mathrm{Re}\,\nu-1}{2}-k}$$

$$\leq Cy^{-\frac{|\mathrm{Re}\,\rho-1|}{2}+1-k}. \qquad (4.4.55)$$

Proof. In view of (4.4.49), one has

$$
S \leq \sum \left(\frac{|n + \frac{1}{2} - mz|^2}{my} \right)^{1-k} \left|\frac{m_1}{m_2}\right|^{\frac{1-\operatorname{Re}\rho}{2}} \left(\frac{|n + \frac{1}{2} - mz|^2}{my} \right)^{\frac{\operatorname{Re}\nu - 3}{2}}
$$

$$
\leq C \sum (|m|y)^{1-k} \left|\frac{m_1}{m_2}\right|^{\frac{1-\operatorname{Re}\rho}{2}} \left(\frac{(n + \frac{1}{2})^2}{my} + |m|y \right)^{\frac{\operatorname{Re}\nu - 3}{2}}
$$

$$
= Cy^{1-k}(S_1 + S_2), \tag{4.4.56}
$$

where

$$
S_1 + S_2 = \sum |m|^{1-k} \left|\frac{m_1}{m_2}\right|^{\frac{1-\operatorname{Re}\rho}{2}} \left(\frac{(n + \frac{1}{2})^2}{my} + |m|y \right)^{\frac{\operatorname{Re}\nu - 3}{2}}, \tag{4.4.57}
$$

and S_1 is obtained by keeping only the terms such that $|n + \frac{1}{2}| \leq |m|y$. One thus has

$$
S_1 \leq C \sum |m|^{1-k} \left|\frac{m_1}{m_2}\right|^{\frac{1-\operatorname{Re}\rho}{2}} (|m|y)^{\frac{\operatorname{Re}\nu - 3}{2}}
$$

$$
\leq Cy^{-\frac{|\operatorname{Re}\rho - 1|}{2}} \sum |m|^{1-k} \left|\frac{m_1}{m_2}\right|^{\frac{1-\operatorname{Re}\rho}{2}} |m|^{-\frac{|\operatorname{Re}\rho - 1|}{2}} (|m|y)^{\frac{\operatorname{Re}\nu - 3 + |\operatorname{Re}\rho - 1|}{2}}
$$

$$
\leq Cy^{-\frac{|\operatorname{Re}\rho - 1|}{2}} \sum |m|^{1-k} |n + \frac{1}{2}|^{\frac{\operatorname{Re}\nu - 3 + |\operatorname{Re}\rho - 1|}{2}} : \tag{4.4.58}
$$

the (double) series converges since $\frac{\operatorname{Re}\nu - 3 + |\operatorname{Re}\rho - 1|}{2} < -1$ and, for $n \neq 0, -1$, the number of integers m dividing $n(n+1)$ is a $O(|n|^\varepsilon)$ for every $\varepsilon > 0$. On the other hand,

$$
S_2 \leq C \sum |m|^{1-k} \left|\frac{m_1}{m_2}\right|^{\frac{1-\operatorname{Re}\rho}{2}} \left(\frac{(n + \frac{1}{2})^2}{my} \right)^{\frac{\operatorname{Re}\nu - 3}{4}}
$$

$$
\leq Cy^{\frac{3-\operatorname{Re}\nu}{2}} \sum |m|^{1-k} |m|^{\frac{3-\operatorname{Re}\nu + |\operatorname{Re}\rho - 1|}{2}} |n + \frac{1}{2}|^{\operatorname{Re}\nu - 3}
$$

$$
\leq Cy^{\frac{3-\operatorname{Re}\nu}{2}} \sum |m|^{1-k} \left(\frac{|n + \frac{1}{2}|}{y} \right)^{\frac{3-\operatorname{Re}\nu + |\operatorname{Re}\rho - 1|}{2}} |n + \frac{1}{2}|^{\operatorname{Re}\nu - 3}
$$

$$
= Cy^{-\frac{|\operatorname{Re}\rho - 1|}{2}} \sum |m|^{1-k} |n + \frac{1}{2}|^{\frac{\operatorname{Re}\nu - 3 + |\operatorname{Re}\rho - 1|}{2}} : \tag{4.4.59}
$$

we arrive at the same sum as in (4.4.58), and the lemma follows. $\qquad\square$

    Coming back to the question of bounding Err, equation (4.4.54) decomposes it into two parts, in view of the last factor $y^{-1} + \frac{|n|}{|m|y}$ on the right-hand side. To the first term, one can apply Lemma 4.4.3 directly, with $k = 1$, giving it a bound $Cy^{-1-\frac{|\operatorname{Re}\rho - 1|}{2}}$. Only the S_2-part of the proof of the lemma does not apply directly to the second term of the right-hand side of (4.4.54): however, writing

$\frac{|n|}{|m|y} \leq \frac{3}{2}(|m|y)^{-\frac{1}{2}} \left(\frac{(n+\frac{1}{2})^2}{|m|y}\right)^{\frac{1}{2}}$, and following the estimate of S_2 in (4.4.59) with $k = \frac{1}{2}$ only and the exponent $\frac{\operatorname{Re}\nu-3}{2}$ replaced by $\frac{\operatorname{Re}\nu-2}{2}$, one still arrives at the estimate $|\mathrm{Err}| \leq Cy^{-\frac{|\operatorname{Re}\rho-1|}{2}}$. Up to an error term majorized by $Cy^{-\frac{|\operatorname{Re}\rho-1|}{2}}$ when $\operatorname{Re}\nu < 1 - |\operatorname{Re}\rho - 1|$, we can thus replace the series $[f_{\rho,\nu}]_{23}(-z^{-1})$ by the series

$$[f_{\rho,\nu}]_{23}^{\sharp}(-z^{-1}) \tag{4.4.60}$$

$$= \sum_{\substack{m\in\mathbb{Z}^\times, n\in\mathbb{Z} \\ m|n(n+1)}} |m_2|^{\rho-1} \left(\frac{|n+\frac{1}{2}-mz|^2}{y}\right)^{\frac{1-\rho}{2}} \chi_{\rho,\nu}^{\mathrm{even}}\left(\frac{|n+\frac{1}{2}-mz|^2}{my} - \frac{1}{4m}\frac{|z|^2}{y}\right).$$

From (4.4.49), the argument t of the function $\chi_{\rho,\nu}^{\mathrm{even}}(t)$ here is bounded away from zero. Then, we substitute for this function its main term

$$2^{\nu-1}\pi^{-\frac{1}{2}}\frac{\Gamma(\frac{\nu}{2})}{\Gamma(\frac{2-\rho+\nu}{2})} \times \frac{1}{2}\left[\left(\frac{-it}{2}\right)_+^{\frac{\rho+\nu-2}{2}} + \left(\frac{-it}{2}\right)_+^{\frac{\rho+\nu-2}{2}}\right]$$

$$= 2^{\frac{\nu-\rho}{2}}\pi^{\frac{1}{2}}\frac{\Gamma(\frac{\nu}{2})}{\Gamma(\frac{2-\rho+\nu}{2})\Gamma(\frac{\rho+\nu}{4})\Gamma(\frac{4-\rho-\nu}{4})}|t|^{\frac{\rho+\nu-2}{2}} \tag{4.4.61}$$

as obtained from (2.3.31) when replacing the hypergeometric function by 1 and powers of $-1 \pm it$ by powers of $\pm it$: the error committed in this approximation is bounded by $C(1+|t|)^{\frac{\operatorname{Re}(\rho+\nu)-4}{2}}$. Also, we write

$$|m_2|^{\rho-1}\left(\frac{|n+\frac{1}{2}-mz|^2}{y}\right)^{\frac{1-\rho}{2}} = \left|\frac{m_1}{m_2}\right|^{\frac{1-\rho}{2}}\left(\frac{|n+\frac{1}{2}-mz|^2}{|m|y}\right)^{\frac{1-\rho}{2}}. \tag{4.4.62}$$

From Lemma 4.4.3, the error induced on the evaluation of $[f_{\rho,\nu}]_{23}^{\sharp}(-z^{-1})$ by the approximation above is bounded, again, by $Cy^{-\frac{|\operatorname{Re}\rho-1|}{2}}$ when $\operatorname{Re}\nu < 1 - |\operatorname{Re}\rho - 1|$. Up to such an error term, we can thus write

$$[f_{\rho,\nu}]_{23}(-z^{-1}) \sim 2^{\frac{\nu-\rho}{2}}\pi^{\frac{1}{2}}$$

$$\times \frac{\Gamma(\frac{\nu}{2})}{\Gamma(\frac{2-\rho+\nu}{2})\Gamma(\frac{\rho+\nu}{4})\Gamma(\frac{4-\rho-\nu}{4})} \sum_{\substack{m\in\mathbb{Z}^\times, n\in\mathbb{Z} \\ m|n(n+1)}} \left|\frac{m_1}{m_2}\right|^{\frac{1-\rho}{2}}\left(\frac{|n+\frac{1}{2}-mz|^2}{|m|y}\right)^{\frac{\nu-1}{2}}$$

$$= 2^{\frac{\nu-\rho+2}{2}}\pi^{\frac{1}{2}}\frac{\Gamma(\frac{\nu}{2})}{\Gamma(\frac{2-\rho+\nu}{2})\Gamma(\frac{\rho+\nu}{4})\Gamma(\frac{4-\rho-\nu}{4})}H_{\rho,\nu}(-z^{-1}). \tag{4.4.63}$$

Contrary to the error terms just analyzed, this main term does not admit a continuation to the full half-plane $\operatorname{Re}\nu < 1 - |\operatorname{Re}\rho - 1|$. According to Theorem 4.3.4, continuation is possible if, besides the usual non-integrality conditions, a certain discrete set of points of an arithmetical nature is avoided.

Using the Fourier expansion (4.3.21) of $H_{\rho,\nu}(-z^{-1})$, one obtains the main term of $[f_{\rho,\nu}]_{23}(-z^{-1})$ as

$$2^{\frac{\nu-\rho+2}{2}}\pi^{\frac{1}{2}}\frac{\Gamma(\frac{\nu}{2})}{\Gamma(\frac{2-\rho+\nu}{2})\Gamma(\frac{4-\rho-\nu}{4})\Gamma(\frac{\rho+\nu}{4})}\pi^{\frac{1}{2}}\frac{\Gamma(-\frac{\nu}{2})}{\Gamma(\frac{1-\nu}{2})}\frac{\zeta(\frac{\rho-\nu}{2})\zeta(\frac{2-\rho-\nu}{2})}{\zeta(1-\nu)}y^{\frac{1+\nu}{2}}. \quad (4.4.64)$$

After elementary manipulations using the link between the functions ζ and ζ^*, as well as the duplication formula of the zeta function, one can write this as

$$2^{1-\rho}\pi^{\frac{1}{2}}\frac{\Gamma(\frac{\nu}{2})\Gamma(\frac{2-\nu}{2})}{\Gamma(\frac{2-\rho+\nu}{2})\Gamma(\frac{2-\rho-\nu}{2})\Gamma(\frac{\rho+\nu}{4})\Gamma(\frac{\rho-\nu}{4})}\frac{\zeta^*(\frac{\rho-\nu}{2})\zeta^*(\frac{\rho+\nu}{2})}{\zeta^*(1-\nu)}y^{\frac{1+\nu}{2}} \quad (4.4.65)$$

or, using (2.3.33),

$$[f_{\rho,\nu}]_{23}(-z^{-1}) = -\frac{C(\rho,\nu)}{\nu}\frac{\zeta^*(\frac{\rho-\nu}{2})\zeta^*(\frac{\rho+\nu}{2})}{\zeta^*(1-\nu)}y^{\frac{1+\nu}{2}}+O(y^{-\frac{|\rho-1|}{2}}). \quad (4.4.66)$$

Coupling this with Theorem 4.4.2, one obtains the following.

Theorem 4.4.4. *Besides* (4.3.1), *assume that* $\mathrm{Re}\,\nu < 1 - |\mathrm{Re}\,\rho - 1|$ *and that* ν *is distinct from all non-trivial zeros of the zeta function, and from all points* $i\lambda_p$ *with* $\frac{1+\lambda_p^2}{4}$ *in the even part of the discrete spectrum of* Δ. *One then has, for* $z = x+iy \in \overline{D}$ *and* $y \to \infty$,

$$f_{\rho,\nu}(z) = I(\rho,\nu)y^{\frac{\rho+1}{2}}+\frac{\Gamma(\frac{2+\rho-\nu}{4})\Gamma(\frac{2+\rho+\nu}{4})}{\Gamma(\frac{4-\rho-\nu}{4})\Gamma(\frac{4-\rho+\nu}{4})}I(2-\rho,\nu)y^{\frac{3-\rho}{2}}$$

$$-\frac{C(\rho,\nu)}{\nu}\frac{\zeta^*(\frac{\rho-\nu}{2})\zeta^*(\frac{\rho+\nu}{2})}{\zeta^*(1-\nu)}y^{\frac{1+\nu}{2}}+O(y^{-\frac{|\rho-1|}{2}}). \quad (4.4.67)$$

One can also take the $\frac{\partial}{\partial y}$-derivative or $\frac{\partial}{\partial x}$-half-derivative (left or right if $x \in \mathbb{Z}$) of this expansion, getting in the first case a remainder $O(y^{-\frac{|\rho-1|}{2}-1})$, in the second case a remainder $O(y^{\frac{|\rho-1|}{2}-1})$.

When $\rho = 1$, the error term $O(y^{-\frac{|\rho-1|}{2}})$ can be improved to $O(y^{-1})$, while the error term in the $\frac{\partial}{\partial y}$ (resp. $\frac{\partial}{\partial x}$-derivative) improves to $O(y^{-1})$ (resp. $O(y^{-2})$).

Corollary 4.4.5. *Noting that both functions* $f_{\rho,\nu}(z)$ *and* $f_{\rho,-\nu}(z)$ *are well-defined as meromorphic functions of* ν *in the strip* $|\mathrm{Re}\,\nu| < 1 - |\mathrm{Re}\,\rho - 1|$, *one has the identity*

$$f_{\rho,\nu} + f_{\rho,-\nu} = c_\rho(\nu)E_{\frac{1+\nu}{2}}, \quad (4.4.68)$$

with

$$c_\rho(\nu) = -\frac{C(\rho,\nu)}{\nu}\frac{\zeta^*(\frac{\rho-\nu}{2})\zeta^*(\frac{\rho+\nu}{2})}{\zeta^*(1-\nu)}. \quad (4.4.69)$$

Proof. Both functions $f_{\rho,\nu}$ and $f_{\rho,-\nu}$ make sense as meromorphic functions in the strip indicated. On the other hand, Theorem 4.2.4 and the observation that $C(\rho,\nu) = -C(\rho,-\nu)$ show that $\left(\Delta - \frac{1-\nu^2}{4}\right)(f_{\rho,\nu} + f_{\rho,-\nu}) = 0$ in the distribution sense in Π. Hence, with the exception of a discrete set of values of ν, the function $f_{\rho,\nu} + f_{\rho,-\nu}$ can only be a multiple of the Eisenstein series $E_{\frac{1+\nu}{2}}$. What remains to be done is computing the coefficient $c_\rho(\nu)$, for which we may assume that $\operatorname{Re}\nu = 0$, also that $\frac{1-\nu^2}{4}$ does not belong to the discrete spectrum of Δ. Since $I(\rho,-\nu) = -I(\rho,\nu)$, we are left with the estimate

$$f_{\rho,\nu}(iy) + f_{\rho,-\nu}(iy) \tag{4.4.70}$$

$$= -\frac{C(\rho,\nu)}{\nu}\,\frac{\zeta^*(\frac{\rho-\nu}{2})\zeta^*(\frac{\rho+\nu}{2})}{\zeta^*(1-\nu)\zeta^*(1+\nu)}\left[\zeta^*(1+\nu)y^{\frac{1+\nu}{2}} + \zeta^*(1-\nu)y^{\frac{1-\nu}{2}}\right] + O(y^{-\frac{|\operatorname{Re}\rho-1|}{2}}),$$

where one recognizes the expression $\zeta^*(1+\nu)y^{\frac{1+\nu}{2}} + \zeta^*(1-\nu)y^{\frac{1-\nu}{2}}$ as the so-called constant term of the function $E^*_{\frac{1+\nu}{2}}(z) = \zeta^*(1+\nu)E_{\frac{1+\nu}{2}}(z)$. This gives the value of $c_\rho(\nu)$.   □

The following proposition is only meant as a verification, since a direct proof of it is the classical way [4, p. 107] of establishing the continuation of the L-function associated to a Maass form. As the residues of the function $H_{\rho,\nu}(z)$, or of the function $f_{\rho,\nu}(z)$, were obtained, in Theorem 4.3.4, as a consequence of the value of the residues of the function $\zeta_k(s,t)$, Proposition 4.4.6 may be regarded as a confirmation of the more computational aspects of Theorem 3.6.2.

Proposition 4.4.6. *Let $\mathcal{M}$ be a cusp-form of even type for the eigenvalue $\frac{1+\lambda_\rho^2}{4}$. The integral of its restriction to D against the measure $ds_\Sigma^{(\rho)}$ introduced in Theorem 4.2.4 is given by the equation*

$$\frac{1}{2}\int_1^\infty \mathcal{M}(iy)\left(y^{\frac{1-\rho}{2}} + y^{\frac{\rho-1}{2}}\right)\frac{dy}{y} = \frac{\pi^{-\frac{\rho}{2}}}{4}\Gamma(\frac{\rho+i\lambda_\rho}{4})\Gamma(\frac{\rho-i\lambda_\rho}{4})L(\frac{\rho}{2},\mathcal{M})$$

$$= \frac{1}{4}L^*(\frac{\rho}{2},\mathcal{M}). \tag{4.4.71}$$

Proof. As a consequence of (4.3.4), one has

$$\operatorname{Res}_{\nu=i\lambda_\rho}f_{\rho,\nu}$$

$$= 2^{\frac{\nu-\rho+2}{2}}\pi^{\frac{1}{2}}\frac{\Gamma(\frac{\nu}{2})}{\Gamma(\frac{2-\rho+\nu}{2})\Gamma(\frac{4-\rho-\nu}{4})\Gamma(\frac{\rho+\nu}{4})}\operatorname{Res}_{\nu=i\lambda_\rho}(z\mapsto H_{\rho,\nu}(-z^{-1})): \tag{4.4.72}$$

using Theorem 4.3.4, one obtains

$$\operatorname{Res}_{\nu=i\lambda_\rho}f_{\rho,\nu} = -2^{\frac{2-\rho+i\lambda_\rho}{2}}\pi^{\frac{\rho}{2}}\frac{\Gamma(\frac{i\lambda_\rho}{2})\Gamma(-\frac{i\lambda_\rho}{2})\Gamma(\frac{2-\rho+i\lambda_\rho}{4})}{\Gamma(\frac{2-\rho+i\lambda_\rho}{2})\Gamma(\frac{\rho+i\lambda_\rho}{4})\Gamma(\frac{4-\rho-i\lambda_\rho}{4})\Gamma(\frac{\rho-i\lambda_\rho}{4})}\mathcal{M}_\rho^\rho$$

$$= -2\pi^{\frac{\rho+1}{2}} \frac{\Gamma(\frac{i\lambda_p}{2})\Gamma(-\frac{i\lambda_p}{2})}{\Gamma(\frac{\rho+i\lambda_p}{4})\Gamma(\frac{\rho-i\lambda_p}{4})\Gamma(\frac{4-\rho+i\lambda_p}{4})\Gamma(\frac{4-\rho-i\lambda_p}{4})} \mathcal{M}_p^\rho, \quad (4.4.73)$$

with

$$\mathcal{M}_p^\rho = \sum_j L(\frac{2-\rho}{2}, \mathcal{M}_{p,j})\mathcal{M}_{p,j} \qquad (4.4.74)$$

(recall that we did not have to put a bar on the functions $\mathcal{M}_{p,j}$ inside the L-function since, taking advantage of Hecke's theory, we chose to have their Fourier coefficients real).

Then, the scalar product $(\mathcal{M}|f_{\rho,\nu}) = (\mathcal{M}|f_{\rho,\nu})_{L^2(\Gamma\backslash\Pi)}$ (recall that the scalar product is antilinear with respect to the function on the left) has a (possible, since $L(\frac{2-\rho}{2}, \mathcal{M})$ could well be zero) simple pole at $\nu = i\lambda_p$, and

$$\mathrm{Res}_{\nu=i\lambda_p}(\overline{\mathcal{M}}|f_{\rho,\nu})$$

$$= -2\pi^{\frac{\rho+1}{2}} \frac{\Gamma(\frac{i\lambda_p}{2})\Gamma(-\frac{i\lambda_p}{2})}{\Gamma(\frac{\rho+i\lambda_p}{4})\Gamma(\frac{\rho-i\lambda_p}{4})\Gamma(\frac{4-\rho+i\lambda_p}{4})\Gamma(\frac{4-\rho-i\lambda_p}{4})} L(\frac{2-\rho}{2}, \mathcal{M}). \quad (4.4.75)$$

According to Theorem 4.2.4, one has

$$(\overline{\mathcal{M}}|(\Delta - \frac{1-\nu^2}{4})f_{\rho,\nu}) = C(\rho,\nu) \int_1^\infty \left(y^{\frac{1-\rho}{2}} + y^{\frac{\rho-1}{2}} \right) \mathcal{M}(iy)\frac{dy}{y} \qquad (4.4.76)$$

with $C(\rho,\nu)$ as given in (2.3.33). On the other hand, this is the same as

$$((\Delta - \frac{1-\bar{\nu}^2}{4})\overline{\mathcal{M}}|f_{\rho,\nu}) = \frac{\lambda_p^2+\nu^2}{4}(\overline{\mathcal{M}}|f_{\rho,\nu}), \qquad (4.4.77)$$

so that

$$(\overline{\mathcal{M}}|f_{\rho,\nu}) = \frac{4}{\lambda_p^2+\nu^2}C(\rho,\nu) \int_1^\infty \left(y^{\frac{1-\rho}{2}} + y^{\frac{\rho-1}{2}} \right) \mathcal{M}(iy)\frac{dy}{y} : \qquad (4.4.78)$$

taking the residue at $\nu = i\lambda_p$, we obtain

$$-2\pi^{\frac{\rho+1}{2}} \frac{\Gamma(\frac{i\lambda_p}{2})\Gamma(-\frac{i\lambda_p}{2})}{\Gamma(\frac{\rho+i\lambda_p}{4})\Gamma(\frac{\rho-i\lambda_p}{4})\Gamma(\frac{4-\rho+i\lambda_p}{4})\Gamma(\frac{4-\rho-i\lambda_p}{4})} L(\frac{2-\rho}{2}, \mathcal{M})$$

$$= \frac{2}{i\lambda_p}C(\rho, i\lambda_p) \int_1^\infty \left(y^{\frac{1-\rho}{2}} + y^{\frac{\rho-1}{2}} \right) \mathcal{M}(iy)\frac{dy}{y} \qquad (4.4.79)$$

where one has, from (2.3.33),

$$-\frac{2}{i\lambda_p}C(\rho, i\lambda_p) = 2^{2-\rho}\pi^{\frac{1}{2}} \frac{\Gamma(\frac{i\lambda_p}{2})\Gamma(-\frac{i\lambda_p}{2})}{\Gamma(\frac{2-\rho+i\lambda_p}{2})\Gamma(\frac{2-\rho-i\lambda_p}{2})\Gamma(\frac{\rho+i\lambda_p}{4})\Gamma(\frac{\rho-i\lambda_p}{4})}, \qquad (4.4.80)$$

which proves (4.4.71), with another use of the duplication formula. $\square$

4.5   The Roelcke-Selberg expansion of $f_{\rho,\nu}$

In this section, we make the Roelcke-Selberg expansion of $f_{\rho,\nu}$ (extending the notion since $f_{\rho,\nu} \notin L^2(D)$) explicit. To make the automorphic function $f_{\rho,\nu}$ more manageable, we subtract from it, if $\rho \neq 1$, a linear combination of the Eisenstein series $E_{\frac{\rho+1}{2}}$ and $E_{\frac{3-\rho}{2}}$; if $\rho = 1$, we replace the inexistent Eisenstein series E_1 by some substitute $E_1^\natural$.

Lemma 4.5.1. *The automorphic function*

$$E_1^\natural(z) = \frac{1}{2} \lim_{\varepsilon \to 0} \left(E_{1+\frac{\varepsilon}{2}}(z) + E_{1-\frac{\varepsilon}{2}}(z) \right) \tag{4.5.1}$$

is given for some constant G^o by the identity

$$E_1^\natural(z) = y - \frac{3}{\pi} \log y + G^o + \frac{6}{\pi} \sum_{n \neq 0} \frac{\sigma_1(|n|)}{|n|} e^{-2\pi|n|y} e^{2i\pi nx}. \tag{4.5.2}$$

Proof. Looking back at the expansion (3.1.6) of $E_s(z)$ (divide both sides of the equation by $\zeta^*(2s)$), one sees that it is only the coefficient of y^{1-s} that becomes infinite when $s = 1$, because of the factor $\Gamma(2 - 2s)$: since a simple pole only is involved, the singularity disappears from the sum $E_s(z) + E_{2-2s}(z)$. Setting the regular terms apart, one thus has

$$E_1^\natural(z) = y + F_1(z) + \frac{6}{\pi} \sum_{n \neq 0} \frac{\sigma_1(|n|)}{|n|} e^{-2\pi|n|y} e^{2i\pi nx}, \tag{4.5.3}$$

with

$$F_1(z) = \frac{1}{2} \lim_{\mu \to 1} \left[\frac{\zeta^*(\mu - 1)}{\zeta^*(3 - \mu)} y^{\frac{\mu-1}{2}} + \frac{\zeta^*(1 - \mu)}{\zeta^*(1 + \mu)} y^{\frac{1-\mu}{2}} \right]$$

$$= \lim_{\mu \to 1} \left(G(\mu) + \frac{\log y}{4} H(\mu) \right), \tag{4.5.4}$$

where

$$G(\mu) = \frac{1}{2} \left[\frac{\zeta^*(\mu - 1)}{\zeta^*(3 - \mu)} + \frac{\zeta^*(1 - \mu)}{\zeta^*(1 + \mu)} \right], \quad H(\mu) = (1 - \mu) \left[\frac{\zeta^*(1 - \mu)}{\zeta^*(1 + \mu)} - \frac{\zeta^*(\mu - 1)}{\zeta^*(3 - \mu)} \right]. \tag{4.5.5}$$

Since $\zeta(0) = -\frac{1}{2}$ and $\zeta^*(2) = \frac{\pi}{6}$, one sees that $H(\mu) \to -\frac{12}{\pi}$, so that the second term of $F_1(z)$ has the limit $-\frac{3}{\pi} \log y$. This proves Lemma 4.5.1.

Note that

$$G^o = G(1) = \frac{d}{d\mu} \left(\frac{(\mu - 1)\zeta^*(1 - \mu)}{\zeta^*(1 + \mu)} \right) (\mu = 1), \tag{4.5.6}$$

an expression useful later.   □

Definition 4.5.2. Assume that the pair ρ, ν satisfies the conditions of Theorem 4.3.4 giving $f_{\rho,\nu}$ a meaning. If $\rho \neq 1$, we define

$$f_{\rho,\nu}^{\sharp}(z) = f_{\rho,\nu}(z) - I(\rho,\nu)E_{\frac{1+\rho}{2}}(z) - \frac{\Gamma(\frac{2+\rho-\nu}{4})\Gamma(\frac{2+\rho+\nu}{4})}{\Gamma(\frac{4-\rho-\nu}{4})\Gamma(\frac{4-\rho+\nu}{4})}I(2-\rho,\nu)E_{\frac{3-\rho}{2}}(z); \quad (4.5.7)$$

we also set

$$f_{1,\nu}^{\sharp}(z) = f_{1,\nu}(z) - 2I(1,\nu)E_1^{\flat}(z). \quad (4.5.8)$$

Note that, since $0 < \operatorname{Re}\rho < 2$, $\zeta^*(1+\rho)$ and $\zeta^*(3-\rho)$ are nonzero, so that $E_{\frac{1+\rho}{2}}$ and $E_{\frac{3-\rho}{2}}$ are meaningful if $\rho \neq 1$. The functions so defined are continuous and automorphic. If $\rho \neq 1$, one has the estimate, as $y \to \infty$,

$$f_{\rho,\nu}^{\sharp}(x+iy) = -\frac{C(\rho,\nu)}{\nu}\frac{\zeta^*(\frac{\rho-\nu}{2})\zeta^*(\frac{\rho+\nu}{2})}{\zeta^*(1-\nu)}y^{\frac{1+\nu}{2}} + O(y^{\frac{|\rho-1|}{2}}); \quad (4.5.9)$$

when $\rho = 1$,

$$f_{1,\nu}^{\sharp}(x+iy) = \frac{6}{\pi}I(1,\nu)\log y - \frac{C(1,\nu)}{\nu}\frac{(\zeta^*(\frac{1-\nu}{2}))^2}{\zeta^*(1-\nu)}y^{\frac{1+\nu}{2}} + O(1); \quad (4.5.10)$$

in both cases, one has $\frac{\partial}{\partial y}f_{\rho,\nu}^{\sharp}(x+iy) = O(y^{\frac{-1+\operatorname{Re}\nu}{2}})$. To see this, one only has to use (4.4.67) and the expansion (3.1.6) of Eisenstein series: it is the second part of the "constant term" of the Fourier expansions of $E_{\frac{1+\rho}{2}}(z)$ and $E_{\frac{3-\rho}{2}}(z)$ that produces the error terms bounded by $Cy^{\frac{|\rho-1|}{2}}$. These expansions are uniform relative to x. In particular, whether $\rho \neq 1$ or not, the function $f_{\rho,\nu}^{\sharp}$ is square-integrable in the fundamental domain when $\operatorname{Re}\nu < 0$, or when $\frac{\zeta^*(\frac{\rho-\nu}{2})\zeta^*(\frac{\rho+\nu}{2})}{\zeta^*(1-\nu)} = 0$.

We wish to introduce substitutes for the inexistent scalar products of a function such as $E_{\frac{1+\rho}{2}}$ (or $E_1^{\flat}$) against Eisenstein series $E_{\frac{1+i\lambda}{2}}$. On the other hand, since $\Delta E_{\frac{1+\rho}{2}} = \frac{1-\rho^2}{4}E_{\frac{1+\rho}{2}}$ and $\Delta E_1^{\flat} = -\frac{3}{\pi}$, it is immediate, using the self-adjointness of Δ in $L^2(\Gamma\backslash\Pi)$, that the scalar products of $E_{\frac{1+\rho}{2}}$ or $E_1^{\flat}$ against a cusp-form exist and are zero.

To avoid divergences, we must introduce as is usual, for every $Y > 1$, the truncated domain

$$D_Y = \{z = x+iy \in \Pi: \ -\frac{1}{2} < x < \frac{1}{2}, |z| > 1, y < Y\}. \quad (4.5.11)$$

We then define

$$\Psi(\rho, i\lambda) = \lim_{Y \to \infty}\left[\int_1^Y y^{\frac{\rho-1}{2}}E_{\frac{1+i\lambda}{2}}(iy)\frac{dy}{y}\right.$$
$$\left. - \int_{D_Y} E_{\frac{1+\rho}{2}}(x+iy)E_{\frac{1+i\lambda}{2}}(x+iy)\frac{dxdy}{y^2}\right] \quad (4.5.12)$$

if $\rho \neq 1$ and

$$\Psi(1, i\lambda) = \lim_{Y \to \infty} \left(\int_1^Y E_{\frac{1+i\lambda}{2}}(iy) \frac{dy}{y} - \int_{D_Y} E_1^\natural(x+iy) E_{\frac{1+i\lambda}{2}}(x+iy) \frac{dxdy}{y^2} \right).$$

(4.5.13)

That these limits exist and are finite follows from (3.1.6) and (4.5.1): we have simply (with a global change of sign) subtracted, up to a quantity with a finite limit as $Y \to \infty$, the integrals on D_Y associated with the first term $y^{\frac{1+\rho}{2}}$, or y, from the Fourier expansion of $E_{\frac{1+\rho}{2}}$ or of $E_1^\natural$: note, since $\mathrm{Re}\,\rho > 0$, that the other part from the "constant term", to wit a multiple of $y^{\frac{1-\rho}{2}}$, or $\log y$, is integrable on D, even after it has been multiplied by $y^{\frac{1}{2}}$.

We now look for the Roelcke-Selberg expansion of $f = f_{\rho,\nu}^\sharp$, with $\mathrm{Re}\,\nu < 0$.

Lemma 4.5.3. *One has*

$$\left(\Delta - \frac{1-\nu^2}{4}\right) f_{\rho,\nu}^\sharp = C(\rho, \nu) \left[2ds_\Sigma^{(\rho)} - E_{\frac{1+\rho}{2}} - E_{\frac{3-\rho}{2}} \right], \qquad \rho \neq 1,$$

$$\left(\Delta - \frac{1-\nu^2}{4}\right) f_{1,\nu}^\sharp = 2C(1, \nu) ds_\Sigma + \frac{6}{\pi} I(1, \nu) + \frac{1-\nu^2}{2} I(1, \nu) E_1^\natural. \qquad (4.5.14)$$

Proof. Starting from the definition (4.5.7) of $f_{\rho,\nu}^\sharp$, we note that the image of $f_{\rho,\nu}$ under $\Delta - \frac{1-\nu^2}{4}$ was given in Theorem 4.2.4, while

$$\left(\Delta - \frac{1-\nu^2}{4}\right) E_{\frac{1+\rho}{2}} = \frac{\nu^2 - \rho^2}{4} E_{\frac{1+\rho}{2}}. \qquad (4.5.15)$$

We also use the expression (2.3.34) of $I(\rho, \nu)$ in terms of $C(\rho, \nu)$, finally the relation

$$C(2-\rho, \nu) \times \frac{\Gamma(\frac{2+\rho-\nu}{4})\Gamma(\frac{2+\rho+\nu}{4})}{\Gamma(\frac{4-\rho-\nu}{4})\Gamma(\frac{4-\rho+\nu}{4})} = C(\rho, \nu), \qquad (4.5.16)$$

a consequence of (2.3.33). The part concerning the function $E_1^\natural$ is proved in the same way, with the help of (4.5.1) and Theorem 4.2.4. One should note that the term $\frac{6}{\pi} I(1, \nu)$ from the second equation (4.5.14), which has no analogue when $\rho \neq 1$, originates from the equation $\Delta E_1^\natural = -\frac{3}{\pi} \neq 0$, while $\Delta E_{\frac{1+\rho}{2}} = \frac{1-\rho^2}{4} E_{\frac{1+\rho}{2}}$ when $\rho \neq 1$: this explains the slight complication that will occur in the special case under consideration, which we shall set apart. $\square$

Lemma 4.5.4. *Together with the standing assumptions* (4.3.1), *assume that* $\mathrm{Re}\,\nu < 0$, *and* $\rho \neq 1$: *the function* $f_{\rho,\nu}^\sharp$ *then admits an expansion*

$$f_{\rho,\nu}^\sharp(z) = \Phi_{\rho,\nu}^0 + \frac{1}{8\pi} \int_{-\infty}^\infty \Phi_{\rho,\nu}(\lambda) E_{\frac{1-i\lambda}{2}}(z) d\lambda + \sum_{p \geq 1} \sum_j \Phi_{\rho,\nu}^{p,j} \mathcal{M}_{p,j}(z), \qquad (4.5.17)$$

some coefficients of which are given by the equations

$$\Phi_{\rho,\nu}(\lambda) = \frac{4C(\rho,\nu)}{\nu^2 + \lambda^2}[\Psi(\rho,i\lambda) + \Psi(2-\rho,i\lambda)],$$

$$\Phi_{\rho,\nu}^{p,j} = (\mathcal{M}_{p,j}|f_{\rho,\nu}). \tag{4.5.18}$$

The expression $\Psi(\rho,i\lambda)$, independent of ν, has been defined in (4.5.12).

Proof. Given any pair f, h of functions, denote as $(f|h)_Y$ the integral $\int_{D_Y} \overline{f}(z)h(z)\frac{dxdy}{y^2}$. Assuming that f and h are automorphic and C^2, Green's formula applies, giving

$$(f|\Delta h)_Y - (\Delta f|h)_Y$$

$$= -\int_{-\frac{1}{2}}^{\frac{1}{2}} \overline{f}(x+iY)\frac{\partial h}{\partial y}(x+iY)dx + \int_{-\frac{1}{2}}^{\frac{1}{2}} \frac{\partial \overline{f}}{\partial y}(x+iY)h(x+iY)dx. \tag{4.5.19}$$

Provided that Δ is applied in the distribution sense, this is valid when $h = f_{\rho,\nu}^\sharp$ as well. Applying it with $h = f_{\rho,\nu}^\sharp$, $f = E_{\frac{1-i\lambda}{2}}$, we note that the one-dimensional integrals on the right-hand side of (4.5.19) go to zero as $Y \to \infty$: this is a consequence of the estimate (4.5.9) and of what has been noted immediately after (4.5.10), together with the estimates $E_{\frac{1+i\lambda}{2}}(x+iy) = O(y^{\frac{1}{2}})$, $\frac{\partial}{\partial y}E_{\frac{1+i\lambda}{2}}(x+iy) = O(y^{-\frac{1}{2}})$. What is left is simply

$$(E_{\frac{1-i\lambda}{2}}|\Delta f_{\rho,\nu}^\sharp) - (\Delta E_{\frac{1-i\lambda}{2}}|f_{\rho,\nu}^\sharp) = 0. \tag{4.5.20}$$

Applying then Lemma 4.5.3 and (4.5.15), this latter one with $-i\lambda$ in place of ρ, we obtain that

$$\frac{\nu^2 + \lambda^2}{4}(E_{\frac{1-i\lambda}{2}}|f_{\rho,\nu}^\sharp)_Y = C(\rho,\nu)\lim_{Y\to\infty}\left(E_{\frac{1-i\lambda}{2}}\left|\left[2ds_\Sigma^{(\rho)} - E_{\frac{1+\rho}{2}} - E_{\frac{3-\rho}{2}}\right]\right)_Y\right. \tag{4.5.21}$$

is $C(\rho,\nu)$ times the limit, as $Y \to \infty$, of

$$\int_1^Y (y^{\frac{\rho-1}{2}} + y^{\frac{1-\rho}{2}})E_{\frac{1+i\lambda}{2}}(iy)\frac{dy}{y} - \int_{D_Y}\left(E_{\frac{1+\rho}{2}}(z) + E_{\frac{3-\rho}{2}}(z)\right)E_{\frac{1+i\lambda}{2}}(z)\frac{dxdy}{y^2}. \tag{4.5.22}$$

Using (4.5.12), one finds the first equation (4.5.18).

The second one is immediate, since

$$(\mathcal{M}_{p,j}|E_{\frac{1+\rho}{2}}) = (\mathcal{M}_{p,j}|E_{\frac{3-\rho}{2}}) = 0. \tag{4.5.23}$$

$\square$

Theorem 4.5.5. Assume that $\rho \neq 1$. The coefficient $\Psi(\rho,i\lambda)$ defined in (4.5.12) satisfies the equation

$$\Psi(\rho,i\lambda) + \Psi(2-\rho,i\lambda) = \frac{\zeta^*(\frac{\rho-i\lambda}{2})\zeta^*(\frac{\rho+i\lambda}{2})}{\zeta^*(1+i\lambda)}. \tag{4.5.24}$$

Also, the constant term $\Phi^0_{\rho,\nu}$ *from the Roelcke-Selberg expansion* (4.5.17) *is zero.*

Proof. Substituting for $i\lambda$ a more general complex number μ, only assuming that $\zeta^*(1+\mu) \neq 0$ so that $E_{\frac{1+\mu}{2}}$ should be well-defined, let us write

$$
\int_1^Y y^{\frac{\rho-1}{2}} E_{\frac{1+\mu}{2}}(iy) \frac{dy}{y} - \int_{D_Y} E_{\frac{1+\rho}{2}}(x+iy) E_{\frac{1+\mu}{2}}(x+iy) \frac{dx\,dy}{y^2}
$$

$$
= \int_1^Y y^{\frac{\rho-1}{2}} E_{\frac{1+\mu}{2}}(iy) \frac{dy}{y} - \int_{D_Y} y^{\frac{\rho-1}{2}} E_{\frac{1+\mu}{2}}(x+iy) \frac{dx\,dy}{y}
$$

$$
+ \int_{D_Y} \left[y^{\frac{1+\rho}{2}} - E_{\frac{1+\rho}{2}}(x+iy) \right] E_{\frac{1+\mu}{2}}(iy) \frac{dx\,dy}{y^2}. \quad (4.5.25)
$$

Since the difference $E_{\frac{1+\mu}{2}}(iy) - \int_{-\frac{1}{2}}^{\frac{1}{2}} E_{\frac{1+\mu}{2}}(x+iy)dx$ is, as $y \to \infty$, a $O(y^{-N})$ for every N, the sum of the first two integrals on the right-hand side has, for every $\mu \in \mathbb{C}$ with $\zeta^*(1+\mu) \neq 0$, a finite limit, and this limit is a holomorphic function of μ. The last integral has a limit as $Y \to \infty$ provided that, moreover, $\mathrm{Re}\left(\frac{1-\rho}{2} + \frac{1\pm\mu}{2}\right) < 1$, i.e., $|\mathrm{Re}\,\mu| < \mathrm{Re}\,\rho$. We may then set

$$
\Psi(\rho,\mu) = \lim_{Y\to\infty} \left[\int_1^Y y^{\frac{\rho-1}{2}} E_{\frac{1+\mu}{2}}(iy) \frac{dy}{y} - \int_{D_Y} E_{\frac{1+\rho}{2}}(x+iy) E_{\frac{1+\mu}{2}}(x+iy) \frac{dx\,dy}{y^2} \right],
$$

$$(4.5.26)$$

under the assumptions that $|\mathrm{Re}\,\mu| < \mathrm{Re}\,\rho$ and $\zeta^*(1+\mu) \neq 0$. Note that, since we need to consider both $\Psi(\rho,\mu)$ and $\Psi(2-\rho,\mu)$, this breaks down when $\mu = \pm 1$, which will make the computation of $\Psi^0(\rho)$ more complicated.

Lemma 4.5.4 has been established under the assumption, among others, that $\mathrm{Re}\,\nu < 0$. Now, as it follows from Theorem 4.3.4, $f_{\rho,\nu}$, and $f^\sharp_{\rho,\nu}$ as a consequence, are meaningful for every value of ν with $\nu \notin \mathbb{Z}, \mathrm{Re}\,\nu < 1 - |\mathrm{Re}\,\rho - 1|, \rho \pm \nu \notin 2\mathbb{Z}$ and ν distinct from non-trivial zeros of zeta and from all numbers $i\lambda_p$ with $\frac{1+\lambda_p^2}{4}$ in the even part of the discrete spectrum of the modular Laplacian. Let $\nu_0 = ia$ with $a > 0$ be such a non-exceptional value. Since, when $\nu = \pm\nu_0$, the function $\Phi_{\rho,\nu}(\lambda)$ made explicit in (4.5.18) has poles at $\lambda = \pm a$, the integral term in the expansion which is the object of Lemma 4.5.4 requires some modification before it can be applied in the case when $\mathrm{Re}\,\nu = 0$: recall that this lemma was established under the assumption that $\mathrm{Re}\,\nu < 0$. Let δ_a and δ_{-a} be two disjoint, closed, half-disks contained in the half-plane $\{\mu \in \mathbb{C} : \mathrm{Re}\,\mu \leq 0\}$ and centered at ia and $-ia$ respectively. Let γ_a and γ_{-a} be the parts of their boundaries consisting of half-circles, and let γ be the contour from $-i\infty$ to $i\infty$ obtained from the straight line after one has substituted the half-circles γ_{-a} and γ_a for the corresponding diameters.

So as to obtain the expansion, similar to (4.5.17), of $f^\sharp_{\rho,ia}(z)$ or $f^\sharp_{\rho,-ia}(z)$, we shall let ν approach ia (*resp.* $-ia$) by values ν such that $\mathrm{Re}\,\nu < 0$. One has

$$\frac{1}{8\pi}\int_{-\infty}^{\infty}\Phi_{\rho,\nu}(\lambda)E_{\frac{1-i\lambda}{2}}(z)d\lambda = \frac{1}{8i\pi}\int_{\mathrm{Re}\,\mu=0}\Phi_{\rho,\nu}(-i\mu)E_{\frac{1-\mu}{2}}(z)d\mu$$

$$= \frac{1}{8i\pi}\int_{\mathrm{Re}\,\mu=0}\frac{4C(\rho,\nu)}{\nu^2-\mu^2}\left(\Psi(\rho,\mu)+\Psi(2-\rho,\mu)\right)E_{\frac{1-\mu}{2}}(z)d\mu. \quad (4.5.27)$$

If ν (with $\mathrm{Re}\,\nu < 0$) is close to ia, the pole ν of the fraction $\mu \mapsto \frac{1}{\nu^2-\mu^2}$ lies within δ_a but the pole $-\nu$ does not lie within δ_{-a}. In this case, one thus has

$$\frac{1}{2i\pi}\int_{\mathrm{Re}\,\mu=0}\frac{C(\rho,\nu)}{\nu^2-\mu^2}\left(\Psi(\rho,\mu)+\Psi(2-\rho,\mu)\right)E_{\frac{1-\mu}{2}}(z)d\mu$$

$$= \frac{1}{2i\pi}\int_\gamma\frac{C(\rho,\nu)}{\nu^2-\mu^2}\left(\Psi(\rho,\mu)+\Psi(2-\rho,\mu)\right)E_{\frac{1-\mu}{2}}(z)d\mu$$

$$- \frac{C(\rho,\nu)}{2\nu}\left(\Psi(\rho,\nu)+\Psi(2-\rho,\nu)\right)E_{\frac{1-\nu}{2}}(z), \quad (4.5.28)$$

an expression the limit of which, as $\nu \to ia$, is obtained by simply substituting ia for ν. When ν is close to $-ia$, it is on the contrary the half-disk δ_{-a} that must be considered, and the same equation as before holds: however, to obtain the limit, we must this time substitute $-ia$ for ν.

We can now compute the sum $f_{\rho,ia}^\sharp(z)+f_{\rho,-ia}^\sharp(z)$. Considerable simplification occurs from the fact that $C(\rho,\nu)$ and $I(\rho,\nu)$ are odd functions of ν: as a start, this sum is not distinct from $f_{\rho,ia}(z)+f_{\rho,-ia}(z)$, hence is a multiple of the Eisenstein series $E_{\frac{1+ia}{2}}(z)$ according to Corollary 4.4.5. In particular, it is orthogonal to cusp-forms. Hence, looking at Lemma 4.5.4, one sees that only the integral terms can contribute to the sum. From (4.5.28), it then follows that

$$f_{\rho,ia}(z)+f_{\rho,-ia}(z) = -\frac{C(\rho,ia)}{2ia}\left[\left(\Psi(\rho,ia)+\Psi(2-\rho,ia)\right)E_{\frac{1-ia}{2}}(z)\right.$$

$$\left. + \left(\Psi(\rho,-ia)+\Psi(2-\rho,-ia)\right)E_{\frac{1+ia}{2}}(z)\right] \quad (4.5.29)$$

or

$$f_{\rho,i\lambda}(z)+f_{\rho,-i\lambda}(z) = -\frac{C(\rho,i\lambda)}{2i\lambda}\left[\left(\Psi(\rho,i\lambda)+\Psi(2-\rho,i\lambda)\right)E_{\frac{1-i\lambda}{2}}(z)\right.$$

$$\left.\left(\Psi(\rho,-i\lambda)+\Psi(2-\rho,-i\lambda)\right)E_{\frac{1+i\lambda}{2}}(z)\right]. \quad (4.5.30)$$

From Corollary 4.4.5,

$$f_{\rho,i\lambda}(z)+f_{\rho,-i\lambda}(z) = -\frac{C(\rho,i\lambda)}{i\lambda}\frac{\zeta^*(\frac{\rho-i\lambda}{2})\zeta^*(\frac{\rho+i\lambda}{2})}{\zeta^*(1-i\lambda)}E_{\frac{1+i\lambda}{2}}(z), \quad (4.5.31)$$

so that

$$\frac{1}{2}\left[\left(\Psi(\rho,i\lambda)+\Psi(2-\rho,i\lambda)\right)E_{\frac{1-i\lambda}{2}}(z)+\left(\Psi(\rho,-i\lambda)+\Psi(2-\rho,-i\lambda)\right)E_{\frac{1+i\lambda}{2}}(z)\right]$$

$$=\frac{\zeta^*(\frac{\rho-i\lambda}{2})\zeta^*(\frac{\rho+i\lambda}{2})}{\zeta^*(1-i\lambda)}E_{\frac{1+i\lambda}{2}}(z)=\frac{\zeta^*(\frac{\rho-i\lambda}{2})\zeta^*(\frac{\rho+i\lambda}{2})}{\zeta^*(1+i\lambda)}E_{\frac{1-i\lambda}{2}}(z).\quad(4.5.32)$$

From (4.5.12), one has

$$\zeta^*(1+i\lambda)\Psi(\rho,i\lambda)=\zeta^*(1-i\lambda)\Psi(\rho,-i\lambda).\qquad(4.5.33)$$

Since $E^*_{\frac{1-i\lambda}{2}}=\zeta^*(1-i\lambda)E_{\frac{1-i\lambda}{2}}$ is again an even function of λ, it follows from this equation that the two terms from the left-hand side of (4.5.32) are identical. This leads to

$$\Psi(\rho,i\lambda)+\Psi(2-\rho,i\lambda)=\frac{\zeta^*(\frac{\rho-i\lambda}{2})\zeta^*(\frac{\rho+i\lambda}{2})}{\zeta^*(1+i\lambda)}.\qquad(4.5.34)$$

We still need to compute the constant term $\Phi^0_{\rho,\nu}$. Now that we know $\Phi_{\rho,\nu}(\lambda)$, Lemma 4.5.4 becomes

$$f_{\rho,\nu}(z)=I(\rho,\nu)E_{\frac{1+\rho}{2}}(z)+\frac{\Gamma(\frac{2+\rho-\nu}{4})\Gamma(\frac{2+\rho+\nu}{4})}{\Gamma(\frac{4-\rho-\nu}{4})\Gamma(\frac{4-\rho+\nu}{4})}I(2-\rho,\nu)E_{\frac{3-\rho}{2}}(z)+\Phi^0_{\rho,\nu}$$

$$+\frac{C(\rho,\nu)}{2\pi}\int_{-\infty}^{\infty}\frac{1}{\nu^2+\lambda^2}\frac{\zeta^*(\frac{\rho-i\lambda}{2})\zeta^*(\frac{\rho+i\lambda}{2})}{\zeta^*(1+i\lambda)}E_{\frac{1-i\lambda}{2}}(z)d\lambda+\sum_{p\geq1}\sum_{j}\Phi^{p,j}_{\rho,\nu}M_{p,j}(z).$$

$$(4.5.35)$$

This leads to an expansion of $f_{\rho,\nu}(iy)$ as $y\to\infty$, which we shall compare to the expansion

$$f_{\rho,\nu}(z)=I(\rho,\nu)y^{\frac{\rho+1}{2}}+\frac{\Gamma(\frac{2+\rho-\nu}{4})\Gamma(\frac{2+\rho+\nu}{4})}{\Gamma(\frac{4-\rho-\nu}{4})\Gamma(\frac{4-\rho+\nu}{4})}I(2-\rho,\nu)y^{\frac{3-\rho}{2}}$$

$$-\frac{C(\rho,\nu)}{\nu}\frac{\zeta^*(\frac{\rho-\nu}{2})\zeta^*(\frac{\rho+\nu}{2})}{\zeta^*(1-\nu)}y^{\frac{1+\nu}{2}}+O\left(y^{-\frac{|\rho-1|}{2}}\right)\quad(4.5.36)$$

from Theorem 4.4.4. Neglecting terms which are a $O(y^{-\frac{|\rho-1|}{2}})$ in (4.5.35), one can first forget the series of cusp-forms. Up to a similar error term, we can also simplify the main integral term in (4.5.35) as follows:

$$L(\rho,\nu;y):=\int_{-\infty}^{\infty}\frac{1}{\nu^2+\lambda^2}\frac{\zeta^*(\frac{\rho-i\lambda}{2})\zeta^*(\frac{\rho+i\lambda}{2})}{\zeta^*(1+i\lambda)}E_{\frac{1-i\lambda}{2}}(z)d\lambda$$

$$\sim\int_{-\infty}^{\infty}\frac{\zeta^*(\frac{\rho-i\lambda}{2})\zeta^*(\frac{\rho+i\lambda}{2})}{\nu^2+\lambda^2}\left[\frac{y^{\frac{1-i\lambda}{2}}}{\zeta^*(1+i\lambda)}+\frac{y^{\frac{1+i\lambda}{2}}}{\zeta^*(1-i\lambda)}\right]d\lambda$$

$$=\frac{2}{i}\int_{\mathrm{Re}\,\mu=0}\frac{\zeta^*(\frac{\rho-\mu}{2})\zeta^*(\frac{\rho+\mu}{2})}{\nu^2-\mu^2}\frac{y^{\frac{1-\mu}{2}}}{\zeta^*(1+\mu)}d\mu.\quad(4.5.37)$$

We move the contour of integration to the line $\operatorname{Re}\mu = 3$: the line integral we arrive at is then bounded, for large y, by Cy^{-1}, but, assuming $\operatorname{Re}\nu < 0$, there are poles at $-\nu$ and $\rho, 2-\rho$ (since $0 < \operatorname{Re}\rho < 2$, the poles at $\rho - 2$ and $-\rho$ do not occur here) to be taken care of. One finds

$$L(\rho,\nu;y) \sim -4\pi \sum_{\alpha=-\nu,\rho,2-\rho} \operatorname{Res}_{\mu=\alpha}\left(\frac{\zeta^*(\frac{\rho-\mu}{2})\zeta^*(\frac{\rho+\mu}{2})}{\nu^2 - \mu^2} \frac{y^{\frac{1-\mu}{2}}}{\zeta^*(1+\mu)} \right). \tag{4.5.38}$$

Taking the coefficient $\frac{C(\rho,\nu)}{2\pi}$ of $L(\rho,\nu;y)$ in (4.5.35) into account, we find

$$\frac{C(\rho,\nu)}{2\pi}L(\rho,\nu;y) \sim -C(\rho,\nu)\frac{\zeta^*(\frac{\rho+\nu}{2})\zeta^*(\frac{\rho-\nu}{2})}{\zeta^*(1-\nu)}\frac{y^{\frac{1+\nu}{2}}}{\nu}$$

$$- 4C(\rho,\nu)\left[\frac{\zeta^*(\rho)}{\zeta^*(1+\rho)}\frac{y^{\frac{1-\rho}{2}}}{\nu^2 - \rho^2} + \frac{\zeta^*(\rho-1)}{\zeta^*(3-\rho)}\frac{y^{\frac{\rho-1}{2}}}{\nu^2 - (2-\rho)^2} \right]. \tag{4.5.39}$$

Using the relation (2.3.34) between $I(\rho,\nu)$ and $C(\rho,\nu)$, one has, up to some rapidly decreasing error term,

$$I(\rho,\nu)E_{\frac{1+\rho}{2}}(z) \sim I(\rho,\nu)y^{\frac{1+\rho}{2}} + \frac{4C(\rho,\nu)}{\nu^2 - \rho^2}\frac{\zeta^*(\rho)}{\zeta^*(1+\rho)}y^{\frac{1-\rho}{2}}. \tag{4.5.40}$$

Using the same, and (4.5.16), one finds

$$\frac{\Gamma(\frac{2+\rho-\nu}{4})\Gamma(\frac{2+\rho+\nu}{4})}{\Gamma(\frac{4-\rho-\nu}{4})\Gamma(\frac{4-\rho+\nu}{4})}I(2-\rho,\nu)E_{\frac{3-\rho}{2}}(z)$$

$$\sim \frac{\Gamma(\frac{2+\rho-\nu}{4})\Gamma(\frac{2+\rho+\nu}{4})}{\Gamma(\frac{4-\rho-\nu}{4})\Gamma(\frac{4-\rho+\nu}{4})}I(2-\rho,\nu)y^{\frac{3-\rho}{2}} + \frac{4C(\rho,\nu)}{\nu^2 - (2-\rho)^2}\frac{\zeta^*(\rho-1)}{\zeta^*(3-\rho)}y^{\frac{\rho-1}{2}}. \tag{4.5.41}$$

With the help of (4.5.39), one may thus rewrite (4.5.35) as

$$f_{\rho,\nu}(z) \sim I(\rho,\nu)y^{\frac{1+\rho}{2}} + \frac{\Gamma(\frac{2+\rho-\nu}{4})\Gamma(\frac{2+\rho+\nu}{4})}{\Gamma(\frac{4-\rho-\nu}{4})\Gamma(\frac{4-\rho+\nu}{4})}I(2-\rho,\nu)y^{\frac{3-\rho}{2}} + \Phi^0_{\rho,\nu}$$

$$- C(\rho,\nu)\frac{\zeta^*(\frac{\rho+\nu}{2})\zeta^*(\frac{\rho-\nu}{2})}{\zeta^*(1-\nu)}\frac{y^{\frac{1+\nu}{2}}}{\nu}. \tag{4.5.42}$$

Comparing this to (4.5.36), we obtain $\Phi^0_{\rho,\nu} = 0$. $\qquad\square$

We consider now the case when $\rho = 1$. The expansion (4.5.17) is still valid, for coefficients $\Phi^0_{1,\nu}$, $\Phi_{1,\nu}(\lambda)$, $\Phi^{p,j}_{1,\nu}$ which we need to determine. There is of course nothing new in the equation for $\Phi^{p,j}_{1,\nu}$.

Theorem 4.5.6. *One has*

$$\Phi_{1,\nu}(\lambda) = \frac{8C(1,\nu)}{\nu^2 + \lambda^2}\Psi(1,i\lambda) \tag{4.5.43}$$

and

$$\Psi(1,i\lambda) = \frac{1}{2}\frac{(\zeta^*(\frac{1-i\lambda}{2}))^2}{\zeta^*(1+i\lambda)}. \tag{4.5.44}$$

On the other hand,

$$\Phi^0_{1,\nu} = \frac{24}{\pi}\frac{I(1,\nu)}{\nu^2 - 1}. \tag{4.5.45}$$

Proof. The peculiarity of this case lies in the extra term $\frac{6}{\pi}I(1,\nu)$ on the right-hand side of (4.5.14). Since constants are orthogonal to Eisenstein series $E_{\frac{1-i\lambda}{2}}$, this has no incidence on the determination of $\Psi(1,i\lambda)$, and the pair of equations (4.5.43), (4.5.44) is just the same as the particular case, when $\rho = 1$, of equations previously obtained. Things are different so far as computations dealing with constant terms from Roelcke-Selberg expansions are concerned.

We start from equation (4.5.37), still valid when $\rho = 1$, at which point differences begin to appear. The integrand now has a simple pole at $\mu = -\nu$ and a double pole at $\mu = 1$. We thus obtain

$$\frac{C(1,\nu)}{2\pi}L(1,\nu;y) \tag{4.5.46}$$

$$\sim -\frac{C(1,\nu)}{\nu}\frac{(\zeta^*(\frac{1-\nu}{2}))^2}{\zeta^*(1-\nu)}y^{\frac{1+\nu}{2}} - 2C(1,\nu)\mathrm{Res}_{\mu=1}\left(\frac{(\zeta^*(\frac{1-\mu}{2}))^2}{\nu^2 - \mu^2}\frac{y^{\frac{1-\mu}{2}}}{\zeta^*(1+\mu)}\right).$$

Next, we rewrite (4.5.36) as

$$f_{1,\nu}(x+iy) = 2I(1,\nu)y - \frac{C(1,\nu)}{\nu}\frac{(\zeta^*(\frac{1-\nu}{2}))^2}{\zeta^*(1-\nu)}y^{\frac{1+\nu}{2}} + \mathrm{O}(y^{-1}). \tag{4.5.47}$$

Recall from Theorem 4.4.4 that the bound on the error term is better in our present case, since the remainder term $\mathrm{O}\left(y^{-\frac{|\rho-1|}{2}}\right)$ in general originated from the error term denoted as Err immediately after (4.4.48), and is no longer present when $\rho = 1$. On the other hand, (4.5.35) is now

$$f_\nu(z) = 2I(1,\nu)E_1^\natural(z) + \Phi^0_{1,\nu} + \frac{C(1,\nu)}{2\pi}\int_{-\infty}^{\infty}\frac{1}{\nu^2 + \lambda^2}\frac{(\zeta^*(\frac{1-i\lambda}{2}))^2}{\zeta^*(1+i\lambda)}E_{\frac{1-i\lambda}{2}}(z)d\lambda$$

$$+ \sum_{p\geq 1}\sum_j \Phi^{p,j}_{1,\nu}\mathcal{M}_{p,j}(z) \tag{4.5.48}$$

or, up to $O(y^{-1})$,

$$f_\nu(z) \sim 2I(1,\nu)\left(y - \frac{3}{\pi}\log y + G^o\right) + \Phi^0_{1,\nu}$$

$$-\frac{C(1,\nu)}{\nu}\frac{(\zeta^*(\frac{1-\nu}{2}))^2}{\zeta^*(1-\nu)}y^{\frac{1+\nu}{2}} - 2C(1,\nu)\mathrm{Res}_{\mu=1}\left(\frac{(\zeta^*(\frac{1-\mu}{2}))^2}{\nu^2-\mu^2}\frac{y^{\frac{1-\mu}{2}}}{\zeta^*(1+\mu)}\right). \quad (4.5.49)$$

Comparing this to (4.5.47), we obtain

$$2I(1,\nu)\left(-\frac{3}{\pi}\log y + G^o\right) + \Phi^0_{1,\nu} - 2C(1,\nu)\mathrm{Res}_{\mu=1}\left(\frac{(\zeta^*(\frac{1-\mu}{2}))^2}{\nu^2-\mu^2}\frac{y^{\frac{1-\mu}{2}}}{\zeta^*(1+\mu)}\right) \sim 0.$$
$$(4.5.50)$$

Using the link between $I(1,\nu)$ and $C(1,\nu)$, one finds, up to a $O(y^{-1})$,

$$\Phi^0_{1,\nu} \sim I(1,\nu)\left[\frac{6}{\pi}\log y - 2G^o + \frac{\nu^2-1}{2}\mathrm{Res}_{\mu=1}\left(\frac{(\zeta^*(\frac{1-\mu}{2}))^2}{\nu^2-\mu^2}\frac{y^{\frac{1-\mu}{2}}}{\zeta^*(1+\mu)}\right)\right].$$
$$(4.5.51)$$

We must now evaluate this residue. With purely temporary notation, we set

$$\psi(\mu) = (\mu-1)\zeta^*(\frac{1-\mu}{2}), \qquad f(\mu) = \frac{1}{\nu^2-\mu^2}$$

$$g(\mu) = y^{\frac{1-\mu}{2}}, \qquad h(\mu) = \frac{1}{\zeta^*(1+\mu)}, \qquad (4.5.52)$$

and

$$F(\mu) = (\psi(\mu))^2 h(\mu) = (\mu-1)^2\frac{(\zeta^*(\frac{1+\mu}{2}))^2}{\zeta^*(1+\mu)}: \qquad (4.5.53)$$

then, the residue under evaluation is

$$\mathrm{Res} = (\psi^2 fgh)'(1) = F(1)(fg)'(1) + F'(1)(fg)(1), \qquad (4.5.54)$$

i.e.,

$$\mathrm{Res} = \frac{24}{\pi}\left[\frac{2}{(\nu^2-1)^2} - \frac{\log y}{2(\nu^2-1)}\right] + \frac{F'(1)}{\nu^2-1}. \qquad (4.5.55)$$

The coefficient of $\log y$ is of course just what is needed, and we finally obtain

$$\Phi^0_{1,\nu} = I(1,\nu)\left[\frac{24}{\pi(\nu^2-1)} - 2G^o + \frac{1}{2}F'(1)\right]. \qquad (4.5.56)$$

What remains to be done is proving that $G^o = \frac{1}{4}F'(1)$, i.e., using (4.5.6) and (4.5.53), that

$$\frac{d}{d\mu}\left(\frac{(\mu-1)\zeta^*(1-\mu)}{\zeta^*(1+\mu)}\right)(\mu=1) = \frac{1}{4}\frac{d}{d\mu}\left(\frac{(\mu-1)^2(\zeta^*(\frac{1-\mu}{2}))^2}{\zeta^*(1+\mu)}\right)(\mu=1):$$
$$(4.5.57)$$

indeed, it follows from a Taylor expansion $s\zeta^*(s) = 1 + \alpha s + O(s^2), s \to 0$, that $(\mu - 1)\zeta^*(1 - \mu)$ and $\frac{1}{4}(\mu - 1)^2(\zeta^*(\frac{1-\mu}{2}))^2$ differ by a $O((\mu - 1)^2)$ as $\mu \to 1$. □

We can now give the full Rankin-Selberg expansion of the function $f_{\rho,\nu}$. This function does not lie in $L^2(D)$ and there are two terms off the spectral line. Observe that, when $\mathrm{Re}\,\nu < 0$, the arithmetic conditions regarding ν stated in Theorem 4.4.4 are void.

Theorem 4.5.7. *Assume that $0 < \mathrm{Re}\,\rho < 2, \rho \neq 1$, that $\nu \notin \mathbb{Z}, \rho \pm \nu \notin 2\mathbb{Z}$, finally that $\mathrm{Re}\,\nu < 0$; recall from (3.1.12) the definition of $L^*(s, \mathcal{M}_{p,j})$. One has the identity*

$$[C(\rho, \nu)]^{-1} f_{\rho,\nu} = \frac{4}{\nu^2 - \rho^2} E_{\frac{1+\rho}{2}} + \frac{4}{\nu^2 - (2 - \rho)^2} E_{\frac{3-\rho}{2}}$$

$$+ \frac{1}{2\pi} \int_{-\infty}^{\infty} \frac{1}{\nu^2 + \lambda^2} \frac{\zeta^*(\frac{\rho - i\lambda}{2})\zeta^*(\frac{\rho + i\lambda}{2})}{\zeta^*(1 + i\lambda)} E_{\frac{1 - i\lambda}{2}} d\lambda$$

$$+ \sum_{p, j \text{ even}} \frac{2}{\nu^2 + \lambda_p^2} L^*(\frac{\rho}{2}, \mathcal{M}_{p,j}) \mathcal{M}_{p,j}. \qquad (4.5.58)$$

In particular, one has $[C(\rho, \nu)]^{-1} f_{\rho,\nu} = [C(2 - \rho, \nu)]^{-1} f_{2-\rho,\nu}$.

Proof. Recall from Proposition 2.3.2 that

$$I(\rho, \nu) = \frac{4}{\nu^2 - \rho^2} C(\rho, \nu) \qquad (4.5.59)$$

and from (3.1.12) that, for Hecke eigenforms of even type (the only ones which can occur in the decomposition of $f_{\rho,\nu}$),

$$L^*(\frac{\rho}{2}, \mathcal{M}_{p,j}) = \pi^{-\frac{\rho}{2}} \Gamma(\frac{\rho + i\lambda_p}{4}) \Gamma(\frac{\rho - i\lambda_p}{4}) L(\frac{\rho}{2}, \mathcal{M}_{p,j}). \qquad (4.5.60)$$

Since, as noted immediately after (2.3.60),

$$C(\rho, \nu) = \frac{\Gamma(\frac{2+\rho-\nu}{4})\Gamma(\frac{2+\rho+\nu}{4})}{\Gamma(\frac{4-\rho-\nu}{4})\Gamma(\frac{4-\rho+\nu}{4})} C(2 - \rho, \nu), \qquad (4.5.61)$$

the equation to be proved is equivalent to

$$f_{\rho,\nu}(z) = I(\rho, \nu) E_{\frac{1+\rho}{2}}(z) + \frac{\Gamma(\frac{2+\rho-\nu}{4})\Gamma(\frac{2+\rho+\nu}{4})}{\Gamma(\frac{4-\rho-\nu}{4})\Gamma(\frac{4-\rho+\nu}{4})} I(2 - \rho, \nu) E_{\frac{3-\rho}{2}}(z)$$

$$+ \frac{C(\rho, \nu)}{2\pi} \int_{-\infty}^{\infty} \frac{1}{\nu^2 + \lambda^2} \frac{\zeta^*(\frac{\rho - i\lambda}{2})\zeta^*(\frac{\rho + i\lambda}{2})}{\zeta^*(1 + i\lambda)} E_{\frac{1 - i\lambda}{2}}(z) d\lambda$$

$$+ 2\pi^{-\frac{\rho}{2}} C(\rho, \nu) \sum_{p, j \text{ even}} \frac{\Gamma(\frac{\rho + i\lambda_p}{4})\Gamma(\frac{\rho - i\lambda_p}{4})}{\nu^2 + \lambda_p^2} L(\frac{\rho}{2}, \mathcal{M}_{p,j}) \mathcal{M}_{p,j}(z). \qquad (4.5.62)$$

All terms with the exception of those in the last series have already been obtained in (4.5.35), to be combined with the last assertion of Theorem 4.5.5. What remains to be done is making the discrete term $\sum_{p,j\text{even}} \Phi^{p,j}_{\rho,\nu} \mathcal{M}_{p,j}(z)$ explicit. One has

$$\mathcal{M}_{p,j} = \frac{4}{\nu^2 + \lambda_p^2} \left(\Delta - \frac{1-\nu^2}{4} \right) \mathcal{M}_{p,j}, \tag{4.5.63}$$

and, using Theorem 4.2.4,

$$\Phi^{p,j}_{\rho,\nu} = (\mathcal{M}_{p,j} | f_{\rho,\nu}) = \frac{4}{\nu^2 + \lambda_p^2} \left(\mathcal{M}_{p,j} \Big| \left(\Delta - \frac{1-\nu^2}{4} \right) f_{\rho,\nu} \right)$$

$$= \frac{8C(\rho,\nu)}{\nu^2 + \lambda_p^2} \int_1^\infty \frac{1}{2} \left(y^{\frac{\rho-1}{2}} + y^{\frac{1-\rho}{2}} \right) \overline{\mathcal{M}_{p,j}}(iy) \frac{dy}{y}, \tag{4.5.64}$$

where the bar over $\mathcal{M}_{p,j}$ in the integral can be dispensed with, after which it suffices to apply Proposition 4.4.6, not forgetting that only Hecke eigenforms of even type occur here. □

That $f_{\rho,\nu}$ and $f_{2-\rho,\nu}$ are proportional was already observed in (4.2.2).

Theorem 4.5.8. *For $\nu \notin \mathbb{Z}, \operatorname{Re}\nu < 0$, one has*

$$[C(1,\nu)]^{-1} f_{1,\nu} = \frac{8}{\nu^2 - 1} E_1^\natural + \frac{96}{\pi} \frac{1}{(\nu^2 - 1)^2} \tag{4.5.65}$$

$$+ \frac{1}{2\pi} \int_{-\infty}^\infty \frac{1}{\nu^2 + \lambda^2} \frac{(\zeta^*(\frac{1-i\lambda}{2}))^2}{\zeta^*(1+i\lambda)} E_{\frac{1-i\lambda}{2}} d\lambda + \sum_{p,j\text{even}} \frac{2}{\nu^2 + \lambda_p^2} L^*(\frac{1}{2}, \mathcal{M}_{p,j}) \mathcal{M}_{p,j}.$$

Proof. The only difference in the proof is that one must add on the right-hand side the extra term $[C(1,\nu)]^{-1} \Phi^0_{1,\nu}$, with $\Phi^0_{1,\nu}$ as made explicit in (4.5.45). □

4.6 Incomplete ρ-series

Classically, incomplete Eisenstein series are defined as integral superpositions of Eisenstein series $E_{\frac{1-\nu}{2}}$ for various values of the parameter ν. The latter ones are built as Poincaré series, starting from powers of $\operatorname{Im} z$ and applying the summation process, which takes place in this case with respect to the generic element of Γ/Γ^0_∞; incomplete Eisenstein series are obtained in the same way, starting from an "arbitrary" function of $\operatorname{Im} z$. We here apply the same idea, replacing Eisenstein series $E_{\frac{1-\nu}{2}}$ by functions $f_{\rho,\nu}$ and fixing the parameter ρ, with $0 < \operatorname{Re}\rho < 2$.

The function $f_{\rho,\nu}$ was built from the function $z \mapsto (\operatorname{Im} z)^{\frac{\rho-1}{2}} \chi^{\text{even}}_{\rho,\nu} \left(\frac{\operatorname{Re} z}{\operatorname{Im} z} \right)$, applying the Poincaré summation process with respect to the group Γ. We consider now, more generally, functions of the kind

$$\phi(z) = (\operatorname{Im} z)^{\frac{\rho-1}{2}} \chi \left(\frac{\operatorname{Re} z}{\operatorname{Im} z} \right), \tag{4.6.1}$$

in other words functions on Π homogeneous of degree $\frac{\rho-1}{2}$: recall from (2.3.15) that, under the dual Radon transform, such functions arise from generalized eigenfunctions, in the plane with coordinates x, ξ, of the operator $\mathcal{B} = \frac{1}{4i\pi}\left(x\frac{\partial}{\partial x} - \xi\frac{\partial}{\partial \xi}\right)$. We shall also assume that χ is even, i.e., that one has $\phi(z) = \phi(-\bar{z})$.

Theorem 2.4.2 provided a fairly general class of such functions ϕ with nice analytical properties: one such example is the function

$$\phi_{\rho,\alpha}(z) = |z|^{\frac{\rho-1}{2}} K_{\frac{\rho-1}{2}}\left(\frac{\alpha|z|}{\operatorname{Im} z}\right), \tag{4.6.2}$$

depending on the free parameter $\alpha > 0$. It has the form (4.6.1) with $\chi(t) = (1+t^2)^{\frac{\rho-1}{4}} K_{\frac{\rho-1}{2}}(\alpha\sqrt{1+t^2})$.

Given any function $\phi(z) = (\operatorname{Im} z)^{\frac{\rho-1}{2}}\chi\left(\frac{\operatorname{Re} z}{\operatorname{Im} z}\right)$ where, as a start, the function χ lies in $\mathcal{S}_{\text{even}}(\mathbb{R})$, we denote as $\mathcal{P}\phi$ the automorphic function — to be called an incomplete ρ-series in analogy with the concept of incomplete Eisenstein (or incomplete theta) series — such that

$$(\mathcal{P}\phi)(z) = \left(\Delta - \frac{1-\rho^2}{4}\right)\left(\Delta - \frac{1-(2-\rho)^2}{4}\right)\cdot\frac{1}{2}\sum_{g\in\Gamma}\phi(g.z). \tag{4.6.3}$$

The series converges for every $z \in \Pi$, as it follows from the methods of Section 3.2, and defines a C^∞ automorphic function. The purpose of the operator $\left(\Delta - \frac{1-\rho^2}{4}\right)\left(\Delta - \frac{1-(2-\rho)^2}{4}\right)$, which could also be applied, instead, to ϕ before the summation process has taken place, is to kill bad terms from the Roelcke-Selberg expansion of $\mathcal{P}\phi$, as will be seen presently.

Note that the summation process destroys the homogeneity of functions but, after it has been performed, something important remains from the parameter ρ anyway, as will be analyzed more thoroughly in the next section. With ϕ as in (4.6.1) (and χ even), one has $\phi(-z^{-1}) = |z|^{1-\rho}\phi(z)$, a function homogeneous of degree $\frac{1-\rho}{2}$, and the space of $\mathcal{P}$-series associated with functions homogeneous of degree $\frac{\rho-1}{2}$ or $\frac{1-\rho}{2}$ coincide. In particular, $\phi_{\rho,\alpha}(-z^{-1}) = \phi_{2-\rho,\alpha}(z)$, so that $\mathcal{P}\phi_{\rho,\alpha} = \mathcal{P}\phi_{2-\rho,\alpha}$. We compute now the Roelcke-Selberg expansion of the function $\mathcal{P}\phi$, with the help of Theorem 2.4.2.

Lemma 4.6.1. *Let Ψ be a bounded holomorphic even function in some strip $\{z \in \mathbb{C}: |\operatorname{Im} z| < \beta_0\}$, such that $\int_{\operatorname{Im} z=\beta} |\Psi(z)|^2 dz < \infty$ whenever $|\beta| < \beta_0$. For $0 < \beta < \beta_0$, one has the identity*

$$\Psi(\lambda) = \frac{1}{i\pi}\int_{\operatorname{Im}\mu=\beta}\frac{\mu}{\lambda^2-\mu^2}\Psi(\mu)d\mu = \frac{1}{\pi}\int_{\operatorname{Re}\nu=-\beta}\frac{i\nu}{\lambda^2+\nu^2}\Psi(i\nu)d\nu. \tag{4.6.4}$$

Proof. For $\lambda \in \mathbb{R}$ and $\beta \in]0, \beta_0[$, one has, since Ψ is even,

$$\frac{1}{2i\pi} \int_{\operatorname{Im}\mu=\beta} \frac{\Psi(\mu)}{\mu - \lambda} d\mu = -\frac{1}{2i\pi} \int_{\operatorname{Im}\mu=-\beta} \frac{\Psi(\mu)}{\mu + \lambda} d\mu, \qquad (4.6.5)$$

so that

$$\frac{1}{2i\pi} \int_{\operatorname{Im}\mu=\beta} \Psi(\mu) \left[\frac{1}{\mu - \lambda} + \frac{1}{\mu + \lambda} \right] d\mu = \frac{1}{2i\pi} \left[\int_{\operatorname{Im}\mu=\beta} \frac{\Psi(\mu)}{\mu - \lambda} - \int_{\operatorname{Im}\mu=-\beta} \frac{\Psi(\mu)}{\mu - \lambda} \right]$$
$$= -\Psi(\lambda). \qquad (4.6.6)$$

$\square$

Theorem 4.6.2. *Let F be an odd function on the real line, such that, for some $\beta_0 > 0$, $\int_{-\infty}^{\infty} |F(\sigma)|^2 e^{4\pi|\beta|\sigma} d\sigma < \infty$ whenever $|\beta| < \beta_0$. Let G be the even function on the real line such that $\frac{1}{2i\pi} G' = F$ and $G(\pm\infty) = 0$, and let ϕ be the function associated to G under equation (2.4.20), in the way introduced in Theorem 2.4.2. The Roelcke-Selberg expansion of the automorphic function $\mathcal{P}\phi$, as defined in (4.6.3), is*

$$(\mathcal{P}\phi)(z) = \frac{1}{8\pi} \int_{-\infty}^{\infty} \Phi(\lambda) E_{\frac{1-i\lambda}{2}}(z) d\lambda + \sum_{p\geq 1} \sum_j \Phi^{p,j} \mathcal{M}_{p,j}(z), \qquad (4.6.7)$$

with

$$\Phi(\lambda) = \widehat{G}(\lambda) \frac{(\rho^2 + \lambda^2)((2-\rho)^2 + \lambda^2)}{16} \frac{\zeta^*(\frac{\rho-i\lambda}{2})\zeta^*(\frac{\rho+i\lambda}{2})}{\zeta^*(1+i\lambda)} \qquad (4.6.8)$$

and, for every $p \geq 1$,

$$\sum_j \Phi^{p,j} \mathcal{M}_{p,j} = \ddot{G}(\lambda_p) \frac{(\rho^2 + \lambda_p^2)((2-\rho)^2 + \lambda_p^2)}{32} \sum_j L^*(\frac{\rho}{2}, \mathcal{M}_{p,j}) \mathcal{M}_{p,j}, \qquad (4.6.9)$$

a sum in which only eigenforms $\mathcal{M}_{p,j}$ of even type are retained.

Proof. It is an elementary matter (use the fact that the convolution by a summable function on the line is an endomorphism of $L^2(\mathbb{R})$) that the indefinite integral G of F vanishing at infinity satisfies, just like F, the condition $\int_{-\infty}^{\infty} |G(\sigma)|^2 e^{4\pi\beta_0\sigma} d\sigma < \infty$. Using Theorem 2.4.2 to express ϕ as an integral superposition of the functions $z \mapsto (\operatorname{Im} z)^{\frac{\rho-1}{2}} \chi_{\rho,\nu}^{\text{even}}(\frac{\operatorname{Re} z}{\operatorname{Im} z})$, then definition (4.2.1) of the function $f_{\rho,\nu}$ and Theorems 4.5.7 and 4.5.8 which give the Roelcke-Selberg expansions of the functions $f_{\rho,\nu}$, finally the pair of equations

$$\left(\Delta - \frac{1-\rho^2}{4}\right) E_{\frac{1-i\lambda}{2}} = \frac{\rho^2 + \lambda^2}{4} E_{\frac{1-i\lambda}{2}}, \qquad \left(\Delta - \frac{1-\rho^2}{4}\right) \mathcal{M}_{p,j} = \frac{\rho^2 + \lambda_p^2}{4} \mathcal{M}_{p,j},$$
$$(4.6.10)$$

we obtain that the Roelcke-Selberg expansion of $\mathcal{P}\phi$ is given by (4.6.7), with

$$\Phi(\lambda) = \frac{(\rho^2 + \lambda^2)((2-\rho)^2 + \lambda^2)}{16\pi} \int_{\operatorname{Re}\nu = -\beta} \widehat{F}(i\nu) \frac{d\nu}{\nu^2 + \lambda^2} \times \frac{\zeta^*(\frac{\rho - i\lambda}{2})\zeta^*(\frac{\rho + i\lambda}{2})}{\zeta^*(1 + i\lambda)},$$

$$\Phi^{p,j} = \frac{(\rho^2 + \lambda_p^2)((2-\rho)^2 + \lambda_p^2)}{32\pi} \int_{\operatorname{Re}\nu = -\beta} \widehat{F}(i\nu) \frac{d\nu}{\nu^2 + \lambda_p^2} \times L^*(\frac{\rho}{2}, \mathcal{M}_{p,j}).$$

$$(4.6.11)$$

Since $\widehat{F}(\lambda) = \lambda \widehat{G}(\lambda)$, the theorem then follows from Lemma 4.6.1. Note the role of the extra operator in front of the right-hand side of (4.6.3): for $\rho \neq 1$, it kills the terms proportional to $E_{\frac{1+\rho}{2}}$ and $E_{\frac{3-\rho}{2}}$ from the right-hand side of (4.5.58); when $\rho = 1$, remembering that $\Delta E_1^\natural$ is a constant, one sees that it kills the constant and the multiple of $E_1^\natural$ from the right-hand side of (4.5.65). $\square$

In particular,

Proposition 4.6.3. *The coefficients of the Roelcke-Selberg expansion of the auto-morphic function $\left(\frac{2\alpha}{\pi}\right)^{-\frac{1}{2}} \mathcal{P}\phi_{\rho,\alpha}$, with $\phi_{\rho,\alpha}$ as recalled in (4.6.2), are given by the equations*

$$\Phi(\lambda) = -\frac{1}{16}(\rho^2 + \lambda^2)((2-\rho)^2 + \lambda^2) K_{\frac{i\lambda}{2}}(\alpha) \frac{\zeta^*(\frac{\rho - i\lambda}{2})\zeta^*(\frac{\rho + i\lambda}{2})}{\zeta^*(1 + i\lambda)},$$

$$\Phi^{p,j} = -\frac{1}{32}(\rho^2 + \lambda_p^2)((2-\rho)^2 + \lambda_p^2) K_{\frac{i\lambda_p}{2}}(\alpha) L^*(\frac{\rho}{2}, \mathcal{M}_{p,j}). \qquad (4.6.12)$$

4.7   The automorphic measures $ds_\Sigma^{(\rho)}$; related work

In this section, we pay a renewed attention to the union Σ of lines congruent to the hyperbolic line from 0 to $i\infty$, with more emphasis on the parameter ρ than on the parameter ν. Also, we briefly discuss some analogues, or possible analogues, of the whole theory, obtained by changing Σ to the union of another set of lines. We first clarify the present constructions by linking them to the automorphic Green's kernel theory.

Let $k_{\frac{1-\nu}{2}}(z, z')$ be the integral kernel, as recalled in (2.4.1), of the resolvent $\left(\Delta - \frac{1-\nu^2}{4}\right)^{-1}$ in $L^2(\Pi)$. The automorphic Green's kernel $G_{\frac{1-\nu}{2}}(z, z')$ is defined as

$$G_{\frac{1-\nu}{2}}(z, z') = \frac{1}{2} \sum_{g \in \Gamma} k_{\frac{1-\nu}{2}}(z, g.z'), \qquad (4.7.1)$$

a series absolutely convergent for $\operatorname{Re}\nu < -1$. Unless z is congruent to i or to $\frac{1+i\sqrt{3}}{2}$, the stabilizer of z in Γ reduces to the pair $\{I, -I\}$, so that $\left(\Delta - \frac{1-\nu^2}{4}\right) G_{\frac{1-\nu}{2}}(z, z')$, considered as a distribution $\mathfrak{S}_z$ with respect to z', coincides in general with the series of unit masses at points in the Γ'-orbit of z, where $\Gamma' = \Gamma/\{\pm I\}$ is the image of Γ in $PSL(2, \mathbb{R})$.

Now, the measure $ds_\Sigma^{(\rho)}(z')$, in other words the sum of g-transforms, with $g \in \Gamma'$, of the measure $\delta_{(0,i\infty)}^{(\rho)} = y'^{\frac{\rho-1}{2}} \frac{dy'}{y'}$ (or of the measure $\frac{1}{2}\left(\delta_{(0,i\infty)}^{(\rho)} + \delta_{(0,i\infty)}^{(2-\rho)}\right)$) considered on the half-line from i to $i\infty$, is just the integral with respect to $\delta_{(0,i\infty)}^{(\rho)}(z)$, taken on this half-line, of the measure $\mathfrak{S}_z$ just considered. In view of Theorem 4.2.4, one might thus expect that a formula such as

$$f_{\rho,\nu}(z') \qquad \text{similar to} \qquad C(\rho, \nu) \int_1^\infty G_{\frac{1-\nu}{2}}(iy, z')\left(y^{\frac{\rho-1}{2}} + y^{\frac{1-\rho}{2}}\right)\frac{dy}{y} \qquad (4.7.2)$$

could be valid. However, this is not possible, since such an integral is divergent: but we show now that, after we have taken care of the divergence in a natural way — removing infinities — the formula will become true.

Theorem 4.7.1. *Assume that* $0 < \operatorname{Re}\rho < 2$, *and* $\nu \notin \mathbb{Z}, \operatorname{Re}\nu < 0$. *One has, if* $\rho \neq 1$,

$$[C(\rho, \nu)]^{-1}f_{\rho,\nu}(z') = \frac{4}{\nu^2 - \rho^2}E_{\frac{1+\rho}{2}}(z') + \frac{4}{\nu^2 - (2-\rho)^2}E_{\frac{3-\rho}{2}}(z')$$

$$+ \lim_{Y \to \infty}\left[\int_1^Y G_{\frac{1-\nu}{2}}(iy, z')\left(y^{\frac{\rho-1}{2}} + y^{\frac{1-\rho}{2}}\right)\frac{dy}{y}\right.$$

$$\left. - \int_{D_Y} G_{\frac{1-\nu}{2}}(z, z')\left[E_{\frac{1+\rho}{2}}(x + iy)y^{\frac{\rho-1}{2}} + E_{\frac{3-\rho}{2}}(x + iy)y^{\frac{1-\rho}{2}}\right]\frac{dxdy}{y^2}\right], \quad (4.7.3)$$

while

$$[C(1, \nu)]^{-1}f_{1,\nu}(z') = \frac{8}{\nu^2 - 1}E_1^\natural(z') + \frac{96}{\pi}\frac{1}{(\nu^2 - 1)^2}$$

$$+ 2\lim_{Y \to \infty}\left[\int_1^Y G_{\frac{1-\nu}{2}}(iy, z')\frac{dy}{y} - \int_{D_Y} G_{\frac{1-\nu}{2}}(z, z')E_1^\natural(x + iy)\frac{dxdy}{y^2}\right]. \quad (4.7.4)$$

Proof. Consider first the case when $\rho \neq 1$, and break the main bracket on the right-hand side of (4.7.3) into three pieces, according to the decomposition [55, p. 271]

$$G_{\frac{1-\nu}{2}}(z, z') = \frac{12}{\pi}(\nu^2 - 1)^{-1} + \frac{1}{2\pi}\int_{-\infty}^\infty E_{\frac{1+i\lambda}{2}}(z)E_{\frac{1-i\lambda}{2}}(z')\frac{d\lambda}{\nu^2 + \lambda^2}$$

$$+ 4\sum_{p,j}\frac{\mathcal{M}_{p,j}(z)\mathcal{M}_{p,j}(z')}{\nu^2 + \lambda_p^2} \qquad (4.7.5)$$

of the automorphic Green's kernel. In view of Theorem 4.5.7, equation (4.7.3) is equivalent to a set of three identities, the first of which is that

$$\int_1^Y \left(y^{\frac{\rho-1}{2}} + y^{\frac{1-\rho}{2}}\right) \frac{dy}{y} - \int_{D_Y} \left(E_{\frac{1+\rho}{2}}(z) + E_{\frac{3-\rho}{2}}(z)\right) \frac{dxdy}{y^2} \qquad (4.7.6)$$

goes to zero as $Y \to \infty$. To see this, we apply (4.5.19) with $f = 1$ and $h = f^\sharp_{\rho,\nu}$, using Lemma 4.5.3 to the effect that

$$\Delta f^\sharp_{\rho,\nu} = \frac{1-\nu^2}{4} f^\sharp_{\rho,\nu} + C(\rho,\nu) \left[2ds_\Sigma^{(\rho)} - E_{\frac{1+\rho}{2}} + E_{\frac{3-\rho}{2}}\right] \qquad (4.7.7)$$

and the fact that, according to the second part of Theorem 4.5.5, $(1|f^\sharp_{\rho,\nu}) = \frac{\pi}{3}\Phi^0_{\rho,\nu} = 0$. This takes care of the identities between the constant terms of the spectral decompositions of the two sides of (4.7.3). The second identity, which expresses that the spectral densities of the two sides are identical too, has already been written, coupling (4.5.12) with the first part of Theorem 4.5.5. The identity between the discrete parts is a consequence of Proposition 4.4.6.

In the case when $\rho = 1$, the only difference lies in the evaluation of the constant term: one must replace (4.7.7) by

$$\Delta f^\sharp_{1,\nu} = \frac{1-\nu^2}{4} f^\sharp_{1,\nu} + \frac{24}{\pi} \frac{C(1,\nu)}{\nu^2 - 1} + 2C(\rho,\nu) \left[ds_\Sigma^{(1)} - E^\flat_1\right], \qquad (4.7.8)$$

recalling from (4.5.45) that $(1|f^\sharp_{1,\nu}) = \frac{\pi}{3} \times \frac{24}{\pi} \frac{I(1,\nu)}{\nu^2-1}$.   □

It is possible, with some prudence of language, to get rid of ν and make what is in effect the Roelcke-Selberg expansion of the measure $ds_\Sigma^{(\rho)}$ explicit.

Theorem 4.7.2. *Assume that* $0 < \text{Re}\,\rho < 2$. *In the weak sense when integrated on D against automorphic functions f in Π such that $f(x+iy)$ and $\Delta f(x+iy)$ are for some $\varepsilon > 0$ a $O\left(y^{-\frac{|\text{Re}\,\rho-1|}{2}-\varepsilon}\right)$ as $y \to \infty$, one has if $\rho \neq 1$ the decomposition*

$$ds_\Sigma^{(\rho)} = \frac{1}{2}\left(E_{\frac{1+\rho}{2}} + E_{\frac{3-\rho}{2}}\right) + \frac{1}{16\pi} \int_{-\infty}^\infty \frac{\zeta^*(\frac{\rho-i\lambda}{2})\zeta^*(\frac{\rho+i\lambda}{2})}{\zeta^*(1+i\lambda)} E_{\frac{1-i\lambda}{2}} d\lambda$$

$$+ \frac{1}{4} \sum_{p,j\,even} L^*(\frac{\rho}{2}, \mathcal{M}_{p,j})\mathcal{M}_{p,j}; \quad (4.7.9)$$

when $\rho = 1$, one has the similar decomposition

$$ds_\Sigma = E^\flat_1 + \frac{1}{16\pi} \int_{-\infty}^\infty \frac{(\zeta^*(\frac{1-i\lambda}{2}))^2}{\zeta^*(1+i\lambda)} E_{\frac{1-i\lambda}{2}} d\lambda + \frac{1}{4} \sum_{p,j\,even} L^*(\frac{1}{2}, \mathcal{M}_{p,j})\mathcal{M}_{p,j}. \quad (4.7.10)$$

Proof. One evaluates the integral

$$\left(\left(\Delta - \frac{1-\nu^2}{4}\right)f|f_{\rho,\nu}\right) = \int_D \left(\Delta - \frac{1-\bar{\nu}^2}{4}\right)\bar{f}(x+iy)f_{\rho,\nu}(x+iy)\frac{dxdy}{y^2} \quad (4.7.11)$$

with the help of (4.5.58), after which it suffices to integrate by parts and to apply Theorem 4.2.4. In the second case, the extra term originating from the term $\frac{96}{\pi}\frac{1}{(\nu^2-1)^2}$ from (4.5.65) cancels off thanks to the second term on the right-hand side of the relation

$$\left(\Delta - \frac{1-\nu^2}{4}\right)E_1^{\natural} = \frac{\nu^2-1}{4}E_1^{\natural} - \frac{3}{\pi}. \quad (4.7.12)$$

$\square$

In analogy with Definition 4.5.2, we set

$$\left[ds_{\Sigma}^{(\rho)}\right]^{\sharp} = ds_{\Sigma}^{(\rho)} - \frac{1}{2}\left(E_{\frac{1+\rho}{2}} + E_{\frac{3-\rho}{2}}\right), \quad \rho \neq 1, \quad (4.7.13)$$

with an obvious modification when $\rho = 1$. With the help of Theorem 2.4.2, we now reformulate Theorem 4.6.2: this will clarify the role of ρ.

Theorem 4.7.3. *Let H be an even function satisfying the conditions in Theorem 2.4.2, and define the operator $H\left(2\sqrt{\Delta - \frac{1}{4}}\right)$ accordingly, with the difference, however, that Δ now stands for the modular Laplacian. Set*

$$f = H\left(2\sqrt{\Delta - \frac{1}{4}}\right)\left[ds_{\Sigma}^{(\rho)}\right]^{\sharp}. \quad (4.7.14)$$

One has, whenever $\beta < \beta_0$, the estimate $f(z) = O\left((\mathrm{Im}\,z)^{\frac{1-\beta}{2}}\right)$ as $\mathrm{Im}\,z \to \infty$ while z lies in the fundamental domain, and the spectral decomposition

$$f(z) = \frac{1}{8\pi}\int_{-\infty}^{\infty} \Phi(\lambda)F_{\frac{1-i\lambda}{2}}(z)d\lambda + \sum_{p\geq 1}\sum_j \Phi^{p,j}\mathcal{M}_{p,j}(z), \quad (4.7.15)$$

with

$$\Phi(\lambda) = \frac{1}{2}H(\lambda)\frac{\zeta^*(\frac{\rho-i\lambda}{2})\zeta^*(\frac{\rho+i\lambda}{2})}{\zeta^*(1+i\lambda)} \quad (4.7.16)$$

and, for every $p \geq 1$,

$$\sum_j \Phi^{p,j}\mathcal{M}_{p,j} = \frac{1}{4}H(\lambda_p)\sum_j L^*(\frac{\rho}{2}, \mathcal{M}_{p,j})\mathcal{M}_{p,j} \quad (4.7.17)$$

(a sum in which only eigenforms $\mathcal{M}_{p,j}$ of even type are considered).

Proof. If one writes, as in Theorem 2.4.2, but addressing oneself, now, to the automorphic Laplacian rather than the free one in the half-plane,

$$H\left(2\sqrt{\Delta - \frac{1}{4}}\right) = -\frac{1}{4i\pi}\int_{\operatorname{Re}\nu=-\beta} \nu H(i\nu)\left(\Delta - \frac{1}{4}\right)^{-1} d\nu, \tag{4.7.18}$$

one finds

$$f(z) = -\frac{1}{8i\pi}\int_{\operatorname{Re}\nu=-\beta} \nu H(i\nu)$$
$$\times \left[\frac{f_{\rho,\nu}(z)}{C(\rho,\nu)} - \frac{4}{\nu^2 - \rho^2}E_{\frac{1+\rho}{2}}(z) - \frac{4}{\nu^2 - (2-\rho)^2}E_{\frac{3-\rho}{2}}(z)\right] d\nu, \tag{4.7.19}$$

where the bracket on the right-hand side is a $O\left((\operatorname{Im} z)^{\frac{1+\operatorname{Re}\nu}{2}}\right)$ as a consequence of Theorem 4.4.4 (in which the implied constant does not depend on the imaginary part of ν), so that f lies in $L^2(\Gamma\backslash\Pi)$ in view of the integrability assumptions made about H. The rest of the proof is just the same as Theorem 4.6.2, with the difference that, having already subtracted the two "bad terms" from $ds_\Sigma^{(\rho)}$, we no longer need to apply the operator $\left(\Delta - \frac{1-\rho^2}{4}\right)\left(\Delta - \frac{1-(2-\rho)^2}{4}\right)$ to kill them. □

Remarks 4.7.a. Theorem 4.7.3 calls for a number of remarks and clarifications. We assume here that the number ρ, with $0 < \operatorname{Re}\rho < 2$, has been fixed.

(i) From the estimate of $f(z)$ at infinity in the fundamental domain, it follows that the density $\Phi(\lambda) = (E_{\frac{1-i\lambda}{2}}|f)$ is a C^∞ function of λ, an essential point in the following remarks, in which pointwise values of Φ and of some of its derivatives need to be considered.

(ii) The discrete part (relative to its spectral decomposition) of the function f is entirely determined by its continuous part: the projection of f on the eigenspace of the (even part of the) modular Laplacian corresponding to the eigenvalue $\frac{1+\lambda_p^2}{4}$ is a multiple of the cusp-form $\sum_j L^*(\frac{\rho}{2}, \mathcal{M}_{p,j})\mathcal{M}_{p,j}$, and the global coefficient is the product of $\Phi(\lambda_p)$ by an explicit function of the pair (ρ, λ_p).

(iii) For any given p, the cusp-forms $\sum_j L^*(\frac{\rho}{2}, \mathcal{M}_{p,j})\mathcal{M}_{p,j}$ will generate the totality of the discrete eigenspace under consideration if one allows ρ to vary as well. If the eigenspace of the even part of Δ corresponding to some eigenvalue $\frac{1+\lambda_p^2}{4}$ is simple, taking just one value of ρ such that $L(\frac{\rho}{2}, \mathcal{M}_p) \neq 0$ would suffice.

(iv) Let $L^{2,\infty}(\Gamma\backslash\Pi)$ be the space of automorphic functions f satisfying the property that the product of $f(z)$ by an arbitrary power of $\log(\operatorname{Im} z)$ is square-integrable in the fundamental domain, so that, as a consequence, the spectral density Φ of f is C^∞: there is on this space a natural Fréchet topology. From the closure, in this space, of the space of functions f introduced in (4.7.14), it is clear, just looking at the zeros of the spectral density $\Phi(\lambda)$, that one can determine whether $\mathrm{Re}\,\rho = 1$ or not; also, in the first case, the value of ρ. But knowing whether this closure coincides, for every given value of ρ with $\mathrm{Re}\,\rho \neq 1$, with the full space $L^{2,\infty}(\Gamma\backslash\Pi)$, is another matter. Indeed, so far as the continuous part of spectral decompositions is concerned, it depends on whether the Riemann hypothesis is true or not for zeta; so far as the discrete part is concerned, it depends on the validity of the same hypothesis for L-functions associated to cusp-forms of even type, and on the simplicity, or not, of the discrete eigenspaces of even type.

(v) If λ is a zero of order κ of the product $\zeta^*(\frac{\rho-i\lambda}{2})\zeta^*(\frac{\rho+i\lambda}{2})$, the density Φ of f (as defined by (4.7.14)) must vanish to the order κ (at least) at λ. In other words, the scalar product of f against all functions $(\frac{\partial}{\partial\lambda})^j E_{\frac{1-i\lambda}{2}}$ with $0 \leq j \leq \kappa$ must be zero there. The space generated by the functions $(\frac{\partial}{\partial s})^j E_s(z)$ with j less than the multiplicity of a given zero of zeta appears in Zagier's paper [69]: this coincidence will be explained at the end of this section, after we have made a short survey of some partial analogue [60, Sec. 19-20] of the present work, developed in relation to real quadratic extensions of the rationals.

Let us indicate that, besides the measures $ds_{\Sigma}^{(\rho)}$, there is a considerable variety of other possible one-dimensional automorphic objects, since nothing — save our desire to deal with a situation as simple as possible, from the point of view of arithmetic if not from that of analysis — forces us to consider the set Σ, among other possibilities. As a starting point, let us consider a nonzero real quadratic form $R(x, \xi) = px^2 + 2qx\xi + r\xi^2$ on $\mathbb{R}^2$: it is the Weyl symbol of the differential operator $\mathrm{Op}(R) = pQ^2 + q(QP + PQ) + rP^2$, with Q and P as defined in (1.1.4) (this is the one-dimensional case). The symbol of the commutator $[\mathrm{Op}(R), \mathrm{Op}(h)]$ of $\mathrm{Op}(R)$ with an arbitrary pseudo-differential operator is then, as a consequence of the relations (1.2.6), the Poisson bracket

$$\frac{1}{2i\pi}\{R, h\} = \frac{1}{i\pi}[(qx + r\xi)\frac{\partial h}{\partial x} - (px + q\xi)\frac{\partial h}{\partial \xi}]$$

$$= \frac{1}{i\pi}\frac{d}{dt}\Big|_{t=0}(h \circ \exp(t(\begin{smallmatrix} q & r \\ -p & -q \end{smallmatrix}))), \qquad (4.7.20)$$

involving the element $X = (\begin{smallmatrix} q & r \\ -p & -q \end{smallmatrix})$ of the Lie algebra $\mathfrak{sl}(2, \mathbb{R})$. Decomposing operators according to the way the commutation with $\mathrm{Op}(R)$ acts on them is thus equivalent with the decomposition of their symbols as generalized eigenvectors of the first-order operator $\{R, .\}$.

Now, the quadratic form $R(x, \xi)$ can be positive-definite (or negative-definite, a case which needs not be considered on its own) or indefinite, or degenerate, i.e., it has to be equivalent, under conjugation by some matrix in $SO(2)$, to a multiple of $x^2 + \xi^2$, $x\xi$ or ξ^2: then, X lies in the orbit in $\mathfrak{sl}(2, \mathbb{R})$, under the adjoint action of $SL(2, \mathbb{R})$, of a multiple of $(\begin{smallmatrix} 0 & 1 \\ -1 & 0 \end{smallmatrix})$, $(\begin{smallmatrix} \frac{1}{2} & 0 \\ 0 & -\frac{1}{2} \end{smallmatrix})$ or $(\begin{smallmatrix} 0 & 1 \\ 0 & 0 \end{smallmatrix})$, and the one-parameter subgroup of $SL(2, \mathbb{R})$ generated by X is conjugate to one of the three subgroups K, A, N from the Iwasawa decomposition of $SL(2, \mathbb{R})$.

Given $X \in \mathfrak{sl}(2, \mathbb{R})$ let us transfer the operator $\{R, \bullet\}$ to Π, getting as a result the operator $\mathcal{L}_X$ defined as $\mathcal{L}_X \phi = \frac{d}{dt}\Big|_{t=0} (\phi \circ \exp(tX))$: the (generalized) eigenfunctions of this operator, solutions of the equation $\mathcal{L}_X \phi = \frac{\rho-1}{2}\phi$, are thus characterized by the equation $\phi \circ \exp(tX) = t^{\frac{\rho-1}{2}}\phi$. We may then ask for the study of the Poincaré series built from such functions ϕ, and for their Roelcke-Selberg decompositions.

This may involve serious difficulties in general, as the one-parameter group G_X (infinitesimally) generated by X may combine with the arithmetic group under consideration (we here specialize in Γ, the full modular group) in a variety of ways. To start with, defining the Poincaré series as

$$z \mapsto \frac{1}{2} \sum_{g \in (G_X \cap \Gamma) \backslash \Gamma} \phi \circ g, \qquad (4.7.21)$$

so as to avoid, in particular, repeating infinitely many times the same term in the case when $G_X \cap \Gamma$ is infinite, it is clear that the way this intersection lies within Γ is all-important: it would also be hard to describe it in general. We shall satisfy ourselves, here, with showing how certain constructions, more or less classical, fit within this scheme.

The best-known case is that for which $X = \left(\begin{smallmatrix} 0 & 1 \\ 0 & 0 \end{smallmatrix}\right)$, so that the associated one-parameter group G_X is N, and $\frac{\rho-1}{2} = 0$: this means that we start from general functions of $\operatorname{Im} z$ alone, and the Poincaré series obtained in this way are exactly the incomplete Eisenstein (or theta-) series. The case when $X = \left(\begin{smallmatrix} 0 & 1 \\ -1 & 0 \end{smallmatrix}\right), G_X = SO(2)$ and $\frac{\rho-1}{2} = 0$ is involved in the construction of the automorphic Green's kernel. We wish to spend some more time reviewing cases in which the group G_X is conjugate to A within $SL(2, \mathbb{R})$. The case when $X = \left(\begin{smallmatrix} \frac{1}{2} & 0 \\ 0 & -\frac{1}{2} \end{smallmatrix}\right), G_X = A$ and $0 < \operatorname{Re}\rho < 2$ has been detailed in the present chapter. When G_X is conjugate to A, i.e., when $X = \left(\begin{smallmatrix} q & r \\ -p & -q \end{smallmatrix}\right)$ with $q^2 - pr > 0$, the set of fixed points of G_X consists of a pair (α, β) of points on $\mathbb{R} \cup \{\infty\}$, and the hyperbolic line through (α, β) is globally invariant under the action of G_X. Let us denote as $\Sigma^{(\alpha,\beta)}$ the union of lines congruent, under Γ, to the line just considered: the set $\Sigma^{(\alpha,\beta)}$ takes the place of the set $\Sigma = \Sigma^{(0,\infty)}$ used in the present work.

It is very likely that, were one willing to face the arithmetic complications associated with the necessary consideration of congruence subgroups of Γ (even when automorphy is defined in terms of Γ), one could study along the same lines the case when (α, β) is any pair of rationals.

In [60, Sec. 19-20], we considered a pair $\tau, \bar{\tau}$ of conjugate irrational numbers in a real quadratic extension of the rationals (this pair was denoted as $\rho, \bar{\rho}$ there, a notation preempted here). We then studied automorphic functions (let us denote them as $f_{1,\nu}^{\tau}$ here) which, in view of the present work, should be considered as the analogues of the functions $f_{1,\nu}$. There are of course many arithmetic differences; also, the analysis was simpler, there, on two accounts. Starting from the function

$$\phi(z) = \chi_{1,\nu}^{\text{even}}\left(\frac{-z+\overline{\tau}}{z-\tau}\right), \qquad (4.7.22)$$

a function invariant under the group $\Gamma_\tau = \{g = \left(\begin{smallmatrix} a & b \\ c & d \end{smallmatrix}\right) : \frac{a\tau+b}{c\tau+d} = \tau\}$, the summation involved in the Poincaré series takes place with respect to $g \in \Gamma_\tau \backslash \Gamma$: this is already a simplification since the group Γ_τ is infinite, which reduces the convergence problems. Then, the theory developed in a way fairly comparable — with the exception of the arithmetic developments, somewhat more involved there — to that which concerns the function $f_{1,\nu}$ in the present work: in particular, the singularities of the Poincaré series obtained are supported in $\Sigma^\tau = \Sigma^{\tau,\overline{\tau}}$ and still consist of discontinuities of the normal derivative. There is a major difference, however, in that the image $\mathfrak{K}$ of Σ^τ in $\Gamma\backslash\Pi$ is compact. A related fact is that there is no need to subtract linear combinations of Eisenstein series from the automorphic function $f_{1,\nu}^{\tau}$ to make it square-integrable: in other words, in this theory, one has $\left[f_{1,\nu}^{\tau}\right]^{\sharp} = f_{1,\nu}^{\tau}$. We had been partly motivated towards this construction by Hejhal's construction of "pseudo-cusp forms" [16], in which imaginary quadratic extensions of the rationals had been considered. With the experience acquired from the present work, it is clear that we might (and should) have considered also the parameter-dependent theory involving the number denoted as ρ here.

The Roelcke-Selberg expansion of the functions $f_{1,\nu}^{\tau}$ was obtained (*loc. cit.*, Prop. 20.7 and Theorem 20.5). We need only to recall here the expression of the spectral density $\Phi(\lambda)$: it was found to be $\frac{2C(1,\nu)}{\nu^2+\lambda^2}c_0(-i\lambda)$, a function obtained by analytic continuation from the function

$$c_0(\mu) = \frac{(\Gamma(\frac{1-\mu}{4})^2}{\Gamma(\frac{1-\mu}{2})}(\tau-\overline{\tau})^{\frac{1-\mu}{2}} \sum_{\left(\begin{smallmatrix} n_1 & n_2 \\ m_1 & m_2 \end{smallmatrix}\right)\in\Gamma_\rho\backslash\Gamma/(\Gamma\cup N)} |N(m_1\tau-n_1)|^{\frac{\mu-1}{2}}, \qquad (4.7.23)$$

involving the norm from $\mathbb{Q}$ to the extension under consideration, and so defined for $\operatorname{Re}\mu < -1$. There is on Σ^τ a canonical line element and, up to multiplication by some constant with a geometric signification, this spectral density can be obtained as the product of $\frac{C(1,\nu)}{\nu^2+\lambda^2}$ by the integral on $\mathfrak{K}$ of the Eisenstein series $E_{\frac{1+i\lambda}{2}}$: recall that $\mathfrak{K}$ is compact. Calculations which lead to the link between this integral and the zeta-function (4.7.23) of some ideal class in the quadratic field under consideration can already be found in Siegel [48, p. 107-128] and in Hecke [15, p. 201], a citation taken from [69]. In this latter reference, Zagier considered in a systematic way linear forms associated with Eisenstein series E_s or their derivatives vanishing in the case when s is a zero of some appropriate order of the zeta function. One of these linear forms was obtained as a sum, leading to the zeta function of some real quadratic field, of Hecke's zeta functions of ideal classes of that field, and as such as a linear combination of integrals taken on compact curves of the type of $\Re$. Despite the fact that, in that work, these integrals did not arise from the calculation of the spectral density of some automorphic function, this explains — in the quadratic case — the coincidence noted as point (v) in the list of remarks which followed the proof of Theorem 4.7.3.

4.8   A duality

The measures $ds_{\Sigma}^{(\rho)}$ may be regarded as the generalized eigenfunctions of some operator on a space of one-dimensional measures supported in Σ, in the following sense: identifying (up to a point: *cf. infra*) such a measure, under the assumption that it is absolutely continuous with respect to $ds_{\Sigma} = ds_{\Sigma}^{(1)}$, with its Radon-Nikodym density, define, at each point z of Σ, the unordered pair $(\nabla, -\nabla)$ as that consisting of the two normalized tangent vectors to the unique (*cf.* beginning of proof of Theorem 4.2.4) line through z belonging to the set Σ is made of. Then, $ds_{\Sigma}^{(\rho)}$ is a generalized eigenfunction of ∇^2 for the eigenvalue $(\frac{\rho-1}{2})^2$. We specialize from the start in the automorphic situation (so that the measure $ds_{\Sigma}^{(\rho)}$ or its density qualifies), defining the scalar product of two such densities as being their integral with respect to $\frac{dy}{y}$, taken on the half-hyperbolic line from i to $i\infty$. We denote as $L^2(\Gamma\backslash\Sigma)$ the corresponding Hilbert space of functions on Σ, and note that the operator ∇^2 is both Γ-invariant and formally self-adjoint on $L^2(\Gamma\backslash\Sigma)$. It is still useful to make a distinction between this space of functions on Σ and the space $L^2_{\mathrm{meas}}(\Gamma\backslash\Sigma)$ of associated measures supported in Σ.

The duality we have in mind is that which makes it possible, on one hand, to expand an element $d\sigma$ of the Hilbert space $L^2_{\mathrm{meas}}(\Gamma\backslash\Sigma)$ as an integral, to be completed by a series, of generalized eigenfunctions of Δ in $L^2(\Gamma\backslash\Pi)$: Theorem 4.7.2 gave the solution to this problem in the (sufficient) case when $d\sigma = ds_{\Sigma}^{(\rho)}$. In the other direction, we wish to expand the restrictions to Σ of elements f of $L^2(\Gamma\backslash\Pi)$ as integrals of (generalized) eigenfunctions of ∇^2 in $L^2(\Gamma\backslash\Sigma)$. This much simpler problem amounts to decomposing the restriction of f to the hyperbolic line from 0 to $i\infty$ into its homogeneous parts, so that a Mellin transform gives theoretically the solution: the question remains of making the coefficients of this decomposition explicit, in the case when f is an Eisenstein series or a Maass form.

Before we do so, let us note that an automorphic function is far from being characterized, in general, by its restriction to Σ or, what amounts to the same, its restriction to the line $(0, i\infty)$: for this leaves completely indeterminate its restriction to any line congruent to some hyperbolic line from α to β when the pair (α, β) does not lie in the Γ-orbit of the pair $(0, i\infty)$. But if it is a generalized eigenfunction of Δ, of even type for the symmetry $z \mapsto -\bar{z}$, for a known eigenvalue, it is indeed characterized by its restriction to $(0, i\infty)$, as a consequence of the Cauchy-Kowalewska theorem: making instead its L-function explicit is certainly better adapted to the present context, and will be done later.

Denote as $\frac{ds^{(\rho)}}{ds}$ the Radon-Nikodym derivative of the measure $ds_\Sigma^{(\rho)}$ with respect to ds_Σ, which coincides with $\frac{1}{2}\left(y^{\frac{\rho-1}{2}} + y^{\frac{1-\rho}{2}}\right)$ on the line from 0 to $i\infty$. If $\mathcal{M}$ is a cusp-form, its restriction to Σ is given (using the Mellin, or Fourier, inversion formula) by the equation

$$\mathcal{M} = \frac{1}{2i\pi} \int_{\operatorname{Re}\rho=1} \left(ds_\Sigma^{(\rho)}|\mathcal{M}\right) \frac{ds^{(\rho)}}{ds} d\rho \qquad (4.8.1)$$

or, restricting it further without loss of information and taking advantage of the symmetry $\rho \mapsto 2 - \rho$,

$$\mathcal{M}(iy) = \frac{1}{2i\pi} \int_{\operatorname{Re}\rho=1} \left(ds_\Sigma^{(\rho)}|\mathcal{M}\right) y^{\frac{\rho-1}{2}} d\rho. \qquad (4.8.2)$$

More preparation is needed in the case when $f = E_{\frac{1-\nu}{2}}$, in which the scalar product $\left(ds_\Sigma^{(\rho)}|f\right)$ does not make sense. However, define the truncated version of this restricted Eisenstein series as

$$\left(E_{\frac{1-\nu}{2}}\right)_{\text{trunc}}(iy) = \begin{cases} E_{\frac{1-\nu}{2}}(z) - y^{\frac{1-\nu}{2}} - \frac{\zeta^*(-\nu)}{\zeta^*(\nu)} y^{\frac{1+\nu}{2}} & \text{if } y \geq 1, \\ E_{\frac{1-\nu}{2}}(z) - y^{\frac{-1+\nu}{2}} - \frac{\zeta^*(-\nu)}{\zeta^*(\nu)} y^{\frac{-1-\nu}{2}} & \text{if } y \leq 1. \end{cases} \qquad (4.8.3)$$

The function so defined is invariant under the symmetry $y \mapsto y^{-1}$, and rapidly decreasing as $y \to \infty$. Asking for its decomposition into homogeneous components, proportional to functions $\frac{1}{2}(y^{\frac{1-\rho}{2}} + y^{\frac{\rho-1}{2}})$ with $\operatorname{Re}\rho = 1$, certainly has an answer, even though its derivative has a discontinuity at $y = 1$. One has

$$\left(E_{\frac{1-\nu}{2}}\right)_{\text{trunc}}(iy) = 4 \sum_{n\geq 1} n^{-\frac{\nu}{2}} \sigma_\nu(n) y^{\frac{1}{2}} K_{\frac{\nu}{2}}(2\pi n y) \qquad \text{if } y \geq 1 \qquad (4.8.4)$$

while, if $y \leq 1$, the same function is given by the expression

$$y^{\frac{1-\nu}{2}} - y^{\frac{-1+\nu}{2}} + \frac{\zeta^*(-\nu)}{\zeta^*(\nu)}\left(y^{\frac{1+\nu}{2}} - y^{\frac{-1-\nu}{2}}\right) + \frac{4}{\zeta^*(\nu)} \sum_{n\geq 1} n^{-\frac{\nu}{2}} \sigma_\nu(n) y^{\frac{1}{2}} K_{\frac{\nu}{2}}(2\pi n y).$$

$$(4.8.5)$$

We compute now, with $\operatorname{Re}\rho = 1$, the integral

$$\left(ds_\Sigma^{(\rho)}|\left(E_{\frac{1-\nu}{2}}\right)_{\text{trunc}}\right) = \frac{1}{2}\int_0^\infty y^{\frac{\rho-1}{2}}\left(E_{\frac{1-\nu}{2}}\right)_{\text{trunc}}(iy)\frac{dy}{y}. \qquad (4.8.6)$$

More generally, we consider for $\mu \in \mathbb{C}$ the convergent integral

$$I(\mu) = \frac{1}{2}\int_0^\infty y^\mu \left(E_{\frac{1-\nu}{2}}\right)_{\text{trunc}}(iy)\frac{dy}{y}. \qquad (4.8.7)$$

Assuming $\operatorname{Re}\mu > \frac{1}{2}(1+|\operatorname{Re}\nu|)$ makes it possible to take advantage of (4.8.4) and (4.8.5), since then

$$\int_0^1 y^\mu \left[y^{\frac{1-\nu}{2}} - y^{\frac{-1+\nu}{2}} + \frac{\zeta^*(-\nu)}{\zeta^*(\nu)}\left(y^{\frac{1+\nu}{2}} - y^{\frac{-1-\nu}{2}}\right)\right]\frac{dy}{y}$$

$$= \frac{1}{\mu + \frac{1-\nu}{2}} - \frac{1}{\mu + \frac{\nu-1}{2}} + \frac{\zeta^*(-\nu)}{\zeta^*(\nu)}\left[\frac{1}{\mu + \frac{1+\nu}{2}} - \frac{1}{\mu - \frac{1+\nu}{2}}\right]. \qquad (4.8.8)$$

On the other hand [36, p. 91], for $n > 0$,

$$\frac{1}{2}\int_0^\infty y^{\mu+\frac{1}{2}} K_{\frac{\nu}{2}}(2\pi ny)\frac{dy}{y} = \frac{1}{8}\Gamma(\frac{2\mu+1+\nu}{4})\Gamma(\frac{2\mu+1-\nu}{4})(\pi n)^{-\mu-\frac{1}{2}}, \qquad (4.8.9)$$

so that the integral

$$\frac{1}{2}\int_0^\infty y^\mu . \frac{4}{\zeta^*(\nu)}\sum_{n\geq 1} n^{-\frac{\nu}{2}}\sigma_\nu(n)y^{\frac{1}{2}}K_{\frac{\nu}{2}}(2\pi ny)\frac{dy}{y} \qquad (4.8.10)$$

can be written as

$$\frac{1}{2}\frac{\pi^{-\mu-\frac{1}{2}}}{\zeta^*(\nu)}\Gamma(\frac{2\mu+1+\nu}{4})\Gamma(\frac{2\mu+1-\nu}{4})\sum_{n\geq 1} n^{-\mu-\frac{1}{2}-\frac{\nu}{2}}\sigma_\nu(n)$$

$$= \frac{1}{2}\frac{\pi^{-\mu-\frac{1}{2}}}{\zeta^*(\nu)}\Gamma(\frac{2\mu+1+\nu}{4})\Gamma(\frac{2\mu+1-\nu}{4})\sum_{d\geq 1, k\geq 1}(kd)^{-\mu-\frac{1}{2}-\frac{\nu}{2}}d^\nu$$

$$= \frac{1}{2}\frac{\pi^{-\mu-\frac{1}{2}}}{\zeta^*(\nu)}\Gamma(\frac{2\mu+1+\nu}{4})\Gamma(\frac{2\mu+1-\nu}{4})\zeta(\mu + \frac{1+\nu}{2})\zeta(\mu + \frac{1-\nu}{2})$$

$$= \frac{1}{2}\frac{\zeta^*(\mu + \frac{1+\nu}{2})\zeta^*(\mu + \frac{1-\nu}{2})}{\zeta^*(\nu)}. \qquad (4.8.11)$$

When $\rho = 1 + ir$ and $\nu = i\lambda$, we have thus obtained

$$\left(ds_\Sigma^{(\rho)}\,\middle|\,\left(E_{\frac{1-i\lambda}{2}}\right)_{\text{trunc}}\right) = \frac{1}{2}\frac{\zeta^*(\frac{\rho+i\lambda}{2})\zeta^*(\frac{\rho-i\lambda}{2})}{\zeta^*(i\lambda)}$$

$$+ \frac{1}{1 - i(r+\lambda)} + \frac{1}{1 + i(r-\lambda)} + \frac{\zeta^*(-i\lambda)}{\zeta^*(i\lambda)}\left(\frac{1}{1 - i(r-\lambda)} + \frac{1}{1 + i(r+\lambda)}\right). \qquad (4.8.12)$$

Using the equations

$$\frac{1}{2\pi}\int_{-\infty}^\infty (1 + ir)^{-1}y^{\frac{ir}{2}}\,dr = y^{-\frac{1}{2}}\operatorname{char}(y \geq 1),$$

$$\frac{1}{2\pi}\int_{-\infty}^\infty (1 - ir)^{-1}y^{\frac{ir}{2}}\,dr = y^{\frac{1}{2}}\operatorname{char}(y \leq 1), \qquad (4.8.13)$$

we find

$$\int_{-\infty}^{\infty}\left[\frac{1}{1-i(r+\lambda)}+\frac{1}{1+i(r-\lambda)}+\frac{\zeta^*(-i\lambda)}{\zeta^*(i\lambda)}\left(\frac{1}{1-i(r-\lambda)}+\frac{1}{1+i(r+\lambda)}\right)\right]y^{\frac{ir}{2}}\,dr$$

$$= 2\pi \times \begin{cases} y^{\frac{-1+i\lambda}{2}}+\frac{\zeta^*(-i\lambda)}{\zeta^*(i\lambda)}y^{\frac{-1-i\lambda}{2}} & \text{if } y \geq 1, \\ y^{\frac{1-i\lambda}{2}}+\frac{\zeta^*(-i\lambda)}{\zeta^*(i\lambda)}y^{\frac{1+i\lambda}{2}} & \text{if } y \geq 1. \end{cases} \qquad (4.8.14)$$

Now, equation (4.8.2) is still valid with $\left(E_{\frac{1-i\lambda}{2}}\right)_{\text{trunc}}$ substituted for $\mathcal{M}$. Using it to obtain the decomposition of the function $\left(E_{\frac{1-i\lambda}{2}}\right)_{\text{trunc}}$ into homogeneous components, finally the link (4.8.3) between this function and the full Eisenstein series, we obtain the following.

Theorem 4.8.1. *One has the decomposition*

$$E_{\frac{1-i\lambda}{2}}(iy) = \frac{1}{2i\pi}\int_{\mathrm{Re}\,\rho=1}\frac{1}{2}\frac{\zeta^*(\frac{\rho+i\lambda}{2})\zeta^*(\frac{\rho-i\lambda}{2})}{\zeta^*(i\lambda)}y^{\frac{\rho-1}{2}}\,d\rho$$

$$+ y^{\frac{-1+i\lambda}{2}}+y^{\frac{1-i\lambda}{2}}+\frac{\zeta^*(-i\lambda)}{\zeta^*(i\lambda)}\left(y^{\frac{-1-i\lambda}{2}}+y^{\frac{1+i\lambda}{2}}\right). \qquad (4.8.15)$$

If $\mathcal{M}$ is a Hecke eigenform of even type, one has

$$\mathcal{M}(iy) = \frac{1}{8i\pi}\int_{\mathrm{Re}\,\rho=1}L^*\left(\frac{\rho}{2},\mathcal{M}\right)y^{\frac{\rho-1}{2}}\,d\rho. \qquad (4.8.16)$$

Proof. The proof of the first equation was completed just before the statement of the theorem, and the second equation is a consequence of (4.8.2) and of the identity $\left(ds_\Sigma^{(\rho)}|\mathcal{M}\right) = \frac{1}{4}L^*(\frac{\rho}{2},\mathcal{M})$ recalled in Proposition 4.4.6. $\qquad\square$

Remarks 4.8.a. (i) If, just as done for cusp-forms, one defines, when $\mathrm{Re}\,s$ is large, the L-function associated to the Eisenstein series $E^*_{\frac{1-i\lambda}{2}}$ as $L(s, E^*_{\frac{1-i\lambda}{2}}) = \sum_{n\geq1} b_n n^{-s}$ with $b_n = 2n^{\frac{i\lambda}{2}}\sigma_{-i\lambda}(n)$ as it follows from the Fourier expansion (3.1.6), one finds

$$L(s, E_{\frac{1-i\lambda}{2}}) = \frac{1}{\zeta^*(i\lambda)} \times 2\sum_{n\geq1}\sigma_{-i\lambda}(n)n^{-s+\frac{i\lambda}{2}} = 2\frac{\zeta(s-\frac{i\lambda}{2})\zeta(s+\frac{i\lambda}{2})}{\zeta^*(i\lambda)}. \qquad (4.8.17)$$

On the other hand, even though the scalar product $\left(ds_\Sigma^{(\rho)}|E_{\frac{1-i\lambda}{2}}\right)$ is undefined, it becomes so after a suitable truncation has been applied to the Eisenstein series: in view of this equation, the main term $\frac{1}{2}\frac{\zeta^*(\frac{\rho+i\lambda}{2})\zeta^*(\frac{\rho-i\lambda}{2})}{\zeta^*(i\lambda)}$ from the scalar product $\left(ds_\Sigma^{(\rho)}|\left(E_{\frac{1-i\lambda}{2}}\right)_{\text{trunc}}\right)$ appears as the natural extension of the scalar product obtained in the case of a cusp-form.

(ii) There are really two automorphic theories at work here: the usual one defined in terms of the pair $(L^2(\Gamma\backslash\Pi), \Delta)$ and that associated to the pair $(L^2(\Gamma\backslash\Sigma), \nabla^2)$, which is essentially trivial and in which only a continuous spectrum occurs. What is non-trivial, however, is the relation between the two, and there is some duality between the operations involved in Theorem 4.7.2 and in the much simpler Theorem 4.8.1. In the first one, we decomposed automorphic measures supported in Σ into Eisenstein series, with the contribution of a series of cusp-forms; in the latter theorem, we decomposed the restriction to the line from 0 to $i\infty$ of Eisenstein series or cusp-forms into their homogeneous components. This restriction operator R, considered as a densely defined operator from $L^2(\Gamma\backslash\Pi)$ to $L^2((0, i\infty), \frac{dy}{y})$, is far from being closed since, in the Hilbert space sense, the domain of its adjoint reduces to $\{0\}$. However, it becomes closed if multiplied, on the right, by $\left(\Delta - \frac{1-\nu^2}{4}\right)^{-1}$ for some ν with $\operatorname{Re}\nu \neq 0$: then, the operator R and that of making from a function on $(0, i\infty)$ invariant under the map $iy \mapsto iy^{-1}$ an automorphic measure supported in Σ become adjoint to each other in some weak sense. It is therefore not too surprising that the main coefficients appearing in the two decompositions should be conjugate to each other. It would probably be interesting to generalize the two theorems at least to the case when the line from 0 to $i\infty$ is replaced by an arbitrary line with rational endpoints. For instance, the next simplest case concerns the line from $\frac{1}{2}$ to $\frac{1}{2} + i\infty$: in this case, the L-series $\sum_{n\geq 1} b_n n^{-s}$ built from the Fourier coefficients of the modular (Eisenstein or Hecke) form under consideration must be replaced by the series $\sum_{n\geq 1}(-1)^n b_n n^{-s}$. It seems likely that the study of the more general case should lead to the occurrence, in place of the product $\frac{1}{2}\zeta^*(\frac{\rho+i\lambda}{2})\zeta^*(\frac{\rho-i\lambda}{2})$ (the L-function associated to $E^*_{\frac{1-i\lambda}{2}}$), of more complicated quadratic expressions involving values of Dirichlet L-functions in place of zeta. It is clear that using the full spectral theory of the Laplacian for congruence subgroups of Γ, as developed in [21], will be necessary here.

Chapter 5

Spectral decomposition of the Poincaré summation process

The following object $\mathfrak{P}$, almost a distribution in $\mathbb{R}^4$ as will be seen, defined by the equation

$$\langle \mathfrak{P}, h \otimes f \rangle = \sum_{g \subset \Gamma} \int_{\mathbb{R}^2} (h \circ g)(x, \xi) f(x, \xi) dx d\xi,$$

is virtually undistinguishable from the Poincaré summation process itself. It thus encompasses, in a way, the whole automorphic theory associated with the group Γ and its action on $\mathbb{R}^2$ by means of linear transformations. It is bi-automorphic in the sense that $\langle \mathfrak{P}, h \otimes f \rangle$ does not change if h or f is transformed under an arbitrary linear transformation of coordinates in Γ. In this chapter, we give a proof of the simple formula

$$\mathfrak{P} = \frac{1}{2\pi} \int_{-\infty}^{\infty} \mathfrak{E}_{i\lambda} \otimes \mathfrak{E}_{-i\lambda} \frac{d\lambda}{\zeta(i\lambda)\zeta(-i\lambda)}$$

$$+ 2 \sum_{p \neq 0} \Gamma\left(\frac{i\lambda_p}{2}\right) \Gamma\left(-\frac{i\lambda_p}{2}\right) \sum_j \epsilon_{p,j} \mathfrak{M}_{p,j} \otimes \mathfrak{M}_{-p,j},$$

where $\epsilon_{p,j} = \pm 1$ indicates the parity of the Hecke eigenform $\mathcal{M}_{p,j}$ under the symmetry $z \mapsto -\bar{z}$. In particular — this is not a priori obvious — the scalar product $\langle \mathfrak{P}, \bar{h} \otimes f \rangle$ is positive semi-definite (Corollary 5.3.4).

One of the advantages of using automorphic distribution theory, as opposed to automorphic function theory, is that one can dispense with the spectral theory of the modular Laplacian, substituting for it the simpler spectral theory of the automorphic Euler operator and avoiding the use of special functions. Still, we chose to let the proof depend on known results regarding series of Kloosterman sums, rather than working in a completely independent way so as to find a new approach to this theory. Besides avoiding repetition, this certainly helped (especially the version [21] based on the spectral expansion of the automorphic Green's operator) in the more technical part, where estimates of Iwaniec's function $Z_s(m, n)$ were required. This is the function advocated by Iwaniec [21, p. 134] or [23, p. 413] (noting that the notation is different in the second reference) as having nicer properties (a functional equation) than the Selberg series to which it reduces when only the first term of the Taylor expansion of some Bessel function is retained. The function $Z_s(m, n)$ is exactly the one needed here.

5.1   A universal Poincaré series

Given $h \in L^2(\mathbb{R}^2)$, the classical formula

$$2\pi h_{i\lambda} = \lim_{\delta \to 0} \left[(2i\pi\mathcal{E} + \delta + i\lambda)^{-1} - (2i\pi\mathcal{E} - \delta + i\lambda)^{-1} \right] h \qquad (5.1.1)$$

makes it possible to recover the decomposition $h = \int_{-\infty}^{\infty} h_{i\lambda} d\lambda$ of h into homogeneous components with the help of the resolvent of the Euler operator. This operator has a continuous spectrum only and what this equation really means is that, given $a < b$, the integral

$$\int_a^b \left[(2i\pi\mathcal{E} + \delta + i\lambda)^{-1} - (2i\pi\mathcal{E} - \delta + i\lambda)^{-1} \right] h \, d\lambda \qquad (5.1.2)$$

converges strongly towards $2\pi \int_a^b h_{i\lambda} d\lambda$ as $\delta \to 0$.

To make an operator such as $(2i\pi\mathcal{E} + \mu)^{-1}$ explicit, we first observe that, given $h \in \mathcal{S}_{\text{even}}(\mathbb{R}^2)$, the functions

$$u(x, \xi) = \int_0^1 h(tx, t\xi) t^\mu dt, \quad v(x, \xi) = - \int_1^\infty h(tx, t\xi) t^\mu dt, \qquad (5.1.3)$$

the first one well defined if $\mathrm{Re}\,\mu > -1$, the second one well defined for all values of (x, ξ) if $\mathrm{Re}\,\mu < -1$, or for $(x, \xi) \neq (0, 0)$ and all values of μ, satisfy the equations

$$(2i\pi\mathcal{E} + \mu)u = h, \quad (2i\pi\mathcal{E} + \mu)v = h, \qquad (5.1.4)$$

as is easily seen with the help of an integration by parts. To show that the two maps $h \mapsto u$ and $h \mapsto v$ define the resolvent $(2i\pi\mathcal{E} + \mu)^{-1}$ of the operator $2i\pi\mathcal{E}$ (i times a self-adjoint one in $L^2(\mathbb{R}^2)$), the first one for $\mathrm{Re}\,\mu > 0$ and the second one for $\mathrm{Re}\,\mu < 0$, it suffices to show that both maps extend as continuous endomorphisms of $L^2(\mathbb{R}^2)$. This is a particular case of the so-called Hardy's inequalities, here a simple consequence of the Cauchy-Schwarz inequality since, say in the first case, given another function $u_1 \in L^2(\mathbb{R}^2)$, one has

$$
|(u_1|u)| = \left| \int_{\mathbb{R}^2} \bar{u}_1(x,\xi) dx d\xi \int_0^1 h(tx, t\xi) t^\mu dt \right|
$$

$$
\leq \int_0^1 t^{\mathrm{Re}\,\mu} dt \int_{\mathbb{R}^2} |u_1(x,\xi)| |h(tx, t\xi)| dx d\xi
$$

$$
\leq \int_0^1 t^{\mathrm{Re}\,\mu} \|u_1\| . t^{-1} \|h\| dt. \tag{5.1.5}
$$

The seemingly dual (formally identical) equation

$$
\mathfrak{G} = \frac{1}{2\pi} \lim_{\delta \to 0} \int_{-\infty}^{\infty} \left[(\delta + i\lambda + 2i\pi\mathcal{E})^{-1} - (-\delta + i\lambda + 2i\pi\mathcal{E})^{-1} \right] \mathfrak{G} d\lambda \tag{5.1.6}
$$

does not apply, in general, in the tempered distribution case, not even in a weak sense. This is due to the fact that, given $h \in \mathcal{S}(\mathbb{R}^2)$, the function (where $\mathrm{Re}\,\mu > 0$)

$$
\left((2i\pi\mathcal{E} + \mu)^{-1} h \right) (x, \xi) = \int_0^1 t^\mu h(tx, t\xi) dt \tag{5.1.7}
$$

does not, generally, lie in $\mathcal{S}(\mathbb{R}^2)$: the problem lies in the behaviour of the function so defined for $|x| + |\xi|$ large, and there is a similar problem for $\mathrm{Re}\,\mu < 0$, this time for small values of $|x| + |\xi|$. However, the space $\mathcal{S}(\mathbb{R}^2)$ is the intersection of the spaces $\mathcal{S}_N(\mathbb{R}^2)$, where h lies in the space associated to $N = 0, 1, \dots$ if and only if the image of h under the product of any number $\leq N$ of operators in the set $\left((x), (\xi), \frac{\partial}{\partial x}, \frac{\partial}{\partial \xi} \right)$ is a bounded function. Then, it is true that, given $N = 0, 1, \dots$ and $h \in \mathcal{S}(\mathbb{R}^2)$, the function $(2i\pi\mathcal{E} + \mu)^{-1} h$ lies in $\mathcal{S}_N(\mathbb{R}^2)$ if $\mathrm{Re}\,\mu$ is large enough. Something entirely similar holds, for $-\mathrm{Re}\,\mu$ large enough, with the function

$$
\left((2i\pi\mathcal{E} + \mu)^{-1} h \right) (x, \xi) = - \int_1^\infty t^\mu h(tx, t\xi) dt, \tag{5.1.8}
$$

where increasing $-\mathrm{Re}\,\mu$ makes it possible to bound, even when $|x| + |\xi|$ is small, any given number of derivatives of the function so defined.

Note that, even when $\mathfrak{G}$ is replaced by a function $h \in \mathcal{S}(\mathbb{R}^2)$, it is impossible, for reasons of convergence, to consider the integrals on the line of $(\delta \pm i\lambda + 2i\pi\mathcal{E})^{-1} h$ separately: one can, however, introduce instead the symmetric-type integral

$$
\int_\mathbb{R}^{\mathrm{sym}} (\delta + i\lambda + 2i\pi\mathcal{E})^{-1} h d\lambda = \lim_{A \to \infty} \int_{-A}^A (\delta + i\lambda + 2i\pi\mathcal{E})^{-1} h d\lambda, \tag{5.1.9}
$$

an integral also to be denoted as $\frac{1}{i} \int_{\mathrm{Re}\,\mu = \delta}^{\mathrm{sym}} (\mu + 2i\pi\mathcal{E})^{-1} h d\mu$.

Given $h \in \mathcal{S}(\mathbb{R}^2)$, the equation, in which $|x| + |\xi| \neq 0$,

$$
h_{i\rho}(x, \xi) = \frac{1}{2\pi} \int_0^\infty s^{i\rho} h(sx, s\xi) ds \tag{5.1.10}
$$

yields, after an integration by parts,

$$-\rho^2 h_{i\rho}(x,\xi) = \frac{1}{2\pi} \int_0^\infty s^{i\rho} \left[(2i\pi\mathcal{E})^2 h \right] (sx, s\xi) ds, \tag{5.1.11}$$

from which the inequality

$$|h_{i\rho}(x,\xi)| \le C(1+\rho^2)^{-1} \int_0^\infty \left[1 + s^2(|x| + |\xi|)^2 \right]^{-1} ds$$
$$\le C(1+\rho^2)^{-1}(|x| + |\xi|)^{-1} \tag{5.1.12}$$

follows. Then, given $\delta > 0$ and $\lambda \in \mathbb{R}$, the double integral

$$\int_0^1 t^{\delta + i\lambda} dt \int_{-\infty}^\infty h_{i\rho}(tx, t\xi) d\rho \tag{5.1.13}$$

is convergent and, still assuming $|x| + |\xi| \ne 0$, one can transform the integral

$$I_1: \; = \frac{1}{2\pi} \int_{\mathbb{R}}^{\text{sym}} \left[(\delta + i\lambda + 2i\pi\mathcal{E})^{-1} h \right] (x,\xi) d\lambda$$

$$= \frac{1}{2\pi} \int_{\mathbb{R}}^{\text{sym}} d\lambda \int_0^1 t^{\delta + i\lambda} h(tx, t\xi) dt$$

$$= \frac{1}{2\pi} \int_{\mathbb{R}}^{\text{sym}} d\lambda \int_0^1 t^{\delta + i\lambda} dt \int_{-\infty}^\infty h_{i\rho}(tx, t\xi) d\rho$$

$$= \frac{1}{2\pi} \int_{\mathbb{R}}^{\text{sym}} d\lambda \int_{-\infty}^\infty d\rho \int_0^1 t^{\delta + i\lambda} h_{i\rho}(tx, t\xi) dt$$

$$= \frac{1}{2\pi} \int_{\mathbb{R}}^{\text{sym}} d\lambda \int_{-\infty}^\infty d\rho \int_0^1 t^{\delta + i\lambda - 1 - i\rho} h_{i\rho}(x, \xi) dt$$

$$= \frac{1}{2\pi} \int_{\mathbb{R}}^{\text{sym}} d\lambda \int_{-\infty}^\infty (\delta + i\lambda - i\rho)^{-1} h_{i\rho}(x, \xi) d\rho. \tag{5.1.14}$$

The same goes for I_2, defined in the same way after δ has been replaced by $-\delta$. Then,

$$I_1 - I_2: \; = \frac{1}{2\pi} \int_{\mathbb{R}}^{\text{sym}} \left[(\delta + i\lambda + 2i\pi\mathcal{E})^{-1} h - (-\delta + i\lambda + 2i\pi\mathcal{E})^{-1} h \right] (x, \xi) d\lambda$$

$$= \frac{1}{2\pi} \int_{\mathbb{R}}^{\text{sym}} d\lambda \int_{-\infty}^\infty \left[(\delta + i\lambda - i\rho)^{-1} - (-\delta + i\lambda - i\rho)^{-1} \right] h_{i\rho}(x, \xi) d\rho$$

$$= \frac{1}{2\pi} \int_{-\infty}^\infty h_{i\rho}(x, \xi) d\rho \int_{\mathbb{R}}^{\text{sym}} \frac{2\delta}{\delta^2 + (\lambda - \rho)^2} d\lambda$$

$$= \int_{-\infty}^\infty h_{i\rho}(x, \xi) d\rho = h(x, \xi). \tag{5.1.15}$$

Note that the regrouping of terms makes the superscript "sym" optional in the last few lines.

While equation (5.1.6) is invalid in the distribution case, equation (5.1.15) has an analogue, obtained by duality, to wit the formally identical decomposition

$$\mathfrak{S} = \frac{1}{2\pi} \int_{-\infty}^{\infty} \left[(\delta + i\lambda + 2i\pi\mathcal{E})^{-1} - (-\delta + i\lambda + 2i\pi\mathcal{E})^{-1} \right] \mathfrak{S} d\lambda, \qquad (5.1.16)$$

valid for any given tempered distribution $\mathfrak{S}$ provided that δ is chosen large enough. With $\mu = \delta + i\lambda$, it will be much better, in calculations, to use $(-\mu + 2i\pi\mathcal{E})^{-1}$, rather than $(\bar{\mu} + 2i\pi\mathcal{E})^{-1}$, as a second term within brackets, but this is possible only if one uses, again, the symmetric-type integral, writing

$$\mathfrak{S} = \frac{1}{2i\pi} \int_{\mathrm{Re}\,\mu=\delta}^{\mathrm{sym}} \left[(\mu + 2i\pi\mathcal{E})^{-1} - (-\mu + 2i\pi\mathcal{E})^{-1} \right] \mathfrak{S} d\mu. \qquad (5.1.17)$$

After this decomposition has been made as explicit as possible, one may, when feasible, make contour deformations and see what happens when δ goes to zero. The limiting formula, if any should exist, will then provide a decomposition of $\mathfrak{S}$ as a superposition of distributions homogeneous of degrees lying on the line $-1 + i\mathbb{R}$.

Given $h \in \mathcal{S}_{\mathrm{even}}(\mathbb{R}^2)$, the would-be distribution $\mathfrak{S} = \sum_{g\in\Gamma} h \circ g$ is the one defined weakly by the equation

$$\langle \mathfrak{S}, f \rangle = \sum_{g\in\Gamma} \int_{\mathbb{R}^2} h(g.(x, \xi)) f(x, \xi) dx d\xi \qquad (5.1.18)$$

for every function $f \in \mathcal{S}_{\mathrm{even}}(\mathbb{R}^2)$. If the convergence of this series of integrals were asserted, $\mathfrak{S}$ would be defined as a (tempered) automorphic distribution: this is almost the case. Note that in the proposition below, an integration by parts would make it possible to replace the assumptions made about h and f by the requirement that h should lie in the image of $\mathcal{S}_{\mathrm{even}}(\mathbb{R}^2)$ under the operator $\pi^2\mathcal{E}^2 \left(\pi^2\mathcal{E}^2 + \frac{1}{4} \right)$, in which case the condition $f \in \mathcal{S}_{\mathrm{even}}(\mathbb{R}^2)$ would suffice so far as f is concerned: then, $\mathfrak{S}$ would become a genuine distribution.

Proposition 5.1.1. *The series* (5.1.18) *is convergent under the assumption that both h and f are functions lying in the image of $\mathcal{S}_{\mathrm{even}}(\mathbb{R}^2)$ under the operator* $2i\pi\mathcal{E}(1 + 2i\pi\mathcal{E})$.

Proof. Assume that $h = 2i\pi\mathcal{E}h_1 = (1 + 2i\pi\mathcal{E})h_2$ and $f = 2i\pi\mathcal{E}f_1 = (1 + 2i\pi\mathcal{E})f_2$ with $h_1, h_2, f_1, f_2 \in \mathcal{S}_{\mathrm{even}}(\mathbb{R}^2)$. Then, for every pair (α, β) of real numbers, one has for $r = 0$ or 1 the identity

$$h(\alpha t, \beta t) = \left[t\frac{d}{dt} + r + 1 \right] (h_r(\alpha t, \beta t)), \qquad (5.1.19)$$

so that

$$t^r h(\alpha t, \beta t) = \frac{d}{dt}(t^{r+1} h_r(\alpha t, \beta t)) : \tag{5.1.20}$$

hence,

$$\int_{-\infty}^{\infty} h(\alpha t, \beta t)dt = \int_{-\infty}^{\infty} th(\alpha t, \beta t)dt = 0 \quad \text{if } |\alpha| + |\beta| \neq 0, \tag{5.1.21}$$

and the same holds with f in place of h.

Given $g = \left(\begin{smallmatrix} a & b \\ c & d \end{smallmatrix}\right) \in \Gamma$, one has, assuming $c \neq 0$ and trading x for the new variable $t = cx + d\xi$,

$$\langle h \circ g, f \rangle = |c|^{-1} \int_{\mathbb{R}^2} h(c^{-1}(at - \xi), t) f(c^{-1}(t - d\xi), \xi) dt d\xi. \tag{5.1.22}$$

Then, a Taylor expansion implies

$$h(c^{-1}(at - \xi), t) = h(c^{-1}at, t) - c^{-1}\xi h_1'(c^{-1}at, t) + R_1(t, \xi),$$
$$f(c^{-1}(t - d\xi), \xi) = f(-c^{-1}d\xi, \xi) + c^{-1}t f_1'(-c^{-1}d\xi, \xi) + R_2(t, \xi), \tag{5.1.23}$$

with

$$|R_1(t, \xi)| \leq C\frac{\xi^2}{c^2}(1 + t^2)^{-2}, \quad |R_2(t, \xi)| \leq C\frac{t^2}{c^2}(1 + \xi^2)^{-2}. \tag{5.1.24}$$

In view of the pair of equations

$$\int_{-\infty}^{\infty} h(c^{-1}at, t)dt = \int_{-\infty}^{\infty} th(c^{-1}at, t)dt = 0,$$
$$\int_{-\infty}^{\infty} f(-c^{-1}d\xi, \xi)d\xi = \int_{-\infty}^{\infty} \xi f(-c^{-1}d\xi, \xi)d\xi = 0 \tag{5.1.25}$$

proved in the beginning of the present proof, one obtains the inequality $|\langle h \circ g, f \rangle| \leq C|c|^{-3}$ and, more generally,

$$|\langle h \circ g, f \rangle| \leq C(\max(|a|, |b|, |c|, |d|))^{-3}. \tag{5.1.26}$$

The absolute convergence of the series $\sum_{g \in \Gamma} \langle h \circ g, f \rangle$ is then a consequence of Lemma 4.1.4. We shall use later (just after (5.1.65)) the fact that the power $|c|^{-3}$ which shows up in the estimate leaves some elbow room: $|c|^{-2-\varepsilon}$ with $\varepsilon > 0$ would have sufficed. $\square$

Before stating and proving the next lemma, let us emphasize that, in several instances, we shall have to deal with series of integrals, or integrals of series, in which the various summation operations could not be permuted, or could be permuted only with difficulty: all superpositions of such operations must be read from the right to the left. In the proof of this lemma and of the following proposition, we make the assumption that $h = 2i\pi \mathcal{E} h_1$ and $f = 2i\pi \mathcal{E} f_1$ with $h_1, f_1 \in \mathcal{S}_{\text{even}}(\mathbb{R}^2)$.

Lemma 5.1.2. *Assume that h, f lie in the image of $\mathcal{S}_{\text{even}}(\mathbb{R}^2)$ under the operator $2i\pi\mathcal{E}$. Then,*

$$T(h, f): = \sum_{k\in\mathbb{Z}} \int_{\mathbb{R}^2} h(x,\xi)f(x-k\xi,\xi)dxd\xi$$

$$= \sum_{n\in\mathbb{Z}} \int_{-\infty}^{\infty} (\mathcal{F}_1^{-1}h)\left(-\frac{n}{\xi},\xi\right)(\mathcal{F}_1^{-1}f)\left(\frac{n}{\xi},\xi\right)\frac{d\xi}{|\xi|}$$

$$+ \int_{-\infty}^{\infty} (\mathcal{F}_1^{-1}h)(\xi,0)(\mathcal{F}_1^{-1}f)(\xi,0)\frac{d\xi}{|\xi|}. \quad (5.1.27)$$

Proof. For $k \neq 0$, one has (a consequence of (5.1.21))

$$\int_{\mathbb{R}^2} h(x,0)f(x-k\xi,0)dxd\xi = \frac{1}{|k|}\int_{\mathbb{R}^2} h(x,0)f(\eta,0)dxd\eta = 0: \quad (5.1.28)$$

but this is not true when $k = 0$. Hence,

$$T(h,f) = \int_{\mathbb{R}^2} h(x,\xi)f(x,\xi)dxd\xi$$

$$+ \sum_{k\neq 0} \int_{\mathbb{R}^2} [h(x,\xi)f(x-k\xi,\xi) - h(x,0)f(x-k\xi,0)]dxd\xi. \quad (5.1.29)$$

The integral on the second line can be written as

$$\frac{1}{|k|}\int_{\mathbb{R}^2} \left[h\left(x,\frac{\xi}{k}\right)f\left(x-\xi,\frac{\xi}{k}\right) - h(x,0)f(x-\xi,0)\right]dxd\xi, \quad (5.1.30)$$

where the new bracket is bounded for every N by

$$C\frac{|\xi|}{|k|}(1+|x|)^{-N}(1+|x-\xi|)^{-N} \quad (5.1.31)$$

for some $C > 0$. It follows that the series of integrals on the second line of (5.1.29) converges, and that the order of the summation and integration can be changed. Hence,

$$T(h,f) = \int_{\mathbb{R}^2} h(x,\xi)f(x,\xi)dxd\xi$$

$$+ \int_{\mathbb{R}^2} dxd\xi \sum_{k\neq 0}[h(x,\xi)f(x-k\xi,\xi) - h(x,0)f(x-k\xi,0)]. \quad (5.1.32)$$

We shall now show that this can be written as

$$T(h,f) = \int_{\mathbb{R}^2} dxd\xi \sum_{k\in\mathbb{Z}} h(x,\xi)f(x-k\xi,\xi) - \int_{\mathbb{R}^2} dxd\xi \sum_{k\neq 0} h(x,0)f(x-k\xi,0),$$

$$(5.1.33)$$

for which we have to show that each of the two terms is a convergent integral of series. Let us start with the first one.

By Poisson's formula, for $\xi \neq 0$,

$$\sum_{k \in \mathbb{Z}} h(x, \xi) f(x - k\xi, \xi) = \frac{1}{|\xi|} \sum_{n \in \mathbb{Z}} e^{-2i\pi n \frac{x}{\xi}} h(x, \xi) \left(\mathcal{F}_1^{-1} f\right) \left(\frac{n}{\xi}, \xi\right), \qquad (5.1.34)$$

where $\mathcal{F}_1^{-1} f$ denotes the partial Fourier transform of f with respect to the first variable, so that

$$\int_{\mathbb{R}^2} dx d\xi \sum_{k \in \mathbb{Z}} h(x, \xi) f(x - k\xi, \xi) = \int_{\mathbb{R}^2} dx \frac{d\xi}{|\xi|} \sum_{n \in \mathbb{Z}} e^{-2i\pi n \frac{x}{\xi}} h(x, \xi) \left(\mathcal{F}_1^{-1} f\right) \left(\frac{n}{\xi}, \xi\right).$$

$$(5.1.35)$$

It is not difficult to permute the series and the (double) integral in this instance since, if $n \neq 0$, one has for every N the estimate

$$\left| \left(\mathcal{F}_1^{-1} f\right) \left(\frac{n}{\xi}, \xi\right) \right| \leq C(1 + |n|)^{-N} \left(|\xi| + |\xi|^{-1}\right)^{-N} : \qquad (5.1.36)$$

note that the term with $n = 0$ on the right-hand side of (5.1.35) is integrable too, though an argument is needed so far as the integrability near $\xi = 0$ is concerned. This follows from the equation

$$\left(\mathcal{F}_1^{-1} f\right)(\eta, \xi) = \left(-\eta \frac{\partial}{\partial \eta} + \xi \frac{\partial}{\partial \xi}\right) \left(\mathcal{F}_1^{-1} f_1\right)(\eta, \xi) \qquad (5.1.37)$$

(a consequence of $f = 2i\pi \mathcal{E} f_1$) and its special case obtained when $\eta = 0$. The first term on the right-hand side of (5.1.33) is thus

$$\sum_{n \in \mathbb{Z}} \int_{-\infty}^{\infty} \left(\mathcal{F}_1^{-1} h\right) \left(-\frac{n}{\xi}, \xi\right) \left(\mathcal{F}_1^{-1} f\right) \left(\frac{n}{\xi}, \xi\right) \frac{d\xi}{|\xi|}. \qquad (5.1.38)$$

What remains to be treated is the second term in (5.1.32), to wit

$$R = \int_{\mathbb{R}^2} dx d\xi \sum_{k \neq 0} h(x, 0) f(x - k\xi, 0). \qquad (5.1.39)$$

First, we show that this integral is convergent, and would remain so even if $d\xi$ were to be replaced by $|\xi|^\mu d\xi$ for any μ with $\operatorname{Re} \mu > -1$. To do so, we write

$$\int_{-\infty}^{\infty} |\xi|^\mu d\xi \sum_{k \geq 1} f(x - k\xi, 0) = \int_{-\infty}^{\infty} |\xi|^\mu d\xi \sum_{k \geq 1} \left[f(x - k\xi, 0) - \int_k^{k+1} f(x - t\xi, 0) dt \right]$$

$$+ \int_{-\infty}^{\infty} |\xi|^\mu d\xi \int_1^{\infty} f(x - t\xi, 0) dt. \qquad (5.1.40)$$

Using the mean value theorem and the elementary inequality (Peetre's inequality from interpolation theory: it will have to be used time and again)

$$(1 + |x - s\xi|)^{-N} \leq C(1 + |x|)^N(1 + |s\xi|)^{-N}, \qquad (5.1.41)$$

one can bound the first term by

$$C \int_{-\infty}^{\infty} |\xi|^{\operatorname{Re}\mu}(1 + |x|)^N \sum_{k \geq 1} |\xi|(1 + |k\xi|)^{-N} d\xi$$

$$= C(1 + |x|)^N \sum_{k \geq 1} k^{-\operatorname{Re}\mu - 2} \int_{-\infty}^{\infty} |\xi|^{\operatorname{Re}\mu}(1 + |\xi|)^{-N} d\xi, \quad (5.1.42)$$

an expression majorized for large N, provided that $\operatorname{Re}\mu > -1$, by $C(1 + |x|)^N$, which can thus be integrated against $|h(x, 0)|dx$. To majorize the second integral on the right-hand side of (5.1.40), we first add to (5.1.40) the analogous expression arising from the consideration of terms with $k \leq -1$, finding the integral

$$\int_{-\infty}^{\infty} |\xi|^\mu d\xi \int_{|t| \geq 1} f(x - t\xi, 0)dt = \int_{-\infty}^{\infty} |\xi|^{\mu - 1} d\xi \int_{|t| \geq |\xi|} f(x - t, 0)dt$$

$$= -\int_{-\infty}^{\infty} |\xi|^{\mu - 1} d\xi \int_{|t| \leq |\xi|} f(x - t, 0)dt \quad (5.1.43)$$

as a result: we have taken advantage of the equation (5.1.21) $\int_{-\infty}^{\infty} f(x - t, 0)dt = 0$. We then write, for some large N,

$$\left| \int_{|t| \geq |\xi|} f(x - t, 0)dt \right| \leq \begin{cases} C(1 + |x|)^{2N}(1 + |\xi|)^{-N} & \text{if } |\xi| \geq 1, \\ C|\xi| & \text{if } |\xi| \leq 1, \end{cases} \quad (5.1.44)$$

obtaining a function of (x, ξ) integrable against $|\xi|^{\operatorname{Re}\mu - 1}|h(x, 0)|dxd\xi$ if $\operatorname{Re}\mu > -1$.

Finally, we compute the expression

$$R(\mu) = \int_{\mathbb{R}^2} |\xi|^\mu dxd\xi \sum_{k \neq 0} h(x, 0)f(x - k\xi, 0)$$

$$= \int_{-\infty}^{\infty} h(x, 0)dx \int_{-\infty}^{\infty} |\xi|^\mu d\xi \sum_{k \neq 0} f(x - k\xi, 0), \quad (5.1.45)$$

under the assumption that $0 < \operatorname{Re}\mu < 1$. Since

$$|f(x - k\xi, 0)| \leq C(1 + |x|)^N(1 + |k\xi|)^{-N} \qquad (5.1.46)$$

and

$$\sum_{k \neq 0} \int_{-\infty}^{\infty} |\xi|^{\operatorname{Re}\mu}(1 + |k\xi|)^{-N} d\xi = \sum_{k \neq 0} |k|^{-\operatorname{Re}\mu - 1} \int_{-\infty}^{\infty} |\xi|^{\operatorname{Re}\mu}(1 + |\xi|)^{-N} d\xi < \infty$$

$$(5.1.47)$$

for large N, one can, again, commute the k-series and the $d\xi$ - integration on the right-hand side of (5.1.45). Now, one has

$$\mathcal{F}^{-1}\left(|\xi|^\mu\right)(y) = \pi^{-\frac{1}{2}-\mu}\frac{\Gamma(\frac{1+\mu}{2})}{\Gamma(-\frac{\mu}{2})}|y|^{-1-\mu} = \frac{\zeta(-\mu)}{\zeta(1+\mu)}|y|^{-1-\mu} : \qquad (5.1.48)$$

it is only when $-1 < \mathrm{Re}\,\mu < 0$ that this is an identity between two locally integrable functions but, using analytic continuation, one can still regard this as an identity between two tempered distributions (more will be said about this later), provided that, on one hand, $\nu \neq -1, -3, \ldots$, on the other hand that $\nu \neq 0, 2, \ldots$; the "integrals" below should be assigned this distribution meaning, only observing at the end ((5.1.37) again) that the function $\left(\mathcal{F}_1^{-1}f\right)\left(\frac{\xi}{k}, 0\right)$ is "divisible" by ξ as a function. We may then, using analytic continuation, start from values of μ such that $0 < \mathrm{Re}\,\mu < 1$. Also,

$$(\mathcal{F}\left[\xi \mapsto f(x - k\xi, 0)\right])(y) = \frac{1}{|k|}\exp\left(-2i\pi\frac{xy}{k}\right)\left(\mathcal{F}_1^{-1}f\right)\left(\frac{y}{k}, 0\right) : \qquad (5.1.49)$$

it follows (reverting to the variable ξ in place of y) that

$$\int_{-\infty}^{\infty}|\xi|^\mu f(x - k\xi, 0)d\xi$$

$$= \frac{1}{|k|}\frac{\zeta(-\mu)}{\zeta(1+\mu)}\int_{-\infty}^{\infty}\exp\left(-2i\pi\frac{x\xi}{k}\right)|\xi|^{-1-\mu}\left(\mathcal{F}_1^{-1}f\right)\left(\frac{\xi}{k}, 0\right)d\xi$$

$$= \frac{\zeta(-\mu)}{\zeta(1+\mu)}\int_{-\infty}^{\infty}e^{-2i\pi x\xi}|k\xi|^{-1-\mu}\left(\mathcal{F}_1^{-1}f\right)(\xi, 0)d\xi. \qquad (5.1.50)$$

As a consequence,

$$R(\mu) = 2\zeta(-\mu)\int_{-\infty}^{\infty}|\xi|^{-1-\mu}\left(\mathcal{F}_1^{-1}h\right)(-\xi, 0)\left(\mathcal{F}_1^{-1}f\right)(\xi, 0)d\xi. \qquad (5.1.51)$$

One obtains $R = R(0)$, the term with a minus sign on the right-hand side of (5.1.33), from the equation $\zeta(0) = -\frac{1}{2}$. □

Proposition 5.1.3. *Recall definition (3.3.12) of Kloosterman sums. Assume that h and f lie in the image of $\mathcal{S}_{\mathrm{even}}(\mathbb{R}^2)$ under the operator $2i\pi\mathcal{E}(1 + 2i\pi\mathcal{E})$. One has*

$$\langle \mathfrak{P}, h \otimes f \rangle = \langle \mathfrak{P}_{\mathrm{min}}, h \otimes f \rangle + \langle \mathfrak{P}^{(11)}, h \otimes f \rangle$$
$$+ \langle \mathfrak{P}^{(01)}, h \otimes f \rangle + \langle \mathfrak{P}^{(01)}, \mathcal{F}^{\mathrm{symp}}h \otimes \mathcal{F}^{\mathrm{symp}}f \rangle$$
$$+ \langle \mathfrak{P}^{(10)}, h \otimes f \rangle + \langle \mathfrak{P}^{(10)}, \mathcal{F}^{\mathrm{symp}}h \otimes \mathcal{F}^{\mathrm{symp}}f \rangle$$
$$+ \langle \mathfrak{P}^{(00)}, h \otimes f \rangle + \langle \mathfrak{P}^{(00)}, \mathcal{F}^{\mathrm{symp}}h \otimes \mathcal{F}^{\mathrm{symp}}f \rangle, \qquad (5.1.52)$$

with

$$
\langle \mathfrak{P}_{\min}, h \otimes f \rangle = 2 \sum_{n \in \mathbb{Z}} \int_{-\infty}^{\infty} (\mathcal{F}_1^{-1} h) \left(-\frac{n}{\xi}, \xi \right) (\mathcal{F}_1^{-1} f) \left(\frac{n}{\xi}, \xi \right) \frac{d\xi}{|\xi|}
$$

$$
+ 2 \int_{-\infty}^{\infty} (\mathcal{F}_1^{-1} h) \, (\xi, 0) \, (\mathcal{F}_1^{-1} f) \, (\xi, 0) \frac{d\xi}{|\xi|} \quad (5.1.53)
$$

and

$$
\langle \mathfrak{P}^{(01)}, h \otimes f \rangle = 2 \sum_{c \geq 1} c^{-1} \sum_{n \neq 0} S(0, n; c) \int_{\mathbb{R}^2} \cos \left(\frac{2\pi n x}{c\xi} \right)
$$

$$
\times (\mathcal{F}_1^{-1} h) \, (0, x) \, (\mathcal{F}_1^{-1} f) \left(\frac{n}{\xi}, \xi \right) \frac{dx \, d\xi}{|x| \, |\xi|} \quad (5.1.54)
$$

and

$$
\langle \mathfrak{P}^{(10)}, h \otimes f \rangle = 2 \sum_{c \geq 1} c^{-1} \sum_{m \neq 0} S(m, 0; c) \int_{\mathbb{R}^2} \cos \left(\frac{2\pi m \xi}{c x} \right)
$$

$$
\times (\mathcal{F}_1^{-1} h) \left(-\frac{m}{x}, x \right) (\mathcal{F}_1^{-1} f) \, (0, \xi) \frac{dx \, d\xi}{|x| \, |\xi|} \quad (5.1.55)
$$

and

$$
\langle \mathfrak{P}^{(11)}, h \otimes f \rangle = 2 \sum_{c \geq 1} c^{-1} \sum_{\substack{m, n \in \mathbb{Z} \\ mn \neq 0}} S(m, n; c)
$$

$$
\times \int_{\mathbb{R}^2} (\mathcal{F}_1^{-1} h) \left(-\frac{m}{x}, x \right) (\mathcal{F}_1^{-1} f) \left(\frac{n}{\xi}, \xi \right) \cos \left(\frac{2\pi}{c} \left(\frac{m\xi}{x} + \frac{nx}{\xi} \right) \right) \frac{dx \, d\xi}{|x| \, |\xi|} \quad (5.1.56)
$$

and, finally,

$$
\langle \mathfrak{P}^{(00)}, h \otimes f \rangle = 2 \sum_{c \geq 1} c^{-1} S(0, 0; c)
$$

$$
\times \int_{\mathbb{R}^2} (\mathcal{F}_1^{-1} h) \, (0, x) \, (\mathcal{F}_1^{-1} f) \, (\xi, 0) \left[\cos \left(\frac{2\pi x \xi}{c} \right) - 1 \right] \frac{dx \, d\xi}{|x| \, |\xi|}. \quad (5.1.57)
$$

Proof. The proof is based on Lemma 5.1.2, to be applied several times. The "minor and major" terms $\mathfrak{S}_{\min}$ and $\mathfrak{S}_{\text{maj}}$ of the decomposition of $\mathfrak{S} = \sum_{g \in \Gamma} h \circ g$ are defined by limiting the summation to the matrices $g = \left(\begin{smallmatrix} a & b \\ c & d \end{smallmatrix} \right)$ such that $c = 0$ or $c \neq 0$: denote as Γ' the subset of Γ defined by the second condition. In view of the equation

$$
\left(\begin{smallmatrix} 1 & j \\ 0 & 1 \end{smallmatrix} \right) g \left(\begin{smallmatrix} 1 & k \\ 0 & 1 \end{smallmatrix} \right) = g + \left(\begin{smallmatrix} jc & * \\ 0 & kc \end{smallmatrix} \right), \quad g = \left(\begin{smallmatrix} a & b \\ c & d \end{smallmatrix} \right), \quad (5.1.58)
$$

the class $(\Gamma \cap N) g (\Gamma \cap N)$ can be parametrized, when $c \neq 0$, by $\mathbb{Z} \times \mathbb{Z}$, under the map $j, k \mapsto \left(\begin{smallmatrix} 1 & j \\ 0 & 1 \end{smallmatrix} \right) g \left(\begin{smallmatrix} 1 & k \\ 0 & 1 \end{smallmatrix} \right)$, and the quotient set $(\Gamma \cap N) \backslash \Gamma' / (\Gamma \cap N)$ can be parametrized by the set of pairs a, c with $c \neq 0$, a mod c and $(a, c) = 1$, associating to such a pair the class of an arbitrary matrix in Γ with $\left(\begin{smallmatrix} a \\ c \end{smallmatrix}\right)$ as a left column. On the other hand, one has $\pm \left(\begin{smallmatrix} 1 & b \\ 0 & 1 \end{smallmatrix}\right)\left(\begin{smallmatrix} 1 & k \\ 0 & 1 \end{smallmatrix}\right) = \pm \left(\begin{smallmatrix} 1 & b+k \\ 0 & 1 \end{smallmatrix}\right)$ so that, given $g \in \Gamma$ with $c = 0$, the class $g(\Gamma \cap N)$ can be parametrized by the sole number k under this correspondence; also, one has in this case $a = \pm 1$, and this number characterizes the class mod $\Gamma \cap N$ under consideration. One thus has

$$\langle \mathfrak{S}_{\mathrm{maj}}, f \rangle = 2 \sum_{\substack{c \geq 1, a \bmod c \\ (a,c)=1}} \sum_{j,k \in \mathbb{Z}} \langle h \circ \left(\left(\begin{smallmatrix} 1 & j \\ 0 & 1 \end{smallmatrix}\right) g \left(\begin{smallmatrix} 1 & k \\ 0 & 1 \end{smallmatrix}\right) \right), f \rangle, \qquad (5.1.59)$$

where g is an arbitrary matrix in Γ with $\left(\begin{smallmatrix} a \\ c \end{smallmatrix}\right)$ as a left column, and

$$\langle \mathfrak{S}_{\mathrm{min}}, f \rangle = 2 \sum_{k \in \mathbb{Z}} \langle h \circ \left(\begin{smallmatrix} 1 & k \\ 0 & 1 \end{smallmatrix}\right), f \rangle. \qquad (5.1.60)$$

In other words, $\langle \mathfrak{S}_{\mathrm{min}}, f \rangle = 2T(h, f)$ with the notation of Lemma 5.1.2, which is the same as $\langle \mathfrak{P}_{\mathrm{min}}, h \otimes f \rangle$, as defined in (5.1.53).

Next, we prove the convergence of the series of integrals on the right-hand sides of equations (5.1.54) to (5.1.57), under the sole assumption, invariant under $\mathcal{F}^{\mathrm{symp}}$, that h and f lie in the image of $\mathcal{S}_{\mathrm{even}}(\mathbb{R}^2)$ under $2i\pi\mathcal{E}$: this suffices to ensure the first of the pair of equations (5.1.21), but not the second one. We still write $h = 2i\pi\mathcal{E}h_1$, $f = 2i\pi\mathcal{E}f_1$. The estimate (5.1.36) or the identity (5.1.37) (in which h may also take the place of f) takes care of the summation with respect to m, n as well as of the integration: the only difficulty lies with the summation with respect to c. Recall Weil's sharp bound [21, p. 52]

$$|S(m, n; c)| \leq (m, n, c)^{\frac{1}{2}} c^{\frac{1}{2}} \sigma_0(c) \quad \text{if } mn \neq 0 \qquad (5.1.61)$$

and, in particular, $|S(m, n; c)| \leq C(m, n)c^{\frac{1}{2}+\varepsilon}$ for every $\varepsilon > 0$, and a constant $C(m, n)$. When $mn = 0$, one has instead [21, p. 52] or [23, p. 44], recalling that Möb denotes the Möbius indicator function,

$$S(0, n; c) = \sum_{1 \leq d \mid (n,c)} \mathrm{M\ddot{o}b} \left(\frac{c}{d} \right) d, \qquad (5.1.62)$$

from which one obtains $|S(0, n; c)| \leq \sigma_1(|n|)$ if $n \neq 0$, an estimate independent of c, while the estimate $|S(0, 0; c)| \leq c$ will suffice. To prove that the right-hand side of (5.1.56) is meaningful, it thus suffices to "save" from the integral a factor $c^{-\delta}$ with some $\delta > \frac{1}{2}$: so far as the the right-hand sides of (5.1.54) and (5.1.55) are concerned, saving a factor $c^{-\delta}$ with $\delta > 0$ would be sufficient. Finally, we need to save $c^{-\delta}$ with $\delta > 1$ from the integral on the right-hand side of (5.1.57), and it is immediate, in this case, that we can actually save the factor c^{-2}.

Let us consider (5.1.54) first. Using (5.1.37) (twice), we may transform the integral on the second line of (5.1.56) to

$$\int_{\mathbb{R}^2} (\text{sign } x) \frac{d}{dx} \left[(\mathcal{F}_1^{-1} h_1)(0, x) \right] \xi \frac{\partial}{\partial \xi} \left[(\mathcal{F}_1^{-1} f_1)\left(\frac{n}{\xi}, \xi\right) \right] \left(\cos \frac{2\pi n x}{c\xi} \right) dx \frac{d\xi}{|\xi|} :$$

(5.1.63)

performing an integration by parts. Instead of letting the operator $\xi \frac{\partial}{\partial \xi}$ act on $(\mathcal{F}_1^{-1} f_1)\left(\frac{n}{\xi}, \xi\right)$, we may let its transpose act on the remaining part of the integral: but

$$\left(1 + \xi \frac{\partial}{\partial \xi}\right) \left(|\xi|^{-1} \cos \frac{2\pi n x}{c\xi}\right) = |\xi|^{-1} \frac{2\pi n x}{c\xi} \left(\sin \frac{2\pi n x}{c\xi}\right),$$

(5.1.64)

so we are done since arbitrary powers of $|\xi|^{-1}$ are taken care of by the function $(\mathcal{F}_1^{-1} f_1)\left(\frac{n}{\xi}, \xi\right)$, in which $n \neq 0$. The case of (5.1.54) is totally similar, while that of (5.1.56) is easier, since in view of (5.1.36), we do not care in this case about powers either of $|x|^{-1}$ or $|\xi|^{-1}$. Note that, while we needed to save in this case from the integral a factor $c^{-\delta}$ with $\delta > \frac{1}{2}$, we actually saved a factor c^{-1}, and we did even better when dealing with (5.1.54) or (5.1.57): this will help in a moment.

To facilitate the proof of the identity (5.1.56) and the three analogous ones, we insert an extra factor $c^{-\mu}$ on both sides. Starting from (5.1.59), we thus set

$$F(\mu) = 2 \sum_{\substack{c \geq 1, a \bmod c \\ (a,c)=1}} c^{-\mu} \sum_{j,k \in \mathbb{Z}} \langle h \circ \left(\begin{smallmatrix} 1 & j \\ 0 & 1 \end{smallmatrix}\right) \circ g, f \circ \left(\begin{smallmatrix} 1 & -k \\ 0 & 1 \end{smallmatrix}\right) \rangle :$$

(5.1.65)

from the proof of Proposition 5.1.1 (cf. last sentence there), it is clear that convergence still holds provided that $\text{Re}\,\mu > -1$, and that the function so defined is analytic in this domain. We do a similar operation with the series on the right-hand sides of (5.1.54), (5.1.55), (5.1.56) and (5.1.57), replacing in each case the factor c^{-1} by $c^{-1-\mu}$ and obtaining as a result a function well-defined and analytic in the domain $\text{Re}\,\mu > -\frac{1}{2}$, as it follows from the considerations just made. The use of analytic continuation will make it possible, assuming that $\text{Re}\,\mu$ is large, not to worry about the summability with respect to c, as all estimates which will appear below are uniform, up to the possible loss of powers of c.

In the computation of $F(0) = \langle \mathfrak{S}_{\text{maj}}, f \rangle$, a number of terms will show up: in order not to get lost, we shall denote these successive contributions as $T_1, T_2, \ldots$: we shall have to add them at the end. Let us write

$$F(\mu) = 2 \sum_{\substack{c \geq 1, a \bmod c \\ (a,c)=1}} c^{-\mu} \sum_{j \in \mathbb{Z}} A_j(a, c),$$

(5.1.66)

with

$$A_j(a, c) = \sum_{k \in \mathbb{Z}} \int_{\mathbb{R}^2} h((a + jc)x + (b + jd)\xi, cx + d\xi) f(x - k\xi, \xi) dx d\xi.$$

(5.1.67)

To compute this, we shall apply Lemma 5.1.2 and, later, the method of proof of this lemma, again. One has

$$\left(\mathcal{F}_1^{-1}\left[(x,\xi) \mapsto h((a+jc)x+(b+jd)\xi, cx+d\xi)]\right)\right)(\eta,\xi)$$

$$= c^{-1}\exp\left(-2i\pi\frac{d\xi\eta}{c}\right)\int_{-\infty}^{\infty}\exp\left(2i\pi\frac{x\eta}{c}\right)h\left(\left(\frac{a}{c}+j\right)x-\frac{\xi}{c},x\right)dx. \quad (5.1.68)$$

It follows from the lemma that

$$A_j(a,c) = A_j^\bullet(a,c) + B_j(a,c) \quad (5.1.69)$$

with

$$A_j^\bullet(a,c) = c^{-1}\sum_{n\in\mathbb{Z}}e^{2i\pi\frac{nd}{c}}\int_{-\infty}^{\infty}(\mathcal{F}_1^{-1}f)\left(\frac{n}{\xi},\xi\right)\frac{d\xi}{|\xi|}$$

$$\times\int_{-\infty}^{\infty}\exp\left(-2i\pi\frac{nx}{c\xi}\right)h\left(\left(\frac{a}{c}+j\right)x-\frac{\xi}{c},x\right)dx, \quad (5.1.70)$$

and

$$B_j(a,c) = c^{-1}\int_{-\infty}^{\infty}(\mathcal{F}_1^{-1}f)(\xi,0)\frac{d\xi}{|\xi|}\cdot\int_{-\infty}^{\infty}e^{-2i\pi\frac{x\xi}{c}}h\left(\left(\frac{a}{c}+j\right)x,x\right)dx. \quad (5.1.71)$$

Next, we must compute the sum as j runs through $\mathbb{Z}$ of each of the last two expressions: we start with the second one. In the sum $\sum_{j\in\mathbb{Z}}B_j(a,c)$, it is impossible to permute the j-series and the integral, and we manage in the same way as that used in the proof of Lemma 5.1.2, subtracting when $j \neq 0$ from the integrand $e^{-2i\pi\frac{x\xi}{c}}h\left(\left(\frac{a}{c}+j\right)x,x\right)$ the function $h(jx,0)$, the integral of which over the real line is zero (5.1.21): since, for $j \neq 0$,

$$\int_{-\infty}^{\infty}\left|e^{-2i\pi\frac{x\xi}{c}}h\left(\left(\frac{a}{c}+j\right)x,x\right)-h(jx,0)\right|dx \leq Cj^{-2}(1+|\xi|), \quad (5.1.72)$$

this makes the permutation possible and leads to

$$\sum_{j\in\mathbb{Z}}B_j(a,c) = c^{-1}\int_{\mathbb{R}^2}dx\frac{d\xi}{|\xi|}(\mathcal{F}_1^{-1}f)(\xi,0)e^{-2i\pi\frac{x\xi}{c}}\sum_{j\in\mathbb{Z}}h\left(\left(\frac{a}{c}+j\right)x,x\right)$$

$$- c^{-1}\int_{\mathbb{R}^2}dx\frac{d\xi}{|\xi|}(\mathcal{F}_1^{-1}f)(\xi,0)\sum_{j\neq 0}h(jx,0) \quad (5.1.73)$$

as soon as the convergence of each of the two terms of this decomposition has been asserted. According to Poisson's formula, one has for $x \neq 0$,

$$\sum_{j\in\mathbb{Z}}h\left(\left(\frac{a}{c}+j\right)x,x\right) = \frac{1}{|x|}\sum_{m\in\mathbb{Z}}e^{2i\pi\frac{am}{c}}(\mathcal{F}_1^{-1}h)\left(-\frac{m}{x},x\right), \quad (5.1.74)$$

so that the first term on the right-hand side of (5.1.73) is, after an easy permutation
of the integral and the m-series (use (5.1.36) in the x-variable if $m \neq 0$ and (5.1.37)
in the ξ-variable)

$$c^{-1} \sum_{m \in \mathbb{Z}} e^{2i\pi \frac{am}{c}} \int_{\mathbb{R}^2} \exp\left(-2i\pi \frac{x\xi}{c}\right) (\mathcal{F}_1^{-1} h) \left(-\frac{m}{x}, x\right) (\mathcal{F}_1^{-1} f) (\xi, 0) \frac{dx \, d\xi}{|x| \, |\xi|}.$$

(5.1.75)

Dealing now with the second term on the right-hand side of (5.1.73), we note that
the integral

$$\int_{\mathbb{R}^2} dx \frac{d\xi}{|\xi|} (\mathcal{F}_1^{-1} f) (\xi, 0) \sum_{j \neq 0} h(jx, 0)$$

(5.1.76)

is similar to the one in (5.1.39): that jx, not $\xi - jx$, is the first argument of the
function h here brings very few changes, and leads to the fact that the value of
this integral is

$$- \int_{\mathbb{R}^2} (\mathcal{F}_1^{-1} h) (x, 0) (\mathcal{F}_1^{-1} f) (\xi, 0) \frac{dx \, d\xi}{|x| \, |\xi|}.$$

(5.1.77)

We have found

$$\sum_{j \in \mathbb{Z}} B_j(a, c) = c^{-1} \int_{\mathbb{R}^2} (\mathcal{F}_1^{-1} h) (x, 0) (\mathcal{F}_1^{-1} f) (\xi, 0) \frac{dx \, d\xi}{|x| \, |\xi|}$$

(5.1.78)

$$+ c^{-1} \sum_{m \in \mathbb{Z}} e^{2i\pi \frac{am}{c}} \int_{\mathbb{R}^2} \exp\left(-2i\pi \frac{x\xi}{c}\right) (\mathcal{F}_1^{-1} h) \left(-\frac{m}{x}, x\right) (\mathcal{F}_1^{-1} f) (\xi, 0) \frac{dx \, d\xi}{|x| \, |\xi|}.$$

To compute the contribution of this sum to $F(\mu)$, as defined in (5.1.66), we recall
that, whether $m = 0$ or not,

$$\sum_{\substack{a \bmod c \\ (a,c)=1}} e^{2i\pi \frac{am}{c}} = S(m, 0; c).$$

(5.1.79)

Since the function

$$\sum_{c \geq 1} c^{-1-\mu} S(0, 0; c) = \frac{\zeta(\mu)}{\zeta(\mu + 1)}$$

(5.1.80)

vanishes at $\mu = 0$, the first term on the right-hand side of (5.1.78) will not contribute to the part of $F(0)$ we are currently dealing with, which will thus reduce
to $T_1 = T_1(0)$, with

$$T_1(\mu) = 2 \sum_{c \geq 1} c^{-1-\mu} \sum_{m \in \mathbb{Z}} S(m, 0; c)$$

$$\times \int_{\mathbb{R}^2} (\mathcal{F}_1^{-1} h) \left(-\frac{m}{x}, x\right) (\mathcal{F}_1^{-1} f) (\xi, 0) \exp\left(-2i\pi \frac{x\xi}{c}\right) \frac{dx \, d\xi}{|x| \, |\xi|}. \quad (5.1.81)$$

Going back to (5.1.66) and (5.1.69), we must now compute the value at $\mu = 0$ of the sum

$$2 \sum_{\substack{c \geq 1, a \bmod c \\ (a,c)=1}} c^{-\mu} \sum_{j \in \mathbb{Z}} A_j^{\bullet}(a, c) \tag{5.1.82}$$

and, to start with, the sum

$$\sum_{j \in \mathbb{Z}} A_j^{\bullet}(a, c) = c^{-1} \sum_{j \in \mathbb{Z}} \sum_{n \in \mathbb{Z}} e^{2i\pi \frac{nd}{c}} \int_{-\infty}^{\infty} (\mathcal{F}_1^{-1} f) \left(\frac{n}{\xi}, \xi \right) \frac{d\xi}{|\xi|}$$

$$\times \int_{-\infty}^{\infty} \exp \left(-2i\pi \frac{nx}{c\xi} \right) h \left(\left(\frac{a}{c} + j \right) x - \frac{\xi}{c}, x \right) dx. \tag{5.1.83}$$

When $n \neq 0$, the function $|\xi|^{-1} (\mathcal{F}_1^{-1} f) \left(\frac{n}{\xi}, \xi \right) \exp \left(-2i\pi \frac{nx}{c\xi} \right)$ lies in $\mathcal{S}_{\text{even}}(\mathbb{R}^2)$. Setting $\xi = ax - cn$, one transforms the double integral just written to

$$c \int_{\mathbb{R}^2} (\mathcal{F}_1^{-1} f) \left(\frac{n}{ax - cn}, ax - cn \right) \exp \left(-2i\pi \frac{nx}{c(ax - cn)} \right) h(\eta + jx, x) dx \frac{d\eta}{|ax - cn|}, \tag{5.1.84}$$

so that one can apply Lemma 5.1.2 directly, exchanging the variables x and ξ: much care is needed when substituting the proper arguments. It is on the functions, within brackets in the following two equations, regarded as functions of η that the inverse Fourier transformation is to be applied:

$$\mathcal{F}^{-1} \left[\frac{c}{|ax - cn|} (\mathcal{F}_1^{-1} f) \left(\frac{n}{ax - cn}, ax - cn \right) \exp \left(-2i\pi \frac{nx}{c(ax - cn)} \right) \right] \left(\frac{m}{x}, x \right)$$

$$= e^{2i\pi \frac{am}{c}} \int_{-\infty}^{\infty} (\mathcal{F}_1^{-1} f) \left(\frac{n}{\xi}, \xi \right) \exp \left(-2i\pi \left(\frac{m\xi}{cx} + \frac{nx}{c\xi} \right) \right) \frac{d\xi}{|\xi|} \tag{5.1.85}$$

and

$$\mathcal{F}^{-1} \left[|\eta|^{-1} (\mathcal{F}_1^{-1} f) \left(-\frac{n}{c\eta}, -c\eta \right) \right] (x, 0) = \int_{-\infty}^{\infty} e^{-2i\pi \frac{x\xi}{c}} (\mathcal{F}_1^{-1} f) \left(\frac{n}{\xi}, \xi \right) \frac{d\xi}{|\xi|}. \tag{5.1.86}$$

Applying Lemma 5.1.2, we obtain that the contribution to $\sum_{j \in \mathbb{Z}} A_j^{\bullet}(a, c)$ of the terms with $n \neq 0$ is

$$c^{-1} \sum_{n \neq 0} e^{2i\pi \frac{am + dn}{c}} \sum_{m \in \mathbb{Z}} \int_{\mathbb{R}^2} (\mathcal{F}_1^{-1} h) \left(-\frac{m}{x}, x \right) (\mathcal{F}_1^{-1} f) \left(\frac{n}{\xi}, \xi \right)$$

$$\times \exp \left(-2i\pi \left(\frac{m\xi}{cx} + \frac{nx}{c\xi} \right) \right) \frac{dx \, d\xi}{|x| \, |\xi|} \tag{5.1.87}$$

$$+ c^{-1} \sum_{n \neq 0} e^{2i\pi \frac{dn}{c}} \int_{\mathbb{R}^2} (\mathcal{F}_1^{-1} h) (x, 0) (\mathcal{F}_1^{-1} f) \left(\frac{n}{\xi}, \xi \right) \exp \left(-2i\pi \frac{x\xi}{c} \right) \frac{dx \, d\xi}{|x| \, |\xi|}.$$

Taking into account the definition of Kloosterman sums, we obtain that the contribution to the value at $\mu = 0$ of the sum (5.1.83) limited to the terms under investigation (those with $n \neq 0$) is $T_2(0) + T_3(0)$, with

$$T_2(\mu) = 2 \sum_{c \geq 1} c^{-1-\mu} \sum_{n \neq 0} \sum_{m \in \mathbb{Z}} S(m, n; c)$$ (5.1.88)

$$\times \int_{\mathbb{R}^2} (\mathcal{F}_1^{-1}h)\left(-\frac{m}{x}, x\right) (\mathcal{F}_1^{-1}f)\left(\frac{n}{\xi}, \xi\right) \exp\left(-2i\pi\left(\frac{m\xi}{cx} + \frac{nx}{c\xi}\right)\right) \frac{dx\, d\xi}{|x|\, |\xi|}$$

and (changing ξ to $\frac{n}{\xi}$ in the integral)

$$T_3(\mu) = 2 \sum_{c \geq 1} c^{-1-\mu} \sum_{n \neq 0} S(0, n; c)$$

$$\times \int_{\mathbb{R}^2} (\mathcal{F}_1^{-1}h)(x, 0)(\mathcal{F}_1^{-1}f)\left(\xi, \frac{n}{\xi}\right) \exp\left(-2i\pi\frac{nx}{c\xi}\right) \frac{dx\, d\xi}{|x|\, |\xi|}.$$ (5.1.89)

The only term that remains to be considered arises from the contribution to $\sum_{j \in \mathbb{Z}} A_j^\bullet(a, c)$ of the terms with $n = 0$, to wit (starting from (5.1.83)), the sum

$$c^{-1} \sum_{j \in \mathbb{Z}} \int_{\mathbb{R}^2} (\mathcal{F}_1^{-1}f)(0, \xi) h\left(\left(\frac{a}{c} + j\right)x - \frac{\xi}{c}, x\right) dx \frac{d\xi}{|\xi|}.$$ (5.1.90)

We can no longer apply Lemma 5.1.2, but the same method will work. Namely, subtracting from the integrand of the dx-integral, when $j \neq 0$, the function $h\left(jx - \frac{\xi}{c}, 0\right)$, the integral of which is zero, we can transform this to

$$c^{-1} \int_{\mathbb{R}^2} (\mathcal{F}_1^{-1}f)(0, \xi) \sum_{j \in \mathbb{Z}} h\left(\left(\frac{a}{c} + j\right)x - \frac{\xi}{c}, x\right) dx \frac{d\xi}{|\xi|}$$

$$- c^{-1} \int_{\mathbb{R}^2} (\mathcal{F}_1^{-1}f)(0, \xi) \sum_{j \neq 0} h\left(jx - \frac{\xi}{c}, 0\right) dx \frac{d\xi}{|\xi|}.$$ (5.1.91)

Applying Poisson's formula again, the first line on the right-hand side can be written

$$c^{-1} \sum_{m \in \mathbb{Z}} e^{\frac{2i\pi am}{c}} \int_{\mathbb{R}^2} (\mathcal{F}_1^{-1}h)\left(-\frac{m}{x}, x\right) (\mathcal{F}_1^{-1}f)(0, \xi) \exp\left(-2i\pi\frac{m\xi}{cx}\right) \frac{dx\, d\xi}{|x|\, |\xi|}.$$ (5.1.92)

It contributes to the value at $\mu = 0$ of the sum (5.1.83) the term $T_4(0)$, with

$$T_4(\mu) = 2 \sum_{c \geq 1} c^{-1-\mu} \sum_{m \in \mathbb{Z}} S(m, 0; c)$$

$$\times \int_{\mathbb{R}^2} (\mathcal{F}_1^{-1}h)\left(-\frac{m}{x}, x\right) (\mathcal{F}_1^{-1}f)(0, \xi) \exp\left(-2i\pi\frac{m\xi}{cx}\right) \frac{dx\, d\xi}{|x|\, |\xi|}.$$ (5.1.93)

Finally, we must consider the second term on the right-hand side of (5.1.91). Calling it $-R$, we may treat it in exactly the same way as that denoted by the same letter in (5.1.39), in the proof of Lemma 5.1.2. Introducing

$$R(\nu) = c^{-1} \int_{\mathbb{R}^2} |x|^\nu dx \frac{d\xi}{|\xi|} \left(\mathcal{F}_1^{-1}f\right)(0, \xi) \sum_{j \neq 0} h\left(jx - \frac{\xi}{c}, 0\right), \qquad (5.1.94)$$

we note that this is an analytic function of ν for $\mathrm{Re}\,\nu > -1$. Following the computation done in the given reference, we obtain

$$R(\nu) = 2\zeta(-\nu) \int_{-\infty}^{\infty} \left(\mathcal{F}_1^{-1}f\right)(0, \xi) \frac{d\xi}{|\xi|} \int_{-\infty}^{\infty} |x|^{-\mu-1} \exp\left(2i\pi \frac{x\xi}{c}\right) \left(\mathcal{F}_1^{-1}h\right)(x, 0)dx.$$

$$(5.1.95)$$

The corresponding contribution to $F(0)$ (this is the last one) is thus $T_5(0)$, with

$$T_5(\mu) = 2 \sum_{c \geq 1} c^{-1-\mu} S(0, 0; c)$$

$$\times \int_{\mathbb{R}^2} \left(\mathcal{F}_1^{-1}h\right)(x, 0) \left(\mathcal{F}_1^{-1}f\right)(0, \xi) \exp\left(2i\pi \frac{x\xi}{c}\right) \frac{dx\, d\xi}{|x|\, |\xi|}. \qquad (5.1.96)$$

What remains to be done is checking that $T_1(0) + \cdots + T_5(0)$ coincides with the sum on the right-hand side of (5.1.52), once the first term (already taken care of right after (5.1.60)) has been removed. We first note the identity, valid for every $h \in \mathcal{S}_{\mathrm{even}}(\mathbb{R}^2)$,

$$\left[\mathcal{F}_1^{-1}\left(\mathcal{F}^{\mathrm{symp}}h\right)\right](\eta, \xi) = \left(\mathcal{F}_1^{-1}h\right)(\xi, \eta): \qquad (5.1.97)$$

it makes it a trivial task to rewrite the right-hand sides of equations (5.1.54), (5.1.55) and (5.1.57) when, as invited by (5.1.52), the expression $\mathcal{F}^{\mathrm{symp}}h \otimes \mathcal{F}^{\mathrm{symp}}f$ has been substituted for $h \otimes f$. This already provides us with some measure of verification since the bilinear form $\mathfrak{P}$ is invariant by construction under such a change while, from (5.1.97) and a look at equations (5.1.53) and (5.1.56), so are the bilinear forms $\mathfrak{P}_{\mathrm{min}}$ and $\mathfrak{P}^{(11)}$: but this verification is valid only under the more stringent assumptions (invariant under the change of h, f to $\mathcal{F}^{\mathrm{symp}}h, \mathcal{F}^{\mathrm{symp}}f$) that h and f lie in the image of $\mathcal{S}_{\mathrm{even}}(\mathbb{R}^2)$ under the operator $2i\pi\mathcal{E}(1 + 2i\pi\mathcal{E})(1 - 2i\pi\mathcal{E})$.

In the observations which follow, one must keep in mind that in $\mathbb{Z}$, as a range of values of m or n, one must always set apart 0 and use when needed (as is the case when dealing with $T_1(\mu)$) the change of variable $x \mapsto -\frac{m}{x}$ (or $\xi \mapsto \frac{n}{\xi}$), which preserves the measure $\frac{dx}{|x|}$ (resp. $\frac{d\xi}{|\xi|}$); also, dealing with $\mathcal{S}_{\mathrm{even}}(\mathbb{R}^2)$ will make it possible to replace exponentials by cosines.

The part of $T_1(0)$ obtained in this way, from the terms with $m \neq 0$ of (5.1.81) is (after this change of variable) exactly $\langle \mathfrak{P}^{(10)}, \mathcal{F}^{\mathrm{symp}}h \otimes \mathcal{F}^{\mathrm{symp}}f \rangle$. In the part of $T_1(\mu)$ obtained from the term with $m = 0$, we cannot make $\mu = 0$ in a brutal way since this would create a divergence: however, we can do so after having subtracted

1 from $\exp\left(-2i\pi\frac{x\xi}{c}\right)$, finding as a result the term $\langle\mathfrak{P}^{(00)}, h\otimes f\rangle$. At the same time, what we would obtain from the replacement of the exponential by 1 would not contribute ultimately to $F(0)$, in view of (5.2.52).

Again $T_2(0)$, as computed in (5.1.88), must be split, setting apart the term with $m = 0$: this term is exactly $\langle\mathfrak{P}^{(01)}, h\otimes f\rangle$. The sum of terms of $T_2(0)$ with $m \neq 0$ is $\langle\mathfrak{P}^{(11)}, h\otimes f\rangle$. The term $T_3(0)$ in (5.1.89) agrees with $\langle\mathfrak{P}^{(01)}, \mathcal{F}^{\mathrm{symp}}h\otimes\mathcal{F}^{\mathrm{symp}}f\rangle$. The sum of terms with $m \neq 0$ from $T_4(\mu)$, as computed in (5.1.93), agrees with $\langle\mathfrak{P}^{(10)}, h\otimes f\rangle$. The term with $m = 0$ from the same sum will not contribute, in view of (5.2.52) again, because c does not occur in the integrand. Finally, the term $T_5(0)$ is made explicit in just the same way as the part of $T_1(0)$ originating from the term with $m = 0$ in (5.1.81): we obtain this time $\langle\mathfrak{P}^{(00)}, \mathcal{F}^{\mathrm{symp}}h\otimes\mathcal{F}^{\mathrm{symp}}f\rangle$. The proof of Proposition 5.1.3 is over. $\qquad\square$

Remark 5.1.a. Recall from the proof that the right-hand side of (5.1.52) is meaningful as soon as h and f lie in the image of $\mathcal{S}_{\mathrm{even}}(\mathbb{R}^2)$ under $2i\pi\mathcal{E}$.

5.2 Spectral decomposition of the bilinear form $\mathfrak{P}$

We now look for a decomposition of $\mathfrak{S} = \sum_{g\in\Gamma} h\circ g$ (not quite a distribution *a priori*: but the map $h\mapsto\mathfrak{S}$ can be extended as a map from the image of $\mathcal{S}_{\mathrm{even}}(\mathbb{R}^2)$ under the operator $\pi^2\mathcal{E}^2(\pi^2\mathcal{E}^2 + \frac{1}{4})$ to the space of automorphic distributions) as a superposition of homogeneous distributions, automorphic as it will turn out (of course, this will not be the case when considering the minor and major parts of $\mathfrak{S}$ separately). It is useful, at this point, to review some notions regarding automorphic distributions.

Definition 5.2.1. Recall that the equation

$$|t|_{\varepsilon}^{-\nu-1} = -\frac{1}{\nu}\frac{d}{dt}|t|_{1-\varepsilon}^{-\nu}, \qquad \varepsilon = 0 \text{ or } 1, \tag{5.2.1}$$

makes it possible, by induction, to give the left-hand side a meaning, as a tempered distribution, provided that $\nu \neq \varepsilon, \varepsilon + 2, \ldots$. We then define, for $k \in \mathbb{Z}$ and $\nu \neq 0, 2, \ldots$, the pair of tempered distributions (still denoted as if they were integrals)

$$\mathfrak{L}_{\nu,k}(h) = \int_{-\infty}^{\infty} |t|^{-\nu-1}\left(\mathcal{F}_1^{-1}h\right)\left(\frac{k}{t}, t\right) dt,$$

$$\mathfrak{R}_{\nu,k}(h) = \int_{-\infty}^{\infty} |t|^{-\nu-1}\left(\mathcal{F}_1^{-1}h\right)\left(t, \frac{k}{t}\right) dt : \tag{5.2.2}$$

one may note that, when $k \neq 0$, this even makes sense when ν takes any of the generally excluded values $0, 2, \ldots$.

If $k \neq 0$, one has $\mathfrak{R}_{\nu,k}(h) = |k|^{-\nu}\mathfrak{L}_{-\nu,k}(h)$, or

$$\sigma_{\nu}(|k|)\mathfrak{R}_{\nu,k}(h) = \sigma_{-\nu}(|k|)\mathfrak{L}_{-\nu,k}(h), \qquad k \neq 0. \tag{5.2.3}$$

There is another link between $\mathfrak{L}_{\nu,k}$ and $\mathfrak{R}_{\nu,k}$, to wit the identity

$$\mathfrak{R}_{\nu,k}(h) = \mathfrak{L}_{\nu,k}\left(\mathcal{F}^{\mathrm{symp}}h\right),\qquad(5.2.4)$$

a consequence of (5.1.97). A last useful elementary property is the identity

$$\mathfrak{L}_{\nu,k}(\bar{h}) = \overline{\mathfrak{L}_{\bar{\nu},-k}(h)},\qquad(5.2.5)$$

which is a consequence of the fact that the complex conjugate of $\left(\mathcal{F}_1^{-1}\bar{h}\right)(\eta,\xi)$ is $\left(\mathcal{F}_1^{-1}h\right)(-\eta,\xi)$.

If $\operatorname{Re}\nu < 0, \nu \neq 0, -2, \ldots$, one has

$$\int_{-\infty}^{\infty}|t|^{-\nu}h(t,0)\,dt$$
$$= \pi^{-\frac{1}{2}+\nu}\frac{\Gamma(\frac{1-\nu}{2})}{\Gamma(\frac{\nu}{2})}\int_{-\infty}^{\infty}|t|^{\nu-1}\left(\mathcal{F}_1^{-1}h\right)(t,0)\,dt = \frac{\zeta(\nu)}{\zeta(1-\nu)}\mathfrak{R}_{-\nu,0}(h):\quad(5.2.6)$$

the second expression is not a genuine integral in general; recall that the notation (5.2.2) deals with the distribution generalization of the notion. Still, if $-1 < \operatorname{Re}\nu < 0$ and $h \in (2i\pi\mathcal{E})\mathcal{S}_{\mathrm{even}}(\mathbb{R}^2)$, this is a genuine integral anyway since, as a consequence of (5.1.37), the function $t^{-1}\left(\mathcal{F}_1^{-1}h\right)(t,0)$ is bounded. Equation (3.2.7) can then be rephrased as

$$\langle\mathfrak{E}_{\nu},h\rangle = \zeta(-\nu)\mathfrak{L}_{\nu,0}(h) + \zeta(\nu)\mathfrak{R}_{-\nu,0}(h) + \sum_{k\neq 0}\sigma_\nu(|k|)\mathfrak{L}_{\nu,k}(h):\qquad(5.2.7)$$

this decomposition is only valid for $\nu \notin 2\mathbb{Z}, \nu \neq \pm 1$, but $\mathfrak{E}_\nu$ is meaningful for $\nu \neq \pm 1$. From (5.2.3) and (5.2.4), one may recheck, looking at the last equation, that $\mathcal{F}^{\mathrm{symp}}\mathfrak{E}_\nu = \mathfrak{E}_{-\nu}$.

The same linear forms can be used to define the cusp-distributions associated to Maass-Hecke cusp-forms. Recall that the conditions $p \geq 1$ and $\lambda_p > 0$ are part of the convention regarding the orthonormal set $(\mathcal{M}_{p,j})$ of Maass-Hecke cusp-forms. But, when defining the associated cusp-distributions $\mathfrak{M}_{p,j}$, we must take $p \neq 0$ instead: it will be useful to define $\lambda_p = -\lambda_{-p}$ if $p < 0$. Then, (3.2.21) becomes simply, for all pairs p, j,

$$\langle\mathfrak{M}_{p,j},h\rangle = \frac{1}{2}\sum_{k\neq 0}|k|^{\frac{i\lambda_p}{2}}b_{k;p,j}\mathfrak{L}_{i\lambda_p,k}(h),\qquad(5.2.8)$$

where our convention is that $b_{k;-p,j} = b_{k;p,j}$.

Recall that a tempered distribution $\mathfrak{T}$ is homogeneous of degree α if it satisfies the identity $\langle\mathfrak{T}, s^{-2i\pi\mathcal{E}}h\rangle = s^{1+\alpha}\langle\mathfrak{T}, h\rangle$ for every $h \in \mathcal{S}(\mathbb{R}^2)$. In this sense, $\mathfrak{L}_{\nu,k}$ is homogeneous of degree $-1 - \nu$ and $\mathfrak{R}_{\nu,k}$ is homogeneous of degree $-1 + \nu$. All terms of the decomposition (5.2.7) are thus homogeneous of degreee $-1 - \nu$, and all terms of the decomposition (5.2.8) are thus homogeneous of degreee $-1 - i\lambda_p$.

We begin now our computation of the decomposition of $\mathfrak{S}$ into homogeneous components, which will require several lemmas. So far as the minor term $\mathfrak{S}_{\min}$ is concerned, one may simply write

$$\mathfrak{S}_{\min} = \int_{-\infty}^{\infty} (\mathfrak{S}_{\min})_{i\lambda}\, d\lambda \tag{5.2.9}$$

with $\langle (\mathfrak{S}_{\min})_{i\lambda}, f \rangle = \langle \mathfrak{S}_{\min}, f_{-i\lambda} \rangle$. From (5.1.53), it follows that $\langle (\mathfrak{S}_{\min})_{i\lambda}, f \rangle$ is the sum of two terms, the first of which is

$$\frac{1}{\pi} \sum_{n \in \mathbb{Z}} \int_0^{\infty} t^{-1-i\lambda} dt \int_{-\infty}^{\infty} (\mathcal{F}_1^{-1} h) \left(-\frac{n}{\xi}, \xi \right) (\mathcal{F}_1^{-1} f) \left(\frac{n}{t\xi}, t\xi \right) \frac{d\xi}{|\xi|} \tag{5.2.10}$$

$$= \frac{1}{2\pi} \sum_{n \in \mathbb{Z}} \int_{-\infty}^{\infty} |\xi|^{-1+i\lambda} (\mathcal{F}_1^{-1} h) \left(-\frac{n}{\xi}, \xi \right) d\xi \times \int_{-\infty}^{\infty} |s|^{-1-i\lambda} (\mathcal{F}_1^{-1} f) \left(\frac{n}{s}, s \right) ds :$$

note that these double integrals are absolutely convergent in a trivial way when $m \neq 0$, and it is also the case when $m = 0$ in view of equation (5.1.37), which eliminates an apparent pair of singularities. The second term is

$$\frac{1}{\pi} \int_0^{\infty} \int_{-\infty}^{\infty} |\xi|^{-1-i\lambda} (\mathcal{F}_1^{-1} h) (\xi, 0) d\xi \times \int_{-\infty}^{\infty} |s|^{-1+i\lambda} (\mathcal{F}_1^{-1} f) (s, 0) ds. \tag{5.2.11}$$

Using Definition 5.2.1, one may write

$$\langle \mathfrak{S}_{\min}, f \rangle$$
$$= \frac{1}{2\pi} \int_{-\infty}^{\infty} \sum_{m \in \mathbb{Z}} \mathcal{L}_{i\lambda, -m}(h) \mathcal{L}_{-i\lambda, m}(f) d\lambda + \frac{1}{2\pi} \int_{-\infty}^{\infty} \mathfrak{R}_{-i\lambda, 0}(h) \mathfrak{R}_{i\lambda, 0}(f) d\lambda. \tag{5.2.12}$$

To obtain a decomposition of $\mathfrak{S}_{\mathrm{maj}}$ into homogeneous components, we shall make use of (5.1.16), which demands computing the image of $\mathfrak{S}_{\mathrm{maj}}$ under $(\mu + 2i\pi\mathcal{E})^{-1}$ for $|\mathrm{Re}\,\mu|$ large, and $\mathrm{Re}\,\mu$ positive or negative. At this point, it is necessary to decompose the linear form $\mathfrak{S}_{\mathrm{maj}}$ in a way paralleling the decomposition of $\langle \mathfrak{P}_{\mathrm{maj}}, h \otimes f \rangle$ in Proposition 5.1.3. With $(p, q) = (0, 1), (1, 0), (1, 1)$ or $(0, 0)$, we set

$$\langle \mathfrak{S}^{(pq)}, \rangle = \langle \mathfrak{P}^{(pq)}, h \otimes f \rangle, \tag{5.2.13}$$

and, except in the case when $(p, q) = (1, 1)$, in which this is not necessary, we set

$$\langle \mathfrak{T}^{(pq)}, f \rangle = \langle \mathfrak{P}^{(pq)}, \mathcal{F}^{\mathrm{symp}} h \otimes \mathcal{F}^{\mathrm{symp}} f \rangle. \tag{5.2.14}$$

Finding the decomposition into homogeneous components of the linear forms $\mathfrak{S}^{(pq)}$ will suffice: the missing ones are related to these in a trivial way.

Let us denote as Σ_{01} (*resp.* Σ_{10}, Σ_{11}) the set of pairs m, n with $(m = 0, n \neq 0)$ (*resp.* $(m \neq 0, n = 0), (m \neq 0, n \neq 0)$). For $t > 0$, one has

$$\mathcal{F}_1^{-1}\left[(y, \eta) \mapsto h(ty, t\eta)\right]\left(-\frac{m}{x}, x\right) = t^{-1}\left(\mathcal{F}_1^{-1}h\right)\left(-\frac{m}{tx}, tx\right): \qquad (5.2.15)$$

when $\operatorname{Re}\mu$ is large, applying the equation

$$\left[(\mu + 2i\pi\mathcal{E})^{-1}\mathfrak{S}^{(pq)}\right](x, \xi) = \int_0^1 t^\mu \mathfrak{S}^{(pq)}(tx, t\xi)dt \qquad (5.2.16)$$

to the various terms from the decomposition of $\mathfrak{S}$ given in Proposition 5.1.3, we obtain if $(p, q) = (0, 1), (1, 0)$ or $(1, 1)$, after having performed the change of variable $x \mapsto t^{-1}x$, the equation

$$\langle(\mu + 2i\pi\mathcal{E})^{-1}\mathfrak{S}^{(pq)}, f\rangle = 2\sum_{c \geq 1} c^{-1} \sum_{(m,n) \in \Sigma_{pq}} S(m, n; c) \qquad (5.2.17)$$

$$\times \int_{\mathbb{R}^2} \left(\mathcal{F}_1^{-1}h\right)\left(-\frac{m}{x}, x\right)\left(\mathcal{F}_1^{-1}f\right)\left(\frac{n}{\xi}, \xi\right) F^+\left(\mu; \frac{nx}{c\xi}, \frac{m\xi}{cx}\right) \frac{dx\,d\xi}{|x|\,|\xi|}, \quad \operatorname{Re}\mu \text{ large,}$$

if we define

$$F^+(\mu; \alpha, \beta) = \int_0^1 t^{\mu-1} \cos\left(2\pi(\alpha t^{-1} + \beta t)\right) dt, \quad \operatorname{Re}\mu > 0. \qquad (5.2.18)$$

Finding the analogue of (5.2.17) in the case when $-\operatorname{Re}\mu$ is large requires defining in the same way

$$F^-(\mu; \alpha, \beta) = -\int_1^\infty t^{\mu-1} \cos\left(2\pi(\alpha t^{-1} + \beta t)\right) dt \quad \operatorname{Re}\mu < 0: \qquad (5.2.19)$$

but it is immediate that $F^-(\mu; \alpha, \beta) = -F^+(-\mu; \beta, \alpha)$ for $\operatorname{Re}\mu < 0$. From (5.1.17), one thus obtains that, for δ large enough,

$$\langle\mathfrak{S}^{(pq)}, f\rangle = \frac{1}{i\pi}\int_{\operatorname{Re}\mu=\delta}^{\text{sym}} d\mu \sum_{c \geq 1} c^{-1} \sum_{m,n \in \Sigma_{pq}} S(m, n; c) \int_{\mathbb{R}^2}\left(\mathcal{F}_1^{-1}h\right)\left(-\frac{m}{x}, x\right)$$

$$\times\left(\mathcal{F}_1^{-1}f\right)\left(\frac{n}{\xi}, \xi\right)\left[F^+\left(\mu; \frac{nx}{c\xi}, \frac{m\xi}{cx}\right) + F^+\left(\mu; \frac{m\xi}{cx}, \frac{nx}{c\xi}\right)\right]\frac{dx\,d\xi}{|x|\,|\xi|}. \qquad (5.2.20)$$

To study the function F^+, one must make use of the incomplete Gamma functions [36, p. 337]. In the equations below, it is assumed that $x > 0$; on the other hand, one has to assume that $\operatorname{Re}\nu > 0$ in the definition of the first function only:

$$\gamma(\nu, x) = \int_0^x t^{\nu-1}e^{-t}dt,$$

$$\Gamma(\nu, x) = \int_x^\infty t^{\nu-1}e^{-t}dt = \Gamma(\nu) - \gamma(\nu, x). \qquad (5.2.21)$$

They extend as holomorphic functions of x in the Riemann surface above the plane punctured at 0. We cut the plane along the interval $]-\infty, 0]$ both for a proper definition of these two functions as for that of power functions. Then, for $z \notin]-\infty, 0]$, one has

$$\int_0^1 t^{\nu-1}e^{-zt}\,dt = z^{-\nu}\gamma(\nu, z), \qquad \int_1^\infty t^{\nu-1}e^{-zt}\,dt = z^{-\nu}\Gamma(\nu, z). \qquad (5.2.22)$$

The function defined as

$$\psi(\nu, z): \ = z^{-\nu}\gamma(\nu, z) = \sum_{r\geq 0}\frac{(-1)^r}{r!}\frac{z^r}{\nu+r}, \qquad \mathrm{Re}\,\nu > 0, \qquad (5.2.23)$$

extends as an entire function of z for all values of $\nu \neq 0, -1, \ldots$.

Lemma 5.2.2. *For* $\mathrm{Re}\,\mu > 0, \mu \notin \mathbb{Z}$, *the function* F^+ *defined in* (5.2.18) *is made explicit as follows. If* $\alpha, \beta \neq 0$, *one has*

$$F^+(\mu; \alpha, \beta) = \frac{1}{2}\Gamma(-\frac{\mu}{2})\Gamma(\frac{2+\mu}{2})|\frac{\alpha}{\beta}|^{\frac{\mu}{2}} \times \begin{cases} J_\mu(4\pi\sqrt{\alpha\beta}) & \text{if } \alpha\beta > 0, \\ I_\mu(4\pi\sqrt{|\alpha\beta|}) & \text{if } \alpha\beta < 0 \end{cases}$$

$$+ \sum_{\substack{\ell,r\geq 0 \\ \ell+r\,\mathrm{even}}} \frac{1}{\ell!r!}\frac{1}{\mu+\ell-r}(2i\pi\beta)^\ell(2i\pi\alpha)^r. \qquad (5.2.24)$$

On the other hand, for $\alpha \neq 0$, *one has*

$$F^+(\mu; 0, \alpha) = \sum_{r\,\mathrm{even}\geq 0} \frac{1}{r!}\frac{(2i\pi\alpha)^r}{\mu+r},$$

$$F^+(\mu; \alpha, 0) = \frac{1}{2}\frac{\zeta(1+\mu)}{\zeta(-\mu)}|\alpha|^\mu + \sum_{r\,\mathrm{even}\geq 0}\frac{1}{r!}\frac{(2i\pi\alpha)^r}{\mu-r}. \qquad (5.2.25)$$

Proof. Set

$$\theta^+(\mu; \alpha, \beta) = \int_0^1 t^{\mu-1}e^{-2i\pi(\alpha t^{-1}+\beta t)}\,dt, \qquad \mathrm{Re}\,\mu > 0, \qquad (5.2.26)$$

so that

$$F^+(\mu; \alpha, \beta) = \frac{1}{2}\left[\theta^+(\mu; \alpha, \beta) + \theta^+(\mu; -\alpha, -\beta)\right]. \qquad (5.2.27)$$

Starting from the first equation (5.2.26), expanding one exponential into a series and performing the change of variable $t \mapsto t^{-1}$ in the remaining integral, one obtains

$$\theta^+(\mu; \alpha, \beta) = \sum_{\ell\geq 0}\frac{(-1)^\ell}{\ell!}(2i\pi\beta)^\ell\int_1^\infty t^{-\mu-\ell-1}e^{-2i\pi\alpha t}\,dt, \qquad (5.2.28)$$

or

$$\theta^+(\mu;\alpha,\beta) = \sum_{\ell\geq 0}\Gamma(-\mu-\ell)\frac{(-1)^\ell}{\ell!}(2i\pi\beta)^\ell(2i\pi\alpha)^{\mu+\ell}$$

$$-\sum_{\ell\geq 0}\frac{(-1)^\ell}{\ell!}(2i\pi\beta)^\ell\psi(-\mu-\ell,2i\pi\alpha). \quad (5.2.29)$$

The first term can be resummed as

$$\Gamma(-\mu)\Gamma(1+\mu)(i\alpha)^\mu|\alpha\beta|^{-\frac{\mu}{2}}\times \begin{cases}J_\mu(4\pi\sqrt{\alpha\beta}) & \text{if } \alpha\beta > 0, \\ I_\mu(4\pi\sqrt{|\alpha\beta|}) & \text{if } \alpha\beta < 0.\end{cases} \quad (5.2.30)$$

Using (5.2.27) and the relation

$$\Gamma(-\mu)\Gamma(1+\mu)\cos\frac{\pi\mu}{2} = \frac{1}{2}\Gamma(-\frac{\mu}{2})\Gamma(\frac{2+\mu}{2}), \quad (5.2.31)$$

one obtains the first term on the right-hand side of (5.2.24). The second term can be computed with the help of (5.2.23): the constraint that $\ell + r$ should be even in the corresponding sum originates from the fact that one must take the average of two values computed at (α, β), and at $(-\alpha, -\beta)$.

The equations concerning the case when $\alpha\beta = 0$ are obtained from the equations, in which $\alpha \neq 0$,

$$\theta^+(\mu;0,\alpha) = \psi(\mu,2i\pi\alpha),$$
$$\theta^+(\mu;\alpha,0) = \Gamma(-\mu)(2i\pi\alpha)^\mu - \psi(-\mu,2i\pi\alpha) \quad (5.2.32)$$

and

$$F^+(\mu;0,\alpha) = \frac{1}{2}[\psi(\mu,2i\pi\alpha) + \psi(\mu,-2i\pi\alpha)]: \quad (5.2.33)$$

in view of (5.2.23), the first one leads to the first equation (5.2.25), while the second one gives

$$F^+(\mu;\alpha,0) = \Gamma(-\mu)(2\pi|\alpha|)^\mu\cos\frac{\pi\mu}{2} - \frac{1}{2}[\psi(-\mu,2i\pi\alpha) + \psi(-\mu,-2i\pi\alpha)], \quad (5.2.34)$$

which leads to the second equation (5.2.25) if one uses the relation

$$\Gamma(-\mu)(2\pi)^\mu\cos\frac{\pi\mu}{2} = \frac{1}{2}\pi^{\frac{1}{2}+\mu}\frac{\Gamma(-\frac{\mu}{2})}{\Gamma(\frac{1+\mu}{2})} = \frac{1}{2}\frac{\zeta(1+\mu)}{\zeta(-\mu)}. \quad (5.2.35)$$

$\square$

Remark 5.2.a. Each of the two terms of the decomposition (5.2.24) has poles at $\mu = 2, 4, \ldots$, which are not poles of their sum.

We begin now our study of the decomposition into homogeneous components of each of the expressions $\langle \mathfrak{G}^{(pq)}, f \rangle$, as computed with the help of (5.2.20): there is a slight difference in the case when $(p, q) = (0, 0)$, which will be considered later. It is essential to note that the $d\mu$ - integration (of a symmetric type) is to be performed as the last operation. We shall work, in the present section, under the assumption that $\mathcal{F}_1^{-1}h$ and $\mathcal{F}_1^{-1}f$ are compactly supported: in view of (5.1.97), this is preserved if the transformation $\mathcal{F}^{\mathrm{symp}}$ is applied to both h and f; it is also preserved under the application of $2i\pi\mathcal{E}$. Then, the summation with respect to m, n will actually reduce to a finite sum: this will be crucial when studying $\mathfrak{G}^{(11)}$, and there is no reason not to take advantage of this assumption even when not really needed. There is no difficulty with the integration with respect to $dx d\xi$ since, in view of (5.1.37), the function $\left(\mathcal{F}_1^{-1}h \right)(0, x)$ is "divisible" by x as a function while, if $m \neq 0$, the function $\left(\mathcal{F}_1^{-1}h \right)\left(-\frac{m}{x}, x \right)$ is flat at zero. The true problems concern only the c-summability and the $d\mu$ - summability of a symmetric type: we know that there is convergence in this sense for δ large enough, and we wish to move δ to lower values, at least to values < 2 as a starting point, so that the poles $2, 4 \ldots$ brought by the decomposition in Lemma 5.2.2 should not be a nuisance.

We shall be helped by the fact that adding any constant ($-\frac{2}{\mu}$ will be our choice) to the quantity within brackets in the integral on the right-hand side of (5.2.17) does not change the integral. This follows from the equation

$$\int_{-\infty}^{\infty} (\mathcal{F}_1^{-1}f) \left(\frac{n}{\xi}, \xi \right) \frac{d\xi}{|\xi|} = \int_{-\infty}^{\infty} \xi \frac{d}{d\xi} \left[(\mathcal{F}_1^{-1}f_1) \left(\frac{n}{\xi}, \xi \right) \right] \frac{d\xi}{|\xi|}, \tag{5.2.36}$$

observed in (5.1.37) and its consequence

$$\int_{-\infty}^{\infty} (\mathcal{F}_1^{-1}f) \left(\frac{n}{\xi}, \xi \right) \frac{d\xi}{|\xi|} = 0 \quad \text{if } n \neq 0 \tag{5.2.37}$$

(since in this case $(\mathcal{F}_1^{-1}f_1) \left(\frac{n}{\xi}, \xi \right)$ vanishes at 0^+ and 0^-) and the similar one involving h in terms of f: recall that there is no term with $m = n = 0$.

Lemma 5.2.3. *Assume that h and f lie in the image under $2i\pi\mathcal{E}$ of the subspace of $\mathcal{S}_{\mathrm{even}}(\mathbb{R}^2)$ consisting of functions the Fourier transform of which with respect to the first variable has compact support. Then, provided that $\mathrm{Re}\,\mu > \frac{1}{2}$, the series (with respect to c, m, n) of $dx d\xi$ - integrals on the right-hand side of (5.2.20) can be transformed to*

$$\sum_{m,n \in \Sigma_{pq}} \int_{\mathbb{R}^2} \sum_{c \geq 1} c^{-1} S(m, n; c) \left(\mathcal{F}_1^{-1}h \right) \left(-\frac{m}{x}, x \right) \left(\mathcal{F}_1^{-1}f \right) \left(\frac{n}{\xi}, \xi \right)$$

$$\times \left[F^+ \left(\mu; \frac{nx}{c\xi}, \frac{m\xi}{cx} \right) + F^+ \left(\mu; \frac{m\xi}{cx}, \frac{nx}{c\xi} \right) - \frac{2}{\mu} \right] \frac{dx \, d\xi}{|x| \, |\xi|}. \tag{5.2.38}$$

Proof. We have just explained why it is possible to subtract $\frac{2}{\mu}$ from the integrand of (5.2.20) without changing the integral. Because of the assumption of compact support, we may replace the set Σ_{pq} by a single pair m, n. If $mn \neq 0$, Weil's sharp bound (5.1.61) gives in particular, $|S(m, n; c)| \leq C(m, n)c^{\frac{1}{2}+\varepsilon}$ for every $\varepsilon > 0$, and a constant $C(m, n)$. When $m = 0, n \neq 0$, one has instead (5.1.62) $|S(0, n; c)| \leq \sigma_1(|n|)$, an estimate independent of c.

To show that the c-series and the integral can be permuted when $\operatorname{Re}\mu > \frac{1}{2}$, it thus suffices to show that, for some $\kappa > \frac{1}{2}$, one can "save" from the (new) expression within brackets a factor $c^{-\kappa}$ without destroying the $dxd\xi$ - integrability. To do so, we write

$$\theta^+(\mu; \alpha, \beta) - \frac{1}{\mu} = \int_0^1 t^{\mu-1} \left[e^{-2i\pi(\alpha t^{-1} + \beta t)} - 1 \right] dt : \qquad (5.2.39)$$

then, using the fact that the quantity within brackets in the integrand is, in absolute value, less than 2 and less than $2\pi|\alpha t^{-1} + \beta t|$, we write

$$\left| e^{-2i\pi(\alpha t^{-1} + \beta t)} - 1 \right| \leq 2^{1-\kappa}(2\pi|\alpha t^{-1} + \beta t|)^\kappa$$
$$\leq 2\pi^\kappa \left[|\alpha|^\kappa t^{-\kappa} + |\beta|^\kappa t^\kappa \right]. \qquad (5.2.40)$$

Integrating this inequality, on $(0, 1)$, against the measure $t^{\operatorname{Re}\mu-1}dt$, we obtain if $\operatorname{Re}\mu > \kappa$ the inequality

$$\left| F^+(\mu; \alpha, \beta) - \frac{1}{\mu} \right| \leq 2\pi^\kappa \left[\frac{|\alpha|^\kappa}{\operatorname{Re}\mu - \kappa} + \frac{|\beta|^\kappa}{\operatorname{Re}\mu + \kappa} \right], \qquad (5.2.41)$$

which is to be applied with $(\alpha, \beta) = \left(\frac{m\xi}{cx}, \frac{nx}{c\xi} \right)$ or $\left(\frac{nx}{c\xi}, \frac{m\xi}{cx} \right)$. The lemma follows. $\square$

The preceding lemma does not yet prove that, when applying (5.2.20), we can move δ from its large initial value to any value $> \frac{1}{2}$. There are certainly no poles to be taken care of since, as already emphasized, the function $F^+(\mu; \alpha, \beta)$ is analytic for $\operatorname{Re}\mu > 0$, contrary to the two terms of its decomposition given in Lemma 5.2.2. So as to perform a change of contour, we must, however, bound (5.2.38) by an expression going to zero as $|\operatorname{Im}\mu| \to \infty$ whenever $\operatorname{Re}\mu > \frac{1}{2}$: this will be done after the next two lemmas have been proved. Since the $d\mu$-integral is an integral of symmetric type only (5.1.17), we must not expect to obtain bounds better than $\operatorname{O}\left(|\operatorname{Im}\mu|^{-1} \right)$.

To study the function

$$H_{m,n}(\mu; x, \xi) = \sum_{c \geq 1} c^{-1} S(m, n; c) \left[F^+\left(\mu; \frac{nx}{c\xi}, \frac{m\xi}{cx} \right) + F^+\left(\mu; \frac{m\xi}{cx}, \frac{nx}{c\xi} \right) - \frac{2}{\mu} \right], \qquad (5.2.42)$$

which occurs in the integrand of (5.2.38), we introduce when $mn \neq 0$, following [21, p. 81], the function

$$Z_s(m,n) = \frac{1}{2\sqrt{|mn|}} \sum_{c \geq 1} c^{-1} S(m,n;c) \times \begin{cases} J_{2s-1}\left(\frac{4\pi\sqrt{mn}}{c}\right) & \text{if } mn > 0, \\[2mm] I_{2s-1}\left(\frac{4\pi\sqrt{|mn|}}{c}\right) & \text{if } mn < 0, \end{cases}$$

(5.2.43)

which extends, from a study initiated by Selberg, as a meromorphic function in the plane. For clarity, we separate the cases when $mn = 0$ or not in what follows: do not confuse "principal" and "residual" with "major" and "minor".

Lemma 5.2.4. *Assume that* $mn \neq 0$. *For* $\operatorname{Re}\mu > \frac{1}{2}$, *one has*

$$H_{m,n}(\mu; x, \xi) = H_{m,n}^{\mathrm{princ}}(\mu; x, \xi) + H_{m,n}^{\mathrm{res}}(\mu; x, \xi), \tag{5.2.44}$$

with

$$H_{m,n}^{\mathrm{princ}}(\mu; x, \xi) = \Gamma\left(-\frac{\mu}{2}\right)\Gamma\left(\frac{2+\mu}{2}\right)$$

$$\times \left[|m|^{\frac{1+\mu}{2}} |n|^{\frac{1-\mu}{2}} \left|\frac{\xi}{x}\right|^{\mu} + |m|^{\frac{1-\mu}{2}} |n|^{\frac{1+\mu}{2}} \left|\frac{x}{\xi}\right|^{\mu} \right] Z_{\frac{1+\mu}{2}}(m,n) \tag{5.2.45}$$

and

$$H_{m,n}^{\mathrm{res}}(\mu; x, \xi) := \sum_{c \geq 1} c^{-1} S(m,n;c)$$

$$\times \sum_{\substack{\ell, r \text{ even} \geq 0 \\ \ell+r \geq 2}} \frac{1}{\ell! r!} \left(2i\pi \frac{m\xi}{cx}\right)^{\ell} \left(2i\pi \frac{nx}{c\xi}\right)^{r} \left[\frac{1}{\mu+\ell-r} + \frac{1}{\mu-\ell+r}\right]. \tag{5.2.46}$$

The residue at 0 of the function $H_{m,n}^{\mathrm{res}}(\mu; x, \xi)$ *is given as*

$$\operatorname{Res}_{\mu=0} H_{m,n}^{\mathrm{res}}(\mu; x, \xi) = 2 \times \begin{cases} J_0\left(\frac{4\pi\sqrt{mn}}{c}\right) - 1 & \text{if } mn > 0, \\[2mm] I_0\left(\frac{4\pi\sqrt{|mn|}}{c}\right) - 1 & \text{if } mn < 0, \end{cases} \tag{5.2.47}$$

an expression independent of (x, ξ).

Proof. The decomposition of $F^{+}(\mu; \alpha, \beta)$ as a sum of two terms provided by Lemma 5.2.2 leads to a decomposition of the function $H_{m,n}(\mu; x, \xi)$. That $H_{m,n}^{\mathrm{res}}(\mu; x, \xi)$ can be written in the way indicated is just a consequence of the observation that the term, deleted from the right-hand side of (5.2.46), that would correspond to $\ell = r = 0$, is $\frac{2}{\mu}$, the quantity subtracted from $F^{+}\left(\mu; \frac{nx}{c\xi}, \frac{m\xi}{cx}\right) + F^{+}\left(\mu; \frac{m\xi}{cx}, \frac{nx}{c\xi}\right)$ when defining $H_{m,n}(\mu; x, \xi)$. The residue at $\mu = 0$ of $H_{m,n}^{\mathrm{res}}(\mu; x, \xi)$ is obtained by collecting the terms with $\ell = r$ and comparing the result to the Taylor series expansion of the function J_0 or I_0. Then, one must set

$$H_{m,n}^{\mathrm{princ}}(\mu; x, \xi) = \frac{1}{2}\Gamma(-\frac{\mu}{2})\Gamma(\frac{2+\mu}{2})\sum_{c\geq 1} c^{-1}S(m,n;c)$$

(5.2.48)

$$\times \left[\left|\frac{\xi}{x}\right|^{\mu} \left|\frac{m}{n}\right|^{\frac{\mu}{2}} + \left|\frac{x}{\xi}\right|^{\mu} \left|\frac{n}{m}\right|^{\frac{\mu}{2}} \right] \times \begin{cases} J_\mu\left(\frac{4\pi\sqrt{mn}}{c}\right) & \text{if } mn > 0, \\ I_\mu\left(\frac{4\pi\sqrt{|mn|}}{c}\right) & \text{if } mn < 0. \end{cases}$$

Using (5.2.43), one obtains Lemma 5.2.4. $\square$

Lemma 5.2.5. *Assume* $n \neq 0$. *For* $\mathrm{Re}\,\mu > \frac{1}{2}$, *one has*

$$H_{0,n}(\mu; x, \xi) = H_{0,n}^{\mathrm{princ}}(\mu; x, \xi) + H_{0,n}^{\mathrm{res}}(\mu; x, \xi),$$

(5.2.49)

with

$$H_{0,n}^{\mathrm{princ}}(\mu; x, \xi) = \frac{1}{2}\frac{\sigma_\mu(|n|)}{\zeta(-\mu)}\left|\frac{x}{\xi}\right|^\mu,$$

$$H_{0,n}^{\mathrm{res}}(\mu; x, \xi) = \sum_{r\,\mathrm{even}\geq 2} \frac{1}{r!}\left(\frac{1}{\mu+r} + \frac{1}{\mu-r}\right)\frac{\sigma_r(|n|)}{\zeta(r+1)}\left(2i\pi\frac{x}{\xi}\right)^r.$$

(5.2.50)

Proof. From the second part of Lemma 5.2.2, one has

$$H_{0,n}(\mu; x, \xi) = \sum_{c\geq 1} c^{-1}S(0,n;c)$$

$$\times \left[\frac{\zeta(1+\mu)}{\zeta(-\mu)}\left|\frac{nx}{c\xi}\right|^\mu + \sum_{r\,\mathrm{even}\geq 2} \frac{1}{r!}\left(2i\pi\frac{nx}{c\xi}\right)^r\left(\frac{1}{\mu+r} + \frac{1}{\mu-r}\right) \right].$$

(5.2.51)

Using the equation [21, p. 52]

$$\sum_{c\geq 1} c^{-1-\mu}S(0,n;c) = \frac{\sigma_{-\mu}(|n|)}{\zeta(\mu+1)}$$

(5.2.52)

and the fact that $|n|^\mu\sigma_{-\mu}(|n|) = \sigma_\mu(|n|)$, one obtains the lemma. $\square$

The two lemmas which precede first show that the parameter δ of the line of integration (of symmetric type) $\mathrm{Re}\,\mu = \delta$ can be taken to arbitrary values $> \frac{1}{2}$ (or even $\delta > 0$ in the case when $mn = 0$, as a consequence of the estimate $|S(0,n;c)| \leq \sigma_1(|n|)$ mentioned above). Indeed, the required estimates as $|\mathrm{Im}\,\mu| \to \infty$ of the term $H_{m,n}^{\mathrm{res}}(\mu; x, \xi)$ are immediately apparent, whether $mn \neq 0$ or $mn = 0$. On the other hand, so far as the first term $H_{m,n}^{\mathrm{princ}}(\mu; x, \xi)$ is concerned, arbitrary powers of μ^{-1} can be gained by means of integrations by parts. For in the case when

$mn \neq 0$, applying $\left(x\frac{\partial}{\partial x}\right)^2$ to the quantity within brackets on the right-hand side of (5.2.45) has the same effect as multiplying it by μ^2, and the same holds with the factor $\left|\frac{x}{\xi}\right|^{\mu}$ in the case when $m = 0$.

Now that we can choose $\delta < 1$, it is possible, still under the assumption that $(p,q) \neq (0,0)$ to split the right-hand side of the equation (*cf.* (5.2.20) and (5.2.42))

$$\langle \mathfrak{S}^{(pq)}, f\rangle = \frac{1}{i\pi} \sum_{m,n\in\Sigma_{pq}} \int_{\mathrm{Re}\,\mu=\delta}^{\mathrm{sym}} d\mu$$

$$\times \int_{\mathbb{R}^2} (\mathcal{F}_1^{-1}h)\left(-\frac{m}{x},x\right)(\mathcal{F}_1^{-1}f)\left(\frac{n}{\xi},\xi\right)H_{m,n}(\mu;x,\xi)\frac{dx\,d\xi}{|x|\,|\xi|} \quad (5.2.53)$$

into a principal and a residual part, without having to worry about the poles of each of the two parts at $\mu = 2, 4, \ldots$.

Lemma 5.2.6. *Assume that* $0 < \delta < 1$. *For every pair* $(m,n) \neq (0,0)$ *one has*

$$\int_{\mathrm{Re}\,\mu=\delta}^{\mathrm{sym}} d\mu \int_{\mathbb{R}^2} (\mathcal{F}_1^{-1}h)\left(-\frac{m}{x},x\right)(\mathcal{F}_1^{-1}f)\left(\frac{n}{\xi},\xi\right)H_{m,n}^{\mathrm{res}}(\mu;x,\xi)\frac{dx\,d\xi}{|x|\,|\xi|} = 0.$$
$$(5.2.54)$$

Proof. Consider the case when $mn \neq 0$ first, using (5.2.46). The factor $c^{-\ell-r}$, with $\ell + r \geq 2$, which appears there, ensures the convergence of the c-series. Despite the fact that the function $H_{m,n}^{\mathrm{res}}(\mu;x,\xi)$ has a (simple) pole at $\mu = 0$, its integral against $(\mathcal{F}_1^{-1}h)\left(-\frac{m}{x},x\right)(\mathcal{F}_1^{-1}f)\left(\frac{n}{\xi},\xi\right)\frac{dx\,d\xi}{|x|\,|\xi|}$ is regular there in view of (5.2.37) since the residue of $H_{m,n}^{\mathrm{res}}(\mu;x,\xi)$ is independent of (x,ξ). Finally, as this function goes to zero, as $|\mathrm{Im}\,\mu| \to \infty$, in a way uniform with respect to $\mathrm{Re}\,\mu$, the integral (of symmetric type) under consideration is the same as the corresponding one, taken on the line $\mathrm{Re}\,\mu = 0$: as $H_{m,n}^{\mathrm{res}}(\mu;x,\xi)$ is an odd function of μ, this integral is zero. The case when $mn = 0$, say $(m = 0, n \neq 0)$ is also based on the fact that $H_{0,n}^{\mathrm{res}}(\mu;x,\xi)$ is an odd function: the sole differences are that the c-summation has already been performed explicitly, and there is no pole at $\mu = 0$ to be taken care of. $\qquad\square$

From now on, we may use on lines $\mathrm{Re}\,\mu = \delta$ genuine integrals rather than integrals of symmetric type only: this is because, as observed between (5.2.52) and (5.2.53), arbitrary powers of μ^{-1} can be gained, by means of integrations by parts, in the integrals where principal parts only of the functions $H_{m,n}$ occur as factors in the integrand.

Lemma 5.2.7. *The contribution of* $\mathfrak{S}^{(01)}$ *to* $\mathfrak{S}$ *is given as*

$$\langle \mathfrak{S}^{(01)}, f\rangle = \frac{1}{2\pi} \int_{-\infty}^{\infty} \sum_{n\neq 0} (\zeta(i\lambda))^{-1}\mathcal{L}_{i\lambda,0}(h)\sigma_{-i\lambda}(|n|)\mathcal{L}_{-i\lambda,n}(f)d\lambda. \quad (5.2.55)$$

Similarly,

$$\langle \mathfrak{S}^{(10)}, f \rangle = \frac{1}{2\pi} \int_{-\infty}^{\infty} \sum_{m \neq 0} \sigma_{i\lambda}(|m|) \mathfrak{L}_{i\lambda,m}(h)(\zeta(-i\lambda))^{-1} \mathfrak{L}_{-i\lambda,0}(f) d\lambda. \qquad (5.2.56)$$

Proof. From (5.2.53) and Lemmas 5.2.5 and 5.2.6, one has

$$\langle \mathfrak{S}^{(01)}, f \rangle = \frac{1}{2i\pi} \sum_{n \neq 0} \int_{\operatorname{Re}\mu=\delta} \frac{\sigma_\mu(|n|)}{\zeta(-\mu)} d\mu$$

$$\times \int_{\mathbb{R}^2} (\mathcal{F}_1^{-1}h)(0,x)(\mathcal{F}_1^{-1}f)\left(\frac{n}{\xi},\xi\right) \frac{dx \, d\xi}{|x| \, |\xi|}, \qquad (5.2.57)$$

which can be rewritten, using (5.2.2), as

$$\langle \mathfrak{S}^{(01)}, f \rangle = \frac{1}{2i\pi} \sum_{n \neq 0} \int_{\operatorname{Re}\mu=\delta} \frac{\sigma_\mu(|n|)}{\zeta(-\mu)} \mathfrak{L}_{-\mu,0}(h) \mathfrak{L}_{\mu,n}(f) d\mu. \qquad (5.2.58)$$

The limit of this expression as $\delta \to 0$ is the right-hand side of (5.2.55). The second equation asserted in Lemma 5.2.7 follows as well, since the expression $\langle \mathfrak{S}, h \rangle$ depends on h, f in a symmetric way. $\qquad \square$

Lemma 5.2.8. *The contribution of $\mathfrak{S}^{(00)}$ to $\mathfrak{S}$ is given as*

$$\langle \mathfrak{S}^{(00)}, f \rangle = \frac{1}{2\pi} \int_{-\infty}^{\infty} \frac{\zeta(i\lambda)}{\zeta(-i\lambda)} \mathfrak{L}_{-i\lambda,0}(h) \mathfrak{R}_{-i\lambda,0}(f) d\lambda. \qquad (5.2.59)$$

Proof. Starting this time from (5.1.57), one obtains, applying (5.1.16) as before, that for $\delta \in]0,1[$, one has

$$\langle \mathfrak{S}^{(00)}, f \rangle = \frac{1}{i\pi} \int_{\operatorname{Re}\mu=\delta} d\mu \sum_{c \geq 1} c^{-1} S(0,0;c) \int_{\mathbb{R}^2} (\mathcal{F}_1^{-1}h)(0,x)$$

$$\times (\mathcal{F}_1^{-1}f)(\xi,0) \left[F^+\left(\mu; \frac{x\xi}{c},0\right) + F^+\left(\mu; 0, \frac{x\xi}{c}\right) - \frac{2}{\mu} \right] \frac{dx \, d\xi}{|x| \, |\xi|}. \qquad (5.2.60)$$

The term $-\frac{2}{\mu}$ originates from the fact that, in (5.1.57), 1 has been subtracted from $\cos \frac{2\pi x\xi}{c}$. Then, the proof is similar to that of Lemma 5.2.7, only replacing $\frac{nx}{c\xi}$ by $\frac{x\xi}{c}$, and using in place of (5.2.52) the equation

$$\sum_{c \geq 1} c^{-1-\mu} S(0,0;c) = \frac{\zeta(\mu)}{\zeta(\mu+1)}. \qquad (5.2.61)$$

$\qquad \square$

When $mn \neq 0$, the study of the function $\mu \mapsto H^{\text{princ}}_{m,n}(\mu; x, \xi)$ depends on some facts [21, p. 137] regarding the analytic continuation of the series $Z_s(m, n)$, initially defined for $\operatorname{Re} s > 1$, or even $\operatorname{Re} s > \frac{3}{4}$ as can be seen with the use of Weil's sharp estimate (5.1.61). Recall that we are at present working under the assumption that the partial Fourier transforms of h and f with respect to the first variable have compact support. We recall from a look at the arguments of these two functions in Lemma 5.2.6 that it is sufficient to study $Z_s(m, n)$ for bounded values of $|m| + |n|$: no uniformity with respect to this pair of integers is required.

The series $Z_s(m, n)$ can be continued as a meromorphic function of s in the full complex plane, holomorphic for $\operatorname{Re} s > \frac{1}{2}$. Its possible poles are $\frac{1}{2}$, the numbers $s = \frac{1 \pm i \lambda_p}{2}$ such that $\frac{1 + \lambda_p^2}{4}$ lies in the discrete spectrum of the modular Laplacian, and the numbers $\frac{\omega}{2}$, where ω is a non-trivial zero of zeta. Let us recall, with Iwaniec's notation, the definition of the functions

$$\phi(s) = \pi^{\frac{1}{2}} \frac{\Gamma(s - \frac{1}{2})}{\Gamma(s)} \sum_{c \geq 1} c^{-2s} S(0, 0; c) = \pi^{\frac{1}{2}} \frac{\Gamma(s - \frac{1}{2})}{\Gamma(s)} \frac{\zeta(2s - 1)}{\zeta(2s)}, \qquad (5.2.62)$$

$$\phi(n, s) = \frac{\pi^s}{\Gamma(s)} |n|^{s-1} \sum_{c \geq 1} c^{-2s} S(0, n; c) = \frac{\pi^s}{\Gamma(s)} |n|^{s-1} \frac{\sigma_{1-2s}(|n|)}{\zeta(2s)}, \qquad n \neq 0.$$

In particular, $\phi(n, \frac{1}{2}) = 0$: since [21] the polar part of $Z_s(m, n)$ at $s = \frac{1}{2}$ is, in the case of the full modular group Γ (Iwaniec also treats the case of congruence groups), a multiple of $\frac{\bar{\phi}(m, \frac{1}{2}) \phi(n, \frac{1}{2})}{s - \frac{1}{2}}$, the pole (possible in general) $s = \frac{1}{2}$ drops from the picture entirely.

We switch to our present notation, in which $s = \frac{1 + \mu}{2}$: each pole of the function $Z_{\frac{1+\mu}{2}}(m, n)$ is of the kind $\mu = \omega - 1$, where ω is a non-trivial zero of zeta, or of the form $\pm i \lambda_p$ for some p. Also,

$$\phi \left(\frac{1 + \mu}{2} \right) = \frac{\zeta^*(\mu)}{\zeta^*(-\mu)}, \quad \phi \left(n, \frac{1 + \mu}{2} \right) = \frac{|n|^{\frac{\mu - 1}{2}} \sigma_{-\mu}(|n|)}{\zeta^*(-\mu)} = \frac{|n|^{\frac{-\mu - 1}{2}} \sigma_{\mu}(|n|)}{\zeta^*(-\mu)}.$$
$$(5.2.63)$$

The set of points denoted as (s_j) in [21], associated to the set $(s_j(1 - s_j))$ of discrete eigenvalues of Δ, is characterized by the further condition $t_j = \operatorname{Im} s_j > 0$: in this reference, each eigenvalue is repeated according to its multiplicity. Our convention with $\lambda_p (p \geq 1)$ is not to repeat eigenvalues, but using for each λ_p a complete orthonormal set $(\mathcal{M}_{p,j})$ of Maass-Hecke forms corresponding to this eigenvalue. We must thus denote as $\rho_{p,j}(n)$ the numbers denoted as $\rho_j(n)$ in [21]. These are just half the nth Fourier coefficients of the corresponding cusp-forms: with our notation,

$$\rho_{p,j}(n) = \frac{1}{2} |n|^{-\frac{1}{2}} b_{n;p,j}. \qquad (5.2.64)$$

The study of the function $Z_s(m, n)$ is based on the following identity [21, p. 136], obtained from an analysis of the automorphic Green function. Assume that

$y \geq 4\pi^2|mn|$ and $\operatorname{Re}\mu > 1$. Then,

$$\frac{1}{4|n|}\delta_{mn}I_{\frac{\mu}{2}}(y)K_{\frac{\mu}{2}}(y) + Z_{\frac{1+\mu}{2}}(m,n)\left[K_{\frac{\mu}{2}}(y)\right]^2$$

$$= |mn|^{-\frac{1}{2}}\sum_{p\geq 1}\sum_j b_{m;p,j}b_{n;p,j}(\mu+i\lambda_p)^{-1}(\mu-i\lambda_p)^{-1}\left[K_{\frac{i\lambda_p}{2}}(y)\right]^2 \qquad (5.2.65)$$

$$+ \frac{1}{\pi}\int_{-\infty}^{\infty}(\mu-i\lambda)^{-1}(\mu+i\lambda)^{-1}\overline{\phi\left(m,\frac{1+i\lambda}{2}\right)}\phi\left(n,\frac{1+i\lambda}{2}\right)\left[K_{\frac{i\lambda}{2}}(y)\right]^2 d\lambda.$$

Actually, we made a global change of sign in this quotation: for the use of the resolvent equation right before Theorem 7.5 in [21] demands that the right inverse $R_s = (-\Delta + s(1-s))^{-1}$ (with our Δ) be used, and (cf. [[21], (5.4)], with the notation there), the integral kernel of R_s is $-G_s(z/z')$, not $G_s(z/z')$.

Still quoting (and taking this change of sign into account), the residue of the function $Z_s(m,n)$ at $s_p = \frac{1+i\lambda_p}{2}$ or $s_p = \frac{1-i\lambda_p}{2}$ is the same, to wit

$$\frac{1}{2s_p-1}\sum_j \overline{\rho_{p,j}(m)}\rho_{p,j}(n): \qquad (5.2.66)$$

or

$$\operatorname{Res}_{\mu=i\lambda_p}Z_{\frac{1+\mu}{2}}(m,n) = \frac{1}{2i\lambda_p}|mn|^{-\frac{1}{2}}\sum_j b_{m;p,j}b_{n;p,j}. \qquad (5.2.67)$$

Finally,

$$Z_s(m,n) - Z_{1-s}(m,n) = -\frac{1}{2\pi|n|}\delta_{m,n}\cos\pi s + \frac{1}{2s-1}\phi(m,1-s)\phi(n,s): \quad (5.2.68)$$

in view of (5.2.62), the right-hand side has poles at points $\frac{\omega}{2}$ and $\frac{2-\omega}{2}$ (with $\zeta^*(\omega) = 0$), which confirms, since $Z_{1-s}(m,n)$ is holomorphic for $\operatorname{Re}s < \frac{1}{2}$, that $Z_s(m,n)$ is singular at points $\frac{\omega}{2}$.

Rewrite (5.2.53) as

$$\langle \mathfrak{S}^{(11)}, f\rangle = \int_{\operatorname{Re}\mu=\delta}\Psi(\mu)d\mu, \qquad (5.2.69)$$

with $\Psi = \sum_{m,n\neq 0}\Psi_{m,n}$ and

$$\Psi_{m,n}(\mu) = \frac{1}{i\pi}\int_{\mathbb{R}^2}(\mathcal{F}_1^{-1}h)\left(-\frac{m}{x},x\right)(\mathcal{F}_1^{-1}f)\left(\frac{n}{\xi},\xi\right)H_{m,n}^{\mathrm{princ}}(\mu;x,\xi)\frac{dx\,d\xi}{|x|\,|\xi|}. $$
$$(5.2.70)$$

Equations (5.2.46) and (5.2.45) show that, in the strip $-1 < \operatorname{Re}\mu < 1$, the singularities of $H_{m,n}^{\mathrm{princ}}(\mu;x,\xi)$ arise from the poles of $Z_{\frac{1+\mu}{2}}(m,n)$. These are either points $\omega - 1$ (one may prefer to say: points $-\omega$) with $\zeta^*(\omega) = 0$ or points $\pm i\lambda_p$ with $\frac{1+\lambda_p^2}{4}$ in the discrete spectrum of the modular Laplacian. In particular, the function Ψ is holomorphic in the half-strip $0 < \operatorname{Re}\mu < 1$.

Lemma 5.2.9. *One has the identity*

$$\frac{1}{2}\left(\Psi(\mu) + \Psi(-\mu)\right) = \frac{1}{4i\pi}\sum_{m,n\neq 0}(\zeta(\mu)\zeta(-\mu))^{-1}$$

$$\times\left[\sigma_\mu(|m|)\sigma_{-\mu}(|n|)\mathfrak{L}_{\mu,m}(h)\mathfrak{L}_{-\mu,n}(f) + \sigma_{-\mu}(|m|)\sigma_\mu(|n|)\mathfrak{L}_{-\mu,m}(h)\mathfrak{L}_{\mu,n}(f)\right]$$

$$-\frac{1}{4i\pi}\sum_{m\neq 0}\left[\mathfrak{L}_{\mu,-m}(h)\mathfrak{L}_{-\mu,m}(f) + \mathfrak{L}_{-\mu,-m}(h)\mathfrak{L}_{\mu,m}(f)\right]. \quad (5.2.71)$$

Proof. In view of the factor $\Gamma(-\frac{\mu}{2})\Gamma(\frac{2+\mu}{2})$ which is apparent in (5.2.45), the sum $H^{\text{princ}}_{m,n}(\mu; x, \xi) + H^{\text{princ}}_{m,n}(-\mu; x, \xi)$ involves the difference $Z_{\frac{1+\mu}{2}}(m, n) - Z_{\frac{1-\mu}{2}}(m, n)$, obtained from (5.2.68). Since

$$\Gamma(-\frac{\mu}{2})\Gamma(\frac{2+\mu}{2})\cos\frac{\pi(1+\mu)}{2} = \pi, \quad (5.2.72)$$

we obtain

$$H^{\text{princ}}_{m,n}(\mu; x, \xi) + H^{\text{princ}}_{m,n}(-\mu; x, \xi) = \left[|m|^{\frac{1+\mu}{2}}|n|^{\frac{1-\mu}{2}}\left|\frac{\xi}{x}\right|^\mu + |m|^{\frac{1-\mu}{2}}|n|^{\frac{1+\mu}{2}}\left|\frac{x}{\xi}\right|^\mu\right]$$

$$\times\left[-\frac{1}{2|n|}\delta_{m,n} + \frac{1}{2}\Gamma(\frac{\mu}{2})\Gamma(-\frac{\mu}{2})\phi\left(m, \frac{1}{2}-\frac{\mu}{2}\right)\phi\left(n, \frac{1}{2}+\frac{\mu}{2}\right)\right]. \quad (5.2.73)$$

One has

$$\phi\left(m, \frac{1-\mu}{2}\right)\phi\left(n, \frac{1+\mu}{2}\right) = \frac{\pi}{\Gamma(\frac{1+\mu}{2})\Gamma(\frac{1-\mu}{2})}\frac{|m|^{\frac{-1-\mu}{2}}|n|^{\frac{-1+\mu}{2}}}{\zeta(1-\mu)\zeta(1+\mu)}\sigma_\mu(|m|)\sigma_{-\mu}(|n|)$$

$$= \frac{1}{\Gamma(\frac{\mu}{2})\Gamma(-\frac{\mu}{2})}\frac{|m|^{\frac{-1-\mu}{2}}|n|^{\frac{-1+\mu}{2}}}{\zeta(\mu)\zeta(-\mu)}\sigma_\mu(|m|)\sigma_{-\mu}(|n|). \quad (5.2.74)$$

Hence,

$$H^{\text{princ}}_{m,n}(\mu; x, \xi) + H^{\text{princ}}_{m,n}(-\mu; x, \xi) = \frac{1}{2}\delta_{mn}\left(\left|\frac{\xi}{x}\right|^\mu + \left|\frac{x}{\xi}\right|^\mu\right)$$

$$+ \frac{1}{2\zeta(\mu)\zeta(-\mu)}\left[\sigma_\mu(|m|)\sigma_{-\mu}(|n|)\left|\frac{\xi}{x}\right|^\mu + \sigma_{-\mu}(|m|)\sigma_\mu(|n|)\left|\frac{x}{\xi}\right|^\mu\right]: \quad (5.2.75)$$

we have used the trivial identity $|m|^{-\mu}\sigma_\mu(|m|) = \sigma_{-\mu}(|m|)$. Recalling (5.2.70) and using (5.2.2), one obtains (5.2.71). $\square$

In order to evaluate the integral in (5.2.69), we need the (easy to guess) following lemma. Unfortunately, the proof of it is quite technical, and we shall postpone it to the next section, mostly devoted to just such a task.

Lemma 5.2.10. *Let $\delta \in]\frac{1}{2}, 1[$ be given. For every pair m, n of nonzero integers, one has*

$$
\int_{\mathrm{Re}\,\mu=\delta} \Psi_{m,n}(\mu)d\mu
$$

$$
= \int_{\mathrm{Re}\,\mu=0} \frac{1}{2} \left(\Psi_{m,n}(\mu) + \Psi_{m,n}(-\mu) \right) d\mu + i\pi \sum_{p\geq 1} \sum_{\pm} \mathrm{Res}_{\mu=\pm i\lambda_p} \Psi_{m,n}(\mu), \quad (5.2.76)
$$

where λ_p runs through the set of positive numbers such that $\frac{1+\lambda_p^2}{4}$ belongs to the discrete spectrum of the modular Laplacian.

Lemma 5.2.11. *Assume that h and f lie in the image, under the operator $2i\pi\mathcal{E}(1 + 2i\pi\mathcal{E})$, of the space of functions in $\mathcal{S}_{\mathrm{even}}(\mathbb{R}^2)$ the partial Fourier transform of which under the first variable has compact support. Then, assuming that Lemma 5.2.10 has been proved, one has*

$$
\langle \mathfrak{S}^{11}, f \rangle = \frac{1}{2\pi} \int_{-\infty}^{\infty} \sum_{m,n\neq 0} \frac{\sigma_{i\lambda}(|m|)\sigma_{-i\lambda}(|n|)}{\zeta(i\lambda)\zeta(-i\lambda)} \mathcal{L}_{i\lambda,m}(h)\mathcal{L}_{-i\lambda,n}(f)d\lambda
$$

$$
- \frac{1}{2\pi} \int_{-\infty}^{\infty} \sum_{m\neq 0} \mathcal{L}_{i\lambda,m}(h)\mathcal{L}_{-i\lambda,-m}(f)d\lambda
$$

$$
+ 2\sum_{p\neq 0} \Gamma\left(\frac{i\lambda_p}{2}\right) \Gamma\left(-\frac{i\lambda_p}{2}\right) \sum_{j} \epsilon_{p,j} \langle \mathfrak{M}_{p,j}, h \rangle \langle \mathfrak{M}_{-p,j}, f \rangle, \quad (5.2.77)
$$

where $\epsilon_{p,j} = 1$ or -1 according to whether the cusp-form $\mathcal{M}_{p,j}$ is of even or odd type.

Proof. Lemma 5.2.9 made the function $\frac{1}{2}\left(\Psi(\mu) + \Psi(-\mu)\right)$ explicit. The integral of this function, taken on the pure imaginary line, coincides with the sum of the first two terms on the right-hand side of (5.2.77).

Next, we must compute the residue of $\Psi(\mu)$ at $\mu = \pm i\lambda_p$. Since the points $\pm i\lambda_p$ are simple poles of the function $Z_{\frac{1+\mu}{2}}(m, n)$, equations (5.2.45) and (5.2.67) yield

$$
\mathrm{Res}_{\mu=i\lambda_p} H_{m,n}^{\mathrm{princ}}(\mu; x, \xi) = \frac{1}{4}|mn|^{-\frac{1}{2}} \Gamma\left(-\frac{i\lambda_p}{2}\right) \Gamma\left(\frac{i\lambda_p}{2}\right) \left(\sum_{j} b_{m;p,j}b_{n;p,j} \right)
$$

$$
\times \left[|m|^{\frac{1+\mu}{2}} |n|^{\frac{1-\mu}{2}} \left|\frac{\xi}{x}\right|^{\mu} + |m|^{\frac{1-\mu}{2}} |n|^{\frac{1+\mu}{2}} \left|\frac{x}{\xi}\right|^{\mu} \right]. \quad (5.2.78)
$$

From definition (5.2.70) of the function Ψ and definition (5.2.2) of the linear form $\mathcal{L}_{\nu,k}$, one obtains

$$\mathrm{Res}_{\mu=i\lambda_p}\Psi(\mu) = \frac{1}{4i\pi}\sum_{m,n\neq 0}\Gamma\left(-\frac{i\lambda_p}{2}\right)\Gamma\left(\frac{i\lambda_p}{2}\right)\left(\sum_j b_{m;p,j}b_{n;p,j}\right) \tag{5.2.79}$$

$$\times\left[|m|^{\frac{i\lambda_p}{2}}|n|^{-\frac{i\lambda_p}{2}}\mathcal{L}_{i\lambda_p,-m}(h)\mathcal{L}_{-i\lambda_p,n}(f) + |m|^{-\frac{i\lambda_p}{2}}|n|^{\frac{i\lambda_p}{2}}\mathcal{L}_{-i\lambda_p,-m}(h)\mathcal{L}_{i\lambda_p,n}(f)\right]:$$

note that, this time, we must be careful to distinguish $-m$ from m, in view of the existence of cusp-forms of odd type relative to the symmetry $z \mapsto -\bar{z}$. The residue at the point $-i\lambda_p$ is just the same. Since extending, as in Lemma 5.2.10, a sum of residues to the set of pairs $(\lambda_p, \pm)$ with $p \geq 1$ is the same as extending it to the set of numbers λ_p with $p \neq 0$, we obtain in this way (multiplying by $i\pi$) that the second term on the right-hand side of (5.2.76) is

$$\frac{1}{2}\sum_{m,n\neq 0}\sum_{p\neq 0}\Gamma\left(-\frac{i\lambda_p}{2}\right)\Gamma\left(\frac{i\lambda_p}{2}\right)|m|^{\frac{i\lambda_p}{2}}|n|^{-\frac{i\lambda_p}{2}}\left(\sum_j b_{-m;p,j}b_{n;p,j}\right)$$

$$\times\mathcal{L}_{i\lambda_p,m}(h)\mathcal{L}_{-i\lambda_p,n}(f). \tag{5.2.80}$$

Using (5.2.8), this agrees with the last line of the right-hand side of (5.2.77). □

Theorem 5.2.12. *Assume that h and f lie in the image, under the operator $2i\pi\mathcal{E}(1 + 2i\pi\mathcal{E})$, of the space of functions in $\mathcal{S}_{\mathrm{even}}(\mathbb{R}^2)$ the partial Fourier transform of which under the first variable has compact support. Then (assuming that Lemma 5.2.10 has been proved), one has*

$$\langle\mathfrak{P}, h\otimes f\rangle = \frac{1}{2\pi}\int_{-\infty}^{\infty}\langle\mathcal{E}_{i\lambda}, h\rangle\langle\mathcal{E}_{-i\lambda}, f\rangle\frac{d\lambda}{\zeta(i\lambda)\zeta(-i\lambda)}$$

$$+2\sum_{p\neq 0}\Gamma\left(\frac{i\lambda_p}{2}\right)\Gamma\left(-\frac{i\lambda_p}{2}\right)\sum_j \epsilon_{p,j}\langle\mathfrak{M}_{p,j}, h\rangle\langle\mathfrak{M}_{-p,j}, f\rangle. \tag{5.2.81}$$

Proof. It is just a matter of collecting the decompositions into bihomogeneous components of the various terms on the right-hand side of (5.1.52), as obtained in (5.2.12) and in Lemmas 5.2.7, 5.2.8 and 5.2.11: we may change the variable of integration λ to $-\lambda$ here and there. Adding the first and the last of the results just given a reference to, we observe that the first term on the right-hand side of (5.2.12) and the second term on the right-hand side of (5.2.77) cancel out, and we obtain

$$\langle\mathfrak{P}_{\mathrm{min}}, h\otimes f\rangle + \langle\mathfrak{P}^{(11)}, h\otimes f\rangle$$

$$= \frac{1}{2\pi}\int_{-\infty}^{\infty}\mathcal{L}_{i\lambda,0}(h)\mathcal{L}_{-i\lambda,0}(f)d\lambda + \frac{1}{2\pi}\int_{-\infty}^{\infty}\mathfrak{R}_{-i\lambda,0}(h)\mathfrak{R}_{i\lambda,0}(f)d\lambda$$

$$+ \frac{1}{2\pi}\int_{-\infty}^{\infty}\sum_{m,n\neq 0}\frac{\sigma_{i\lambda}(|m|)\sigma_{-i\lambda}(|n|)}{\zeta(i\lambda)\zeta(-i\lambda)}\mathcal{L}_{i\lambda,m}(h)\mathcal{L}_{-i\lambda,n}(f)d\lambda$$

$$+ 2 \sum_{p \neq 0} \Gamma \left(\frac{i\lambda_p}{2} \right) \Gamma \left(-\frac{i\lambda_p}{2} \right) \sum_j \epsilon_{p,j} \langle \mathfrak{M}_{p,j}, h \rangle \langle \mathfrak{M}_{-p,j}, f \rangle. \tag{5.2.82}$$

Next, from Lemma 5.2.7, and (5.2.4) together with (5.2.3)

$$\langle \mathfrak{P}^{(01)}, h \otimes f \rangle + \langle \mathfrak{P}^{(01)}, \mathcal{F}^{\text{symp}} h \otimes \mathcal{F}^{\text{symp}} f \rangle$$

$$= \frac{1}{2\pi} \int_{-\infty}^{\infty} \sum_{n \neq 0} (\zeta(i\lambda))^{-1} \mathcal{L}_{i\lambda,0}(h) \sigma_{-i\lambda}(|n|) \mathcal{L}_{-i\lambda,n}(f) d\lambda$$

$$+ \frac{1}{2\pi} \int_{-\infty}^{\infty} \sum_{n \neq 0} (\zeta(-i\lambda))^{-1} \mathfrak{R}_{-i\lambda,0}(h) \sigma_{-i\lambda}(|n|) \mathcal{L}_{-i\lambda,n}(f) d\lambda \tag{5.2.83}$$

and

$$\langle \mathfrak{P}^{(10)}, h \otimes f \rangle + \langle \mathfrak{P}^{(10)}, \mathcal{F}^{\text{symp}} h \otimes \mathcal{F}^{\text{symp}} f \rangle$$

$$= \frac{1}{2\pi} \int_{-\infty}^{\infty} \sum_{m \neq 0} \sigma_{i\lambda}(|m|) \mathcal{L}_{i\lambda,m}(h) (\zeta(-i\lambda))^{-1} \mathcal{L}_{-i\lambda,0}(f) d\lambda$$

$$+ \frac{1}{2\pi} \int_{-\infty}^{\infty} \sum_{m \neq 0} \sigma_{i\lambda}(|m|) \mathcal{L}_{i\lambda,m}(h) (\zeta(i\lambda))^{-1} \mathfrak{R}_{i\lambda,0}(f) d\lambda. \tag{5.2.84}$$

Finally, from Lemma 5.2.8,

$$\langle \mathfrak{P}^{(00)}, h \otimes f \rangle + \langle \mathfrak{P}^{(00)}, \mathcal{F}^{\text{symp}} h \otimes \mathcal{F}^{\text{symp}} f \rangle \tag{5.2.85}$$

$$= \frac{1}{2\pi} \int_{-\infty}^{\infty} \frac{\zeta(-i\lambda)}{\zeta(i\lambda)} \mathcal{L}_{i\lambda,0}(h) \mathfrak{R}_{i\lambda,0}(f) d\lambda + \frac{1}{2\pi} \int_{-\infty}^{\infty} \frac{\zeta(i\lambda)}{\zeta(-i\lambda)} \mathfrak{R}_{-i\lambda,0}(h) \mathcal{L}_{-i\lambda,0}(f) d\lambda.$$

Adding all these results and using (5.2.7), we obtain Theorem 5.2.12.   □

5.3   Technicalities and complements

In this section, we prove Lemma 5.2.10, by means of a deformation of contour. Even though we start from an integral of $\Psi_{m,n}(\mu)$ on a line contained in the half-plane $\text{Re}\,\mu > 0$, having to decompose this function into its even and odd parts will force us to cross the pure imaginary line at some points. It is there that (small) difficulties will accumulate when estimating the function $Z_s(m,n)$, for which we rely on Iwaniec's identity (5.2.65), here recalled:

$$\frac{1}{4|n|} \delta_{mn} I_{\frac{\mu}{2}}(y) K_{\frac{\mu}{2}}(y) + Z_{\frac{1+\mu}{2}}(m,n) \left[K_{\frac{\mu}{2}}(y) \right]^2 \tag{5.3.1}$$

$$= |mn|^{-\frac{1}{2}} \sum_{p \geq 1} \sum_j b_{m;p,j} b_{n;p,j} (\mu + i\lambda_p)^{-1} (\mu - i\lambda_p)^{-1} \left[K_{\frac{i\lambda_p}{2}}(y) \right]^2$$

$$+ \frac{1}{\pi} \int_{-\infty}^{\infty} (\mu - i\lambda)^{-1} (\mu + i\lambda)^{-1} \phi \left(m, \frac{1 - i\lambda}{2} \right) \phi \left(n, \frac{1 + i\lambda}{2} \right) \left[K_{\frac{i\lambda}{2}}(y) \right]^2 d\lambda.$$

To make the best of this identity, we must choose paths in the μ-plane keeping away, as much as is possible, from the points $\pm i\lambda_p$. Let us rename the eigenvalues of the modular Laplacian as $\frac{1+\lambda'^2_q}{4}$ with $0 < \lambda'_q \to \infty$, with the convention (contrary to that used when eigenvalues are denoted $\frac{1+\lambda^2_p}{4}$) that eigenvalues are repeated according to multiplicity. This makes it possible to apply the fact (a consequence of Selberg's trace formula) that λ'_q is of the order of $q^{\frac{1}{2}}$ as $q \to \infty$: then, one can construct a sequence (A_q) going to infinity such that the distance from A_q to the closest λ'_r is $\geq C^{-1}A_q^{-\frac{1}{2}}$ for some $C > 0$.

Next, in view of (5.2.71), we must also be able, on the paths to be chosen, to have a good control over $(\zeta(\pm\mu))^{-1}$. Classical results regarding zero-free regions of zeta [54, p. 160-161] provide the answer. There exists a positive constant c such that all zeros $\sigma + it$ of zeta satisfy the condition

$$\sigma < 1 - \frac{c}{\log(2 + |t|)} : \tag{5.3.2}$$

moreover, for some choice of c, there exists another constant $C > 0$ such that

$$\frac{1}{|\zeta(\sigma + it)|} \leq C \log(2 + |t|) \quad \text{if } \sigma \geq 1 - \frac{c}{\log(2 + |t|)} : \tag{5.3.3}$$

this applies in particular if $\sigma + it$ lies to the right of the critical strip, or to the left (use the functional equation of zeta).

Also, we need some bounds for the Fourier coefficients $b_{n;p,j}$ of the Maass-Hecke cusp-forms $\mathcal{M}_{p,j}$. On one hand, a result of Smith [49] gives the bound $C\lambda_p^{\frac{1}{2}} \exp\left(\frac{\pi\lambda_p}{4}\right)$ for the first Fourier coefficient of a cusp-form for the eigenvalue $\frac{1+\lambda^2_p}{4}$, $\lambda_p > 0$, normalized in $L^2(\Gamma\backslash\Pi)$. On the other hand, the ratio $\frac{b_{n;p,j}}{b_{1;p,j}}$, which can be interpreted directly as an eigenvalue of a Hecke operator, is bounded by $C|n|^\tau$ for some τ: which τ is best at present depends on progress towards the Ramanujan conjecture for non-holomorphic cusp-forms, but exponents much better than 1 are known. Hence,

$$|b_{n;p,j}| \leq C|n|\lambda_p^{\frac{1}{2}} \exp\left(\frac{\pi\lambda_p}{4}\right) : \tag{5.3.4}$$

actually, since m and n are bounded in our present study, Smith's result would suffice here; (5.3.4) will be helpful in the proof of Theorem 5.3.3, where we get rid of the assumption that $\mathcal{F}_1^{-1}h$ and $\mathcal{F}_1^{-1}f$ have compact support.

Some elementary estimates regarding Bessel functions are required. In (5.3.1), the number y must be large enough ($> 2\pi|mn|^{\frac{1}{2}}$), but can be considered as fixed since m, n are bounded in the present study. Assume that $\mu = \sigma + i\rho$ with $-1 < \sigma < 1$. The Taylor series expansion

$$I_{\frac{\mu}{2}}(y) = \frac{(\frac{y}{2})^{\frac{\mu}{2}}}{\Gamma\left(\frac{2+\sigma+i\rho}{2}\right)} \sum_{m \geq 0} \frac{1}{m!} \frac{(\frac{y}{2})^{2m}}{\left(\frac{2+\sigma+i\rho}{2}\right)_m} \tag{5.3.5}$$

(involving a Pochhammer symbol) yields the inequality

$$\frac{\left(\frac{y}{2}\right)^{\frac{\sigma}{2}}}{\left|\Gamma\left(\frac{2+\sigma+i\rho}{2}\right)\right|}\left[2-\exp\left(\frac{y^2}{2|2+\sigma+i\rho|}\right)\right]\leq\left|I_{\frac{\sigma+i\rho}{2}}(y)\right|$$

$$\leq\frac{\left(\frac{y}{2}\right)^{\frac{\sigma}{2}}}{\left|\Gamma\left(\frac{2+\sigma+i\rho}{2}\right)\right|}\exp\left(\frac{y^2}{2|2+\sigma+i\rho|}\right). \quad (5.3.6)$$

Then, recalling asymptotics for the Gamma function on vertical lines [36, p. 13],

$$\left|\Gamma\left(\frac{2+\sigma+i\rho}{2}\right)\right|\sim(2\pi)^{\frac{1}{2}}e^{-\frac{\pi|\rho|}{4}}\left|\frac{\rho}{2}\right|^{\frac{1+\sigma}{2}}, \quad (5.3.7)$$

one sees that in the equation

$$K_{\frac{\sigma+i\rho}{2}}(y)=\frac{1}{2}\Gamma\left(\frac{\sigma+i\rho}{2}\right)\Gamma\left(\frac{2-\sigma-i\rho}{2}\right)\left[I_{\frac{-\sigma-i\rho}{2}}(y)-I_{\frac{\sigma+i\rho}{2}}(y)\right], \quad (5.3.8)$$

the first term dominates the second if $\sigma>0$ and $|\rho|$ is large enough, in a way depending on y. Using (5.3.6), one obtains for $|\rho|$ large enough the estimate, valid for some $C>0$,

$$C^{-1}e^{-\frac{\pi|\rho|}{4}}|\rho|^{\frac{|\sigma|-1}{2}}\leq\left|K_{\frac{\sigma+i\rho}{2}}(y)\right|\leq Ce^{-\frac{\pi|\rho|}{4}}|\rho|^{\frac{|\sigma|-1}{2}}, \quad (5.3.9)$$

while

$$\left|I_{\frac{\sigma+i\rho}{2}}(y)\right|\leq Ce^{\frac{\pi|\rho|}{4}}|\rho|^{\frac{-\sigma-1}{2}}. \quad (5.3.10)$$

We need to obtain good bounds for the function $\Psi_{m,n}(\mu)$. Recall that, as already observed just after Lemma 5.2.5, integrations by parts in the integral (5.2.70) defining this function make it possible to gain arbitrary powers of μ^{-1}: with $\mu=\sigma+i\rho$, it will not be necessary to keep track of factors bounded by powers of $1+|\rho|$. In the equation just mentioned, $\Psi_{m,n}(\mu)$ was defined in terms of $H_{m,n}^{\mathrm{princ}}(\mu;x,\xi)$, the definition of which was given in (5.2.45). Observe that, there, the function $Z_{\frac{1+\mu}{2}}(m,n)$ is accompanied by the factor $\Gamma(-\frac{\mu}{2})\Gamma(\frac{2+\mu}{2})$, which has for large $|\rho|$ the same size, up to some power of $1+|\rho|$, as the factor $\left[K_{\frac{\mu}{2}}(y)\right]^2$ which accompanies $Z_{\frac{1+\mu}{2}}(m,n)$ in (5.3.1): this will dispense us with having to chase the factor $e^{\frac{\pi|\rho|}{2}}$ up and down several times. Finally, we just need, roughly speaking, to bound by powers of $1+|\rho|$ the three terms in that identity not involving the function $Z_{\frac{1+\mu}{2}}$. Note that we shall have to settle for a little less, especially so far as the series is concerned, when μ is close to the pure imaginary line: that we can still manage with the results of the next two lemmas will be seen later.

Warning. Even though we do not care about powers of $1+|\rho|$ in our estimates of $Z_{\frac{1+\mu}{2}}(m,n)$, we must be very careful with powers of $1+|\lambda_p|$ or of $1+|\lambda|$ in the series and integral on the right-hand side of (5.3.1), where no integration by parts could come to our help.

In view of (5.3.9), (5.3.10), the extra term on the left-hand side of this identity is bounded by a constant for $-1 < \operatorname{Re}\mu < 1$. Let us consider now the series on the right-hand side.

Lemma 5.3.1. *Let $\mu = \sigma + i\rho$, with $|\sigma| < 1$. Assume that, for some C, one has $|\mu \pm i\lambda_p| \geq C^{-1}\lambda_p^{-\frac{1}{2}}$ for every $p \geq 1$. For any number α such that $\frac{2|\sigma|+1}{3} < \alpha < 1$, the series on the right-hand side of (5.3.1) is bounded by $C|\sigma|^{-\alpha}(1+|\rho|)^\alpha$ for some new constant C.*

Proof. Let us revert to the notation (λ'_q) rather than $(\lambda_{p,j})$ so as to have to sum over the index q only. As recalled in (5.3.4), the first factor $b_{m;q}b_{n;q}$ of the general term of the series under examination is bounded by $C\lambda'_q e^{-\frac{\pi\lambda'_q}{2}}$. The last factor $\left[K_{\frac{i\rho}{2}}(y)\right]^2$ is bounded by $C\lambda_q'^{|\sigma|-1}e^{-\frac{\pi\lambda'_q}{2}}$ according to (5.3.9). Since the set $\{\lambda'_q\}$ is bounded away from zero, one at least of the two factors $(\mu \pm i\lambda'_q)^{-1}$, say $(\mu + i\lambda'_q)^{-1}$, is bounded by $C\lambda_q'^{-1}$. Finally, one has on one hand

$$|\mu - i\lambda'_q|^{-1} = [\sigma^2 + (\rho - \lambda'_q)^2]^{-\frac{1}{2}} \leq C|\sigma|^{-1}(1+|\rho|)\lambda_q'^{-1}, \tag{5.3.11}$$

on the other hand $|\mu - i\lambda'_q|^{-1} \leq C\lambda_q'^{\frac{1}{2}}$: taking advantage of the two inequalities, one may write

$$|\mu - i\lambda'_q|^{-1} \leq C|\sigma|^{-\alpha}(1+|\rho|)^\alpha \lambda_q'^{-\alpha}.\lambda_q'^{\frac{1-\alpha}{2}} = C|\sigma|^{-\alpha}(1+|\rho|)^\alpha \lambda_q'^{\frac{1-3\alpha}{2}}. \tag{5.3.12}$$

Piecing together the four inequalities just mentioned, we obtain that the general term of the series is bounded by $C(1+|\rho|)^\alpha \lambda_q'^{|\sigma|-\frac{1+3\alpha}{2}}$. The lemma follows. $\square$

Lemma 5.3.2. *The integral term on the right-hand side of (5.3.1), initially defined for $\sigma > 0$ (with $\mu = \sigma + i\rho$), extends as an analytic function for $\sigma > -\frac{c}{2\log(2+|\rho|)}$, where c is the constant defined in (5.3.3). For every $\beta \in]0, 1[$, this integral term is bounded by $C|\sigma|^{-\beta}(1+|\rho|)^{1-\beta}$ for some constant $C > 0$ when μ lies in the domain just defined.*

Proof. Let us note first, recalling (5.2.63), that

$$\phi\left(m, \frac{1-i\lambda}{2}\right)\phi\left(n, \frac{1+i\lambda}{2}\right) = \frac{|m|^{\frac{i\lambda-1}{2}}\sigma_{-i\lambda}(|n|)}{\zeta^*(i\lambda)} \times \frac{|n|^{\frac{-i\lambda-1}{2}}\sigma_{i\lambda}(|n|)}{\zeta^*(-i\lambda)}. \tag{5.3.13}$$

According to (5.3.3) and (5.3.7), this product is bounded by $Ce^{\frac{\pi|\lambda|}{2}}|\lambda|[\log(2+|\lambda|)]^2$. Taking into account (5.3.9) as well, one obtains

$$\left|\phi\left(m, \frac{1-i\lambda}{2}\right)\phi\left(n, \frac{1+i\lambda}{2}\right)\left[K_{\frac{i\lambda}{2}}(y)\right]^2\right| \leq C[\log(2+|\lambda|)]^2 : \tag{5.3.14}$$

this function can be integrated against $|(\mu - i\lambda)^{-1}(\mu + i\lambda)^{-1}|$, but not against one of the two factors only. This is the reason for the following splitting:

$$(\mu - i\lambda)^{-1}(\mu + i\lambda)^{-1} = \frac{1}{2\mu}\left[\frac{1}{\mu - i\lambda} + \frac{1}{i\lambda}\right] + \frac{1}{2\mu}\left[\frac{1}{\mu + i\lambda} - \frac{1}{i\lambda}\right]. \qquad (5.3.15)$$

The apparent singularity at $\lambda = 0$ so introduced is harmless, since the two ϕ-factors in the integral under examination vanish at $\lambda = 0$.

This integral term is currently meaningful when $\operatorname{Re}\mu > 0$. So as to make a crossing of the pure imaginary line possible, we use a change of contour in each of the two integrals obtained after the splitting (5.3.15) has been made, writing the second term on the right-hand side of (5.3.1) as

$$\sum_{\varepsilon = \pm 1} \frac{1}{2i\pi\mu} \int_{C_\varepsilon} \left[\frac{1}{\mu - \varepsilon\nu} + \frac{1}{\varepsilon\nu}\right] \phi\left(m, \frac{1-\nu}{2}\right) \phi\left(n, \frac{1+\nu}{2}\right) \left[K_{\frac{\nu}{2}}(y)\right]^2 d\nu, \qquad (5.3.16)$$

where the two paths C_ε are chosen in the following way. First, to take benefit from this change, we must have $\operatorname{Re}\nu < 0$ on C_1 and $\operatorname{Re}\nu > 0$ on C_{-1}. Next, so as to keep away from the poles of the two ϕ-factors, we must manage so that no zeros of $\zeta^*(\nu)$ will be encountered on the way from the pure imaginary line to any of the lines C_1 and $-C_{-1}$. Using (5.3.3) and the functional equation of zeta again, we see that a possible choice consists in defining C_1 by the equation (in which $\nu = \tau + i\lambda$) $\tau = -\frac{c}{\log(2+|\lambda|)}$, and C_{-1} by the equation $\tau = \frac{c}{\log(2+|\lambda|)}$.

Equation (5.3.1) becomes a valid one for $\operatorname{Re}\mu > -\frac{c}{\log(2+|\operatorname{Im}\mu|)}$ as soon as we have replaced the integral term there by (5.3.16). To obtain good bounds, let us assume that $\sigma > -\frac{1}{2}\frac{c}{\log(2+|\rho|)}$, recalling that $\mu = \sigma + i\rho$. Consider for instance the integral corresponding to the choice $\varepsilon = 1$. Noting that, when $\tau + i\lambda$ lies on one of the lines $C_{\pm 1}$, the expression

$$|\lambda|^{\pm\frac{\tau}{2}} = \exp\left(\pm\frac{c}{2}\frac{\log|\lambda|}{\log(2 + |\lambda|)}\right), \qquad (5.3.17)$$

which occurs in the bound for $K_{\frac{\nu}{2}}(y)$ or on that for the inverse of $\Gamma(\pm\frac{\tau}{2})$, is bounded, we observe that the estimate (5.3.14) remains valid, with the same right-hand side, after $i\lambda$ has been replaced by $\nu = \tau + i\lambda$ in the left-hand side. Write

$$\frac{1}{\mu}\left[\frac{1}{\mu - \nu} + \frac{1}{\nu}\right] = \nu^{-1}(\mu - \nu)^{-1}, \qquad (5.3.18)$$

already saving $(1 + |\lambda|)^{-1}$ (for $|\lambda|$ large) from the first factor. On the other hand, when $\mu = \sigma + i\rho$ lies in the domain indicated and $\nu = \tau + i\lambda$ lies on C_1, one always has

$$\text{either } |\mu - \nu| \geq C(1 + |\lambda|) \quad \text{or } |\sigma - \tau| \geq \frac{|\sigma|}{3}. \qquad (5.3.19)$$

Indeed, if $\sigma \geq 0$, one has $|\sigma - \tau| > \sigma$. If $\sigma < 0$ and the second inequality is violated, one has $\sigma \leq \tau + |\sigma - \tau| < \tau + \frac{|\sigma|}{3} = \tau - \frac{\sigma}{3}$, i.e., $\sigma < \frac{3\tau}{4}$. Then, $-\frac{1}{2\log(2+|\rho|)} < -\frac{3}{4\log(2+|\lambda|)}$ and $2 + |\lambda| > (2 + |\rho|)^{\frac{3}{2}}$, so that $|\lambda - \rho|$ is of the same order as $|\lambda|$; of course, $|\mu - \nu| \geq |\lambda - \rho|$.

In the first case, the term with $\varepsilon = 1$ in (5.3.16) is bounded by an absolute constant. In the second case, one can majorize it by

$$C \int_{-\infty}^{\infty} (1 + |\lambda|)^{-1}[|\sigma| + |\rho - \lambda|]^{-1}[\log(2 + |\lambda|)]^2 d\lambda$$

$$\leq C|\sigma|^{-\beta} \int_{-\infty}^{\infty} (1 + |\lambda|)^{-1}|\rho - \lambda|^{\beta-1}[\log(2 + |\lambda|)]^2 d\lambda \leq C|\sigma|^{-\beta} \quad (5.3.20)$$

(just split the last integral into two terms, according to whether $|\lambda - \rho| \geq 1$ or not, and use the inequality $(1 + |\rho - \lambda|)^{\beta-1} \leq C(1 + |\rho|)^{1-\beta}(1 + |\lambda|)^{\beta-1})$. This concludes the proof of the lemma. $\qquad \square$

PROOF OF LEMMA 5.2.10. We shall verify it in two steps. First, let us prove that

$$\int_{\text{Re}\,\mu - \delta} \Psi_{m,n}(\mu) d\mu = \int_{\gamma_1} \Psi_{m,n}(\mu) d\mu, \quad (5.3.21)$$

where $\gamma_+ = \{\sigma + i\rho \colon \sigma = \frac{c}{2\log(2+|\rho|)}\}$. On one hand, the move takes place entirely within the right half-plane, so that we can apply Lemma 5.3.2. Points $\mu = \sigma + i\rho$ between γ_+ and the line $\text{Re}\,\mu = \delta$ lie also, for some C, within the range of applicability of Lemma 5.3.1: if this were not the case, there would exist some pair $(p, \pm)$ satisfying the inequalities $\sigma \leq C^{-1}\lambda_p^{-\frac{1}{2}}$ and $|\rho \pm \lambda_p| \leq C^{-1}\lambda_p^{-\frac{1}{2}}$. But this is impossible if C is large enough, for the second inequality implies that λ_p is of the same order as $|\rho|$, which makes the first inequality incompatible with the condition $\sigma \geq \frac{c}{2\log(2+|\rho|)}$. The results of the last two lemmas make the move from the line $\text{Re}\,\mu = \delta$ to γ_+ possible: note that, though the factors $|\sigma|^{-\alpha}$ in Lemma 5.3.1 and $|\sigma|^{-\beta}$ in Lemma 5.3.2 lessen to some extent the quality of the estimates, this is harmless here since at most a factor of the order of a power of $\log(2 + |\rho|)$ is lost in this way.

At this point, it is possible to write

$$\int_{\gamma_+} \Psi_{m,n}(\mu) d\mu = \int_{\gamma_+} \frac{\Psi_{m,n}(\mu) + \Psi_{m,n}(-\mu)}{2} d\mu + \int_{\gamma_+} \frac{\Psi_{m,n}(\mu) - \Psi_{m,n}(-\mu)}{2} d\mu,$$

$$(5.3.22)$$

because both terms make sense, as a consequence of Lemma 5.3.2. Transforming the first term is easy: as the region between γ_+ and the pure imaginary line is contained in the zero-free region of both functions $\mu \mapsto \zeta(\pm\mu)$, as defined in (5.3.3), it is immediate (remember that arbitrary powers of μ^{-1} can be obtained with the help of integrations by parts in the integrals making up the expression within brackets on the right-hand side of (5.2.71)) that the term under examination can be transformed to $\int_{\mathrm{Re}\,\mu=0} \frac{\Psi_{m,n}(\mu)+\Psi_{m,n}(-\mu)}{2}d\mu$.

We finally come to the slightly more complicated study of the second term. After another look at the definition of the sequence (A_q) in the beginning of this section, we shall substitute for the line γ_+ the sequence of (oriented) contours $(\gamma_q)_{q\geq 1}$ defined as follows. Set $\psi(\rho) = \frac{c}{2\log(2+|\rho|)}$. A point describes the contour γ_q if moving first from $-iA_q$ to $\psi(A_q) - iA_q$ along a horizontal piece of line, next from $\psi(A_q) - iA_q$ to $\psi(A_q) + iA_q$ along the corresponding part of γ_+, finally from $\psi(A_q) + iA_q$ to iA_q along a horizontal piece of line again. Let us now show that, as $q \to \infty$, the integral of $\frac{\Psi_{m,n}(\mu)-\Psi_{m,n}(-\mu)}{2}$ on the contour γ_q has the corresponding integral taken on γ_+ as a limit. To do so, we must verify that the integrals over the remaining parts (the parts of γ_+ which have been discarded and the parts on the new rectilinear small segments which have been added) go to zero as $q \to \infty$. The contributions of the first two parts have actually already been treated, when we established (5.3.21). The sole difficulty with the small segments is that σ moves between 0 and $\psi(A_q)$ there, and the estimates in Lemmas 5.3.1 and (5.3.2) blow up as $\sigma \to 0$: however, $\sigma^{-\alpha}$ (or $\sigma^{-\beta}$) there remains integrable near 0 since $\alpha, \beta < 1$, so we are done.

For the very last step, let us introduce the image $\widetilde{\gamma}_q$ of the oriented path γ_q under the map $\mu \mapsto -\mu$. Then, with $\Xi(\mu) = \frac{1}{2}\left[\Psi_{m,n}(\mu) - \Psi_{m,n}(-\mu)\right]$, one has

$$\int_{\gamma_q} \Xi(\mu)d\mu = \frac{1}{2}\left[\int_{\gamma_q} \Xi(\mu)d\mu - \int_{\widetilde{\gamma}_q} \Xi(-\mu)d\mu\right]: \qquad (5.3.23)$$

but Ξ is an odd function, so that this is the same as $i\pi$ times the sum of residues of Ξ, or of Ψ, inside the closed path $\gamma_q + \widetilde{\gamma}_q$. The proof of Lemma 5.2.10 is complete, provided that we understand the sum of the series of residues there in the following way: as the limit, as $q \to \infty$, of the sum of residues at poles between $-iA_q$ and iA_q. This restriction will be lifted next. $\qquad\square$

Theorem 5.3.3. *The right-hand side of (5.2.81) extends as a continuous bilinear form of the pair h, f, each function lying in the Fréchet space which is the image of $\mathcal{S}_{\mathrm{even}}(\mathbb{R}^2)$ under the operator $2i\pi\mathcal{E}$. Equation (5.2.81) is still valid whenever h and f lie in the image of $\mathcal{S}_{\mathrm{even}}(\mathbb{R}^2)$ under the operator $2i\pi\mathcal{E}(1+2i\pi\mathcal{E})$.*

Proof. The second part is a consequence of the first, together with Theorem 5.2.12 and Proposition 5.1.1. To prove the first part, we shall use (5.2.7) and (5.2.8). With $h = 2i\pi\mathcal{E}h_1$, one has

$$\begin{aligned}
(1+i\lambda)(2+i\lambda)\mathcal{L}_{i\lambda,0}(h) &= \left[(1 - 2i\pi\mathcal{E})(2 - 2i\pi\mathcal{E})\mathcal{L}_{i\lambda,0}\right](h) \\
&= \mathcal{L}_{i\lambda,0}\left[(1 + 2i\pi\mathcal{E})(2 + 2i\pi\mathcal{E})h\right] \\
&= \mathcal{L}_{i\lambda,0}\left[2i\pi\mathcal{E}(1 + 2i\pi\mathcal{E})(2 + 2i\pi\mathcal{E})h_1\right] \qquad (5.3.24) \\
&= \int_{-\infty}^{\infty} |t|^{-i\lambda-1}t\frac{d}{dt}\left[\mathcal{F}_1^{-1}(1 + 2i\pi\mathcal{E})(2 + 2i\pi\mathcal{E})h_1\right](0,t)dt,
\end{aligned}$$

so that

$$\left|(1+i\lambda)(2+i\lambda)\mathcal{L}_{i\lambda,0}(h)\right| \le \int_{-\infty}^{\infty} \left|\frac{d}{dt}\left[\mathcal{F}_1^{-1}(1+2i\pi\mathcal{E})(2+2i\pi\mathcal{E})h_1\right](0,t)\right| dt,$$

(5.3.25)

where the right-hand side is a continuous linear form on $h_1 \in \mathcal{S}_{\text{even}}(\mathbb{R}^2)$. This already shows the convergence, in an appropriate norm, of the part of the integral in (5.2.81) obtained from keeping only the first two "cuspidal" terms from the decomposition (5.2.7) of $\langle \mathcal{E}_{i\lambda}, h \rangle$, and doing the same with f. To bound the terms from the series (5.2.7) with $k \ne 0$, it is no longer necessary to assume that h lies in the image of $\mathcal{S}_{\text{even}}(\mathbb{R}^2)$ under $2i\pi\mathcal{E}$: that $h \in \mathcal{S}_{\text{even}}(\mathbb{R}^2)$ suffices. We write

$$(1+i\lambda)(2+i\lambda)\mathcal{L}_{i\lambda,k}(h) = \mathcal{L}_{i\lambda,k}\left[(1+2i\pi\mathcal{E})(2+2i\pi\mathcal{E})h\right]$$

$$= \int_{-\infty}^{\infty} |t|^{-i\lambda}(\text{sign}\,t).t^{-1}\left[\mathcal{F}_1^{-1}\left((1+2i\pi\mathcal{E})(2+2i\pi\mathcal{E})h\right)\right]\left(\frac{k}{t},t\right) dt. \quad (5.3.26)$$

Setting

$$g(\eta,t) = \eta^2 t\left[\mathcal{F}_1^{-1}\left((1+2i\pi\mathcal{E})(2+2i\pi\mathcal{E})h\right)\right](\eta,t), \quad (5.3.27)$$

one may write

$$(1+i\lambda)(2+i\lambda)\mathcal{L}_{i\lambda,k}(h) = k^{-2}\int_{-\infty}^{\infty} |t|^{-i\lambda}(\text{sign}\,t)g\left(\frac{k}{t},t\right) dt : \quad (5.3.28)$$

this ensures the sought-after convergence with respect to both the $d\lambda$ - integration and the k-series.

Treating the series on the right-hand side of (5.2.81) can be done in just the same way, only replacing the logarithmic bound for $(\zeta(i\lambda)\zeta(-i\lambda))^{-1}$ (which we have not made explicit since we just showed how to gain the factor $(1+|\lambda|)^{-2}$) by the estimate (5.3.4). The factor $\lambda_p \exp\left(\frac{\pi\lambda_p}{2}\right)$ arising from a product of two coefficients of the decomposition (5.2.8)) of $\langle \mathfrak{M}_{p,j}, \cdot \rangle$ is immediately taken care of by the extra coefficient $\Gamma\left(\frac{i\lambda_p}{2}\right)\Gamma\left(-\frac{i\lambda_p}{2}\right)$. The factor $|n|$ in (5.3.4) (which could be improved, but this is not necessary) only requires that, besides the function $g(\eta,t)$, we introduce the same function with an outside factor $\eta^3 t^2$ in place of $\eta^2 t$. This concludes the proof of Theorem 5.3.3.   □

Corollary 5.3.4. *The scalar product* $(h,f) \mapsto \langle \mathfrak{P}, \bar{h} \otimes f \rangle$ *on the space of functions considered in Theorem 5.3.3 is semi-definite non-negative.*

Proof. As a consequence of (5.2.5) and of the equation $b_{-k;p,j} = \epsilon_{p,j} b_{k;p,j}$, one obtains the pair of equations

$$\langle \mathcal{E}_\nu, \bar{h} \rangle = \overline{\langle \mathcal{E}_{\bar{\nu}}, h \rangle}, \quad \langle \mathfrak{M}_{p,j}, \bar{h} \rangle = \epsilon_{p,j}\overline{\langle \mathfrak{M}_{-p,j}, h \rangle} : \quad (5.3.29)$$

the corollary follows, yielding

$$\langle \mathfrak{P}, \bar{h} \otimes h \rangle = \frac{1}{2\pi} \int_{-\infty}^{\infty} |\langle \mathfrak{E}_{i\lambda}, h \rangle|^2 |\zeta(i\lambda)|^{-2} d\lambda$$

$$+ 2 \sum_{p \neq 0} \sum_{j} |\Gamma \left(\frac{i\lambda_p}{2} \right)|^2 |\langle \mathfrak{M}_{p,j}, h \rangle|^2. \quad (5.3.30)$$

$\square$

The scalar product which is the object of the last corollary is far from being positive definite, because distinct functions h in the image of $\mathcal{S}_{\text{even}}(\mathbb{R}^2)$ under $2i\pi\mathcal{E}(1 + 2i\pi\mathcal{E})$ can lead to the same automorphic distribution after having been applied the Poincaré summation process: to start with, the result depends only on the restriction of the function $(\mathcal{F}_1^{-1}h)(\eta, \xi)$ to the set where $\eta\xi \in \mathbb{Z}$ (Proposition 5.1.3). As a matter of fact, as will be seen in the next corollary, if $\mathfrak{S} = \sum_{g \in \Gamma} h \circ g$, the number $\langle \mathfrak{P}, \bar{h} \otimes h \rangle$ depends only on $\mathfrak{S}$ and is indeed the square of some Hilbert norm on the (incomplete) space of automorphic distributions so defined. Moreover, pseudo-differential analysis will make it possible to understand this norm in terms depending on $\mathfrak{S}$ rather than h. The following corollary is an automorphic analogue of the pair of equations (1.1.43) and (1.1.44) in Remark 1.1.b.

Corollary 5.3.5. *Recall definition* (2.1.1) *of the* Θ - *transformation. Let h lie in the image of $\mathcal{S}_{\text{even}}(\mathbb{R}^2)$ under the operator $(2i\pi\mathcal{E})^2(1 + 2i\pi\mathcal{E})(1 - 2i\pi\mathcal{E})$, and let $\mathfrak{S} = \sum_{g \in \Gamma} h \circ g$, a genuine distribution under these assumptions (cf. Proposition 5.1.1). The number*

$$\|\mathfrak{S}\|_{L^2(\Gamma \backslash \mathbb{R}^2)}^2 : = \langle \mathfrak{P}, \bar{h} \otimes h \rangle \quad (5.3.31)$$

depends only on $\mathfrak{S}$ and defines the square of a norm of Hilbert type on the (incomplete with respect to it) space of automorphic distributions so defined. Moreover, one has the following pair of identities. If h is $\mathcal{G}$-invariant,

$$\|\Theta_0 \mathfrak{S}\|_{L^2(\Gamma \backslash \Pi)} = 2\|\Gamma(i\pi\mathcal{E})\mathfrak{S}\|_{L^2(\Gamma \backslash \mathbb{R}^2)}, \quad (5.3.32)$$

while, if h changes to $-h$ under $\mathcal{G}$,

$$\|\Theta_1 \mathfrak{S}\|_{L^2(\Gamma \backslash \Pi)} = 4\|\Gamma(1 + i\pi\mathcal{E})\mathfrak{S}\|_{L^2(\Gamma \backslash \mathbb{R}^2)}. \quad (5.3.33)$$

Proof. Using (3.2.8) and (3.2.21), not forgetting the role of the rescaling operator $2^{-\frac{1}{2} + i\pi\mathcal{E}}$ one has

$$\Theta_0 \mathfrak{E}_{-i\lambda} = 2^{\frac{1-i\lambda}{2}} E_{\frac{1-i\lambda}{2}}^*, \qquad \Theta_1 \mathfrak{E}_{-i\lambda} = 2^{\frac{1-i\lambda}{2}} i\lambda E_{\frac{1-i\lambda}{2}}^*,$$

$$\Theta_0 \mathfrak{M}_{p,j} = 2^{\frac{1-i\lambda_p}{2}} M_{|p|,j}, \qquad \Theta_1 \mathfrak{M}_{p,j} = 2^{\frac{1-i\lambda_p}{2}} i\lambda_p M_{|p|,j}. \quad (5.3.34)$$

It follows from these equations and (5.2.81) that

$$\Theta_0 \mathfrak{S} = \frac{1}{2\pi} \int_{-\infty}^{\infty} 2^{\frac{1-i\lambda}{2}} \langle \mathfrak{E}_{i\lambda}, h \rangle E_{\frac{1-i\lambda}{2}}^* \frac{d\lambda}{|\zeta(i\lambda)|^2}$$

$$+ \sum_{p \neq 0} \sum_{j} 2^{\frac{3-i\lambda_p}{2}} |\Gamma(\frac{i\lambda_p}{2})|^2 \epsilon_{p,j} \langle \mathfrak{M}_{p,j}, h \rangle M_{|p|,j} : \quad (5.3.35)$$

to obtain the Θ_1-transform of $\mathfrak{S}$, it suffices to insert the extra factor $i\lambda$ (*resp.* $i\lambda_p$) in the integral (*resp.* the series), so that considering $\Theta_0\mathfrak{S}$ only will do. Since $\mathcal{G} = 2^{2i\pi\mathcal{E}}\mathcal{F}^{\text{symp}}$ as observed just after (1.1.24), the function $\lambda \mapsto 2^{\frac{1-i\lambda}{2}}\langle\mathfrak{E}_{i\lambda}, h\rangle$ is even when h is $\mathcal{G}$-invariant: similarly, the number $2^{\frac{1-i\lambda_p}{2}}\langle\mathfrak{M}_{p,j}, h\rangle$ is then invariant under the change $p \mapsto -p$. It follows from the first property that the function Φ entering the Roelcke-Selberg expansion (3.1.13) of $\Theta_0\mathfrak{S}$ is

$$\Phi(\lambda) = 2^{\frac{5-i\lambda}{2}}\pi^{-\frac{i\lambda}{2}}\Gamma(\frac{i\lambda}{2})(\zeta(-i\lambda))^{-1}\langle\mathfrak{E}_{i\lambda}, h\rangle, \tag{5.3.36}$$

while the second property makes it possible to reduce the sum for $p \neq 0$ on the right-hand side of (5.3.35) to twice the same sum, restricted to values $p \geq 1$. Using also the equation

$$\langle\mathfrak{E}_{i\lambda}, \Gamma(i\pi\mathcal{E})h\rangle = \langle\Gamma(-i\pi\mathcal{E})\mathfrak{E}_{i\lambda}, h\rangle = \Gamma(\frac{i\lambda}{2})\langle\mathfrak{E}_{i\lambda}, h\rangle \tag{5.3.37}$$

and the similar equation involving cusp-distributions, one obtains the corollary from a comparison of (3.1.14), applied to (5.3.35), and of (5.3.30). □

Chapter 6

The totally radial Weyl calculus and arithmetic

A common theme of several chapters of this book is the search for some spectral interpretation of the real and imaginary parts of the argument of interesting functions: typically, zeta and other L-functions qualify; also, in a simpler non-arithmetic environment, one can consider the Gamma function. For instance, the most important single element of Theorem 4.5.7 was the appearance of the product $\zeta^*(\frac{\rho-i\lambda}{2})\zeta^*(\frac{\rho+i\lambda}{2})$ as a constituent of some spectral density: there, the pair of numbers $\pm\frac{\lambda}{2}$ was clearly related to a generalized eigenvalue of the operator $\sqrt{\Delta - \frac{1}{4}}$ (that corresponding to the generalized eigenfunction $E_{\frac{1-i\lambda}{2}}$). Assuming that ρ is real, it is less obvious, in this instance, how to give it a direct spectral interpretation, besides recalling that the construction of the function $f_{\rho,\nu}$ started from the consideration, in (2.3.5), of the function $\hom_{\rho,\nu}$, a generalized eigenfunction of the operator $4i\pi\mathcal{B}$ for the eigenvalue $\rho - 1$: but this property does not survive the Poincaré summation process. It would be more to the point to recall that, as observed in the beginning of Section 4.8, the measure $ds_\Sigma^{(\rho)}$ is a generalized eigenfunction of the operator ∇^2 for the eigenvalue $(\frac{\rho-1}{2})^2$.

In all examples we have worked with, the imaginary part of the argument of the function under consideration is associated, again, to the operator $\sqrt{\Delta - \frac{1}{4}}$ on Π or to the related operator $\pi\mathcal{E}$ on $\mathbb{R}^2$. The real part of the same argument is often related to a dimension. For instance, the (non-arithmetic) equation (1.1.43), which expresses the norm of the Θ_0-transform of an even-even symbol h, involves the operator $\Gamma(i\pi\mathcal{E})$: then, its generalization to the higher-dimensional totally radial case, to be quoted in (6.2.12), involves the function $\Gamma(\frac{n-1}{2} + i\pi\mathcal{E})$. Note that, alternatively, a shift by k units would occur if considering, in place of radial functions, products of such by a fixed spherical harmonic of degree k (see Remark 1.3.a.(i)): a special case occurs in (1.1.44).

The Θ_0-transformation and higher-dimensional analogues are based on the characterization of operators by means of their matrix elements against diagonal pairs of coherent states, in the sense given in (1.1). More interesting results follow if one replaces the function, typically of Gaussian style, one starts the construction with, by a discrete measure with suitable arithmetic properties: assuming it to be invariant under some representation of $SL(2, \mathbb{Z})$ in place of $SO(2)$, one obtains results which can be compared to, say, (1.1.43), but in which the Gamma function is replaced by an arithmetic function of interest. For instance, the simplest arithmetic analogues of this equation, based on the use of the discrete measures $\eth_{\text{even}}$ and $\eth_{\text{odd}}$ in (7.2.1) in the next chapter, involve the operator $(1 - 2^{2i\pi\mathcal{E}})(1 - 3^{2i\pi\mathcal{E}})\zeta(1 - 2i\pi\mathcal{E})$ or $(1 - 2^{2i\pi\mathcal{E}})\zeta(1 - 2i\pi\mathcal{E})$ in place of $\Gamma(i\pi\mathcal{E})$, as shown in [63, p. 41].

The present short chapter is mostly a reformulation of results in [63, Chap. 3]. After having lectured on this subject, we came to realize that the presentation in the given reference was unsuitable for readers not especially interested in quantization theory, for the following reason: it required the introduction of a maze of symbolic calculi, with symbols living either on $\mathbb{R}^2$ or on Π. Here, besides the totally radial n-dimensional calculus, as discussed in Sections 1.3 and 2.2, we shall need only one unfamiliar-looking symbolic calculus, to be called the "soft" calculus, the reasons for the introduction of which — having to do both with special function theory and arithmetic — will be fully explained at the end of the chapter. There is a small price to pay for this simplification: there was a free real parameter $\tau > -1$ (sometimes $\tau > -\frac{1}{2}$) in the given reference, while we shall have here $\tau = \frac{n-2}{2}$ in terms of the given dimension $n \geq 2$. However, one can, ultimately, forget about this demand, provided that (as we have done), one bases one's computations on formulas from special function theory rather than on algebraic tricks: in this sense, the totally radial pseudo-differential calculus may be regarded as an introduction to the quantization theory associated to the projective discrete series of representations of $SL(2, \mathbb{R})$.

Let us recall that, at the end of Section 1.3, symbols f of totally radial operators were made to live on the hyperbolic half-plane. As experienced several times, however, life is more pleasant in the plane than in the half-plane, and the species of symbol one can really work with when difficult computations of an arithmetic character are involved must be an associate (in the sense of Chapter 2) of the Radon transform of f. One is of course guided, in the choice of such an associate, by the desire to simplify computations as much as possible: this will be done in Section 6.1. But it will also be necessary, in a crucial way, to prepare for the applicability of a certain extension of the Rankin-Selberg unfolding method on which we shall report in Section 6.2: this will dictate our choice of a "soft" calculus.

We shall then be able to state our main results, dealing with spectral resolutions of functions obtained from the consideration of families of coherent states of an arithmetic nature. This will be, actually, just a quotation from [63, Sec.

9-10], where all this was obtained as a result of much more complicated definitions. In Remark 6.2.a., we shall point towards some seemingly difficult arithmetic questions, raised in a natural way from the considerations in this chapter.

6.1 Radial functions and measures on $\mathbb{R}^n$; the soft calculus

Given any continuous function u in $\mathbb{R}^n \backslash \{0\}$, denote as u_{rad} the unique radial function in $\mathbb{R}^n \backslash \{0\}$ such that the result of testing u or u_{rad} against any radial continuous function with compact support in $\mathbb{R}^n \backslash \{0\}$ should be the same. This can be characterized by the set of equations

$$< d\sigma_\alpha, u >=< d\sigma_\alpha, u_{\text{rad}} >, \quad \alpha > 0, \tag{6.1.1}$$

if one denotes as $d\sigma_\alpha$ the rotation-invariant measure on the sphere $\{x \in \mathbb{R}^n : |x| = a\}$, with total mass 1, in other words the one defined by the equation

$$< d\sigma_\alpha, u >= \omega_n^{-1} \alpha^{1-n} \int_{|x|=\alpha} u(x) d\sigma(x) : \tag{6.1.2}$$

we also set $\delta(|x| \quad a) = d\sigma_a(x)$.

Note the relation

$$< d\sigma_r, u >= \frac{1}{\omega_n} r^{1-n} \frac{d}{dr} \int_{|x| \le r} u(x) dx. \tag{6.1.3}$$

Identifying functions on $\mathbb{R}^n$ (*resp. on the half-line*) with measures with the help of the fixed densities dx (*resp. dt*), one can extend the map R introduced in (1.3.1) to radial measures with support disjoint from $\{0\}$. Arithmetic applications demand in particular that one consider (series of) point masses, and we note with the help of the last relation that if $x_0 \in \mathbb{R}^n \backslash \{0\}$ and $u(x) = \delta(x - x_0)$, one has

$$< d\sigma_r, u >= \frac{1}{\omega_n} r^{1-n} \delta(r - |x_0|), \tag{6.1.4}$$

from which it follows that the "functions" (in this case, measures) U and Ru_{rad} associated in the way indicated before (1.3.1) to the radial part of u are

$$U(r) = \frac{1}{\omega_n} r^{1-n} \delta(r - |x_0|),$$

$$(Ru_{\text{rad}})(t) = (2t)^{-\frac{1}{2}} \delta(\sqrt{2t} - |x_0|) = \delta\left(t - \frac{|x_0|^2}{2}\right). \tag{6.1.5}$$

In other words, the R-transform of the measure $d\sigma_\alpha$ is the measure $\delta_{\frac{\alpha^2}{2}}$, where we denote as δ_a, for $a > 0$, the unit mass at a.

As an example which we shall have to generalize following (7.2.1), consider the measure

$$\mathfrak{d}_{\text{even}}(x) = \sum_{m \in \mathbb{Z}} \chi^{(12)}(m)\delta\left(x - \frac{m}{\sqrt{12}}\right) \tag{6.1.6}$$

on the line, where $\chi^{(12)}$ is the unique Dirichlet character mod 12 such that $\chi^{(12)}(\pm 5) = -1$; then, consider its nth symmetric tensor power, to be identified with the measure $\mathfrak{d}^{(n)}$ on $\mathbb{R}^n$ such that

$$\mathfrak{d}^{(n)}(x) = \sum_{\mu \in \mathbb{Z}^n} \chi(\mu)\delta\left(x_1 - \frac{\mu_1}{\sqrt{12}}\right) \ldots \delta\left(x_n - \frac{\mu_n}{\sqrt{12}}\right), \tag{6.1.7}$$

with

$$\chi(\mu) = \prod_{j=1}^{n} \chi^{(12)}(\mu_j). \tag{6.1.8}$$

The R-transform of the radial part of $\mathfrak{d}^{(n)}$ is the measure (on the half-line)

$$\left(R\mathfrak{d}^{(n)}_{\text{rad}}\right)(t) = \sum_{\mu \in \mathbb{Z}^n} \chi(\mu)\delta\left(t - \frac{|\mu|^2}{24}\right): \tag{6.1.9}$$

if $\chi(\mu) \neq 0$, one has for every j either $\mu_j \equiv \pm 1$ or $\mu_j \equiv \pm 5 \bmod 12$, so that $\mu_j^2 \equiv 1$ mod 24 in every case, from which it follows that $|\mu|^2$ is divisible by 24 in the case when n is. In the case when $n = 24$, one may thus write

$$\left(R\mathfrak{d}^{(n)}_{\text{rad}}\right)(t) = \sum_{m \geq 1} b_m \delta(t - m), \tag{6.1.10}$$

with

$$b_m = \sum_{\substack{\mu \in \mathbb{Z}^{24} \\ |\mu|^2 = 24m}} \chi(\mu). \tag{6.1.11}$$

Recall [41, p. 22] that, setting as is usual $q = e^{2i\pi z}$ for $z \in \Pi$, one has

$$\sum_{\mu_1 \in \mathbb{Z}} \chi^{(12)}(\mu_1)q^{\frac{\mu_1^2}{24}} = 2q^{\frac{1}{24}} \sum_{\ell \in \mathbb{Z}} (-1)^\ell q^{\frac{\ell(3\ell-1)}{2}}$$

$$= 2q^{\frac{1}{24}} \prod_{k \geq 1}(1 - q^k)$$

$$= 2\eta(z), \tag{6.1.12}$$

where η is Dedekind's eta function. Then,

$$2^{24}(\eta(z))^{24} = \sum_{\mu \in \mathbb{Z}^{24}} \chi(\mu)q^{\frac{|\mu|^2}{24}}. \tag{6.1.13}$$

The coefficient b_m in (6.1.11) thus coincides with the mth Fourier coefficient of the function $(2\eta)^{24} = \left(\frac{2}{\pi}\right)^{12} \Delta$ in terms of Ramanujan's Δ-function.

We now define the appropriate symbolic calculus of totally radial operators on $L^2(\mathbb{R}^n)$, to be called the "soft" calculus (because of the "softening" effect of the Gamma factor $\Gamma(1+i\pi\mathcal{E})$: the Gamma function is rapidly decreasing at infinity on vertical lines), in which symbols are functions, or distributions, on $\mathbb{R}^2$, changing to their negatives under the transformation $\mathcal{G}$ introduced in (1.1.24). We cannot fully justify our choice in the present section, though one of the advantages of this definition will be immediately apparent: the integral kernel of the operator $R\mathrm{Op}_{\mathrm{soft}}(h)R^{-1}$, in other words the matrix coefficient of the operator $\mathrm{Op}_{\mathrm{soft}}(h)$ against a pair $(d\sigma_\alpha, d\sigma_\beta)$, will be totally explicit, and not too complicated. However, the reason why we do not choose an even simpler-looking $\mathcal{G}$-invariant species of symbols instead (replacing the factor $\Gamma(1+i\pi\mathcal{E})$ in (6.1.14) by $\Gamma(i\pi\mathcal{E})$) is subtler, and will have to wait for the next section, in which it will be explained in connection with the applicability of a certain extension of the Rankin-Selberg unfolding method.

Though this is not really necessary [63, Sec. 11], we here restrict our interest in totally radial operators commuting with the harmonic oscillator, the symbols of which, in the soft calculus or in the calculus $f \mapsto \mathrm{Op}(\Lambda f)$ as introduced in Theorem 1.3.1, are invariant under the action of $SO(2)$ by linear or fractional-linear transformations. It will then be possible to express the results described in the next section in terms of usual automorphic function theory.

Definition 6.1.1. Assume $n \geq 2$. Given a radial function $h \in \mathcal{S}(\mathbb{R}^2)$, changing to its negative under $\mathcal{G}$, we define the totally radial operator $\mathrm{Op}_{\mathrm{soft}}(h)$ in $L^2(\mathbb{R}^n)$ by the equation

$$\mathrm{Op}_{\mathrm{soft}}(h) = 2^{\frac{n+2}{2}} \frac{\Gamma(\frac{n}{2})}{\Gamma(\frac{n+1}{2})} \mathrm{Op}\left(\Lambda V^* T^* \left((2\pi)^{-\frac{1}{2}-i\pi\mathcal{E}}(\Gamma(1+i\pi\mathcal{E})h)\right)\right). \quad (6.1.14)$$

Let us first observe that, setting

$$f = V^* T^* \left[2^{\frac{n+2}{2}} \frac{\Gamma(\frac{n}{2})}{\Gamma(\frac{n+1}{2})}(2\pi)^{-\frac{1}{2}-i\pi\mathcal{E}}(\Gamma(1+i\pi\mathcal{E})h\right], \quad (6.1.15)$$

one has

$$h = 2^{\frac{-n-2}{2}} \frac{\Gamma(\frac{n+1}{2})}{\Gamma(\frac{n}{2})}(2\pi)^{\frac{1}{2}+i\pi\mathcal{E}}(\Gamma(1+i\pi\mathcal{E}))^{-1} T V f. \quad (6.1.16)$$

This is a consequence of Theorem 2.1.2, noting in particular that the function on the right-hand side of the last equation indeed changes to its negative under $\mathcal{G}$, as it follows from this theorem together with the relation $\mathcal{G}(i\pi\mathcal{E}) = (-i\pi\mathcal{E})\mathcal{G}$.

Given $\alpha, \beta > 0$, let us consider the totally radial operator of rank one built from the pair of measures $d\sigma_\alpha, d\sigma_\beta$, i.e., the operator $P_{\alpha,\beta}$ such that

$$P_{\alpha,\beta} u = < d\sigma_\alpha, u > d\sigma_\beta \quad (6.1.17)$$

or, more explicitly,

$$(P_{\alpha,\beta}u)(x) = \omega_n^{-1}\alpha^{1-n}\left[\int_{|y|=\alpha} u(y)d\sigma(y)\right]\delta(|x|-\beta). \qquad (6.1.18)$$

Recalling definition (2.2.11) of the function ϕ_z, one has

$$(\phi_z|P_{\alpha,\beta}\phi_z) = \left(2\mathrm{Im}\left(-\frac{1}{z}\right)\right)^{\frac{n}{2}} < d\sigma_\alpha, e^{i\pi\bar{z}^{-1}|\cdot|^2} > \left(e^{i\pi\bar{z}^{-1}|\cdot|^2}|d\sigma_\beta\right)$$

$$= \left(2\mathrm{Im}\left(-\frac{1}{z}\right)\right)^{\frac{n}{2}} \exp\left(i\pi(-\beta^2 z^{-1} + \alpha^2 \bar{z}^{-1})\right). \qquad (6.1.19)$$

Denote as $g_{\alpha,\beta}(z)$ the function just made explicit. The function

$$\tilde{g}_{\alpha,\beta}(z) = g_{\alpha,\beta}(-z^{-1}) = (2\mathrm{Im}\, z)^{\frac{n}{2}} e^{i\pi(\beta^2 z - \alpha^2 \bar{z})} \qquad (6.1.20)$$

is of course slightly simpler.

In view of the proof of the next theorem, we need a lemma.

Lemma 6.1.2. *A bounded totally radial operator A is characterized by the family of scalar products $(\phi_w|A\phi_w)$.*

Proof. In view of (1.3.3), the functions ϕ_w are permuted, up to multiplication by phase factors, under any operator $\mathrm{Met}^{(n)}(g)$: since the representation $\mathrm{Met}^{(n)}$, equivalent to $\pi_{\frac{n-2}{2}}$, is known to be irreducible, the space linearly generated by the functions ϕ_w is dense in $L^2_{\mathrm{rad}}(\mathbb{R}^n)$. As the function ϕ_w is the product of $\left(\mathrm{Im}\left(-\frac{1}{w}\right)\right)^{\frac{n}{4}}$ by an antiholomorphic function of w, the scalar product $(\phi_w|A\phi_z)$ is the product of $\left(\mathrm{Im}\left(-\frac{1}{w}\right)\mathrm{Im}\left(-\frac{1}{z}\right)\right)^{\frac{n}{4}}$ by a function holomorphic with respect to w and anti-holomorphic with respect to z: as such, it is characterized by its restriction to the diagonal $z = w$. This completes the proof of Lemma 6.1.2. The one-dimensional case of this argument was already used in the proof of Theorem 1.1.1.   □

Theorem 6.1.3. *Given a radial function $h \in \mathcal{S}(\mathbb{R}^2)$ changing to its negative under $\mathcal{G}$, and $\alpha, \beta > 0$ and distinct, one has*

$$(d\sigma_\alpha|\mathrm{Op}_{\mathrm{soft}}(h)d\sigma_\beta) = \frac{\Gamma(\frac{n}{2})}{(2\pi)^{\frac{n}{2}}}\int_{\mathbb{R}^2} h(x,\xi)\exp\left(i\pi(\alpha^2 - \beta^2)\frac{\xi}{x}\right)|x|^{n-2} \qquad (6.1.21)$$

$$\times \left[x^4 - \left(\frac{\alpha^2-\beta^2}{4}\right)^2\right]\left[\left(x^2 + \frac{(\alpha-\beta)^2}{4}\right)\left(x^2 + \frac{(\alpha+\beta)^2}{4}\right)\right]^{\frac{-n-1}{2}} dx d\xi.$$

On the other hand, one has

$$(d\sigma_\alpha|\mathrm{Op}_{\mathrm{soft}}(h)d\sigma_\alpha) = \frac{\Gamma(\frac{n}{2})}{(2\pi)^{\frac{n}{2}}}\left[\int_{\mathbb{R}^2} h(x,\xi)|x|(x^2+\alpha^2)^{\frac{-n-1}{2}}dx d\xi \qquad (6.1.22)\right.$$

$$\left. - \int_{-\infty}^{\infty} h(0,\xi)d\xi \int_{-\infty}^{\infty} e^{-4i\pi y\xi}|y|(y^2+\alpha^2)^{\frac{-n-1}{2}}dy\right].$$

Proof. If $f \in L^2(\Pi)$, and B is the totally radial operator of rank one defined as $Bu = (\psi|u)\phi$, where ψ and ϕ are radial functions in $L^2(\mathbb{R}^n)$, one has

$$\mathrm{Tr}(\mathrm{Op}(\Lambda f)^* B) = 2^{-n-\frac{1}{2}} \int_\Pi \overline{f}(z)b(z)dm(z) \qquad (6.1.23)$$

if one sets

$$b(z) = (\phi_z|B\phi_z). \qquad (6.1.24)$$

Indeed, in view of Lemma 6.1.2, it suffices to prove (6.1.23) in the case when $\psi = \phi = \phi_w$ for some $w \in \Pi$. In such a case, using (2.2.12), one finds

$$\mathrm{Tr}(\mathrm{Op}(\Lambda f)^* B) = (\phi_w|\mathrm{Op}(\Lambda f)^* \phi_w)$$
$$= (\mathrm{Op}(\Lambda f)\phi_w|\phi_w) = 2^{-n-\frac{1}{2}} \int_\Pi \overline{f}(z)b(z)dm(z). \qquad (6.1.25)$$

Now, when $f \in L^2(\Pi)$, the operator $\mathrm{Op}(\Lambda f)$ has a C^∞ integral kernel and the operator $(\mathrm{Op}(\Lambda f)^* B$ is trace-class even when the rank-one operator B has quite poor continuity properties. It is the case, in particular, when $B = P_{\alpha,\beta}$: then,

$$\mathrm{Tr}(\mathrm{Op}(\Lambda f)^* P_{\alpha,\beta}) = \mathrm{Tr}(\mathrm{Op}(\Lambda f)^* (<d\sigma_\alpha, \bullet > d\sigma_\beta >))$$
$$= (\mathrm{Op}(\Lambda f)d\sigma_\alpha|d\sigma_\beta). \qquad (6.1.26)$$

With $g_{\alpha,\beta}$ as defined in (6.1.19), we have proved the identity

$$(\mathrm{Op}(\Lambda f)d\sigma_\alpha|d\sigma_\beta) = 2^{-n-\frac{1}{2}} \int_\Pi \overline{f}(z)g_{\alpha,\beta}(z)dm(z). \qquad (6.1.27)$$

Note, from (2.2.12), that the adjoint of $\mathrm{Op}(\Lambda f)$ is the same as $\mathrm{Op}(\Lambda \overline{f})$, so that the last identity can also be written as

$$(d\sigma_\alpha|\mathrm{Op}(\Lambda f)d\sigma_\beta) = 2^{-n-\frac{1}{2}} \int_\Pi f(z)g_{\alpha,\beta}(z)dm(z). \qquad (6.1.28)$$

Using Definition 6.1.1, one obtains

$$(d\sigma_\alpha|\mathrm{Op}_{\mathrm{soft}}(h)d\sigma_\beta)$$
$$= 2^{\frac{1-n}{2}} \frac{\Gamma(\frac{n}{2})}{\Gamma(\frac{n+1}{2})} \int_\Pi V^* T^* \left((2\pi)^{-\frac{1}{2}-i\pi\mathcal{E}} \Gamma(1 + i\pi\mathcal{E})h \right)(z)g_{\alpha,\beta}(z)dm(z)$$
$$= 2^{\frac{1-n}{2}} \frac{\Gamma(\frac{n}{2})}{\Gamma(\frac{n+1}{2})} \int_{\mathbb{R}^2} h(x,\xi) \left[(2\pi)^{-\frac{1}{2}+i\pi\mathcal{E}} \Gamma(1 - i\pi\mathcal{E})TV g_{\alpha,\beta} \right](x,\xi)dxd\xi. \qquad (6.1.29)$$

Since h is an even function and

$$\begin{pmatrix} 0 & 1 \\ -1 & 0 \end{pmatrix} \begin{pmatrix} x & b \\ \xi & d \end{pmatrix} = \begin{pmatrix} \xi & d \\ -x & -b \end{pmatrix}, \qquad (6.1.30)$$

one can also write, in terms of the function in (6.1.20),

$$(d\sigma_\alpha | \mathrm{Op}_{\mathrm{soft}}(h) d\sigma_\beta) \qquad (6.1.31)$$

$$= 2^{\frac{1-n}{2}} \frac{\Gamma(\frac{n}{2})}{\Gamma(\frac{n+1}{2})} \int_{\mathbb{R}^2} h(x,\xi) \left[(2\pi)^{-\frac{1}{2}+i\pi\mathcal{E}} \Gamma(1 - i\pi\mathcal{E}) TV\tilde{g}_{\alpha,\beta} \right] (\xi, -x) dx d\xi.$$

Let us first assume that $\alpha \neq \beta$. One has

$$(TV\tilde{g}_{\alpha,\beta})^{\flat}_\lambda (s)$$

$$= \frac{1}{2}(2\pi)^{-\frac{3}{2}} \frac{\Gamma(\frac{1+i\lambda}{2})}{\Gamma(\frac{i\lambda}{2})} \int_\Pi \left(\frac{|z - s|^2}{\mathrm{Im}\, z} \right)^{-\frac{1}{2}-\frac{i\lambda}{2}} (2\mathrm{Im}\, z)^{\frac{n}{2}} e^{i\pi(\beta^2 z - \alpha^2 \bar{z})} dm(z)$$

$$= 2^{\frac{n-5}{2}} \pi^{-\frac{3}{2}} \frac{\Gamma(\frac{1+i\lambda}{2})}{\Gamma(\frac{i\lambda}{2})} \int_0^\infty y^{\frac{n-3+i\lambda}{2}} dy \qquad (6.1.32)$$

$$\times \int_{-\infty}^\infty [(x - s)^2 + y^2]^{-\frac{1}{2}-\frac{i\lambda}{2}} e^{i\pi[(\beta^2 - \alpha^2)x + i(\alpha^2 + \beta^2)y]} dx.$$

Using the equation [36, p. 401]

$$\int_{-\infty}^\infty (x^2 + y^2)^{-\frac{1}{2}-\frac{i\lambda}{2}} e^{i\pi(\beta^2 - \alpha^2)x} dx = \frac{2^{\frac{2-i\lambda}{2}} \pi^{\frac{1+i\lambda}{2}}}{\Gamma(\frac{1+i\lambda}{2})} |\alpha^2 - \beta^2|^{\frac{i\lambda}{2}} y^{-\frac{i\lambda}{2}} K_{\frac{i\lambda}{2}} (\pi|\alpha^2 - \beta^2|y)$$

$$(6.1.33)$$

and the equation [36, p. 92]

$$\int_0^\infty y^{\frac{n-3}{2}} e^{-\pi(\alpha^2 + \beta^2)y} K_{\frac{i\lambda}{2}} (\pi|\alpha^2 - \beta^2|y) dy = 2^{\frac{1-n}{2}} \pi^{\frac{2-n}{2}} \Gamma\left(\frac{n - 1 - i\lambda}{2} \right)$$

$$\times \Gamma\left(\frac{n - 1 + i\lambda}{2} \right) (\alpha\beta)^{\frac{2-n}{2}} |\alpha^2 - \beta^2|^{-\frac{1}{2}} \mathfrak{P}^{\frac{2-n}{2}}_{-\frac{1}{2}+\frac{i\lambda}{2}} \left(\frac{\alpha^2 + \beta^2}{|\alpha^2 - \beta^2|} \right), \quad (6.1.34)$$

one obtains

$$(TV\tilde{g}_{\alpha,\beta})^{\flat}_\lambda (s) = 2^{\frac{-2-i\lambda}{2}} \pi^{\frac{-n+i\lambda}{2}} \frac{\Gamma\left(\frac{n-1-i\lambda}{2} \right) \Gamma\left(\frac{n-1+i\lambda}{2} \right)}{\Gamma(\frac{i\lambda}{2})} e^{i\pi(\beta^2 - \alpha^2)s}$$

$$\times (\alpha\beta)^{\frac{2-n}{2}} |\alpha^2 - \beta^2|^{-\frac{1}{2}+\frac{i\lambda}{2}} \mathfrak{P}^{\frac{2-n}{2}}_{-\frac{1}{2}+\frac{i\lambda}{2}} \left(\frac{\alpha^2 + \beta^2}{|\alpha^2 - \beta^2|} \right). \quad (6.1.35)$$

With $k = (2\pi)^{-\frac{1}{2}+i\pi\mathcal{E}} \Gamma(1 - i\pi\mathcal{E}) TV\tilde{g}_{\alpha,\beta}$, one has, using (2.1.12),

$$k^{\flat}_{i\lambda}(s) = (2\pi)^{\frac{-1-i\lambda}{2}} \Gamma(\frac{2 + i\lambda}{2}) (TV\tilde{g}_{\alpha,\beta})^{\flat}_\lambda (s). \qquad (6.1.36)$$

On the other hand,

$$k_{i\lambda}(\xi, -x) = |\xi|^{-1-i\lambda} k^{\flat}_{i\lambda} \left(-\frac{\xi}{x} \right). \qquad (6.1.37)$$

Then,

$$\left[(2\pi)^{-\frac{1}{2}+i\pi\mathcal{E}}\Gamma(1-i\pi\mathcal{E})TV\check{g}_{\alpha,\beta}\right]_{i\lambda}(\xi,-x)$$

$$= 2^{-\frac{3}{2}-i\lambda}\pi^{\frac{-n-1}{2}}\frac{i\lambda}{2}\Gamma\left(\frac{n-1-i\lambda}{2}\right)\Gamma\left(\frac{n-1+i\lambda}{2}\right)$$

$$\times(\alpha\beta)^{\frac{2-n}{2}}|\alpha^2-\beta^2|^{-\frac{1}{2}+\frac{i\lambda}{2}}\mathfrak{P}^{\frac{2-n}{2}}_{-\frac{1}{2}+\frac{i\lambda}{2}}\left(\frac{\alpha^2+\beta^2}{|\alpha^2-\beta^2|}\right)$$

$$\times|x|^{-1-i\lambda}\exp\left(\frac{i\pi(\alpha^2-\beta^2)\xi}{x}\right). \qquad (6.1.38)$$

From [36, p. 409], one has, for every $\gamma > 0$,

$$\int_{-\infty}^{\infty}\gamma^{\frac{i\lambda}{2}}\Gamma\left(\frac{n-1-i\lambda}{2}\right)\Gamma\left(\frac{n-1+i\lambda}{2}\right)\mathfrak{P}^{\frac{2-n}{2}}_{-\frac{1}{2}+\frac{i\lambda}{2}}\left(\frac{\alpha^2+\beta^2}{|\alpha^2-\beta^2|}\right)d\lambda$$

$$= 2^{\frac{3}{2}}\pi^{\frac{1}{2}}\Gamma(\frac{n-1}{2})\left(\frac{2\alpha\beta}{|\alpha^2-\beta^2|}\right)^{\frac{n-2}{2}}\left(\frac{\alpha^2+\beta^2}{|\alpha^2-\beta^2|}+\frac{\gamma+\gamma^{-1}}{2}\right)^{\frac{1-n}{2}}: \qquad (6.1.39)$$

with an extra factor $\frac{i\lambda}{2}$ in the integrand, one obtains instead the result

$$2^{\frac{3}{2}}\pi^{\frac{1}{2}}\Gamma(\frac{n+1}{2})\frac{\gamma^{-1}-\gamma}{2}\left(\frac{2\alpha\beta}{|\alpha^2-\beta^2|}\right)^{\frac{n-2}{2}}\left(\frac{\alpha^2+\beta^2}{|\alpha^2-\beta^2|}+\frac{\gamma+\gamma^{-1}}{2}\right)^{\frac{-n-1}{2}}. \qquad (6.1.40)$$

With $\gamma = \frac{|\alpha^2-\beta^2|}{4x^2}$, one has

$$\frac{\alpha^2+\beta^2}{|\alpha^2-\beta^2|}+\frac{\gamma+\gamma^{-1}}{2} = \frac{2\left(x^2+\frac{(\alpha-\beta)^2}{4}\right)\left(x^2+\frac{(\alpha+\beta)^2}{4}\right)}{|\alpha^2-\beta^2|x^2}. \qquad (6.1.41)$$

Using (6.1.31) and (6.1.38), one obtains finally

$$(d\sigma_\alpha|\mathrm{Op}_{\mathrm{soft}}(h)d\sigma_\beta) = \frac{\Gamma(\frac{n}{2})}{(2\pi)^{\frac{n}{2}}}\int_{\mathbb{R}^2}h(x,\xi)\exp\left(i\pi(\alpha^2-\beta^2)\frac{\xi}{x}\right)|x|^{n-2} \qquad (6.1.42)$$

$$\times\left[x^4-\left(\frac{\alpha^2-\beta^2}{4}\right)^2\right]\left[\left(x^2+\frac{(\alpha-\beta)^2}{4}\right)\left(x^2+\frac{(\alpha+\beta)^2}{4}\right)\right]^{\frac{-n-1}{2}}dxd\xi.$$

This is just equation (6.1.21). A limiting argument makes it possible to treat the case when $\beta = \alpha$. Denote as $h_{\alpha,\beta}(x,\xi)$ the integrand of the right-hand side of the last equation, from which the factor $h(x,\xi)$ has been deleted. This function does not have a locally integrable limit as $\beta \to \alpha$, but this becomes true if one replaces the term $x^4 - \left(\frac{\alpha^2-\beta^2}{4}\right)^2$ by its first term x^4, obtaining in this way the first term on the right-hand side of (6.1.22). But it is easily seen that the two parts of

$h_{\alpha,\beta}(x,\xi)$ obtained from the splitting of $x^4 - \left(\frac{\alpha^2-\beta^2}{4}\right)^2$ are just the $\mathcal{G}$-transforms of each other (if needed, details are provided in [63, (7.45)-(7.48)]): this makes it possible to obtain the limit, in a distribution sense, of the second part of $h_{\alpha,\beta}$ as well.   □

Since $Rd\sigma_\alpha = \delta_{\frac{\alpha^2}{2}}$ as observed just after (6.1.5), equation (6.1.21) can be rewritten as

$$(R^{-1}\delta_a|\mathrm{Op}_{\mathrm{soft}}(h)R^{-1}\delta_b)_{L^2(\mathbb{R}^n)} = \frac{\Gamma(\frac{n}{2})}{(2\pi)^{\frac{n}{2}}} \int_{\mathbb{R}^2} h(x,\xi) \exp\left(2i\pi(a-b)\frac{\xi}{x}\right) |x|^{n-2}$$

$$\times \left[x^4 - \frac{(a-b)^2}{4}\right]\left[\left(x^2 + \frac{1}{2}\left(\sqrt{a}-\sqrt{b}\right)^2\right)\left(x^2 + \frac{1}{2}\left(\sqrt{a}+\sqrt{b}\right)^2\right)\right]^{\frac{-n-1}{2}} dxd\xi.$$

$$(6.1.43)$$

This is exactly the transfer, under R^{-1}, of the working definition (with $\tau = \frac{n-2}{2}$)

$$(\delta_a|\mathrm{Op}^\tau_{\mathrm{soft}}\delta_b)_{\tau+1} = \int_{\mathbb{R}^2} h(x,\xi)W^\tau(\delta_a,\delta_b)(x,\xi)dxd\xi \qquad (6.1.44)$$

used in [63, p. 69]. There is no need to explain fully the notation in the last equation, taken from the given reference: only note that the extra factor $\frac{\Gamma(\frac{n}{2})}{(2\pi)^{\frac{n}{2}}}$ present in (6.1.43), but absent from the function denoted as $W^\tau(\delta_a,\delta_b)(x,\xi)$ there (the conjugate under R of this Wigner-style function will occur in the next section), is explained by the fact that R^{-1} is an isometry from $H_{\tau+1} = H_{\frac{n}{2}}$ to $L^2_{\mathrm{rad}}(R^n)$ only up to the exact constant under examination, as seen from (1.3.2). This will make it possible, in the next section, to rely on results already proven in the given reference.

Remark 6.1.a. If, in the definition (6.1.14) of $\mathrm{Op}_{\mathrm{soft}}(h)$, we had replaced $\Gamma(1+i\pi\mathcal{E})$ by $\Gamma(i\pi\mathcal{E})$, the factor $x^4 - \left(\frac{\alpha^2-\beta^2}{4}\right)^2$ on the right-hand side of (6.1.21) would not have appeared (while the exponent $\frac{-1-n}{2}$ of the other factor would have been replaced by $\frac{1-n}{2}$). However, far from being a nuisance, this extra factor will be just what is needed, as explained immediately after the proof of Proposition 6.2.3.

6.2   Totally radial operators and arithmetic measures

This section is mostly expository, as it will consist in transferring results from [63, Sec. 9-10] to the present environment, rid of extra sets of symbolic calculi. The program starts with the consideration of so-called resolutions of the identity: these consist in (non-unique, but canonical) decompositions of elements of the space $\mathcal{S}_{\mathrm{rad}}(\mathbb{R}^n)$ as integral superpositions of functions, or measures, the $\mathrm{Met}^{(n)}(g)$-transforms of some fixed function or measure. For instance, if $n \geq 3$, one has the decomposition, in the weak sense in $L^2_{\mathrm{rad}}(\mathbb{R}^n)$,

$$u = \frac{n-2}{8\pi} \int_\Pi (\phi_w | u)_{L^2(\mathbb{R}^n)} \phi_w \, dm(w), \qquad (6.2.1)$$

where the function ϕ_w was defined in (2.2.11). The identity $\|u\|^2 = \frac{n-2}{8\pi} \int_\Pi |(\phi_w | u)|^2 dm(w)$ of which it is a polarized version is equivalent to the fact that, for $\tau = \frac{n-2}{2} > 0$, the map $v \mapsto f$ in (1.3.7) is an isometry.

The arithmetic counterpart to this starts with the consideration of a holomorphic function

$$\psi(z) = q^\kappa \sum_{m \geq 0} a_m q^m \qquad (6.2.2)$$

in Π, where κ is some fixed number in $]0, 1]$, $q = e^{2i\pi z}$, $q^\kappa = e^{2i\pi\kappa z}$, the sequence (a_m) is bounded by some power of $m + 1$, and we assume, with $\tau = \frac{n-2}{2}$, that the function ψ is invariant, up to multiplication by some complex number of absolute value 1, by the transformation $\pi_{\tau+1}\left(\left(\begin{smallmatrix} 0 & 1 \\ -1 & 0 \end{smallmatrix}\right)\right)$. In other words, since of necessity the same goes with the translation $\pi_{\tau+1}\left(\left(\begin{smallmatrix} 1 & 1 \\ 0 & 1 \end{smallmatrix}\right)\right)$, ψ is essentially a holomorphic modular form of weight $\tau + 1 = \frac{n}{2}$ for the full modular group, except for the holomorphy condition at infinity, if stated as is usual in terms of the local coordinate $e^{2i\pi z}$ (taking $e^{\frac{2i\pi z}{N}}$ instead, when κ is rational, one would still have to pay the heavy price of replacing Γ by some congruence subgroup). Note that we do not pay much attention to phase factors, which will always disappear from our final formulas anyway.

We associate with ψ the radial measure on $\mathbb{R}^n$

$$\mathfrak{s}^{(n)} = \sum_{m \geq 0} a_m d\sigma_{\sqrt{2(m+\kappa)}}, \qquad (6.2.3)$$

or

$$\mathfrak{s}^{(n)}(x) = \sum_{m \geq 0} a_m \delta(|x| - \sqrt{2(m+\kappa)}). \qquad (6.2.4)$$

The R-transform of this measure is the measure, on the half-line,

$$t \mapsto \sum_{m \geq 0} a_m \delta(t - m - \kappa). \qquad (6.2.5)$$

As an example, starting from Dedekind's η-function, one may take

$$\psi(z) = (\eta(z))^n = q^{\frac{n}{24}} \prod_{k \geq 1} (1 - q^k)^n : \qquad (6.2.6)$$

in particular, when $n = 24$, one obtains a multiple of Ramanujan's Δ-function; since $\kappa = 1$ in this case, a_m is the $(m + 1)$th Fourier coefficient of this function. It then follows from (6.1.10) that when $\psi = \eta^{24}$, the measure $\mathfrak{s}^{(24)}$ under consideration is just the radial part of the measure $\mathfrak{d}^{(24)} = \otimes^{24} \mathfrak{d}_{\mathrm{even}}$.

Proposition 6.2.1. *Let ψ be a holomorphic function in Π satisfying the conditions stated just after (6.2.2) and, given $n \geq 2$, let $\mathfrak{s}^{(n)}$ be the radial measure in $\mathbb{R}^n$ defined in (6.2.3). Given $u \in \mathcal{S}_{\mathrm{rad}}(\mathbb{R}^n)$, one has the identity*

$$\int_{\Gamma \backslash SL(2,\mathbb{R})} |(\mathrm{Met}^{(n)}(g^{-1})\mathfrak{s}^{(n)}|u)|^2_{L^2(\mathbb{R}^n)} dg = C(n,\psi)\|u\|^2_{L^2(\mathbb{R}^n)} \tag{6.2.7}$$

with

$$C(n,\psi) = \|z \mapsto (2\mathrm{Im}\,z)^{\frac{n}{4}}\psi(z)\|^2_{L^2(\Gamma \backslash \Pi)}, \tag{6.2.8}$$

and the weak decomposition, in $\mathcal{S}'_{\mathrm{rad}}(\mathbb{R}^n)$, of functions in $\mathcal{S}_{\mathrm{rad}}(\mathbb{R}^n)$ as integral superpositions of measures $\mathrm{Met}^{(n)}(g^{-1})\mathfrak{s}^{(n)}$, obtained by polarization of this identity. The Haar measure on $SL(2,\mathbb{R})$ has been normalized as $dg = \frac{1}{2\pi}dm(z)d\theta$ if

$$g = \begin{pmatrix} y^{\frac{1}{2}} & xy^{-\frac{1}{2}} \\ 0 & y^{-\frac{1}{2}} \end{pmatrix} \begin{pmatrix} \cos\theta & -\sin\theta \\ \sin\theta & \cos\theta \end{pmatrix}.$$

Proof. Let us first remark that the fact that, when n is odd, $\mathrm{Met}^{(n)}(g^{-1})$ is only defined up to multiplication by ± 1, is of course harmless. That the integral on $\Gamma \backslash SL(2,\mathbb{R})$ makes sense on the left-hand side of (6.2.7) comes from the fact that the radial measure $\mathfrak{s}^{(n)}$ is invariant, up to multiplication by phase factors (of absolute value 1), by all transformations $\mathrm{Met}^{(n)}(g)$ with $g \in \Gamma$; finally, the integral on $\Gamma \backslash \Pi$ on the right-hand side of (6.2.8) makes sense because the function $z \mapsto (2\mathrm{Im}\,z)^{\frac{n}{2}}|\psi(z)|^2$ is automorphic. The identity is obtained from a transfer under R^{-1} of [63, p. 85], taking the extra coefficient occurring in (1.3.2) into account. $\square$

The next thing to do, starting with the non-arithmetic case again, is to appreciate up to which point a totally radial operator A can be described by means of its diagonal matrix elements $(\phi_w | A\phi_w)$, as indicated by Lemma 6.1.2: the point is to make clear the exact way in which the map from A to this function of w differs from an isometry. From Theorem 2.2.2, it follows that the Hilbert-Schmidt norm $\|\mathrm{Op}(\Lambda f)\|_{H.S.}$ in $L^2(\mathbb{R}^n)$ of the operator $\mathrm{Op}(\Lambda f)$ is given by the equation

$$\|\mathrm{Op}(\Lambda f)\|^2_{H.S.} = \|h^{\mathrm{iso}}\|^2_{L^2(\mathbb{R}^2)} \tag{6.2.9}$$

if one defines

$$h^{\mathrm{iso}} = \frac{2^{-n}}{\Gamma(\frac{n}{2})}(2\pi)^{\frac{1}{2}+i\pi\mathcal{E}}\Gamma(\frac{n-1}{2} - i\pi\mathcal{E})TVf. \tag{6.2.10}$$

Note that (6.1.16) provides a link between h and h^{iso}. In view of (6.2.9), h^{iso} may be called an isometric species of symbol: such a symbol is invariant under the involution

$$\frac{\Gamma(i\pi\mathcal{E})}{\Gamma(-i\pi\mathcal{E})} \frac{\Gamma(\frac{n-1}{2} - i\pi\mathcal{E})}{\Gamma(\frac{n-1}{2} + i\pi\mathcal{E})}. \tag{6.2.11}$$

No useful isometric species of symbol living on Π, rather than $\mathbb{R}^2$, can be defined: recall from Theorem 2.1.2 that the dual Radon transformation transfers even functions (only) of $i\pi\mathcal{E}$ into functions of Δ.

From [63, p. 69], one obtains

$$\int_\Pi |(\phi_w|\mathrm{Op}(\Lambda f)\phi_w)|^2 dm(w) = \frac{4\pi}{(\Gamma(\frac{n}{2}))^2} \left\| \Gamma\left(\frac{n-1}{2} + i\pi\mathcal{E}\right) h^{\mathrm{iso}} \right\|_{L^2(\mathbb{R}^2)}^2. \quad (6.2.12)$$

This is the generalization of (1.1.43) we had in mind. Theorem 2.1.2 and (6.2.10) would make it easy to get rid of h^{iso} and to transform the right-hand side of the last expression into an expression depending on f in a direct way. Using (2.3.3), one can also express the right-hand side in terms of a certain norm of the operator $\mathrm{Op}(\Lambda f)$ instead, with the help of the obvious self-adjoint extension (in the space of Hilbert-Schmidt operators) of the operator $A \mapsto PAQ - QAP$.

We now look for an arithmetic analogue of the last identity, using the measures $\mathrm{Met}^{(n)}(g^{-1})\mathfrak{s}^{(n)}$ in place of the functions ϕ_w. Results are based, this time, on the much more difficult Theorem 10.1 in [63] and its corollaries.

We must recall first the definition of the Rankin-Selberg L-function

$$L^*(\overline{\psi} \otimes \psi, s) = \zeta^*(2s)L(\overline{\psi} \otimes \psi, s)$$

$$= \zeta^*(2s) \times \sum_{m \geq 0} \frac{|a_m|^2}{(m+\kappa)^{s+\tau}}$$

$$= \zeta^*(2s) \times \int_{\Gamma \backslash \Pi} E_s(z)(\mathrm{Im}\, z)^{\tau+1} |\psi(z)|^2 dm(z). \quad (6.2.13)$$

That the series on the right-hand side converges at least when $\mathrm{Re}\, s > 2$ is due to the (generally not optimal) estimate $|a_m| \leq Cm^{\frac{\tau+1}{2}}$, a standard consequence of the modularity of ψ, while the integral is meaningful whenever $s \neq 0, 1$ (the poles of the function $s \mapsto \zeta^*(2s)E_s(z)$) because ψ is rapidly decreasing at infinity in the fundamental domain. The identity is proved by the usual Rankin-Selberg unfolding method [4, p. 70], and the function so defined, holomorphic for $s \neq 0, 1$, is invariant under the symmetry $s \mapsto 1 - s$.

Given a symbol $h \in \mathcal{S}(\mathbb{R}^2)$, changing to its negative under $\mathcal{G}$ and such that $h(x, \xi)$ depends only on $x^2 + \xi^2$, we interest ourselves in the function

$$g \mapsto \left(\mathrm{Met}^{(n)}(g^{-1})\mathfrak{s}^{(n)}\middle|\mathrm{Op}_{\mathrm{soft}}(h)\mathrm{Met}^{(n)}(g^{-1})\mathfrak{s}^{(n)}\right) = \left(\mathfrak{s}^{(n)}\middle|\mathrm{Op}_{\mathrm{soft}}(h \circ g^{-1})\mathfrak{s}^{(n)}\right): \quad (6.2.14)$$

since $h \circ g^{-1}$ depends only on gK (recall that $K = SO(2)$), the function of g just defined can be considered as a function of $z = g.i$: let us denote it as $F(z)$. On the other hand, since $\mathrm{Met}^{(n)}(g^{-1})\mathfrak{s}^{(n)}$ depends only on the class $\Gamma g \subset SL(2, \mathbb{R})$, the same is true of the function $F(g.i)$ just defined: finally, the function F is an automorphic function in Π. We wish to make its Roelcke-Selberg expansion explicit.

The result is best expressed in the frame of automorphic distribution theory. Recall from Proposition 3.2.1 the definition of Eisenstein distributions $\mathfrak{E}_\nu$, and their relation to usual Eisenstein series. Actually, since the transformation $\mathcal{G}$ is more present when pseudo-differential analysis is concerned than the transformation $\mathcal{F}^{\mathrm{symp}}$, it is preferable to use the rescaled version (3.2.1) $\mathfrak{E}_\nu^{\mathrm{resc}} = 2^{\frac{-1-\nu}{2}}\mathfrak{E}_\nu$, so that $\mathcal{G}\mathfrak{E}_\nu^{\mathrm{resc}} = \mathfrak{E}_{-\nu}^{\mathrm{resc}}$. Also note the equation $\mathcal{G}1 = \frac{\delta}{2}$, where δ is the unit mass at $0 \in \mathbb{R}^2$.

Theorem 6.2.2. *Given a symbol $h \in \mathcal{S}(\mathbb{R}^2)$ satisfying the conditions stated just before (6.2.14), one has*

$$(\mathfrak{s}^{(n)}|\mathrm{Op}_{\mathrm{soft}}(h)\mathfrak{s}^{(n)}) = < h, W^{(n)}(\mathfrak{s}^{(n)}, \mathfrak{s}^{(n)}) >, \tag{6.2.15}$$

where the brackets on the right-hand side stand for the duality between $\mathcal{S}(\mathbb{R}^2)$ and $\mathcal{S}'(\mathbb{R}^2)$, and the decomposition into homogeneous components of the "Wigner distribution" there can be made explicit as

$$W^{(n)}(\mathfrak{s}^{(n)}, \mathfrak{s}^{(n)}) = \frac{1}{8\pi^{\frac{3}{2}}}\frac{\Gamma(\frac{n}{2})}{\Gamma(\frac{n+1}{2})}\left[\int_{-\infty}^{\infty}\frac{L^*(\overline{\psi}\otimes\psi, \frac{1-i\lambda}{2})}{\zeta(i\lambda)\zeta(-i\lambda)}i\lambda\mathfrak{E}_{i\lambda}^{\mathrm{resc}}d\lambda\right.$$

$$\left. - 24\pi\|(\mathrm{Im}\, z)^{\frac{n}{4}}\psi\|^2_{L^2(\Gamma\backslash\Pi)}\left(\frac{\delta}{2} - 1\right)\right]. \tag{6.2.16}$$

Proof. This is just a rephrasing of Theorem 11.5 in [63], taking the extra coefficient in front of (6.1.43) into account. One may also transform the coefficient of the non-integral term here with the help of the following analogue of [22, p. 246]:

$$\|(\mathrm{Im}\, z)^{\frac{n}{4}}\psi\|^2_{L^2(\Gamma\backslash\Pi)} = \frac{\pi}{3}\frac{\Gamma(\frac{n}{2})}{(4\pi)^{\frac{n}{2}}}\mathrm{Res}_{s=1}L(\overline{\psi}\otimes\psi, s). \tag{6.2.17}$$
$\square$

Let us now indicate the way this leads to the Roelcke-Selberg decomposition of the function $F(z) = F(g.i)$ defined in (6.2.14). Given the function $h(x, \xi) = H(x^2 + \xi^2)$, we set

$$\chi(\nu) = \frac{1}{2\pi}\int_0^\infty r^{-\nu}H(r^2)dr \tag{6.2.18}$$

and we note, when $|j| + |k| \neq 0$ and $\mathrm{Re}\,\nu < -1$, the equation

$$\int_{-\infty}^\infty |t|^{-\nu}h(tj, tk)dt = 4\pi(j^2 + k^2)^{\frac{\nu-1}{2}}\chi(\nu) : \tag{6.2.19}$$

then,

$$< h, \mathfrak{E}_\nu^{\mathrm{resc}} > = 2^{\frac{-3-\nu}{2}}\zeta(1 - \nu)\sum_{(j.k)=1}\int_{-\infty}^\infty |t|^{-\nu}h(tj, tk)dt$$

$$= 2^{\frac{3-\nu}{2}}\pi\zeta(1 - \nu)\chi(\nu)E_{\frac{1-\nu}{2}}(i) \tag{6.2.20}$$

and more generally, with $z = g.i$,

$$< h \circ g^{-1}, \mathfrak{E}_\nu^{\mathrm{resc}} > = 2^{\frac{3-\nu}{2}} \pi \zeta(1-\nu) \chi(\nu) E_{\frac{1-\nu}{2}}(z). \qquad (6.2.21)$$

The distribution $\frac{\delta}{2} - 1$ on the right-hand side of (6.2.16) can be replaced by δ when tested against functions changing to their negatives under $\mathcal{G}$. It is then an easy matter to obtain the following rephrasing of Theorem 10.1 in the given reference.

Proposition 6.2.3. *Given a symbol $h \in \mathcal{S}(\mathbb{R}^2)$ satisfying the conditions stated just before (6.2.14), the function $F(z) = F(g.i)$ defined in (6.2.14) can be made explicit as*

$$F(z) = \frac{\Gamma(\frac{n}{2})}{(2\pi)^{\frac{n}{2}}} \left[\frac{2\pi^{\frac{1}{2}}}{i} \int_{\frac{1}{2}-i\infty}^{\frac{1}{2}+i\infty} \chi(1-2s) \frac{\Gamma(s-\frac{1}{2})\Gamma(\frac{3}{2}-s)\Gamma(s+\frac{n-2}{2})}{2^{s+\frac{n-2}{2}}\Gamma(s)\Gamma(\frac{n+1}{2})} \right.$$

$$\left. \times \frac{L(\bar{\psi} \otimes \psi, s)}{\zeta^*(2-2s)} E_s^*(s) ds - 2^{-\frac{n}{2}} \pi^{\frac{1}{2}} \frac{\Gamma(\frac{n}{2})}{\Gamma(\frac{n+1}{2})} h(0) \mathrm{Res}_{s=1} L(\bar{\psi} \otimes \psi, s) \right], \quad (6.2.22)$$

and χ is the function introduced in (6.2.18).

Proof. It is just a matter of applying several times the functional equations of zeta and of the function $L(\bar{\psi} \otimes \psi, s)$. $\qquad\square$

Even though we have chosen, in the present exposition, to state Theorem 6.2.2, of which the last proposition is an immediate corollary, first, it is Proposition 6.2.3 that was proved first as [63, Theor. 10.1]. The proof is actually quite complicated and requires the use of an extension of the Rankin-Selberg unfolding method. We now briefly recall this extension, since it will finally explain why we defined the soft symbolic calculus as in Definition 6.1.1, though substituting $\Gamma(i\pi\mathcal{E})$ for $\Gamma(1 + i\pi\mathcal{E})$ in (6.1.14) would have led to an integral kernel simpler than the one in (6.1.21). Note that, after the theorem under discussion has been proved, it would of course still be possible to use, instead of the symbolic calculus $\mathrm{Op}_{\mathrm{soft}}$, the calculus $h \mapsto \mathrm{Op}_{\mathrm{soft}} ((i\pi\mathcal{E})^{-1}h)$: under such a change, the factor $i\lambda$ in the integral on the right-hand side of (6.2.16) would be replaced by 2, leading to the simplest formula conceivable. A remark more to the point concerns the fact, observed in the proof of Theorem 6.1.3, that splitting the factor $x^4 - \left(\frac{\alpha^2 - \beta^2}{4}\right)^2$ in (6.1.21) breaks the integral kernel there into the difference of two terms the $\mathcal{G}$-transforms of each other. In Chapter 2, we already observed the benefit, from the point of view of convergence, of breaking a symmetry: indeed, Proposition 2.3.3 introduced the two functions $\chi_{\rho,\pm\nu}$ which made the construction of the function $f_{\rho,\nu}$ as a Poincaré series convergent at least in a certain domain of values of ν possible. A somewhat similar idea led to the following arithmetic developments.

The classical Rankin-Selberg formula is the equation (in which $z = x + iy$ on the left-hand side)

$$\int_{-\frac{1}{2} < x < \frac{1}{2}} y^s \phi(z) dm(z) = \int_{\Gamma \backslash \Pi} E_s(z) \phi(z) dm(z), \qquad (6.2.23)$$

in which $\operatorname{Re} s > 1$ and ϕ is a C^∞ automorphic function, rapidly decreasing at infinity in the fundamental domain. The Eisenstein series on the right-hand side can be obtained as a series of transforms, over the set $\Gamma_\infty \backslash \Gamma$, with $\Gamma_\infty = \pm(\Gamma \cap N)$, of the power of y occurring on the left-hand side. We would like to generalize the formula, replacing this power of y by the function

$$h_{s,k} \colon z \mapsto y^{\frac{1}{2}} K_{s-\frac{1}{2}}(2\pi k y) e^{2i\pi k x}, \qquad k \geq 1, \tag{6.2.24}$$

which occurs as the kth term, up to some coefficient, in the Fourier expansion of the Eisenstein series under consideration. However, the left-hand side of the equation obtained in such a way in place of (6.2.23) would diverge, and so would the (Poincaré) series of transforms of $h_{s,k}$ substituting for the series defining $E_s(z)$ for $\operatorname{Re} s > 1$.

So as to make the idea work anyway, we take advantage of the decomposition [36, p. 67]

$$K_{s-\frac{1}{2}}(2\pi k y) = \frac{2\pi k y}{1 - 2s} \left[K_{s-\frac{3}{2}}(2\pi k y) - K_{s+\frac{1}{2}}(2\pi k y) \right], \tag{6.2.25}$$

and we avoid using the divergent integral

$$\int_{-\frac{1}{2} < x < \frac{1}{2}} y^{\frac{1}{2}} K_{s-\frac{1}{2}}(2\pi k y) e^{2i\pi k x} \phi(z) d\mu(z) \tag{6.2.26}$$

by substituting for it the expression

$$I(s, \phi) = \frac{2\pi k}{1 - 2s} \left[A(s - 1, \phi) - A(s + 1, \phi) \right], \tag{6.2.27}$$

with (note the exponent of y, distinct from that in (6.2.26)),

$$A(s, \phi) = \int_{-\frac{1}{2} < x < \frac{1}{2}} y^{\frac{3}{2}} K_{s-\frac{1}{2}}(2\pi k y) e^{2i\pi k x} \phi(z) d\mu(z). \tag{6.2.28}$$

It will turn out that each of the two terms of the sum (6.2.27), initially convergent in an appropriate domain of values of s, can be continued analytically so that the sum will eventually make sense.

On the other side, we decompose the function $h_{s,k}$ in (6.2.24) as

$$h_{s,k}(z) = \frac{2\pi k}{1 - 2s} \left[c_{s-1,k}(z) - c_{s+1,k} \right], \tag{6.2.29}$$

with

$$c_{s,k}(z) = y^{\frac{3}{2}} K_{s-\frac{1}{2}}(2\pi k y) e^{2i\pi k x}. \tag{6.2.30}$$

The series

$$\mathfrak{f}_s(z) = \sum_{g \in \Gamma_\infty \backslash \Gamma} c_{s,k}(g.z), \tag{6.2.31}$$

with $k = 1, 2, \ldots$, can be analyzed with the help of the Selberg series on the right-hand side of (3.2.36). The results are the following. Initially convergent for $\frac{1}{2} < \mathrm{Re}\, s < 1$, it extends as a meromorphic function in the complex plane, with the following families of poles: the points $s = n + \frac{3}{2} \pm \frac{i\lambda_j}{2}$ or $s = -n - \frac{1}{2} \pm \frac{i\lambda_j}{2}$ where $n = 0, 1, \ldots$, where $\left(\frac{1+\lambda_j^2}{4}\right)$ is the sequence of eigenvalues of the hyperbolic Laplacian Δ in $L^2(\Gamma\backslash\Pi)$; the points $s = \pm\frac{\omega}{2} + n + 2$ and $s = \pm\frac{\omega}{2} - n - 1$, where ω is a non-trivial zero of zeta; the remaining poles are to be found within the sequence $\{-\frac{1}{2} - n, n = 0, 1, \ldots\}$ or $\{\frac{3}{2} + n, n = 0, 1, \ldots\}$. The function

$$E_k(z, s) = \frac{2\pi k}{1 - 2s} [\mathfrak{f}_{s-1}(z) - \mathfrak{f}_{s+1}(z)] \tag{6.2.32}$$

coincides with $\alpha_k(s) E_s(z)$, with

$$\alpha_k(s) = \frac{1}{2} \frac{\Gamma(s - \frac{1}{2})\Gamma(\frac{1}{2} - s)}{\zeta^*(2 - 2s)} k^{s - \frac{1}{2}} \sigma_{1-2s}(k). \tag{6.2.33}$$

We have just quoted [63, Theor. 8.2]: we forgot to mention there, in the list of poles, those depending on zeros of zeta, which originate from the continuous part of the spectral decomposition of Selberg's series. Since, as proved there, $\mathfrak{f}_s = \mathfrak{f}_{1-s}$ so that they come by pairs with residues the negatives of each other, the argument already used in [63, (10.44)] to show that poles of the first species from the list above do not contribute to the main formula (6.2.22) applies just as well with the poles disregarded in the given reference. A corollary of the study of $\mathfrak{f}_s$ is the following extension of the Rankin-Selberg formula: if neither $s - 1$ nor $s + 1$ coincides with a pole of the function $\mathfrak{f}_s$, and $s \neq \frac{1}{2}$, one has

$$\frac{2\pi k}{1 - 2s} [A(s - 1, \phi) - A(s + 1, \phi)] = \int_{\Gamma\backslash\Pi} E_k(z, s)\phi(z)d\mu(z). \tag{6.2.34}$$

Now, the use of the soft symbolic calculus led, in (6.1.21), to an "integral kernel" appearing as the sum of two terms, obtained from the natural decomposition of the factor $x^4 - \left(\frac{\alpha^2 - \beta^2}{4}\right)^2$. It is precisely this fact which, in the course of the proof of [63, Theor. 10.1], made the use of the modified Rankin-Selberg formula possible, providing just the required splitting of the main term.

An application of Proposition 6.2.3 is the following, stated as Corollary 10.3 in [63].

Corollary 6.2.4. *Let $F(z) = F(g.i)$ be the function defined in (6.2.14), under the assumptions about h which immediately precede this equation: moreover, assume that $h(0) = 0$. Defining the function f in Π by (6.1.15), so that $\mathrm{Op}_{\mathrm{soft}}(h) = \mathrm{Op}(\Lambda f)$, let h^{iso} be the (isometric) symbol of the operator $\mathrm{Op}_{\mathrm{soft}}(h)$, as defined in (6.2.10). One has*

$$\|F\|_{L^2(\Gamma\backslash\Pi)}^2 = \frac{1}{\pi} \left\| \frac{\zeta(1 - 2i\pi\mathcal{E})}{\zeta(-2i\pi\mathcal{E})} L(\overline{\psi} \otimes \psi, \frac{1}{2} - i\pi\mathcal{E})h^{\mathrm{iso}} \right\|_{L^2(\mathbb{R}^2)}^2. \tag{6.2.35}$$

Note that, again, one can dispense with the consideration of the symbol h^{iso}, using (6.2.9) and Theorem 2.1.2 to express the right-hand side of the last equation in terms of f: alternatively, using also (6.1.16) one may just as easily express it in terms of h.

Remark 6.2.a. Zeros of the function $L(\overline{\psi} \otimes \psi, s)$ of the form $s = \frac{1 \pm i\lambda}{2}$ are thus given a spectral interpretation: the corresponding numbers $-\frac{\lambda}{2\pi}$ are points on the spectrum of the automorphic Euler operator in $\mathbb{R}^2$ which, together with the points $\pm\infty$, account for the non-invertibility of the map $h \mapsto F$ (or of the map $h^{\text{iso}} \mapsto F$). From results of Shimura [46] and Iwaniec [22], the function $L(\overline{\psi}\otimes\psi, s)$ is "divisible" by $\zeta(s)$ in a neighbourhood of the half-plane $\operatorname{Re} s \geq \frac{1}{2}$ in the case when ψ is a holomorphic cusp-form of Hecke type (in particular, when ψ corresponds to the radial part of the measure $\mathfrak{d}^{(24)}$ as indicated just after (6.2.5)): in this case, the critical zeros of zeta are thus given an interpretation as well. It would certainly be nice, but it is probably too much to conjecture this, if considering $\mathfrak{d}^{(n)}$ for other values of n, one could make the restriction of zeta to some other lines appear: from [63, p. 41], the restriction of $\zeta(s)$ to the line $\operatorname{Re} s = 1$ does show in this way if $n = 1$ or 3.

Chapter 7

Should one generalize the Weyl calculus to an adelic setting ?

As indicated by the question mark, this chapter is a tentative one. One of its purposes is to point towards one reason for doing pseudo-differential analysis in an adelic setting, beyond the obvious observation that dealing with prime numbers one at a time, when possible, certainly clarifies things. On the other hand, one can also find some reasons not to do so, especially if one restricts one's interest, as will be the case here, to adeles of the field $\mathbb{Q}$. Before we give a short summary of this chapter, let us emphasize again, as done in the introduction of the book, that we do not regard it as an introduction to adelic pseudo-differential analysis: realizing such a program would require, on the number-theoretic side, an expertise much beyond this author's. The idea advocated here will be best explained if starting from a classical distribution point of view.

In the first section, we call attention to a certain automorphic distribution $\mathfrak{T}_\infty$, which appeared in [63, Chap. 4] as the limit, as $N \to \infty$, of a sequence $(\mathfrak{T}_N)$ of automorphic distributions, *almost* the Weyl symbols of some interesting sequence of finite-rank operators from $S(\mathbb{R})$ to $S'(\mathbb{R})$. It is one of the two terms of an automorphic object $\mathfrak{T}_\infty - \mathcal{GM}_\infty$ — not quite an automorphic distribution in the technical sense — which is a series, over the non-trivial zeros of zeta, of Eisenstein distributions $\mathfrak{E}_{-\mu}$ and of some of their $\frac{d}{d\mu}$-derivatives. In view of the role, explained in Theorem 1.1.3, of automorphic distributions as symbols of special operators, one would like to characterize in simple terms the operator $\mathrm{Op}(\mathfrak{T}_N)$ for every N; however, it is not this operator, but the operator with symbol $N^{i\pi\mathcal{E}}\mathfrak{T}_N$ that can really be understood. This does not answer the same question about $\mathfrak{T}_N$, but taking the Θ-transform of this latter distribution is possible, leading to the appearance of theta functions and to most probably untractable questions of boundedness.

Our reason for raising the adelic point of view has to do with the question

of going to the limit as $N \to \infty$. An operator such as $N^{i\pi\mathcal{E}}$, acting on symbols, involves what physicists, and PDE practitioners of so-called "semi-classical
analysis", would call a change of Planck's constant. Contrary to linear changes of
coordinates in $\mathbb{R}^2$ preserving the 2-form $dx \wedge d\xi$ (symplectic transformations), such
a rescaling does not correspond to any transformation on functions defined on the
line. The necessity of dealing with an N-dependent Planck's constant prevents
one from giving a true role as a symbol to the distribution $\mathfrak{T}_\infty$. But such a thing
should be possible in an adelic environment, in which one can use a p-dependent
Planck's constant. The p-adic pseudo-differential analyses so introduced thus differ
from the ones developed in [14, 1]: but we have not approached the question —
a deep one if one insists on not losing the link with the Archimedean problem we
started from — of piecing these various analyses together in an efficient way. We
have made a few local computations, the results of which look quite satisfactory.
Finally, we believe that, whether adelic techniques will eventually be of help, or
not, in problems which seem to dwell as much on analysis as on arithmetic (such
as the Riemann hypothesis), they certainly help in clarifying things: it will be even
more so in situations more intricate arithmetically than the one we have considered, starting with the presence (which we have discarded) of ramified characters,
or the replacement of $\mathbb{Q}$ by a quadratic extension (*cf.* Remark 7.2.b.(iv)).

Despite point (iv) in the list of remarks which followed the proof of Theorem
4.7.3, we do not believe that this constituted a possibly new approach towards
the study of the zeros of zeta, even though it put the L-function associated to
a cusp-form, or that, to wit $\frac{1}{2} \frac{\zeta^*(\frac{\rho+i\lambda}{2})\zeta^*(\frac{\rho-i\lambda}{2})}{\zeta^*(i\lambda)}$, associated to an Eisenstein series,
in a new light. The main point against it — or against Theorem 5.2.12 — seems
to be that zeros of zeta appear there in a "negative" way, i.e., as points where
a spectral density vanishes — or where the continuation of this density becomes
infinite— rather than as points in some discrete spectrum. This phenomenon is
rather general when dealing with zeta, and *may* be connected to the notion of
"absorption spectrum" as suggested by Connes [6] as a way of explaining the sign
of the main correction term to the " $N(\lambda)$-estimate" for the zeta function. In
this direction, let us recall that the Riemann hypothesis is the statement that
all points $i(\omega_j - \frac{1}{2})$, where (ω_j) is the set of non-trivial zeros of zeta, are the
eigenvalues of a self-adjoint operator in some Hilbert space. The "Hilbert-Polya
dream" [39, p. 7] is the conjecture that this can be achieved in an efficient (i.e.,
at least not circuitous) way. Very impressive suggestions in a related (not quite
analogous, though) direction are to be found in the book [25] by Katz and Sarnak,
partly based on deep numerical "experiments" by Odlyzko. On the other hand,
Connes [6] has suggested an adelic approach: our use of adeles is unrelated to this
interesting one-dimensional approach, and linked to pseudo-differential analysis
instead.

The other term $\mathfrak{T}_\infty$ from the sum alluded to above is also interesting on its
own right. Contrary to that of $\mathfrak{M}_\infty$, its study can be done in a usual distribution

setting, and will be made in the first section to come.

7.1 Eisenstein distributions with zeros of zeta for parameters

In this chapter, we shall have to let analytic functions of the operator $2i\pi\mathcal{E}$ act on tempered distributions: these functions, defined by duality, will often be integral superpositions of functions $t^{2i\pi\mathcal{E}}$ with $t > 0$, or discrete versions thereof, in particular Dirichlet series in the argument $2i\pi\mathcal{E}$. For $h \in S(\mathbb{R}^2)$, one has

$$< t^{2i\pi\mathcal{E}}\mathfrak{S}, h >=< \mathfrak{S}, t^{-2i\pi\mathcal{E}}h >= t^{-1} < \mathfrak{S}, (x,\xi) \mapsto h(\frac{x}{t},\frac{\xi}{t}) > . \qquad (7.1.1)$$

A consequence of (5.2.7), which could also be obtained directly from Proposition 3.2.1, is that the family of Eisenstein distributions $\mathfrak{E}_{i\lambda}$, with $\lambda \in \mathbb{R}$, is weakly bounded in $S'(\mathbb{R}^2)$. This gives a meaning, as an automorphic (tempered) distribution, to any integral $\int_{-\infty}^{\infty} a(\lambda)\mathfrak{E}_{i\lambda}d\lambda$ whenever the function a is C^∞ on the real line and there is a common exponent κ such that a derivative of any order of a is bounded by some constant times $(1+|\lambda|)^\kappa$: it suffices indeed to integrate by parts, using the equation $(1 + 2i\pi\mathcal{E})\mathfrak{E}_{i\lambda} = (1 - i\lambda)\mathfrak{E}_{i\lambda}$. This type of integral will occur consistently in what follows.

We have introduced in Proposition 3.2.2 the Dirac comb $\mathfrak{D}$ and given its decomposition into homogeneous (modular) components, here recalled:

$$\mathfrak{D} = 2\pi + \int_{-\infty}^{\infty} \mathfrak{E}_{i\lambda}d\lambda. \qquad (7.1.2)$$

If one considers instead of $\mathfrak{D}$ the automorphic distribution

$$\mathfrak{D}^{\mathrm{prime}}(x, \xi) = 2\pi \sum_{(m,n)=1} \delta(x - n)\delta(\xi - m), \qquad (7.1.3)$$

one finds in the same way that

$$\mathfrak{D}^{\mathrm{prime}} = \frac{12}{\pi} + \int_{-\infty}^{\infty} (\zeta(1 - i\lambda))^{-1} \mathfrak{E}_{i\lambda}d\lambda, \qquad (7.1.4)$$

and the link between $\mathfrak{D}$ and $\mathfrak{D}^{\mathrm{prime}}$ can also be written as

$$\mathfrak{D} = \zeta (1 + 2i\pi\mathcal{E}) \mathfrak{D}^{\mathrm{prime}}. \qquad (7.1.5)$$

One may observe that $\mathfrak{D}$ is quite naturally associated to the additive theory of $\mathbb{Z}$, while $\mathfrak{D}^{\mathrm{prime}}$ is associated to the multiplicative theory of $\mathbb{Z}^\times$, i.e., to the decomposition of integers as products of primes, the zeta function thus providing the link between the two points of view.

Consider now the distribution $\mathfrak{T}_\infty$ in the plane defined by the equation

$$\mathfrak{T}_\infty(x,\xi) = \sum_{\substack{j,k \in \mathbb{Z} \\ |j|+|k| \neq 0}} \Gamma_{jk}^{(\infty)} \delta(x-j)\delta(\xi-k), \tag{7.1.6}$$

where

$$\Gamma_{jk}^{(\infty)} = \prod_{\substack{p\,\text{prime} \\ p|(j,k)}} (1-p). \tag{7.1.7}$$

It is not difficult to see, as proved in [63, Theor.12.5] (an alternative proof will follow from (7.1.27)), that the decomposition into homogeneous components of the distribution $\mathfrak{T}_\infty$ is given by the equation

$$\mathfrak{T}_\infty = \frac{1}{2\pi} \int_{-\infty}^{\infty} (\zeta(-i\lambda))^{-1} \mathfrak{E}_{i\lambda} d\lambda + \sum_j \operatorname{Res}_{\mu=\omega_j} \left(\frac{\mathfrak{E}_{-\mu}}{\zeta(\mu)} \right), \tag{7.1.8}$$

where ω_j is to vary over the set of non-trivial zeros of the Riemann zeta function.

We need to introduce one last automorphic distribution, this time not a series of point-masses but a series of measures carried by lines in $\mathbb{R}^2$, starting from the automorphic object $\mathfrak{M}_\infty$ such that

$$< \mathfrak{M}_\infty, h > = \sum_{(j,k)=1} \int_{-\infty}^{\infty} h(tj,tk) e^{4i\pi t} dt. \tag{7.1.9}$$

We are familiar, by now, with the occurrence of automorphic objects, which become automorphic distributions only after they have been applied to suitable polynomials in the Euler operator. Exactly the same phenomenon occurs here: $\mathfrak{M}_\infty$ is not initially defined as a tempered distribution, but as a continuous linear form on the space image of $S(\mathbb{R}^2)$ by $2i\pi\mathcal{E}(2i\pi\mathcal{E}+1)$ or, to state this in another way, the image of $\mathfrak{M}_\infty$ under $2i\pi\mathcal{E}(2i\pi\mathcal{E}-1)$ is a tempered distribution. Indeed, after an integration by parts, one has, if $h_1 \in S(\mathbb{R}^2)$,

$$\int_{-\infty}^{\infty} [2i\pi\mathcal{E}(2i\pi\mathcal{E}+1)h_1](tj,tk) e^{4i\pi t} dt = (4i\pi)^2 \int_{-\infty}^{\infty} h_1(tj,tk) t^2 e^{4i\pi t} dt. \tag{7.1.10}$$

Using an estimate such as $|h(x,\xi)| \leq C(1+|x|+|\xi|)^{-\delta}$ for some $\delta > 3$, one obtains if $h \in 2i\pi\mathcal{E}(2i\pi\mathcal{E}+1)S(\mathbb{R}^2)$ that

$$\sum_{(j,k)=1} \int_{-\infty}^{\infty} |h(tj,tk)| dt < \infty. \tag{7.1.11}$$

But the right-hand side of (7.1.9) makes sense under the sole assumption that $h \in (2i\pi\mathcal{E})S(\mathbb{R}^2)$: to prove this, we shall relate the automorphic object $\mathfrak{M}_\infty$ to $\mathfrak{D}^{\text{prime}}$. At the same time, this will give us its decomposition into homogeneous components. In the theorem which follows and its proof, all identities concern, rather than tempered distributions, continuous linear forms on the space image of $S(\mathbb{R}^2)$ under $2i\pi\mathcal{E}$.

Theorem 7.1.1. *As introduced in (7.1.9), $\mathfrak{M}_\infty$ is a continuous linear form on the space $(2i\pi\mathcal{E})\mathcal{S}(\mathbb{R}^2)$. One has*

$$\mathfrak{M}_\infty = \frac{1}{2\pi^{\frac{1}{2}}}\cdot(2\pi)^{2i\pi\mathcal{E}}\frac{\Gamma(-i\pi\mathcal{E})}{\Gamma(\frac{1}{2}+i\pi\mathcal{E})}\mathfrak{D}^{\mathrm{prime}}$$

$$= -24 + \frac{1}{2\pi}\int_{-\infty}^{\infty}2^{-i\lambda}(\zeta(i\lambda))^{-1}\mathcal{E}_{i\lambda}d\lambda. \tag{7.1.12}$$

The $\mathcal{G}$-transform of $\mathfrak{M}_\infty$ can be written as the Poincaré series

$$\mathcal{G}\mathfrak{M}_\infty = \sum_{g\in\Gamma/\Gamma_\infty^0} \mathfrak{q}_\infty\circ g^{-1} \quad\text{with } \mathfrak{q}_\infty(x,\xi) = \delta(\xi-1). \tag{7.1.13}$$

One has the identity

$$\mathfrak{T}_\infty - \mathcal{G}\mathfrak{M}_\infty - 12\delta = \sum_j \mathrm{Res}_{\mu=\omega_j}\left(\frac{1}{\zeta(\mu)}\mathcal{E}_{-\mu}\right). \tag{7.1.14}$$

Proof. Let h lie in the image of $\mathcal{S}(\mathbb{R}^2)$ by the operator $2i\pi\mathcal{E}$: since we are only dealing with automorphic objects of even type, it is no loss of generality to assume that h is an even function in what follows. Using (1.2.11), (1.2.12) and a change of contour, we write, with $0 < \varepsilon < 1$,

$$h = \frac{1}{i}\int_{\mathrm{Re}\,\mu=-\varepsilon}h_\mu d\mu, \quad h_\mu(x,\xi) = \frac{1}{2\pi}\int_0^\infty s^\mu h(sx,s\xi)ds. \tag{7.1.15}$$

We temporarily set

$$\mathfrak{S} = \frac{1}{2\pi^{\frac{1}{2}}}\cdot(2\pi)^{2i\pi\mathcal{E}}\frac{\Gamma(-i\pi\mathcal{E})}{\Gamma(\frac{1}{2}+i\pi\mathcal{E})}\mathfrak{D}^{\mathrm{prime}}, \tag{7.1.16}$$

where the function of the Euler operator is defined by duality, in other words

$$< \mathfrak{S}, h > = < \mathfrak{D}^{\mathrm{prime}}, \frac{1}{2\pi^{\frac{1}{2}}}\cdot(2\pi)^{-2i\pi\mathcal{E}}\frac{\Gamma(i\pi\mathcal{E})}{\Gamma(\frac{1}{2}-i\pi\mathcal{E})}h > . \tag{7.1.17}$$

Note that if $h = 2i\pi\mathcal{E}h_1$, one has $\frac{\Gamma(i\pi\mathcal{E})}{\Gamma(\frac{1}{2}-i\pi\mathcal{E})}h = 2\frac{\Gamma(1+i\pi\mathcal{E})}{\Gamma(\frac{1}{2}-i\pi\mathcal{E})}h_1$, which takes care of the pole of Gamma at 0. Hence,

$$< \mathfrak{S}, h > = < \mathfrak{D}^{\mathrm{prime}}, \frac{1}{2i\pi}\int_{\mathrm{Re}\,\mu=-\varepsilon}2^\mu\pi^{\frac{1}{2}+\mu}\frac{\Gamma(-\frac{\mu}{2})}{\Gamma(\frac{1+\mu}{2})}h_\mu d\mu >: \tag{7.1.18}$$

we have used the fact that h_μ is homogeneous of degree $-1-\mu$.

On the other hand, one has $\mathfrak{M}_\infty = \lim_{N\to\infty} \mathfrak{M}_\infty^{(N)}$, with

$$
< \mathfrak{M}_\infty^{(N)}, h > = \sum_{\substack{(j,k)=1 \\ |j|+|k|\le N}} \int_{-\infty}^{\infty} e^{4i\pi t}\,dt. \frac{1}{i} \int_{\mathrm{Re}\,\mu=-\varepsilon} h_\mu(tj, tk)\,d\mu
$$

$$
= \sum_{\substack{(j,k)=1 \\ |j|+|k|\le N}} \lim_{A\to\infty} \int_{-A}^{A} e^{4i\pi t}\,dt. \frac{1}{i} \int_{\mathrm{Re}\,\mu=-\varepsilon} |t|^{-1-\mu} h_\mu(j, k)\,d\mu
$$

$$
= \sum_{\substack{(j,k)=1 \\ |j|+|k|\le N}} \frac{1}{i}. \int_{\mathrm{Re}\,\mu=-\varepsilon} 2^\mu \pi^{\frac{1}{2}+\mu} \frac{\Gamma(-\frac{\mu}{2})}{\Gamma(\frac{1+\mu}{2})} h_\mu(j, k)\,d\mu. \tag{7.1.19}
$$

As $N \to \infty$, this expression converges towards that of $< \mathfrak{S}, h >$ as given in (7.1.18).

This proves the first equation (7.1.12). To prove the second one, we start from (7.1.4), and we first move the contour to a contour γ where $\mathrm{Re}\,\mu > 0$: note that, unless a weak form of Riemann's hypothesis is already known (we do not assume this), to wit the fact that zeros of zeta cannot be arbitrarily close to the boundary of the critical strip, it is impossible to assume that γ is a straight line. Then, since $2i\pi\mathcal{E}\mathfrak{E}_\mu = -\mu\mathfrak{E}_\mu$, we obtain

$$
\mathfrak{M}_\infty = -24 + \frac{1}{2i\pi^{\frac{1}{2}}} \int_\gamma (2\pi)^{-\mu} \frac{\Gamma(\frac{\mu}{2})}{\Gamma(\frac{1-\mu}{2})} (\zeta(1-\mu))^{-1} \mathfrak{E}_\mu\,d\mu
$$

$$
= -24 + \frac{1}{2i\pi} \int_\gamma 2^{-\mu}(\zeta(\mu))^{-1}\mathfrak{E}_\mu\,d\mu
$$

$$
= -24 + \frac{1}{2\pi} \int_{-\infty}^{\infty} 2^{-i\lambda}(\zeta(i\lambda))^{-1}\mathfrak{E}_{i\lambda}\,d\lambda. \tag{7.1.20}
$$

We now take advantage of the similarity between the integral terms which occur in the spectral decompositions of $\mathfrak{M}_\infty$ and $\mathfrak{T}_\infty$. Recall definition (1.1.24) of the operator $\mathcal{G} = 2^{2i\pi\mathcal{E}}\mathcal{F}^{\mathrm{symp}}$. Making the change of variable $\mu \mapsto -\mu$ in the integral term on the last right-hand side of (7.1.20) and remembering that $\mathcal{F}^{\mathrm{symp}}\mathfrak{E}_{-i\lambda} = \mathfrak{E}_{i\lambda}$, one obtains

$$
\mathcal{G}\mathfrak{M}_\infty = -12\delta + \frac{1}{2\pi} \int_{-\infty}^{\infty} (\zeta(-i\lambda))^{-1}\mathfrak{E}_{i\lambda}\,d\lambda : \tag{7.1.21}
$$

subtracting this from equation (7.1.8), one finds (7.1.14).

Finally, the computation of the $\mathcal{G}$-transform of $\mathfrak{M}_\infty$ goes as follows. Given $h \in (-2i\pi\mathcal{E})\mathcal{S}(\mathbb{R}^2)$, one has

$$
< \mathcal{G}\mathfrak{M}_\infty, h > = \sum_{(j,k)=1} \int_{-\infty}^{\infty} (\mathcal{G}h)(tj, tk)e^{4i\pi t}\,dt
$$

$$= 2 \sum_{(j,k)=1} \int_{-\infty}^{\infty} e^{4i\pi t} dt \int_{\mathbb{R}^2} e^{4i\pi t(j\xi - kx)} h(x, \xi) dx d\xi$$

$$= \sum_{(j,k)=1} \int_{-\infty}^{\infty} \delta(1 + j\xi - kx) h(x, \xi) dx d\xi, \tag{7.1.22}$$

which leads to (7.1.13). In the same vein, one has $\mathfrak{M}_\infty = \sum_{g \in \Gamma/\Gamma_\infty^0} \mathfrak{m}_\infty \circ g^{-1}$ with $\mathfrak{m}_\infty(x, \xi) = e^{4i\pi x} \delta(\xi)$.

One should observe the similarity of this last equation with the equation (3.2.30) which defined the "Bezout distribution". Simply changing $\delta(\xi)$ to $\delta(\xi - 1)$ implies considerable changes in the automorphic distribution obtained by the Poincaré series process from such a start, including the appearance of cusp-distributions. □

Remarks 7.1.a. (i) The right-hand side of (7.1.20) is a bona fide tempered distribution: however, if $h \in \mathcal{S}(\mathbb{R}^2)$ does not lie in the image of this space under $2i\pi\mathcal{E}$, one cannot, in general, express the result of testing this distribution on h by means of the series (7.1.9).

(ii) Just as in the so-called explicit formulas from the theory of the zeta function and related L-functions [28, p. 3], the right-hand side of (7.1.14) is a sum over the non-trivial zeros of zeta. The left-hand side is quite different, but it is also, what is the most important, "purely arithmetical" in character, by which we mean that we got rid of all Gamma factors.

We need to introduce a quite specific approximation of the automorphic distribution $\mathfrak{T}_\infty$: it will depend on some positive integer N, actually only on the set of prime divisors of N. It is defined by the equation

$$\mathfrak{T}_N(x, \xi) = \sum_{|j|+|k|\neq 0} \Gamma_{jk}^{(N)} \delta(x - j)\delta(\xi - k) \tag{7.1.23}$$

with

$$\Gamma_{jk}^{(N)} = \prod_{\substack{p \text{ prime} \\ p|(j,k,N)}} (1 - p). \tag{7.1.24}$$

We turn now to the decomposition of $\mathfrak{T}_N$ into homogeneous components. Let A be the set of squarefree divisors of N and, for every $M \in A$, let $d_0(M)$ be the number of prime divisors of M. With $\mathfrak{D}$ as defined in (3.2.2), one has

$$\frac{1}{2\pi} \prod_{p|N} (1 - p^{-2i\pi\mathcal{E}}) \mathfrak{D}(x, \xi) = \sum_{M \in A} (-1)^{d_0(M)} M^{-2i\pi\mathcal{E}} \sum_{|j|+|k|\neq 0} \delta(x - j)\delta(\xi - k)$$

$$= \sum_{M \in A} (-1)^{d_0(M)} M \sum_{|j|+|k|\neq 0} \delta(x - Mj)\delta(\xi - Mk)$$

$$= \sum_{\substack{|j|+|k|\neq 0}} \sum_{\substack{M\in A \\ M|(j,k)}} (-1)^{d_0(M)} M\delta(x-j)\delta(\xi-k)$$

$$= \mathfrak{T}_N(x,\xi). \tag{7.1.25}$$

If one sets

$$\zeta_N(s) = \prod_{p|N}(1-p^{-s})^{-1}, \tag{7.1.26}$$

one can then write

$$\mathfrak{T}_N = \prod_{p|N}\left(1-p^{-2i\pi\mathcal{E}}\right)\left[1 + \frac{1}{2\pi}\int_{-\infty}^{\infty}\mathcal{E}_{i\lambda}d\lambda\right]$$

$$= \prod_{p|N}(1-p^{-1}) + \frac{1}{2\pi}\int_{-\infty}^{\infty}(\zeta_N(-i\lambda))^{-1}\mathcal{E}_{i\lambda}d\lambda$$

$$= \frac{1}{\zeta_N(1)} + \frac{1}{2\pi}\int_{-\infty}^{\infty}(\zeta_N(-i\lambda))^{-1}\mathcal{E}_{i\lambda}d\lambda. \tag{7.1.27}$$

Recall (3.2.6) and observe that, as $N \to \infty$ in such a way that, from a certain point on, every given prime divides N, the automorphic distribution $\mathfrak{T}_N$ converges towards $\mathfrak{T}_\infty$. Comparing the decompositions into homogeneous components of (7.1.27) and (7.1.8), note that, in the limit, the constant term on the last line of (7.1.27) disappears. On the other hand, a series of residues at the non-trivial zeros of zeta makes its appearance: $\zeta(-i\lambda)$ is of course not the limit of $\zeta_N(-i\lambda)$. What one can do, however, is moving first $i\lambda$ to $1 + \varepsilon + i\lambda$, which is immediate since $(\zeta_N(-i\lambda))^{-1}$ is a polynomial: then, one can let N go to ∞ in the required way, so that $(\zeta_N(1 + \varepsilon - i\lambda))^{-1} \to (\zeta(1 + \varepsilon - i\lambda))^{-1}$; applying the residue theorem again to the formula obtained, one can verify (7.1.8).

In the same way, the automorphic object $\mathfrak{M}_\infty$ can be approached by the sequence $(\mathfrak{M}_N)$ defined by the equation

$$< \mathfrak{M}_N, h >= \sum_{\substack{|j|+|k|\neq 0 \\ (j,k,N)=1}} \int_{-\infty}^{\infty} h(tj, tk)e^{4i\pi t}dt, \tag{7.1.28}$$

assuming again that N goes to ∞ in such a way that, from a certain point on, any given prime divides N. This will be apparent from the decomposition of $\mathfrak{M}_N$ into homogeneous components: note the fact that the approximation $\mathfrak{M}_N$ of $\mathfrak{M}_\infty$ has "more" terms than its limit. Consider for $N \geq 1$ the distribution

$$\mathfrak{D}^{(N)} = 2\pi \sum_{\substack{|j|+|k|\neq 0 \\ (j,k,N)=1}} \delta(x-j)\delta(\xi-k). \tag{7.1.29}$$

Again, $\mathfrak{M}_N$ and $\mathfrak{D}^{(N)}$ depend only on the set of prime factors of N, which we may thus assume to be squarefree. One has, for every such N,

$$\mathfrak{D}(x,\xi) = 2\pi \sum_{\substack{1 \le d \mid N \\ (j,k,\frac{N}{d})=1}} \sum_{|j|+|k| \ne 0} \delta(x-dj)\delta(\xi-dk)$$

$$= 2\pi \sum_{\substack{1 \le r \mid N \\ (j,k,r)=1}} \sum_{|j|+|k| \ne 0} \delta\left(x-\frac{N}{r}j\right)\delta\left(\xi-\frac{N}{r}k\right)$$

$$= 2\pi \sum_{\substack{1 \le r \mid N \\ (j,k,r)=1}} \sum_{|j|+|k| \ne 0} \left(\frac{N}{r}\right)^{-1-2i\pi\mathcal{E}} [\delta(x-j)\delta(\xi-k)], \qquad (7.1.30)$$

in other words

$$\mathfrak{D} = N^{-1-2i\pi\mathcal{E}} \sum_{1 \le r \mid N} r^{1+2i\pi\mathcal{E}} \mathfrak{D}^{(r)} : \qquad (7.1.31)$$

it follows, by induction on the number of prime factors of N, that

$$\mathfrak{D}^{(N)} = (\zeta_N(1+2i\pi\mathcal{E}))^{-1} \mathfrak{D}. \qquad (7.1.32)$$

With the same proof as that between (7.1.15) and (7.1.19), one obtains

$$\frac{1}{2\pi^{\frac{1}{2}}} \cdot (2\pi)^{2i\pi\mathcal{E}} \frac{\Gamma(-i\pi\mathcal{E})}{\Gamma(\frac{1}{2}+i\pi\mathcal{E})} \mathfrak{D}^{(N)} = \mathfrak{M}_N. \qquad (7.1.33)$$

From (7.1.32) and (7.1.4), one has if $0 < \varepsilon < 1$ the equation

$$\mathfrak{D}^{(N)} = \frac{2\pi}{\zeta_N(2)} + \frac{1}{i} \int_{\operatorname{Re}\mu=-\varepsilon} (\zeta_N(1-\mu))^{-1} \mathcal{E}_\mu d\mu : \qquad (7.1.34)$$

applying (7.1.33), one obtains the decomposition

$$\mathfrak{M}_N = -\frac{4\pi^2}{\zeta_N(2)} + \frac{1}{2\pi^{\frac{1}{2}}i} \int_{\operatorname{Re}\mu=-\varepsilon} (2\pi)^{-\mu} \frac{\Gamma(\frac{\mu}{2})}{\Gamma(\frac{1-\mu}{2})} (\zeta_N(1-\mu))^{-1} \mathcal{E}_\mu d\mu. \qquad (7.1.35)$$

In the limit as $N \to \infty$ while "absorbing" all primes, one can move the contour of integration to the line $\operatorname{Re}\mu = 0$, provided that, just as in Theorem 7.1.1, we regard the resulting identity as relating two linear forms on $(2i\pi\mathcal{E})\mathcal{S}(\mathbb{R}^2)$ rather than two genuine distributions. Using the functional equation of the zeta function, one obtains again the spectral decomposition (7.1.20) of $\mathfrak{M}_\infty$.

Section 7.2 will be devoted to the study of the operator the Weyl symbol of which is a seemingly harmless (this is actually not the case) modification of the distribution $\mathfrak{T}_N$. Meanwhile, let us make the structure of the operator $\operatorname{Op}(\mathfrak{M}_\infty)$ clear, as is possible in a usual distribution setting. We wish to take advantage of the spectral decomposition (7.1.12) by testing it on a Wigner function $W(v,u)$: the difficulty is that it is only applicable if $W(v,u)$ lies in the image of $\mathcal{S}(\mathbb{R}^2)$ under $2i\pi\mathcal{E}$.

Lemma 7.1.2. *If u, v lie in $S_{odd}(\mathbb{R})$, the Wigner function $W(v, u)$ lies in the image of $S(\mathbb{R}^2)$ under the operator $2i\pi\mathcal{E}$. If $u \neq 0$ lies in $S_{even}(\mathbb{R})$, the Wigner function $W(u, u)$ can never lie in the image of $S(\mathbb{R}^2)$ under $2i\pi\mathcal{E}$.*

Proof. With $h = W(v, u)$, where u and v are assumed to be of the same parity, the condition that $W(v, u)$ lies in the image of $S(\mathbb{R}^2)$ under the operator $2i\pi\mathcal{E}$ expresses itself, in terms of the spectral decomposition (1.2.11) $h = \int_{-\infty}^{\infty} h_{i\lambda} d\lambda$ of the (even) function h, as the condition $h_0 = 0$. Indeed, this condition is the identity $\int_0^{\infty} h(tx, t\xi) dt = 0$: it is clearly necessary and, conversely, assuming it to hold makes it possible to define

$$\left[\left(x\frac{\partial}{\partial x} + \xi\frac{\partial}{\partial \xi} + 1 \right)^{-1} h \right] (x, \xi) = \int_0^1 h(tx, t\xi) dt = -\int_1^{\infty} h(tx, t\xi) dt, \quad (7.1.36)$$

a function in $S(\mathbb{R}^2)$. Starting from

$$h(x, \xi) = 2 \int_{-\infty}^{\infty} \bar{v}(x + r)u(x - r)e^{4i\pi r\xi} dr \quad (7.1.37)$$

and using the fact that h is even, one writes, for $s \neq 0$,

$$h_0^b(s) = h_0(s, 1) = \int_{-\infty}^{\infty} dt \int_{-\infty}^{\infty} \bar{v}(ts + r)u(ts - r)e^{4i\pi tr\xi} dr$$

$$= \frac{1}{2|s|} \int_{\mathbb{R}^2} \bar{v}(y)u(x)e^{\frac{i\pi(y^2 - x^2)}{s}} dx dy = 0. \quad (7.1.38)$$

The lemma follows.   □

To make the most of (7.1.12), it is useful to introduce a large parameter q and consider the automorphic object $q^{2i\pi\mathcal{E}}\mathfrak{M}_{\infty}$.

Theorem 7.1.3. *Set*

$$\psi_{j,k}^{(q)}(x) = \begin{cases} |j|^{-\frac{1}{2}} e^{\frac{i\pi k}{j} x^2} e^{\frac{2i\pi q}{j} x} & \text{if } j \neq 0, \\ \delta\left(x + \frac{q}{k}\right) & \text{if } j = 0, k = \pm 1. \end{cases} \quad (7.1.39)$$

Given $u, v \in S_{odd}(\mathbb{R})$, one has

$$(v|\mathrm{Op}(\mathcal{G}\left(q^{2i\pi\mathcal{E}}\mathfrak{M}_{\infty}\right))u) = \sum_{(j,k)=1} \left(v|\psi_{j,k}^{(q)} \right) \left(\psi_{j,k}^{(q)}|u \right)$$

$$= -24q(v|\check{u}) + \frac{1}{2i\pi} \int_{\mathrm{Re}\,\mu=0} q^{-\mu}(\zeta(\mu))^{-1}(v|\mathrm{Op}(\mathcal{E}_{-\mu})u) d\mu. \quad (7.1.40)$$

Proof. In view of (7.1.21), one has

$$\mathcal{G}\left(q^{2i\pi\mathcal{E}}\mathfrak{M}_{\infty}\right) = -12q\delta + \frac{1}{2i\pi} \int_{\mathrm{Re}\,\mu=0} q^{-\mu}(\zeta(\mu))^{-1}\mathcal{E}_{-\mu} d\mu. \quad (7.1.41)$$

For every function $h \in (2i\pi\mathcal{E})\mathcal{S}(\mathbb{R}^2)$, one has

$$< q^{2i\pi\mathcal{E}}\mathfrak{M}_\infty, h >= q^{-1} \sum_{(j,k)=1} h\left(\frac{tj}{q},\frac{tk}{q}\right) e^{4i\pi t} dt \qquad (7.1.42)$$

so that, for every pair $u,v \in \mathcal{S}_{\mathrm{odd}}(\mathbb{R})$,

$$\left(v|\mathrm{Op}\left(q^{2i\pi\mathcal{E}}\mathfrak{M}_\infty\right)u\right) =< q^{2i\pi\mathcal{E}}\mathfrak{M}_\infty, W(v,u) > \qquad (7.1.43)$$

$$= q^{-1} \sum_{(j,k)=1} \int_{-\infty}^{\infty} e^{4i\pi t} dt. 2 \int_{-\infty}^{\infty} \bar{v}\left(\frac{tj}{q}+s\right) u\left(\frac{tj}{q}-s\right) \exp\left(4i\pi\frac{tk}{q}s\right) ds.$$

Making when $j \neq 0$ the change of variables $(s,t) \mapsto (x,y) = \left(\frac{jt}{q}+s, \frac{jt}{q}-s\right)$, one can write the generic term as

$$|j|^{-1} \int_{-\infty}^{\infty}\int_{-\infty}^{\infty} e^{\frac{2i\pi q}{j}(x+y)} e^{\frac{i\pi k}{j}(x^2-y^2)} \bar{v}(x)u(y) dx dy, \qquad (7.1.44)$$

and one arrives at the equation

$$\left(v|\mathrm{Op}(\mathcal{G}\left(q^{2i\pi\mathcal{E}}\mathfrak{M}_\infty\right))u\right) = \left(v|\mathrm{Op}(q^{2i\pi\mathcal{E}}\mathfrak{M}_\infty)\check{u}\right) = \sum_{(j,k)=1}\left(v|\psi_{j,k}^{(q)}\right)\left(\psi_{j,k}^{(q)}|u\right). \qquad (7.1.45)$$

$\square$

Remark 7.1.b. In particular,

$$\sum_{(j,k)=1} \left|\left(\psi_{j,k}^{(q)}|u\right)\right|^2 = 24q\|u\|^2 + \frac{1}{2i\pi}\int_{\mathrm{Re}\,\mu=0} q^{-\mu}(\zeta(\mu))^{-1}(u|\mathrm{Op}(\mathcal{E}_{-\mu})u) d\mu.$$

$$(7.1.46)$$

The possibility to move the line of integration on the right-hand side of (7.1.40) to any line $\mathrm{Re}\,\mu = \frac{1}{2} - \varepsilon$ is linked to the Riemann hypothesis: such a possibility would indeed lead to an estimate of the integral term as a $O(q^{-\frac{1}{2}+\varepsilon})$, while a weaker converse can be stated too.

The more pleasant features of the identity (7.1.46) are the fact that the left-hand side is a sum of squares, and the fact, of course linked to the use of pseudo-differential analysis in its derivation, that it connects one-dimensional and two-dimensional analyses. One may transform the integral by moving the contour to the line $\mathrm{Re}\,\mu = a, a \to -\infty$. Since $\zeta(-1) = -\frac{1}{12}$ and the residue of $\mathcal{E}_{-\mu}$ at $\mu = -1$ is $-\delta$ (3.2.6), one sees that, when u is an odd function, the first term on the right-hand side of (7.1.40) will disappear when $a < -1$. Then, for $n = 0,1,\ldots$, the residue of $\frac{1}{\zeta(\mu)}$ at $\mu = -2n-2$ is

$$\frac{(-1)^{n+1}}{(n+1)!}\frac{2\pi^{\frac{5}{2}+2n}}{\Gamma(\frac{3}{2}+n)\zeta(3+2n)}. \qquad (7.1.47)$$

When $u \in \mathcal{S}_{\mathrm{odd}}(\mathbb{R})$, one thus has the identity

$$\sum_{(j,k)=1} \left| \left(\psi^{(q)}_{j,k} | u \right) \right|^2 = 2 \sum_{n \geq 0} \frac{(-q^2)^{n+1}}{(n+1)!} \, \frac{\pi^{\frac{5}{2}+2n}}{\Gamma(\frac{3}{2}+n)\zeta(3+2n)} \left(u | \mathrm{Op}(\mathfrak{E}_{2n+2}) u \right). \quad (7.1.48)$$

This generating series provides some understanding of the odd-odd parts of the operators with symbols $\mathfrak{E}_2, \mathfrak{E}_4, \ldots$.

A special case of (7.1.46) is obtained when choosing $u = \phi^1_z$, as introduced in Theorem 1.1.1.

Proposition 7.1.4. *For every $q > 0$, and every $z \in \Pi$, one has the identity*

$$\sum_{(j,k)=1} \left| \left(\psi^{(q)}_{j,k} | \phi^1_z \right) \right|^2 = 2^{\frac{3}{2}} \pi q^2 \sum_{(j,k)=1} \left(\frac{\mathrm{Im}\, z}{|j - kz|^2} \right)^{\frac{3}{2}} \exp\left(-2\pi \frac{q^2 \mathrm{Im}\, z}{|j - kz|^2} \right)$$

$$= 24q - \frac{1}{2i\pi} \int_{\mathrm{Re}\, \mu=0} \frac{2^{\frac{1-\mu}{2}} q^{-\mu} \mu}{\zeta(\mu)} E^*_{\frac{1+\mu}{2}}(z) d\mu. \quad (7.1.49)$$

Proof. Computing the right-hand side of (7.1.46) with this choice of u is immediate, in view of (3.2.8). To compute the left-hand side, one may use the equation

$$\left| \left(\psi^{(q)}_{j,k} | \phi^1_z \right) \right|^2 = < W\left(\psi^{(q)}_{j,k}, \psi^{(q)}_{j,k} \right), W(\phi^1_z, \phi^1_z) > \quad (7.1.50)$$

and write when $j \neq 0$ (if one is not afraid of distributions)

$$W\left(\psi^{(q)}_{j,k}, \psi^{(q)}_{j,k} \right)(x, \xi)$$

$$= 2|j|^{-1} \int_{-\infty}^{\infty} e^{-\frac{i\pi k}{j}(x+t)^2} e^{-\frac{2i\pi q}{j}(x+t)} e^{\frac{i\pi k}{j}(x-t)^2} e^{\frac{2i\pi q}{j}(x-t)} e^{4i\pi t\xi} dt$$

$$= |j|^{-1}\delta\left(\xi - \frac{kx+q}{j} \right), \quad (7.1.51)$$

while, if $j = 0$ and $k = \pm 1$,

$$W\left(\psi^{(q)}_{0,k}, \psi^{(q)}_{0,k} \right)(x, \xi) = 2 \int_{-\infty}^{\infty} \delta(x + t + \frac{q}{k})\delta(x - t + \frac{q}{k}) e^{4i\pi t\xi} dt$$

$$= \delta(x + \frac{q}{k}). \quad (7.1.52)$$

On the other hand, recall from (1.1.38) that

$$W(\phi^1_z, \phi^1_z)(x, \xi) = 2 \left[\frac{4\pi}{\mathrm{Im}\, z} |x - z\xi|^2 - 1 \right] \exp\left(-\frac{2\pi}{\mathrm{Im}\, z} |x - z\xi|^2 \right). \quad (7.1.53)$$

The left-hand side of (7.1.46) can thus be written as $\sum_{(j,k)=1} A_{j,k}$ with, for $j \neq 0$,

$$A_{j,k} = \frac{2}{|j|} \int_{-\infty}^{\infty} \left[\frac{4\pi}{\operatorname{Im} z} \left| x - z \frac{kx+q}{j} \right|^2 - 1 \right] \exp \left(-\frac{2\pi}{\operatorname{Im} z} \left| x - z \frac{kx+q}{j} \right|^2 \right) dx.$$

(7.1.54)

Using the identity

$$\left| x - z \frac{kx+q}{j} \right|^2 = \frac{|j - kz|^2}{j^2} (x - x_0)^2 + q^2 \frac{(\operatorname{Im} z)^2}{|j - kz|^2}$$

(7.1.55)

with $x_0 = \frac{q \operatorname{Re}(\bar{z}(j-kz))}{|j-kz|^2}$, and the equations

$$\int_{-\infty}^{\infty} e^{-2\pi a(x-x_0)^2} dx = 2^{-\frac{1}{2}} a^{-\frac{1}{2}}, \quad \int_{-\infty}^{\infty} (x - x_0)^2 e^{-2\pi a(x-x_0)^2} dx = 2^{-\frac{3}{2}} a^{-\frac{3}{2}} \pi^{-1},$$

(7.1.56)

we obtain the result of Proposition 7.1.4 after we have done a similar computation in the case when $j = 0$. □

Remark 7.1.c. One may check the second of the pair of identities in the proposition, expanding on one side the exponential into a series, applying on the other side the transformed version (7.1.48) of (7.1.49) and using the fact that $E^*_{-\frac{1}{2}-k} = E^*_{\frac{3}{2}+k}$. What is more interesting is the identity (7.1.48)

$$\sum_{(j,k)=1} \left| \left(\psi_{j,k}^{(q)} | \phi_z^1 \right) \right|^2 = 2^{\frac{5}{2}} \pi q^2 \sum_{n \geq 0} \frac{(-2\pi q^2)^n}{n!} E_{\frac{3}{2}+n}(z),$$

(7.1.57)

in which the left-hand side may be thought of as linking two families of coherent states of sorts: the usual family (ϕ_z^1) associated to the odd part of the one-dimensional metaplectic representation and one, of an arithmetic character, well adapted to the restriction of the metaplectic representation to $SL(2,\mathbb{Z})$. The formulas

$$e^{i\pi x^2} \psi_{j,k}^{(q)} = \begin{cases} \psi_{j,k+j}^{(q)} & \text{if } j \neq 0, \\ e^{i\pi q^2} \psi_{j,k}^{(q)} & \text{if } j = 0, k = \pm 1 \end{cases}$$

(7.1.58)

and

$$\mathcal{F}\psi_{j,k}^{(q)} = \begin{cases} \exp\left(\frac{i\pi}{4} \operatorname{sign}(\frac{k}{j})\right) e^{-i\pi \frac{q^2}{jk}} \psi_{k,-j}^{(q)} & \text{if } j \neq 0, k \neq 0 \\ \psi_{k,0}^{(q)} & \text{if } j^2 + k^2 = 1, \end{cases}$$

(7.1.59)

show that, for fixed q, the family of functions (or measures) $\left(\psi_{j,k}^{(q)} \right)$ is globally invariant, up to phase factors, under the restriction of the metaplectic representation to $SL(2,\mathbb{Z})$. It follows from the last equations that one has for $\left(\begin{smallmatrix} a & b \\ c & d \end{smallmatrix} \right) \in SL(2,\mathbb{Z})$ the general formula

$$\pm \operatorname{Met} \left(\left(\begin{smallmatrix} a & b \\ c & d \end{smallmatrix} \right) \right) \psi_{j,k}^{(q)} = w \psi_{aj+bk,cj+dk}^{(q)} \quad \text{for some } w \text{ with } |w| = 1.$$

(7.1.60)

This gives another (not really needed !) construction of a certain set of Eisenstein series. But it would probably be possible, and possibly interesting, to obtain some generalization of (7.1.57), replacing in the definition of the functions $\psi_{j,k}^{(q)}$ the field $\mathbb{Q}$ by a real quadratic extension of class number one, finding on the right-hand side an (automorphic) expression not reducing to a sum of Eisenstein series: then, the Roelcke-Selberg expansion of the result might lead to the occurrence of interesting cusp-forms, in Maass' style.

7.2   The automorphic distribution $N^{i\pi\mathcal{E}}\mathfrak{T}_N$ as a symbol

In Theorem 7.1.1, the difference $\mathfrak{T}_\infty - \mathcal{G}\mathfrak{M}_\infty - 12\delta$ was shown to be just a series of Eisenstein distributions $\mathfrak{E}_{-\mu}$ together, possibly, with some of their $\frac{d}{d\mu}$-derivatives if zeta happens to have multiple zeros, taken only for values of μ in the set of non-trivial zeros of zeta. This explains our desire to understand as deeply as is possible the operators the symbols of which are the two main terms of this decomposition. Such an understanding seems to be out of reach so far as the first one is concerned: however, one can "almost" answer the same question, after one has replaced $\mathfrak{T}_\infty$ by its approximation $\mathfrak{T}_N$. This analysis has already been performed elsewhere [63] (with the exception of Theorem 7.2.1): we shall not review this construction in detail at present, but we shall give, in Section 7.4, a fully parallel construction in an adelic setting. It would actually be an easy matter to "translate" either point of view into the other, starting from an identification of the distribution $\mathfrak{d}_\rho^\sharp$ in (7.4.4) with a function on adeles built from the collection, for all prime divisors p of N, of the functions $\mathfrak{t}_p$ on $\mathbb{Q}_p$ introduced in (7.4.11) below.

Let us consider on the real line the measures

$$\mathfrak{d}_{\text{even}}(x) = \sum_{m\in\mathbb{Z}} \chi^{(12)}(m)\delta\left(x - \frac{m}{\sqrt{12}}\right),$$

$$\mathfrak{d}_{\text{odd}}(x) = \sum_{m\in\mathbb{Z}} \chi^{(4)}(m)\delta\left(x - \frac{m}{2}\right), \tag{7.2.1}$$

where $\chi^{(12)}$ and $\chi^{(4)}$ are the unique Dirichlet characters mod 12 and 4 respectively such that $\chi^{(12)}(\pm 5) = -1$ and $\chi^{(4)}(3) = -1$. These two measures, introduced in [63, Sec. 3], are the simplest even and odd discretely supported distributions invariant, up to multiplication by some phase factor (a 24th root of unity in the first case, an 8th root of unity in the second), by any metaplectic transformation lying above $SL(2,\mathbb{Z})$. It was also shown there that, after some natural quadratic transformation, making a function on the half-line from an even or odd function on the line (*cf.* (3.4.6)), followed by a Laplace transformation (in other words a theta-transformation), the measure $\mathfrak{d}_{\text{even}}$ becomes the Dedekind eta function, and the measure $\mathfrak{d}_{\text{odd}}$ becomes the cube of Dedekind's eta function.

Remark 7.2.a. in [63, p. 24], it was pointed out that one has the identity

$$\int_{SL(2,\mathbb{Z})\backslash SL(2,\mathbb{R})} |<\text{Met}(g^{-1})\mathfrak{d}_{\text{odd}}, u>|^2 dg = \frac{2\pi}{3}\|u\|^2 \qquad (7.2.2)$$

for $u \in \mathcal{S}_{\text{odd}}(\mathbb{R})$, and a similar one, valid for even functions, involving transforms, under the metaplectic representation, of the measure $\mathfrak{d}_{\text{even}}$. There is a kind of complementarity between the family of measures $\text{Met}(g^{-1})\mathfrak{d}_{\text{odd}}, g \in SL(2,\mathbb{R})$, a generalization of which will play a role on the way towards the analysis of the distribution $\mathfrak{T}_\infty$ in the identity (7.1.14), and the family (for given q) of functions or measures $\text{Met}(g^{-1})\psi_{1,0}^{(q)}, g \in SL(2,\mathbb{Z})$, involved in the analysis of the other distribution $\mathcal{G}\mathfrak{M}_\infty$ taken from the same identity.

More generally, we defined in *loc.cit.*, for every integer $N = 2\prod_{p\in S} p$, where S is any finite set of primes including 2, a finite-dimensional space of measures, reducing to the one generated by $\mathfrak{d}_{\text{even}}$ (*resp.* $\mathfrak{d}_{\text{odd}}$) when $N = 12$ (*resp.* $N = 4$). The construction is as follows. Let Λ be the set $\{\mu \bmod N: \mu^2 \equiv N \bmod 2N\}$, in other words, with $\alpha_p = 1$ if $p \in S, p \neq 2$ and $\alpha_2 = 2$, the set of $\mu \in (\mathbb{Z}/N\mathbb{Z})^\times$ such that $\mu \equiv \pm 1 \bmod p^{\alpha_p}$ for every $p \in S$: one defines a character $\chi: \Lambda \to \{\pm 1\}$ by setting $\chi(\mu) = -1$ if one has $\mu \equiv 1 \bmod p^{\alpha_p}$ except for an odd number of p's. Then, fixing a set R_N of representatives of $(\mathbb{Z}/N\mathbb{Z})^\times \bmod \Lambda$, one defines, for every $\rho \in R_N$, the measure

$$\mathfrak{d}_\rho(x) = \sum_{\mu\in\Lambda} \chi(\mu) \sum_{\ell\in\mathbb{Z}} \delta\left(x - \frac{N\ell + \rho\mu}{\sqrt{N}}\right). \qquad (7.2.3)$$

Since $\chi(-\mu) = (-1)^{\gamma(N)}\chi(\mu)$ if $\gamma(N)$ is the number of prime factors of N, $\mathfrak{d}_\rho$ has the parity associated to $\gamma(N)$. Note that the measure $\mathfrak{d}_\rho$ was denoted as ϖ_ρ in [63]: besides generalizing (7.2.1), our present notation may recall that we are dealing with discrete measures, or series of Dirac masses. For a fixed N, the space linearly generated by the measures $\mathfrak{d}_\rho$ is again invariant under all metaplectic transformations lying above matrices in $SL(2,\mathbb{Z})$.

Denote as W_N the Weyl symbol of the map $u \mapsto \sum_\rho <\mathfrak{d}_\rho, u>\mathfrak{d}_\rho$ from $\mathcal{S}(\mathbb{R})$ to $\mathcal{S}'(\mathbb{R})$,

$$W_N = \sum_{\rho\in R_N} W(\mathfrak{d}_\rho, \mathfrak{d}_\rho); \qquad (7.2.4)$$

this is a discrete measure on $\mathbb{R}^2$ and, subtracting from it the mass present at the origin, one finds, denoting as S the set of prime divisors of N,

$$W_N^\times := W_N - N^{-\frac{1}{2}}\prod_{p|N}(1-p) \times \delta$$

$$= N^{-\frac{1}{2}} \sum_{|j|+|k|\neq 0} \prod_{\substack{p \text{ prime} \\ p|(j,k,N)}} (1-p) \times \delta\left(x - \frac{j}{\sqrt{N}}\right)\delta\left(\xi - \frac{k}{\sqrt{N}}\right). \qquad (7.2.5)$$

Then, the distribution $N^{-i\pi\mathcal{E}}W_N^\times$, defined as

$$< N^{-i\pi\mathcal{E}}W_N^\times, h >= N^{\frac{1}{2}} < W_N^\times, (x,\xi) \mapsto h(N^{\frac{1}{2}}x, N^{\frac{1}{2}}\xi) >, \qquad (7.2.6)$$

coincides with the distribution $\mathfrak{T}_N$.

We have thus found an interpretation, as a Weyl symbol, of the automorphic distribution $N^{i\pi\mathcal{E}}\mathfrak{T}_N$. Note that a rescaling operation on symbols such as $N^{-i\pi\mathcal{E}}$ does not admit any clear interpretation as an action on the corresponding operators: recall, however, that (2.3.3) did give such an interpretation to the Euler operator $2i\pi\mathcal{E}$. If one couples the operation $N^{-i\pi\mathcal{E}}$ with the operation $\mathfrak{S} \mapsto \mathfrak{S}_1$, with $\mathfrak{S}_1(x,\xi) = \mathfrak{S}(N^{\frac{1}{2}}x, N^{-\frac{1}{2}}\xi)$, which is a symplectic transformation and, as such, admits an operator-interpretation (1.1.20), one is led to the operation $\mathfrak{S} \mapsto \mathfrak{S}^N$, with $\mathfrak{S}^N(x,\xi) = N^{-\frac{1}{2}}\mathfrak{S}(x, N^{-1}\xi)$. This is strictly an operation on symbols, called a change of Planck's constant by physicists (indeed, when x and η, instead of being pure numbers, represent a length and a momentum, it is necessary to divide by a unit of action products such as $x\eta$ occurring as exponents in equations such as (1.1.14)), or by analysts interested in so-called semi-classical partial differential equations: contrary to symplectic changes of coordinates in $\mathbb{R}^2$, it does not correspond to any operator on functions on the real line. Making such a change of Planck's constant, which amounts to changing the rule Op to a certain N-dependent rule, finally gives the distribution $\mathfrak{T}_N$ an interesting meaning, as a finite sum of Wigner functions, in a sense analogous to (1.1.15). The problem is that it is not the distribution $\mathfrak{T}_N$ we are interested in, but its limit $\mathfrak{T}_\infty$: trying to make something from a limit, as $N \to \infty$, of the rescaling operation $N^{-i\pi\mathcal{E}}$, does not lead anywhere, unless drastic changes are made, making it possible to use a collection of p-dependent Planck's constants.

This is the reason for our going adelic in the next few sections. Taking inspiration from the interpretation of the operator with usual Weyl-type symbol $N^{i\pi\mathcal{E}}\mathfrak{T}_N$, we shall suggest a definition of adelic pseudo-differential analysis incorporating the required change of Planck's constant. When this is done, we shall use the obvious identification of a certain class of automorphic distributions, to be called fractional combs, with functions on adeles, to give a meaning to $\mathfrak{T}_\infty$ as the symbol of some operator, in a symbolic calculus of a more arithmetical nature than Weyl's.

There is, however, still one thing one can do with the help of the usual Weyl calculus, taking $\mathfrak{T}_N$ rather than $N^{i\pi\mathcal{E}}\mathfrak{T}_N$ as a symbol: namely, give an interpretation to its Θ-transform (f_0, f_1) as defined in Theorem 1.1.3. To do so, we need to reconsider the harmonic oscillator L as introduced in (1.1.27): recall that, in $L^2(\mathbb{R})$, it has the purely discrete spectrum $\{\frac{1}{2}, \frac{3}{2}, \dots\}$ and that each eigenvalue is simple. For every j, one can choose in a canonical way a normalized eigenstate ϕ^j for the eigenvalue $j + \frac{1}{2}$, using the so-called creation and annihilation operators to be found in most elementary textbooks on quantum mechanics, and the states ϕ^j

are pairwise orthogonal. We computed in (1.2.9) the Weyl symbol of the operator $e^{-i\pi tL}$: using analytic continuation, one obtains the equation

$$e^{-tL} = \mathrm{Op}\left((x,\xi) \mapsto \frac{1}{\cosh\frac{t}{2}}\exp\left(-2\pi(x^2+\xi^2)\tanh\frac{t}{2}\right)\right), \quad t > 0. \quad (7.2.7)$$

Setting $s = e^{-t}$, one sees that the Weyl symbol $W(\phi^j, \phi^j)$ of the rank-one operator $u \mapsto (\phi^j|u)\phi^j$ can be obtained from the generating series

$$\sum_{j\geq 0} s^j W(\phi^j, \phi^j) = \frac{2}{1+s}\exp\left(-2\pi(x^2+\xi^2)\frac{1-s}{1+s}\right). \quad (7.2.8)$$

In particular, it is invariant under the linear action of the subgroup $SO(2)$ of $SL(2,\mathbb{R})$: it then follows from the equation, in which $\tilde{g}$ is one of the two points of the metaplectic group lying above some $g \in SL(2,\mathbb{R})$,

$$W(\mathrm{Met}(\tilde{g})\phi, \mathrm{Met}(\tilde{g})\phi) = W(\phi, \phi) \circ g^{-1} \quad (7.2.9)$$

(a consequence of (1.1.20)) that, up to some phase factor which we do not need to make explicit here, the eigenfunction ϕ^j remains invariant under the metaplectic transformation associated to any point lying above $SO(2)$. As a consequence, if z is any point of the hyperbolic half-plane, choosing some $g = \left(\begin{smallmatrix} a & b \\ c & d \end{smallmatrix}\right) \in SL(2,\mathbb{R})$ such that $z = g.i$, next some $\tilde{g}$ lying above g in the metaplectic group, one can define the sequence $(\phi_z^j)_{j\geq 0}$, with $\phi_z^j = \mathrm{Met}(\tilde{g})\phi^j$. The functions in $S(\mathbb{R})$ so defined make up an orthonormal basis of eigenstates of the oscillator L_z with symbol $(dx - b\xi)^2 + (-cx + a\xi)^2 = \frac{|x-z\xi|^2}{\mathrm{Im}\,z}$, and we already considered the functions ϕ_z^0 and ϕ_z^1 in Theorem 1.1.3.

Theorem 7.2.1. *Given an integer $N \geq 4$ of the type considered in the present section, one has the equation, in which $\varepsilon = 0$ if the number of prime factors of N is even, $\varepsilon = 1$ in the other case,*

$$(-1)^\varepsilon(\phi_z^0|\mathrm{Op}(\mathfrak{T}_N)\phi_z^0) = -2\prod_{p|N}(p-1) \quad (7.2.10)$$

$$+ \frac{2N^{\frac{1}{2}}}{N+1}\sum_{k\geq 0}\left(\frac{N-1}{N+1}\right)^{2k+\varepsilon}\sum_{\rho\in R_N}|(\partial_\rho|\phi_z^{2k+\varepsilon})|^2.$$

On the other hand,

$$(-1)^\varepsilon(\phi_z^1|\mathrm{Op}(\mathfrak{T}_N)\phi_z^1) = 2\prod_{p|N}(p-1) \quad (7.2.11)$$

$$+ \frac{2N^{\frac{1}{2}}}{(N+1)^3}\sum_{k\geq 0}[(N+1)^2 - 4N(2k+1+\varepsilon)]\left(\frac{N-1}{N+1}\right)^{2k-1+\varepsilon}\sum_{\rho\in R_N}|(\partial_\rho|\phi_z^{2k+\varepsilon})|^2.$$

Proof. Given $t > 0$, the $\mathcal{G}$-transform, as defined in (1.1.24), of the symbol on the right-hand side of (7.2.7) is given by the equation

$$\mathcal{G}\left(\frac{1}{\cosh\frac{t}{2}}\exp\left(-2\pi(x^2+\xi^2)\tanh\frac{t}{2}\right)\right) = \frac{1}{\sinh\frac{t}{2}}\exp\left(-2\pi\frac{x^2+\xi^2}{\tanh\frac{t}{2}}\right). \quad (7.2.12)$$

Set $r = \frac{1}{\tanh\frac{t}{2}}$, so that $e^t = \frac{r+1}{r-1}$ and $\sinh\frac{t}{2} = (r^2-1)^{-\frac{1}{2}}$. Recall that applying the operator $\mathcal{G}$ to an even symbol amounts to taking the composition of the associated operator, on the left or on the right, by the symmetry operator ch: $u \mapsto \check{u}$. With the help of (7.2.7), one obtains that, for $r \geq 1$, $\exp\left(-2\pi r(x^2+\xi^2)\right)$ is the symbol of the operator

$$(r^2-1)^{-\frac{1}{2}}\left(\frac{r-1}{r+1}\right)^L \text{ch} = \frac{1}{r+1}\left(\frac{r-1}{r+1}\right)^{L-\frac{1}{2}}\text{ch}. \quad (7.2.13)$$

On the other hand, a rescaled version of the Wigner function $W(\phi_z^1, \phi_z^1)$ as given in (1.1.38), to wit

$$2\left[4\pi r(x^2+\xi^2)-1\right]e^{-2\pi r(x^2+\xi^2)} = \left(-4r\frac{d}{dr}-2\right)e^{-2\pi r(x^2+\xi^2)}, \quad (7.2.14)$$

is the symbol of the operator

$$\left(-4r\frac{d}{dr}-2\right)\left[(r^2-1)^{-\frac{1}{2}}\left(\frac{r-1}{r+1}\right)^L\right]\text{ch}$$

$$= \frac{2}{(r+1)^3}[r^2+1-4rL]\left(\frac{r-1}{r+1}\right)^{L-\frac{3}{2}}\text{ch}. \quad (7.2.15)$$

Changing r to N and composing the symbols with the transformation g^{-1}, one obtains the equations

$$\text{Op}\left(\exp\left(-2\pi N\frac{|x-z\xi|^2}{\text{Im}\,z}\right)\right)u = \frac{1}{N+1}\left(\frac{N-1}{N+1}\right)^{L_z-\frac{1}{2}}\check{u}$$

$$= \frac{1}{N+1}\sum_{j\geq0}(-1)^j\left(\frac{N-1}{N+1}\right)^j(\phi_z^j|u)\phi_z^j \quad (7.2.16)$$

and

$$\text{Op}\left(2\left[4\pi N\frac{|x-z\xi|^2}{\text{Im}\,z}-1\right]\exp\left(-2\pi N\frac{|x-z\xi|^2}{\text{Im}\,z}\right)\right)u$$

$$= \frac{2}{(N+1)^3}[N^2+1-4NL_z]\left(\frac{N-1}{N+1}\right)^{L_z-\frac{1}{2}}\check{u}. \quad (7.2.17)$$

Then, we write

$$(\phi_z^0|\text{Op}(\mathfrak{T}_N)\phi_z^0) = <\mathfrak{T}_N, W(\phi_z^0, \phi_z^0) >$$
$$= < N^{i\pi\mathcal{E}}\mathfrak{T}_N, N^{i\pi\mathcal{E}}W(\phi_z^0, \phi_z^0) >$$
$$= < N^{i\pi\mathcal{E}}\mathfrak{T}_N, (x, \xi) \mapsto 2N^{\frac{1}{2}} \exp\left(-2\pi N \frac{|x - z\xi|^2}{\text{Im } z}\right) > . \quad (7.2.18)$$

On the other hand, from (7.2.5) and (7.2.6), one obtains

$$N^{i\pi\mathcal{E}}\mathfrak{T}_N = W_N^{\times} = W_N - N^{-\frac{1}{2}} \prod_{p|N}(1 - p) \times \delta. \quad (7.2.19)$$

From (7.2.18), (7.2.4) and (7.2.16), one obtains

$$(\phi_z^0|\text{Op}(\mathfrak{T}_N)\phi_z^0) = -2 \prod_{p|N}(1 - p)$$
$$+ \frac{2N^{\frac{1}{2}}}{N+1} \sum_{j\geq 0}(-1)^j \left(\frac{N-1}{N+1}\right)^j \sum_{\rho\in R_N} |(\mathfrak{d}_\rho|\phi_z^j)|^2. \quad (7.2.20)$$

Equation (7.2.10) follows, since the distributions $\mathfrak{d}_\rho$ have the parity defined by the number of prime factors of N; equation (7.2.11) is obtained in the same way. $\square$

Remarks 7.2.b. (i) The scalar products $(\mathfrak{d}_\rho|\phi_z^j)$ which occur here are linear combinations of theta functions of sorts: for instance,

$$(\mathfrak{d}_\rho|\phi_{-\frac{1}{z}}^0) = (2\text{Im } z)^{\frac{1}{4}} \sum_{\substack{\mu\in\mathbb{Z}/N\mathbb{Z} \\ \mu^2\equiv 1}} \chi(\mu) \sum_{\ell\in\mathbb{Z}} \exp\left(-i\pi\overline{z}\frac{(N\ell + \rho\mu)^2}{N}\right). \quad (7.2.21)$$

(ii) The limit of (7.2.10), as $N \to \infty$ while "absorbing" all primes, has the expression

$$< \mathfrak{T}_\infty, W(\phi_z^0, \phi_z^0) >= 2 \sum_{\substack{|j|+|k|\neq 0}} \prod_{p|(j,k)}(1 - p) . \exp\left(-2\pi\frac{|j - kz|^2}{\text{Im } z}\right). \quad (7.2.22)$$

It is tempting to believe, but we do not know how to do it, that there must exist another expression of this limit, with a structure as rich as the one that led to (7.2.10). Note that all terms on the right-hand side of this latter equation, apart from the first one, are positive: life would be too simple if the sum of these terms, or the first one, had a finite limit as $N \to \infty$. Something similar goes with the difference $(-1)^\varepsilon[(\phi_z^0|\text{Op}(\mathfrak{T}_N)\phi_z^0) - \frac{N-1}{N+1}(\phi_z^1|\text{Op}(\mathfrak{T}_N)\phi_z^1)]$.

(iii) Even though there is no general transformation linking two operators the Weyl symbols of which are related by a rescaling of coordinates (a change of Planck's constant), equation (7.2.16) made it possible to understand fully the operator the symbol of which is an arbitrarily rescaled version of the "basic symbol" $2 \exp\left(-2\pi(x^2 + \xi^2)\right)$. In analysis over the field $\mathbb{Q}_p$, there is also an obvious notion of "basic symbol", to wit the characteristic function of $\mathbb{Z}_p \times \mathbb{Z}_p$. In this case, there is also a family of natural rescaling operations, the rescaling factors being just powers of p (or, which amounts to the same, general ideles after $\mathbb{Q}_p$ has been imbedded into the ring of adeles). We shall see, in Section 7.4, that one can indeed understand fully the operators the symbols of which (in the symbolic calculus to be introduced there) are rescaled versions of the basic symbol.

(iv) If one could deal with powers of $p^{\frac{1}{2}}$ in place of powers of p, there would be no need to "change the rule", which is what is really being done by the change of Planck's constant to be introduced in Section 7.4. A deeper possibility *might be* to replace adeles of the field $\mathbb{Q}$ by a restricted direct product of local fields at points $\sqrt{p}$ in the corresponding sequence of quadratic extensions: we do not know how to deal with it at present.

7.3 Adeles and ideles

The present short section addresses itself primarily to analysts — number theorists know much more about adeles than the present author — and is devoted mainly to definitions, and to a few easy calculations for practice. Also, we wish to implement on one example the principle, well-experimented in a much deeper way than in the present context, that arithmetic facts can often be presented in two equivalent versions, one stated in adelic terms, and the other in a classical distribution setting. More serious matters will be dealt with in the last section of this book.

Let us remind our fellow analysts that adeles first found a basic role in number theory with Tate's thesis, which changed the approach to the zeta function, or to Hecke's L-function (in algebraic number fields), in a drastic way, by showing it — this is especially clear in [29, p. 127] — to be just a factor of proportionality between two "naturally" defined functions or distributions depending on a complex parameter (the unramified part of some quasi-character): see [52, 33]. We shall try later (Proposition 7.3.2) a 2-dimensional approach, as an illustration of the above-stated "principle".

This section is just a subminimal introduction to analysis in an adelic setting, and we consider only the field of rationals and its various completions: also, we dispense with ramified quasi-characters entirely, even though taking them into account would probably not be, up to some point, much more complicated. We do not assume any knowledge of adeles and ideles, beyond the basic definitions, starting with the notion of restricted direct product: besides, of course, Tate's foundational paper [52], our favourite references on this subject are [33, Chap. 7 and 14], [29] and [4, Sec. 3.1]; so far as the metaplectic representation is concerned ([66], [4, Sec. 4.8]), only the one-dimensional case is needed here.

Following usual notation, with the simplification that we are only dealing, here, with the field of rational numbers, we denote for every prime p as $\mathbb{Q}_p$ (*resp.* $\mathbb{Z}_p$) the field of p–adic numbers (*resp.* the compact ring of p–adic integers), setting also $\mathbb{Q}_\infty = \mathbb{R}$. The multiplicative group of non-zero elements of $\mathbb{Q}_v$, where v is any *place* (i.e., a prime number p or ∞), is denoted as $\mathbb{Q}_v^\times$: for every prime p, let $\mathbb{Z}_p^\times$ be the (compact) group of units (i.e., invertible elements) of the ring $\mathbb{Z}_p$. The restricted direct product of the fields $\mathbb{Q}_v$ with respect to the family of subrings $\mathbb{Z}_p$ (undefined for the infinite place) is the ring of adeles $\mathbb{A}$. The restricted direct product of the groups $\mathbb{Q}_v^\times$ with respect to the family of subgroups $\mathbb{Z}_p^\times$ is the group $\mathbb{A}^\times$ of ideles. Recall [33, p. 139] that, under the diagonal embedding, the additive group $\mathbb{Q}$ imbeds as a discrete subgroup of $\mathbb{A}$, and the multiplicative group $\mathbb{Q}^\times$ imbeds as a discrete subgroup of $\mathbb{A}^\times$.

The absolute value $||_p$ on $\mathbb{Q}_p$ is the usual one, such that $|p|_p = p^{-1}$. The so–called self–dual measure dx on the locally compact additive group $\mathbb{Q}_p$ is the Haar measure which gives $\mathbb{Z}_p$ the measure 1 [33, p. 277]. On $\mathbb{Q}_p^\times$, the appropriate Haar measure $d^\times x$ is that [33, p. 279-280] or [52, p. 319] which gives the set of invertible elements of $\mathbb{Z}_p$ the measure 1: in other words $d^\times x = \frac{p}{p-1}\frac{dx}{|x|_p}$. On $\mathbb{Q}_\infty = \mathbb{R}$ (*resp.* $\mathbb{Q}_\infty^\times = \mathbb{R}^\times$), the measures are the usual Lebesgue measure dx and the measure $\frac{dx}{|x|}$. We drop the subscript p or ∞ when the context makes it clear that we are dealing with a fixed place.

The Schwartz–Bruhat space of complex-valued functions on $\mathbb{Q}_p$ is defined in the usual way: a function lies in $\mathcal{S}(\mathbb{Q}_p)$ if it has compact support and it is invariant under translations by vectors in $p^j\mathbb{Z}_p$ for sufficiently large j; on $\mathbb{Q}_\infty$, the usual Schwartz definition of $\mathcal{S}(\mathbb{R})$ applies. On the ring of adeles, one uses the spaces as defined in [66, p. 157-158; 177-178]: a dense subspace $\mathcal{S}_0(\mathbb{A})$ of $\mathcal{S}(\mathbb{A})$ is then made up of the linear combinations of tensor products of Schwartz–Bruhat functions, most (i.e., all but a finite number) of which coincide, for finite p, with the characteristic function of $\mathbb{Z}_p$. Similar definitions apply in the case of $\mathbb{A} \times \mathbb{A}$ in place of $\mathbb{A}$.

Given $x \in \mathbb{Q}_p$, there is a rational number α such that $p^j\alpha \in \mathbb{Z}$ for large j and such that $x + \alpha \in \mathbb{Z}_p$ (note the $+$ sign): we denote as $\kappa(x)$ (or $\kappa_p(x)$ when p is allowed to vary) the uniquely defined class of $\alpha \bmod \mathbb{Z}$: it makes then sense to define the additive character $x \mapsto e^{2i\pi\kappa(x)}$ of $\mathbb{Q}_p$, a notation which takes us close to the familiar Archimedean case. If $x = (x_v)$ is an adele (the *place* v is any prime or ∞), setting also $\kappa_\infty(x_\infty) = x_\infty$, we define $e^{2i\pi\kappa(x)} = \prod_v e^{2i\pi\kappa_v(x_v)}$: recall (this is where the notion of *restricted* product applies) that $x_p \in \mathbb{Z}_p$ for almost every p, so that this is really a finite product. One has $e^{2i\pi\kappa(x)} = 1$ whenever x lies in the image in $\mathbb{A}$ of a rational number under the diagonal embedding: $\mathbb{Q} \to \mathbb{A}$. On the other hand, note that the rational number mod $\mathbb{Z}$ denoted as $\kappa(x)$ is the negative of that denoted as $\lambda(x)$ in [33, p. 276].

Fixing again a prime number p, let us briefly consider the one-dimensional Heisenberg and metaplectic representations in the p-adic environment. Again, we shall satisfy ourselves with defining these as projective representations. The first one is defined with the help of the equation

$$(\tau_{y,\eta}u)(x) = u(x-y)e^{2i\pi\kappa((x-\frac{y}{2})\eta)}. \tag{7.3.1}$$

Just as in the Archimedean case, replacing the domain $\mathbb{Q}_p \times \mathbb{Q}_p$ of (y,η) by the set-theoretic product $\mathbb{Q}_p \times \mathbb{Q}_p \times S^1$, where $S^1 = \{z \in \mathbb{C} : |z| = 1\}$, and providing the result with the appropriate group structure, called the Heisenberg group, makes it possible to regard the map $(y,\eta) \mapsto \tau_{y,\eta}$ as a restriction to the set $\mathbb{Q}_p \times \mathbb{Q}_p \times \{1\}$ of a representation, the factor in S^1 operating by scalar multiplications: the existence and uniqueness of the group structure and Heisenberg representation are immediate [4, p. 525].

The group $SL(2,\mathbb{Q}_p)$, denoted as G_p, can still be characterized as the group of linear transformations of $\mathbb{Q}_p^2$ which preserve the ($\mathbb{Q}_p$-valued) symplectic form defined just as in (1.1.8). Equation (1.1.7) continues to be valid, only replacing the symplectic form in the exponent by its image under κ. The metaplectic (projective) representation Met_p or Met of G_p in the space $L^2(\mathbb{Q}_p)$ is characterized up to multiplication by some *phase factor*: $G_p \to S^1$ by the equation (1.1.19), after we have replaced there the expression $\exp(2i\pi(< \eta, Q > - < y, P >))$, no longer meaningful, by $\tau_{y,\eta}$. More precisely, for $g \in G_p$, the class (mod S^1) of a unitary transformation U associated to g is characterized by the validity of the equation $U\tau_{y,\eta}U^{-1} = \tau_{y',\eta'}$, with $\begin{pmatrix} y' \\ \eta' \end{pmatrix} = g.\begin{pmatrix} y \\ \eta \end{pmatrix}$ for every pair $(y,\eta) \in \mathbb{Q}_p \times \mathbb{Q}_p$. This is to be found in [4, p. 531], together with explicit formulas for the metaplectic representation on generators (just as done, in the Archimedean case, after (1.1.20)). The transformation $u \mapsto v$ corresponding to the matrix $\begin{pmatrix} 1 & 0 \\ c & 1 \end{pmatrix}, \begin{pmatrix} a & 0 \\ 0 & a^{-1} \end{pmatrix}$ with $a \in \mathbb{Q}_p^\times$, or $\begin{pmatrix} 0 & 1 \\ -1 & 0 \end{pmatrix}$ is defined by the equation $v(x) = e^{2i\pi\kappa(\frac{cx^2}{2})}u(x)$, or $v(x) = |a|^{-\frac{1}{2}}u(a^{-1}x)$, or $v = \mathcal{F}u$, where $|a| = |a|_p$ and the Fourier transformation is normalized as

$$(\mathcal{F}u)(y) = \int_{\mathbb{Q}_p} u(x)e^{-2i\pi\kappa(xy)}dy. \tag{7.3.2}$$

There is some minor discrepancy with the formulas in [4, p. 531], which can be taken care of if one takes the conjugate of $g \in G_p$ under the matrix $\begin{pmatrix} 0 & 1 \\ 1 & 0 \end{pmatrix}$: this matrix does not lie in G_p, but Weil's results on the metaplectic representation resist what amounts in effect to a change of sign of Planck's constant. As shown in [66, p. 196], this representation can still be defined as a genuine one of the twofold cover of G_p, but we shall have no need for this here.

Pseudo-differential analysis in a p-adic setting has been studied by Haran [14] and Bechata [1]. The methods of the second reference, which provides a p-adic extension of some (Archimedian) previous results of the present author, in particular regarding the composition formula (Theorem 1.2.2 only: there is no analogue of Moyal's formula in this context), rely to a great extent on the notion, again, of Wigner function; this will play a decisive role here too. With our present notation, the basic formula would be

$$(\mathrm{Op}(h)u)(x) = \int_{\mathbb{Q}_p \times \mathbb{Q}_p} h\left(\frac{x+y}{2}, \eta\right) e^{2i\pi\kappa((x-y)\eta)} u(y) dy d\eta, \qquad (7.3.3)$$

the obvious analogue of (1.1.14). The linear map Op establishes an isometry from the space $L^2(\mathbb{Q}_p^2)$ to the space of Hilbert-Schmidt operators on $L^2(\mathbb{Q}_p)$, and the two covariance formulas (1.1.18) and (1.1.20) extend. However, we shall be led, in Section 7.4, to changing this rule by a p-dependent change of Planck's constant.

Just as in Archimedean analysis (*cf.* Section 1.2), one can decompose complex-valued functions on $\mathbb{Q}_p^2$ into "homogeneous" components. In general, this is slightly more complicated, since ramified characters, i.e., characters the restriction of which to $\mathbb{Z}_p^\times$ is non-trivial, is necessary. However, as this is just a first approach, we shall consider only unramified characters. Recall that for every even function $h \in \mathcal{S}(\mathbb{R}^2)$, one has $h = \int_{-\infty}^{\infty} h_{i\lambda} d\lambda$, with

$$h_{i\lambda}(x, \xi) = \frac{1}{2\pi} \int_0^\infty t^{1+i\lambda} h(tx, t\xi) \frac{dt}{t}, \qquad (x, \xi) \in \mathbb{R}^2 \backslash \{0\}. \qquad (7.3.4)$$

In the p-adic case, we set, provided that convergence is ensured,

$$h_{i\lambda}(x, \xi) = \int_{\mathbb{Q}_p^\times} |t|^{1+i\lambda} h(tx, t\xi) d^\times t \qquad (7.3.5)$$

with $|t| = |t|_p$ and $d^\times t = \frac{p}{p-1} \frac{dt}{|t|}$. Also, more generally, we set

$$h_\nu(x, \xi) = \int_{\mathbb{Q}_p^\times} |t|^{1+\nu} h(tx, t\xi) d^\times t. \qquad (7.3.6)$$

Then, the map $\lambda \mapsto h_{\nu+i\lambda}$ is periodic of period $\frac{2\pi}{\log p}$.

We now verify that the integral (7.3.6) converges for $(x, \xi) \neq (0, 0)$ if h has compact support and $\mathrm{Re}\,\nu > -1$. Indeed, in that case, the function $t \mapsto h(tx, t\xi)$ is also compactly supported, so that $0 < |t| \leq p^k$ for some $k = 0, 1, \ldots$ when $h(tx, t\xi) \neq 0$. In the case when $\mathrm{Re}\,\nu \geq 0$, one has

$$\int_{0<|t|\leq p^k} |t|^{\mathrm{Re}\,\nu} dt = p^k \int_{0<|t|\leq 1} |p^{-k}t|^{\mathrm{Re}\,\nu} dt \leq p^{k(1-\mathrm{Re}\,\nu)}. \qquad (7.3.7)$$

When $\mathrm{Re}\,\nu < 0$, one has on one hand

$$\int_{1\leq|t|\leq p^k} |t|^{\mathrm{Re}\,\nu} dt \leq \int_{|t|\leq p^k} dt = p^k; \qquad (7.3.8)$$

on the other hand,

$$\int_{0<|t|<1} |t|^{\operatorname{Re}\nu}\,dt = \sum_{j\geq 0}\int_{|t|=p^{-j-1}} |t|^{\operatorname{Re}\nu}\,dt$$

$$\leq \sum_{j\geq 0} p^{-(j+1)\operatorname{Re}\nu}\int_{|t|\leq p^{-j-1}} dt$$

$$= \sum_{j\geq 0} p^{-(j+1)(\operatorname{Re}\nu+1)}, \tag{7.3.9}$$

a series convergent when $\operatorname{Re}\nu > -1$.

Next, we note that, provided that convergence is ensured, one has $h_\nu(\rho x, \rho\xi)$ $= h_\nu(x,\xi)$ whenever $(x,\xi)\neq(0,0)$ and $\rho\in\mathbb{Z}_p^\times$. For the time being, we shall only consider the decomposition into homogeneous components of functions h satisfying the condition

$$h(\rho x, \rho\xi) = h(x,\xi), \quad (x,\xi)\in\mathbb{Q}_p^2, \rho\in\mathbb{Z}_p^\times, \tag{7.3.10}$$

and lying in $\mathcal{S}(\mathbb{Q}_p^2)$: we shall denote as $\mathcal{S}_{\mathrm{inv}}(\mathbb{Q}_p^2)$ the linear space of such functions. The consideration of such symbols should suffice for the possible application we have in mind concerning the zeta function, not for the corresponding one dealing with Dirichlet L-functions.

We now examine the way the function $h\in\mathcal{S}_{\mathrm{inv}}(\mathbb{Q}_p^2)$ can be recovered from the family of functions $h_{i\lambda}, \lambda\in\mathbb{R}$, under the initial extra assumption that $\max(|x|,|\xi|)$ is bounded away from zero on the support of h. If such is the case, there exists $k = 0,1,\ldots$ such that, for every $(x,\xi)\neq(0,0)$, one has

$$h_{i\lambda}(x,\xi) = \frac{p}{p-1}\sum_{j=-k}^{k} a_j p^{ij\lambda} \quad \text{with } a_j = \int_{|t|=p^j} h(tx,t\xi)dt. \tag{7.3.11}$$

Now, one has, for $A > 0$,

$$\frac{1}{2A}\int_{-A}^{A} p^{ij\lambda}d\lambda = \begin{cases} 1 & \text{if } j = 0 \\ \dfrac{1}{A}\dfrac{\sin(jA\log p)}{j\log p} & \text{if } j\neq 0: \end{cases} \tag{7.3.12}$$

when $A = \frac{\pi}{\log p}$, the integral is zero unless $j = 0$. Hence,

$$\log p\int_{-\frac{\pi}{\log p}}^{\frac{\pi}{\log p}} h_{i\lambda}(x,\xi)d\lambda = \frac{p}{p-1}\int_{|t|=1} h(tx,t\xi)dt$$

$$= \frac{p}{p-1}h(x,\xi)\times\int_{\mathbb{Z}_p^\times} dt = h(x,\xi): \tag{7.3.13}$$

we have used, of course, the fact that $h(tx,t\xi) = h(x,\xi)$ whenever $t\in\mathbb{Z}_p^\times$.

As has been done in the archimedean case, we set

$$h_\nu^\flat(s) = h_\nu(s,1) = \int_{\mathbb{Q}_p^\times} |t|^{1+\nu} h(ts,t) d^\times t, \qquad (7.3.14)$$

a function on $\mathbb{Q}_p$. We now turn to the analogue of (1.2.15). One has

$$\|h\|_{L^2(\mathbb{Q}_p^2)}^2 = \int_{\mathbb{Q}_p \times \mathbb{Q}_p} |h(x,\xi)|^2 dx d\xi$$

$$= (\log p) \int_{\mathbb{Q}_p \times \mathbb{Q}_p} \overline{h}(x,\xi) dx d\xi \int_{-\frac{\pi}{\log p}}^{\frac{\pi}{\log p}} h_{i\lambda}(x,\xi) d\lambda$$

$$= (\log p) \int_{-\frac{\pi}{\log p}}^{\frac{\pi}{\log p}} d\lambda \int_{\mathbb{Q}_p \times \mathbb{Q}_p} \overline{h}(x,\xi) |\xi|^{-1-i\lambda} h_{i\lambda}^\flat\left(\frac{x}{\xi}\right) dx d\xi, \qquad (7.3.15)$$

since one has if $\xi \neq 0$ the equation $h_{i\lambda}(x,\xi) = |\xi|^{-1-i\lambda} h_{i\lambda}^\flat\left(\frac{x}{\xi}\right)$: setting $x = s\xi$ in the last integral on the right-hand side, so that $dx = |\xi| ds$, one can write it as

$$\int_{\mathbb{Q}_p} h_{i\lambda}^\flat(s) ds \int_{\mathbb{Q}_p^\times} \overline{h}(s\xi,\xi) |\xi|^{-i\lambda} d\xi = \frac{p-1}{p} \int_{\mathbb{Q}_p} h_{i\lambda}^\flat(s) ds \int_{\mathbb{Q}_p^\times} \overline{h}(s\xi,\xi) |\xi|^{1-i\lambda} d^\times \xi$$

$$= \frac{p-1}{p} \int_{\mathbb{Q}_p} |h_{i\lambda}^\flat(s)|^2 ds. \qquad (7.3.16)$$

Finally, we obtain

$$\|h\|_{L^2(\mathbb{Q}_p^2)}^2 = \frac{(p-1)\log p}{p} \int_{-\frac{\pi}{\log p}}^{\frac{\pi}{\log p}} \|h_{i\lambda}^\flat\|_{L^2(\mathbb{Q}_p)}^2 d\lambda, \qquad (7.3.17)$$

an equation to be compared to (1.2.15).

As a consequence, one can extend the collection of transforms $(h \mapsto h_{i\lambda}^\flat)_{\lambda \in \mathbb{R}}$ initially considered only in the case when, on the support of h, the number $\max(|x|,|\xi|)$ is both bounded and bounded away from zero, as follows: given $h \in L^2(\mathbb{Q}_p^2)$, the function $h_{i\lambda}^\flat$ is well-defined, for almost all λ, as an element of $L^2(\mathbb{Q}_p)$, and the equation (7.3.17) continues to hold.

Proposition 7.3.1. *Consider on $\mathbb{Q}_p \times \mathbb{Q}_p$ the function*

$$\Phi_p(x,\xi) = \mathrm{char}((x,\xi) \in \mathbb{Z}_p \times \mathbb{Z}_p): \qquad (7.3.18)$$

one has if $\mathrm{Re}\,\nu > 1$ *the equation*

$$(\Phi_p)_\nu^\flat(s) = (1 - p^{-\nu-1})^{-1} [\max(1,|s|)]^{-\nu-1}, \qquad (7.3.19)$$

in other words

$$(\Phi_p)_\nu(x,\xi) = (1 - p^{-\nu-1})^{-1} [\max(|x|,|\xi|)]^{-\nu-1}. \qquad (7.3.20)$$

The proof of this equation is straightforward. The function Φ_p may be considered as the "fundamental" function on $\mathbb{Q}_p^2$ since all Schwartz-Bruhat functions derive from it by simple constructions: it has competitors in this respect, but not so simple-looking.

Just as an exercise (for analysts), let us consider the following proposition.

Proposition 7.3.2. *Given a prime p, consider the field $\mathbb{Q}_p$ of p-adic numbers and its absolute value $s \mapsto |s|_p$, or simply $s \mapsto |s|$. Given a complex number ν, define on $\mathbb{Q}_p$ the two functions f_ν^p and h_ν^p with*

$$f_\nu^p(s) = [\max(1, |s|)]^{-\nu-1},$$
$$h_\nu^p(s) = (1 - p^{-\nu-1})^{-1} f_\nu^p(s). \tag{7.3.21}$$

The Fourier transform of the second function is given by the equation

$$(\mathcal{F}h_\nu^p)(\sigma) = \frac{1 - p^{-\nu}|\sigma|^\nu}{1 - p^{-\nu}} \operatorname{char}(\sigma \in \mathbb{Z}_p). \tag{7.3.22}$$

Proof. Using the definition (7.3.2) of the p-adic Fourier transformation, one has

$$(1 - p^{-\nu-1})(\mathcal{F}h_\nu^p)(\sigma) = \int_{\mathbb{Q}_p} [\max(1, |s|)]^{-\nu-1} e^{-2i\pi\kappa(\sigma s)} ds \tag{7.3.23}$$

$$= \int_{\mathbb{Z}_p} e^{-2i\pi\kappa(\sigma s)} + \sum_{j \geq 1} p^{-j(\nu+1)} \int_{|s|=p^j} e^{-2i\pi\kappa(\sigma s)} ds.$$

Noting that

$$\{s \in \mathbb{Q}_p : |s| = p^j\} = (p^{-j}\mathbb{Z}_p) \backslash (p^{1-j}\mathbb{Z}_p) \tag{7.3.24}$$

and that

$$\int_{p^{-j}\mathbb{Z}_p} e^{-2i\pi\kappa(\sigma s)} ds = p^j \operatorname{char}(\sigma \in p^j \mathbb{Z}_p), \tag{7.3.25}$$

one obtains

$$(1 - p^{-\nu-1})(\mathcal{F}h_\nu^p)(\sigma) = \operatorname{char}(\sigma \in \mathbb{Z}_p)$$
$$+ \sum_{j \geq 1} p^{-j(\nu+1)} \left[p^j \operatorname{char}(\sigma \in p^j \mathbb{Z}_p) - p^{j-1} \operatorname{char}(\sigma \in p^{j-1} \mathbb{Z}_p) \right]. \tag{7.3.26}$$

The right-hand side is zero unless $\sigma \in \mathbb{Z}_p$: then, if $|\sigma| = p^{-k}$ for some $k \geq 0$, the condition $\sigma \in p^j \mathbb{Z}_p$ holds if and only if $j \leq k$, so that, using the equations

$$\sum_{1 \leq j \leq k} p^{-j\nu} = \frac{p^{-\nu} - p^{-(k+1)\nu}}{1 - p^{-\nu}},$$

$$\sum_{1 \leq j \leq k+1} p^{-j\nu-1} = \frac{1}{p} \frac{p^{-\nu} - p^{-(k+2)\nu}}{1 - p^{-\nu}}, \tag{7.3.27}$$

one obtains

$$(1 - p^{-\nu-1})(\mathcal{F}h_\nu^p)(\sigma) = 1 + \frac{p^{-\nu} - p^{-(k+1)\nu}}{1 - p^{-\nu}} - \frac{1}{p}\frac{p^{-\nu} - p^{-(k+2)\nu}}{1 - p^{-\nu}} \qquad (7.3.28)$$

or, finally,

$$(\mathcal{F}h_\nu^p)(\sigma) = \frac{1 - p^{-\nu}(p^{-k})^\nu}{1 - p^{-\nu}}, \qquad (7.3.29)$$

the same as (7.3.22). □

In the following observation, we consider, instead of the ring $\mathbb{A}$, the ring $\mathbb{A}_f$ (f stands for "finite") obtained in the same way, only forgetting the Archimedean (real) place: by "adeles", we here mean elements of $\mathbb{A}_f$.

Proposition 7.3.3. *Denote adeles as* $s = (s_p)_{p \in P}$ *or* $s = (s_p)$, *recalling that* $s_p \in \mathbb{Q}_p$ *for all* p *and that* $s_p \in \mathbb{Z}_p$ *for almost all* p. *Consider for* $\mathrm{Re}\,\nu > 0$ *the functions (on* $\mathbb{A}_f$)

$$F_\nu(s) = \prod_p f_\nu^p(s_p),$$

$$H_\nu(s) = \prod_p h_\nu^p(s_p)$$

$$= \prod_p (1 - p^{-\nu-1})^{-1} f_\nu^p(s_p) = \zeta(\nu+1)F_\nu(s). \qquad (7.3.30)$$

The Fourier transform of H_ν *is the function on* $\mathbb{A}_f$ *given as*

$$(\mathcal{F}H_\nu)(\sigma) = \prod_p \frac{1 - p^{-\nu}|\sigma|_p^\nu}{1 - p^{-\nu}}\,\mathrm{char}(\sigma_p \in \mathbb{Z}_p). \qquad (7.3.31)$$

The function F_ν, *as a function on* $\mathbb{A}_f$, *is still well-defined for every value of* $\nu \in \mathbb{C}$. *The restriction of the function* $\mathcal{F}H_\nu$ *to ideles (exactly to* $\mathbb{A}_f^\times$) *extends as well to all values of* ν *not in* $\frac{2i\pi}{\log p}\mathbb{Z}$ *for any prime* p.

Proof. Equation (7.3.31) follows from its local versions in Proposition 7.3.2. Clearly, for every adele s, the infinite product defining $F_\nu(s)$ has only finitely many factors distinct from 1 (since $|s_p|_p \le 1$ for almost every p) and so does the infinite product defining $(\mathcal{F}H_\nu)(\sigma)$ if σ is an idele (since, then, $|\sigma_p|_p = 1$ for almost every p). □

It may thus look as if, writing

$$\zeta(\nu+1) = \frac{(\mathcal{F}H_\nu)(\sigma)}{(\mathcal{F}F_\nu)(\sigma)}, \qquad (7.3.32)$$

one had a possible approach towards a study of the zeros of zeta, since each of the two functions F_ν and $\mathcal{F}H_\nu$ is well-defined, as a function on adeles or ideles, except for quite specific values of $\nu \in \mathbb{C}$. Of course, this is hardly more than a joke. What is wrong with the equation $\frac{1}{\zeta(\nu+1)}\mathcal{F}H_\nu = \mathcal{F}F_\nu$, unless $\mathrm{Re}\,\nu > 0$, is that the two sides are really adelic distributions in the sense of [29, p. 127] rather than functions on adeles or ideles: indeed, the local factor $\sigma \mapsto (1-p^{-\nu-1})\frac{1-p^{-\nu}|\sigma|_p^\nu}{1-p^{-\nu}}\mathrm{char}(\sigma_p \in \mathbb{Z}_p)$ takes the value 1 when tested on the characteristic function of $\mathbb{Z}_p$), as it follows from the decomposition

$$\frac{1-p^{-\nu-1}}{1-p^{-\nu}}\int_{\mathbb{Z}_p}(1-p^{-\nu}|\sigma|_p^\nu)d\sigma = \frac{1-p^{-\nu-1}}{1-p^{-\nu}}\sum_{j\geq 0}(p^{-j}-p^{-j-1})(1-p^{-(j+1)\nu}) = 1.$$

$$(7.3.33)$$

Up to a point, one can build a "dictionary" between related objects defined in an adelic or a classical setting, starting from the point of view that, in dimension one, the characteristic function of the ring $\widehat{\mathbb{Z}}$ of adeles s such that $|s_p| \leq 1$ for each p should correspond to the measure $x \mapsto \sum_{n\in\mathbb{Z}}\delta(x-n)$, and going further by finding the natural analogues, in distribution theory, of operations sufficient to generate tensor products of Schwartz-Bruhat spaces of functions on $\mathbb{Q}_p$ from the characteristic function of $\widehat{\mathbb{Z}}$. It is with this in mind that, in the next section, we shall reconsider the p-adic Weyl calculus so as to make room for the family $(\mathfrak{t}_\rho)$ (*cf.* Definition 7.4.2) of p-adic analogues of the distributions $\mathfrak{d}_\rho$ defined in (7.2.3).

As described in introductions to the adelic version of modular form theory [10, 4, 29], the latter theory can be recast in terms starting from the group $GL(2, \mathbb{A})$. Here, with a view comparable to that which led to the consideration of automorphic distributions in the plane, we shall consider only the plane $\mathbb{A}^2$, which is certainly insufficient for arithmetic but just what is needed for pseudo-differential analysis. With a realization that our "dictionary" is far from complete, let us only observe that the calculations which led to (7.3.31) are just a disguised way to obtain, again, the non-cuspidal coefficients of Eisenstein series. Indeed, the function $(\Phi_p)^\flat_\nu$ is identical to the function denoted as h_ν^p in Proposition 7.3.2, which can be compared to the fact that Eisenstein distributions are the homogeneous terms from the decomposition of the Dirac comb $\mathfrak{D}$. On the other hand, if σ is the idele associated (under the diagonal embedding $\mathbb{Q}^\times \to \mathbb{A}^\times$) to an integer $n \in \mathbb{Z}^\times, n = \pm\prod p^{v_p(n)}$, one has

$$\prod_p \frac{1-p^{-\nu}|\sigma|_p^\nu}{1-p^{-\nu}} = \prod_{p|n} \frac{1-p^{-\nu}p^{-\nu v_p(n)}}{1-p^{-\nu}}$$

$$= \prod_{p|n}\sum_{j=0}^{v_p(n)} p^{-j\nu} = \sigma_{-\nu}(|n|) : \qquad (7.3.34)$$

one recognizes the arithmetic part of the nth Fourier coefficient of the Eisenstein serie $E_{\frac{1+\nu}{2}}(z)$ (the extra factor $|n|^{\frac{\nu}{2}}$, absent from (5.2.7) but present in (3.1.6), comes from the required Archimedean factor accompanying the Eisenstein series, which we have neglected, though it could be accounted for, too, by an extension of the same analysis, using $\mathbb{A}$ in place of $\mathbb{A}_f$).

7.4   Renormalizing the *p*-adic Weyl calculus

In Section 7.2, we have defined, for every number N equal to 4 times a squarefree odd integer, a finite set $(\mathfrak{d}_\rho)_{\rho \in R_N}$ of discrete measures on the line with the following properties: (i) given N, the linear space generated by the measures $\mathfrak{d}_\rho$ is invariant under the part of the metaplectic representation lying above $SL(2, \mathbb{Z})$; (ii) the sum $\sum_{\rho \in R_N} W(\mathfrak{d}_\rho, \mathfrak{d}_\rho)$ coincides with the distribution $N^{i\pi\mathcal{E}}\mathfrak{T}_N$, up to the addition of some multiple of δ.

We now reformulate this construction in adelic terms: the main issue is that it will force us to change in a *p*-dependent way the normalization of *p*-adic pseudo-differential analysis as defined in (7.3.3). First, it is essential, so as to find a *p*-adic analogue of (7.2.3), to get rid of the square-root there. Note the general formula

$$
\begin{pmatrix} N^{\frac{1}{2}} & 0 \\ 0 & N^{-\frac{1}{2}} \end{pmatrix} \begin{pmatrix} a & b \\ c & d \end{pmatrix} \begin{pmatrix} N^{-\frac{1}{2}} & 0 \\ 0 & N^{\frac{1}{2}} \end{pmatrix} = \begin{pmatrix} a & Nb \\ N^{-1}c & d \end{pmatrix}. \tag{7.4.1}
$$

On the other hand,

$$
N^{\frac{1}{4}} \mathrm{Met} \left(\begin{pmatrix} N^{-\frac{1}{2}} & 0 \\ 0 & N^{\frac{1}{2}} \end{pmatrix} \right) \delta(x - b) = N^{\frac{1}{2}} \delta(N^{\frac{1}{2}} x - b) = \delta\left(x - \frac{b}{\sqrt{N}} \right), \tag{7.4.2}
$$

so that

$$
\mathfrak{d}_\rho = N^{\frac{1}{4}} \mathrm{Met} \left(\begin{pmatrix} N^{-\frac{1}{2}} & 0 \\ 0 & N^{\frac{1}{2}} \end{pmatrix} \right) \mathfrak{d}_\rho^{\sharp} \tag{7.4.3}
$$

with

$$
\mathfrak{d}_\rho^{\sharp}(x) = \sum_{\mu \in \Lambda} \chi(\mu) \sum_{\ell \in \mathbb{Z}} \delta(x - (N\ell + \rho\mu)). \tag{7.4.4}
$$

The factors $N^{\pm\frac{1}{2}}$ have disappeared from the arguments of δ: only, the space generated by the measures $\mathfrak{d}_\rho^{\sharp}$ is no longer invariant under the metaplectic transformations lying above the group generated by the matrices $\begin{pmatrix} 0 & 1 \\ -1 & 0 \end{pmatrix}$ and $\begin{pmatrix} 1 & 0 \\ 1 & 1 \end{pmatrix}$, but under those lying above the group generated by the matrices $\begin{pmatrix} 0 & N \\ -N^{-1} & 0 \end{pmatrix}$ and $\begin{pmatrix} 1 & 0 \\ N^{-1} & 1 \end{pmatrix}$. To recover $SL(2, \mathbb{Z})$, we would need to modify the definition of the metaplectic representation as well.

We shall not pursue this, in a classical distribution environment, any further, rather reset the transformations involved in a *p*-adic setting. As will be seen, besides much simplification due to the use of localization from the start of the construction, we shall gain in this way a hint towards the solution to our "change of Planck's constant" dilemma as explained just after (7.2.6).

Definition 7.4.1. Fixing a prime p, we set $q = 4$ if $p = 2$, and $q = p$ if $p \neq 2$. We renormalize the definition (7.3.3) of p-adic pseudo-differential analysis by using instead the "bulletted" version defined by

$$\text{Op}^\bullet(h) = \text{Op}(h_1), \quad h_1(x, \xi) = q^{-\frac{1}{2}} h(x, q\xi), \tag{7.4.5}$$

in other words

$$(\text{Op}^\bullet(h)u)(x) = q^{\frac{1}{2}} \int_{\mathbb{Q}_p \times \mathbb{Q}_p} h\left(\frac{x+y}{2}, \eta\right) e^{2i\pi\kappa\left(\frac{(x-y)\eta}{q}\right)} u(y) dy d\eta. \tag{7.4.6}$$

At the same time, we use the modified metaplectic (projective) representation such that

$$\text{Met}^\bullet\left(\begin{pmatrix} a & b \\ c & d \end{pmatrix}\right) = \text{Met}\left(\begin{pmatrix} a & qb \\ q^{-1}c & d \end{pmatrix}\right): \tag{7.4.7}$$

the covariance formula then expresses itself as

$$\text{Met}^\bullet(g)\text{Op}^\bullet(h)\text{Met}^\bullet(g)^{-1} = \text{Op}^\bullet(h \circ g^{-1}), \quad g \in G_p. \tag{7.4.8}$$

The normalization chosen preserves the identity between the Hilbert-Schmidt norm of an operator in $L^2(\mathbb{Q}_p)$ and the norm of its symbol in $L^2(\mathbb{Q}_p \times \mathbb{Q}_p)$: in particular, if one defines the Wigner function $W^\bullet(v, u)$ as the function on $\mathbb{Q}_p \times \mathbb{Q}_p$ making the relation

$$(v|\text{Op}^\bullet(h)u)_{L^2(\mathbb{Q}_p)} = \int_{\mathbb{Q}_p \times \mathbb{Q}_p} h(x, \xi) W^\bullet(v, u)(x, \xi) dx d\xi, \tag{7.4.9}$$

analogous to (1.1.15), true for every $h \in S(\mathbb{Q}_p \times \mathbb{Q}_p)$, the function $W^\bullet(v, u)$ also serves as the $\text{Op}^\bullet$-symbol of the operator $w \mapsto (v|w)u$. It is made explicit by the equation

$$W^\bullet(v, u)(x, \xi) = |2|q^{\frac{1}{2}} \int_{\mathbb{Q}_p} \bar{v}(x+t)u(x-t)e^{2i\pi\kappa\left(\frac{2t\xi}{q}\right)} dt. \tag{7.4.10}$$

There is a small price to pay for this normalization: the symbol of the identity operator is $q^{\frac{1}{2}}$, not 1.

We introduce now the p-adic analogue of the measure in (7.4.4). In analogy with (7.4.4), noting that when $N = q$ ($= 4$ or a prime $p \geq 3$), one has $\Lambda = \{\pm 1 \bmod q\}$ and $\chi(\pm 1 \bmod q) = \pm 1$, we define the following set of functions.

Definition 7.4.2. Given $\rho \in \mathbb{Z}_p^\times$, we define on $\mathbb{Q}_p$ the function t_ρ such that

$$t_\rho(x) = \sum_{\varepsilon = \pm 1} \varepsilon \text{char}(x \in \varepsilon\rho + q\mathbb{Z}_p), \tag{7.4.11}$$

recalling that $q = 4$ if $p = 2$ and $q = p$ otherwise.

Note that the definition of t_ρ only depends on $\rho \bmod q\mathbb{Z}_p$, a class which can be identified in a canonical way with an element of the group $(\mathbb{Z}/q\mathbb{Z})^\times$: we may consider the subscript of t_ρ as being an element of this group. The functions t_ρ are odd, supported in $\mathbb{Z}_p$, and invariant under translations by vectors in $q\mathbb{Z}_p$, hence lie in $\mathcal{S}(\mathbb{Q}_p)$. One has $t_{-\rho} = -t_\rho$, but if one denotes as R_p any set of representatives of $(\mathbb{Z}/q\mathbb{Z})^\times \bmod \{\pm 1\}$, the functions t_ρ with $\rho \in R_p$ are linearly independent. The functions t_ρ are permuted under a natural action of $\mathbb{Z}_p^\times$ on $L^2(\mathbb{Q}_p)$, in the sense that

$$t_\rho(\sigma^{-1}x) = t_{\sigma\rho}(x). \tag{7.4.12}$$

Since the sets $\rho + q\mathbb{Z}_p$ and $-\rho + q\mathbb{Z}_p$ do not intersect, one has

$$\|t_\rho\|_{L^2(\mathbb{Q}_p)}^2 = \int_{\mathbb{Q}_p} |t_\rho(x)|^2 dx = 2\int_{\rho+q\mathbb{Z}_p} dx = \frac{2}{q}. \tag{7.4.13}$$

Definition 7.4.3. We define the unitary transformation $\mathcal{K}$ of $L^2(\mathbb{Q}_p)$ by the equation

$$(\mathcal{K}u)(x) = q^{\frac{1}{2}} \int_{\mathbb{Q}_p} u(y) e^{-2i\pi\kappa(\frac{xy}{q})} dy, \tag{7.4.14}$$

and we denote as $\mathcal{T}$ the transformation, associated under $\mathrm{Met}^\bullet$ to the matrix $\left(\begin{smallmatrix} 1 & 0 \\ 1 & 1 \end{smallmatrix}\right)$, which consists in multiplying a functions of $x \in \mathbb{Q}_p$ by $e^{2i\pi\kappa(\frac{x^2}{2q})}$.

The first transformation is of course a substitute for the Fourier transformation, since it is (up to some phase factor) the image under $\mathrm{Met}^\bullet$ of the matrix $\left(\begin{smallmatrix} 0 & 1 \\ -1 & 0 \end{smallmatrix}\right)$: indeed, $\mathrm{Met}^\bullet\left(\left(\begin{smallmatrix} 0 & 1 \\ -1 & 0 \end{smallmatrix}\right)\right) = \mathrm{Met}\left(\left(\begin{smallmatrix} 0 & q \\ -q^{-1} & 0 \end{smallmatrix}\right)\right)$, and $\left(\begin{smallmatrix} 0 & q \\ -q^{-1} & 0 \end{smallmatrix}\right) = \left(\begin{smallmatrix} q & 0 \\ 0 & q^{-1} \end{smallmatrix}\right)\left(\begin{smallmatrix} 0 & 1 \\ -1 & 0 \end{smallmatrix}\right)$: now, the images, under Met, of the last two matrices, are the unitary transformations of $L^2(\mathbb{Q}_p)$ listed just before (7.3.2): do not forget that $|q|^{-1} = q$. The transformation $\mathcal{K}^{-1}$ is obtained by changing i to $-i$ in the exponent on the right-hand side of (7.4.14).

The following would have been more properly placed right after the definition of the "bulletted" calculus, but we need to rely on properties of the operator $\mathcal{K}$. As in the Archimedean case, we introduce in p-adic analysis the involution $\mathcal{G}_p$ making the identity

$$\mathrm{Op}^\bullet(\mathcal{G}_p h)u = \mathrm{Op}^\bullet(h)\check{u} \tag{7.4.15}$$

valid for any pair $(h, u) \in \mathcal{S}(\mathbb{Q}_p^2) \times \mathcal{S}(\mathbb{Q}_p)$.

Proposition 7.4.4. *The operator $\mathcal{G}_p$ in (7.4.15) is given by the equation*

$$(\mathcal{G}_p h)(x, \xi) = p\int_{\mathbb{Q}_p \times \mathbb{Q}_p} h(y, \eta) e^{2i\pi\kappa(\frac{2(x\eta - y\xi)}{q})} dy d\eta. \tag{7.4.16}$$

Proof. It is immediate that $\mathrm{Op}^\bullet(h)\check{u} = \mathrm{Op}^\bullet(h_1)u$ provided that

$$q^{\frac{1}{2}} \int h(\frac{x-y}{2}, \eta) e^{2i\pi\kappa(\frac{(x+y)\eta}{q})} d\eta = q^{\frac{1}{2}} \int h_1(\frac{x+y}{2}, \eta) e^{2i\pi\kappa(\frac{(x-y)\eta}{q})} d\eta, \tag{7.4.17}$$

i.e., denoting as $\mathcal{K}_2$ the transformation in (7.4.14) when regarded as concerning itself with the second variable of a pair,

$$(\mathcal{K}_2^{-1}h)(\frac{x-y}{2}, x+y) = (\mathcal{K}_2^{-1}h_1)(\frac{x+y}{2}, x-y), \tag{7.4.18}$$

or

$$(\mathcal{K}_2^{-1}h_1)(x, y) = (\mathcal{K}_2^{-1}h)(\frac{y}{2}, 2x). \tag{7.4.19}$$

Then,

$$h_1(x, \xi) = q^{\frac{1}{2}} \int (\mathcal{K}_2^{-1}h)(\frac{y}{2}, 2x) e^{-2i\pi\kappa(\frac{y\xi}{q})} dy$$

$$= q \int h(\frac{y}{2}, \eta) e^{2i\pi\kappa(\frac{2x\eta}{q})} e^{-2i\pi\kappa(\frac{y\xi}{q})} dy d\eta$$

$$= |2|q \int_{\mathbb{Q}_p \times \mathbb{Q}_p} h(y, \eta) e^{2i\pi\kappa(\frac{2(x\eta - y\xi)}{q})} dy d\eta. \tag{7.4.20}$$

$\square$

It will be suggestive to denote as p^{-Eul_p}, or p^{-Eul}, the unitary operator on $L^2(\mathbb{Q}_p \times \mathbb{Q}_p)$ such that

$$(p^{-Eul}f)(x, \xi) = pf(\frac{x}{p}, \frac{\xi}{p}): \tag{7.4.21}$$

the notation does certainly not imply that the exponent Eul has a meaning. One may then observe the relation (analogous to the fact that, in real analysis, the symplectic Fourier transformation on $\mathbb{R}^2$ anticommutes with the Euler operator $2i\pi\mathcal{E}$)

$$p^{-Eul}\mathcal{G}_p = \mathcal{G}_p p^{Eul}; \tag{7.4.22}$$

in view of (7.4.15), one still has $\mathcal{G}_p^2 = I$.

For later use, one should note the following relation:

$$\mathcal{G}_p \Phi_p = p^{-Eul}\Phi_p, \tag{7.4.23}$$

which follows from (7.4.16):

$$(\mathcal{G}_p \Phi_p)(x, \xi) = p \int_{\mathbb{Z}_p \times \mathbb{Z}_p} e^{2i\pi\kappa(\frac{2(x\eta - y\xi)}{q})} dy d\eta$$

$$= p\Phi_p(\frac{2\xi}{q}, -\frac{2x}{q}) = p\Phi_p(\frac{x}{p}, \frac{\xi}{p}). \tag{7.4.24}$$

Theorem 7.4.5. *The linear space generated by the functions* $t_\rho, \rho \in (\mathbb{Z}/q\mathbb{Z})^\times$, *is stable under the metaplectic transformations* $\mathcal{T}$ *and* $\mathcal{K}$. *In the basis made up by the set* $(t_\rho)_{\rho \in R_p}$, *in which* R_p *is an arbitrary set of representatives of* $(\mathbb{Z}/q\mathbb{Z})^\times$

mod ± 1, *the matrices* T *and* K *representing the corresponding operators are the diagonal matrix with entries* $e^{-\frac{i\pi\rho^2}{q}}$ *and the matrix* K *with entries*

$$K(\rho,\sigma) = q^{-\frac{1}{2}}\left(e^{\frac{2i\pi\rho\sigma}{q}} - e^{\frac{-2i\pi\rho\sigma}{q}}\right). \tag{7.4.25}$$

Proof. First note that, when $\rho \in (\mathbb{Z}/q\mathbb{Z})^\times$, ρ^2 is well-defined mod $2q$, which gives $e^{-\frac{i\pi\rho^2}{q}}$ a meaning. One has $e^{2i\pi\kappa(\frac{\rho^2}{2q})} = e^{-i\pi\frac{\rho^2}{q}}$, but note that on the left-hand side of this equation, $\rho \in \mathbb{Z}_p^\times$ while, on the right-hand side, ρ denotes the associated element of $(\mathbb{Z}/q\mathbb{Z})^\times$: that $T t_\rho = e^{-\frac{i\pi\rho^2}{q}} t_\rho$ follows. On the other hand, one has

$$(K t_\rho)(x) = q^{\frac{1}{2}} \int_{\mathbb{Q}_p} \sum_{\varepsilon=\pm 1} \varepsilon \mathrm{char}(y \in \varepsilon\rho + q\mathbb{Z}_p) e^{-2i\pi\kappa(\frac{xy}{q})} dy$$

$$= q^{\frac{1}{2}} \sum_{\varepsilon=\pm 1} \varepsilon \exp\left(-2i\pi\varepsilon\kappa(\frac{x\rho}{q})\right) \int_{q\mathbb{Z}_p} e^{-2i\pi\kappa(\frac{xy}{q})} dy$$

$$= q^{-\frac{1}{2}} \sum_{\varepsilon=\pm 1} \varepsilon \exp\left(-2i\pi\varepsilon\kappa(\frac{x\rho}{q})\right) \int_{\mathbb{Z}_p} e^{-2i\pi\kappa(xy)} dy$$

$$= q^{-\frac{1}{2}} \mathrm{char}(x \in \mathbb{Z}_p) \sum_{\varepsilon=\pm 1} \varepsilon e^{-2i\pi\varepsilon\kappa(\frac{x\rho}{q})}$$

$$= -2iq^{-\frac{1}{2}} \mathrm{char}(x \in \mathbb{Z}_p) \sin(2\pi\kappa(\frac{x\rho}{q})). \tag{7.4.26}$$

We now show that

$$K t_\rho = \frac{1}{2} \sum_{\sigma \in (\mathbb{Z}/q\mathbb{Z})^\times} K(\rho,\sigma) t_\sigma \tag{7.4.27}$$

with $K(\rho,\sigma)$ as indicated in (7.4.25): we need to verify that, for $x \in \mathbb{Z}_p$, one has

$$-2iq^{-\frac{1}{2}} \sin(2\pi\kappa(\frac{x\rho}{q})) = \frac{1}{2} \sum_{\sigma \in (\mathbb{Z}/q\mathbb{Z})^\times} K(\rho,\sigma) \sum_{\varepsilon=\pm 1} \varepsilon \mathrm{char}(x \in \varepsilon\sigma + q\mathbb{Z}_p), \tag{7.4.28}$$

in other words that

$$\sin(2\pi\kappa(\frac{x\rho}{q})) = -\frac{1}{2} \sum_{\sigma \in (\mathbb{Z}/q\mathbb{Z})^\times} \sin\frac{2\pi\rho\sigma}{q} \sum_{\varepsilon=\pm 1} \varepsilon \mathrm{char}(x \in \varepsilon\sigma + q\mathbb{Z}_p): \tag{7.4.29}$$

this follows from the fact that, given $x \in \mathbb{Z}_p^\times$, the condition $x \in \varepsilon\sigma + q\mathbb{Z}_p$ is equivalent to the equation $\frac{\varepsilon\rho\sigma}{q} \equiv -\kappa(\frac{x\rho}{q}) \mod \mathbb{Z}$. $\square$

Remark 7.4.a. Theorem 7.4.5 is fully analogous — with the simplification in the proof brought by localization — to Lemma 3.2 in [63], which concerned distributions on the real line.

The distributions $\mathfrak{d}_\rho$ in (7.2.3) were introduced in connection with the meta-plectic representation: the idea was to define finite-dimensional spaces, depending on appropriate integers N, of discretely supported distributions, globally invariant under the image of (the twofold cover of) $SL(2,\mathbb{Z})$ by the metaplectic represen-tation. The functions t_ρ were introduced in (7.4.11) as a p-adic analogue. In this context, there is a related class of functions, even more natural.

Let $F_0^p \subset L^2(\mathbb{Q}_p)$ consist of all functions supported in $\mathbb{Z}_p$ invariant under translations by vectors in $q\mathbb{Z}_p$. This is a q-dimensional space, an orthogonal basis of which consists of the functions $\mathfrak{r}_\rho$ on $\mathbb{Q}_p$, parametrized by $\rho \in \mathbb{Z}_p$ mod $q\mathbb{Z}_p$, such that

$$\mathfrak{r}_\rho(x) = \mathrm{char}(x \in \rho + q\mathbb{Z}_p). \tag{7.4.30}$$

Note that the quotient set $\mathbb{Z}_p/q\mathbb{Z}_p$ can be identified in a natural way with the additive group $\mathbb{Z}/q\mathbb{Z}$, and that the norm of $\mathfrak{r}_\rho$ is $q^{-\frac{1}{2}}$ for every ρ. Also, for $\rho \in \mathbb{Z}_p^\times$, one has

$$t_\rho = \mathfrak{r}_\rho - \mathfrak{r}_{-\rho}. \tag{7.4.31}$$

The following extends Theorem 7.4.5.

Proposition 7.4.6. *The space F_0^p is invariant under the pair $\mathcal{T}, \mathcal{K}$ of transforma-tions given in Definition 7.4.3.*

Proof. Again, one has $\mathcal{T}\mathfrak{r}_\rho = e^{i\pi\kappa(\frac{\rho^2}{q})}\mathfrak{r}_\rho$. On the other hand, using in the middle the fact that the characteristic function of $\mathbb{Z}_p$ is invariant under the p-Fourier transformation normalized in the usual way, one has

$$(\mathcal{K}\mathfrak{r}_\rho)(x) = q^{\frac{1}{2}} \int_{\mathbb{Q}_p} \mathrm{char}(y \in \rho + q\mathbb{Z}_p)e^{-2i\pi\kappa(\frac{xy}{q})}dy$$

$$= q^{\frac{1}{2}}e^{-2i\pi\kappa(\frac{x\rho}{q})} \int_{q\mathbb{Z}_p} e^{-2i\pi\kappa(\frac{xy}{q})}dy$$

$$= q^{-\frac{1}{2}}e^{-2i\pi\kappa(\frac{x\rho}{q})}\mathrm{char}(x \in \mathbb{Z}_p)$$

$$= q^{-\frac{1}{2}} \sum_{\sigma \in \mathbb{Z}_p/q\mathbb{Z}_p} \mathrm{char}(x \in \sigma + q\mathbb{Z}_p)e^{-2i\pi\kappa(\frac{\sigma\rho}{q})}, \tag{7.4.32}$$

hence

$$\mathcal{K}\mathfrak{r}_\rho = q^{-\frac{1}{2}} \sum_{\sigma \in \mathbb{Z}_p/q\mathbb{Z}_p} e^{-2i\pi\kappa(\frac{\rho\sigma}{q})}\mathfrak{r}_\sigma. \tag{7.4.33}$$

$\square$

Since the (metaplectic) operators $\mathcal{T}$ and $\mathcal{K}$ preserve parity, they also preserve the two subspaces $(F_0^p)_{\mathrm{even}}$ and $(F_0^p)_{\mathrm{odd}}$ of F_0^p.

Theorem 7.4.7. *In the $\mathrm{Op}^\bullet$– symbolic calculus associated with some prime number p, the function Φ_p on $\mathbb{Q}_p \times \mathbb{Q}_p$ which is the characteristic function of $\mathbb{Z}_p \times \mathbb{Z}_p$ is the symbol of $q^{-\frac{1}{2}}$ times the operator of orthogonal projection: $L^2(\mathbb{Q}_p) \to F_0^p$. The* function Ψ_p defined as $\Psi_p = 2^{-\frac{1}{2}}(I - p^{-Eul})\Phi_p$ if $p \geq 3$, and $\Psi_2 = (I - 2^{-Eul})\Phi_2$ when $p = 2$, is the symbol of $(\frac{2}{p})^{\frac{1}{2}}$ times the operator of orthogonal projection: $L^2(\mathbb{Q}_p) \to (F_0^p)_{\text{odd}}$ in all cases.

Proof. First observe that the dimension of F_0^p is q, while that of the odd part of this space is $\frac{p-1}{2}$ if $p \geq 3$, and 1 if $p = 2$. On the other hand,

$$\|(I - p^{-Eul})\Phi_p\|^2_{L^2(\mathbb{Q}_p \times \mathbb{Q}_p)} = \int_{\mathbb{Z}_p \times \mathbb{Z}_p} [1 - 2p\text{char}((x,\xi) \in (p\mathbb{Z}_p) \times (p\mathbb{Z}_p))$$

$$+ p^2\text{char}((x,\xi) \in (p\mathbb{Z}_p) \times (p\mathbb{Z}_p))]dxd\xi = 2(1 - \frac{1}{p}). \quad (7.4.34)$$

Hence, $\|\Phi_p\| = 1$ and $\|\Psi_p\| = (1 - p^{-1})^{\frac{1}{2}}$ when $p \geq 3$, finally $\|\Psi_2\| = 1$: this provides the useful verification that, indeed, the identity between the Hilbert-Schmidt norm of an operator and the L^2-norm of its symbol is respected in this theorem. We now prove it.

Since $\mathfrak{r}_\rho(x) = \mathfrak{r}_0(x - \rho)$, one has, from (7.4.10),

$$W^\bullet(\mathfrak{r}_\rho, \mathfrak{r}_\rho)(x, \xi) = W^\bullet(\mathfrak{r}_0, \mathfrak{r}_0)(x - \rho, \xi), \quad (7.4.35)$$

and since the set $(q^{\frac{1}{2}}\mathfrak{r}_\rho)_{\rho \in \mathbb{Z}_p/q\mathbb{Z}_p}$ is an orthonormal basis of F_0^p, the symbol of the orthogonal projection on F_0^p is the function

$$(x, \xi) \mapsto q \sum_{\rho \in \mathbb{Z}_p/q\mathbb{Z}_p} W^\bullet(\mathfrak{r}_0, \mathfrak{r}_0)(x - \rho, \xi). \quad (7.4.36)$$

We first do the computation in the case when $p \geq 3$, so that $q = p$ and

$$W^\bullet(\mathfrak{r}_0, \mathfrak{r}_0)(x, \xi) = p^{\frac{1}{2}} \int_{\mathbb{Q}_p} \text{char}(x + t \in p\mathbb{Z}_p)\text{char}(x - t \in p\mathbb{Z}_p)e^{2i\pi\kappa(\frac{2t\xi}{p})}dt$$

$$= p^{\frac{1}{2}}\text{char}(x \in p\mathbb{Z}_p) \int_{p\mathbb{Z}_p} e^{2i\pi\kappa(\frac{2t\xi}{p})}dt$$

$$= p^{-\frac{1}{2}}\text{char}(x \in p\mathbb{Z}_p)\text{char}(\xi \in \mathbb{Z}_p). \quad (7.4.37)$$

From (7.4.36), one then obtains the symbol of the orthogonal projection on F_0^p as the function

$$(x, \xi) \mapsto p^{\frac{1}{2}} \sum_{\rho \in \mathbb{Z}_p/p\mathbb{Z}_p} \text{char}(x \in \rho + p\mathbb{Z}_p)\text{char}(\xi \in \mathbb{Z}_p)$$

$$= p^{\frac{1}{2}}\Phi_p(x, \xi). \quad (7.4.38)$$

When $p = 2$, one has

$$W^\bullet(\mathfrak{r}_0, \mathfrak{r}_0)(x, \xi) = \int_{\mathbb{Q}_2} \text{char}(x + t \in 4\mathbb{Z}_2)\text{char}(x - t \in 4\mathbb{Z}_2)e^{2i\pi\kappa(\frac{t\xi}{2})}dt$$

$$= \text{char}(x \in 2\mathbb{Z}_2) \int_{\mathbb{Q}_2} \text{char}(t \in x + 4\mathbb{Z}_2)e^{2i\pi\kappa(\frac{t\xi}{2})}dt$$

$$= \text{char}(x \in 2\mathbb{Z}_2)e^{2i\pi\kappa(\frac{x\xi}{2})} \int_{4\mathbb{Z}_2} e^{2i\pi\kappa(\frac{t\xi}{2})}dt$$

$$= \frac{1}{4}\text{char}(x \in 2\mathbb{Z}_2)\text{char}(\xi \in \frac{1}{2}\mathbb{Z}_2)e^{2i\pi\kappa(\frac{x\xi}{2})}. \qquad (7.4.39)$$

Then,

$$W^\bullet(\mathfrak{r}_0, \mathfrak{r}_0)(x, \xi) + W^\bullet(\mathfrak{r}_0, \mathfrak{r}_0)(x - 2, \xi)$$

$$= \frac{1}{4}\text{char}(x \in 2\mathbb{Z}_2)\text{char}(\xi \in \frac{1}{2}\mathbb{Z}_2)e^{2i\pi\kappa(\frac{x\xi}{2})}[1 + e^{-2i\pi\kappa(\xi)}], \quad (7.4.40)$$

where the last bracket is zero if $\xi \in \frac{1}{2}\mathbb{Z}_2$ but $\xi \notin \mathbb{Z}_2$: hence,

$$W^\bullet(\mathfrak{r}_0, \mathfrak{r}_0)(x, \xi) + W^\bullet(\mathfrak{r}_0, \mathfrak{r}_0)(x - 2, \xi) = \frac{1}{2}\text{char}(x \in 2\mathbb{Z}_2)\text{char}(\xi \in \mathbb{Z}_2). \quad (7.4.41)$$

Finally, when $p = 2$, the symbol of the orthogonal projection on F_0^2 is the function

$$(x, \xi) \mapsto 4 \sum_{\rho \in \mathbb{Z}_2/4\mathbb{Z}_2} W^\bullet(\mathfrak{r}_0, \mathfrak{r}_0)(x - \rho, \xi)$$

$$= 2\text{char}(x \in 2\mathbb{Z}_2)\text{char}(\xi \in \mathbb{Z}_2) + 2\text{char}(x \in 1 + 2\mathbb{Z}_2)\text{char}(\xi \in \mathbb{Z}_2)$$

$$= 2\Phi_p(x, \xi). \qquad (7.4.42)$$

This completes the proof of the first part of Theorem 7.4.7: the second part then follows from (7.4.23) and from the interpretation (7.4.15) of the operator $\mathcal{G}_p$. $\qquad \square$

Setting, just as in the Archimedean case, ch $u = \check{u}$, it follows from Theorem 7.4.7 that, for all values of p, the operator with symbol $p^{-Eul}\Phi_p$ is $q^{-\frac{1}{2}}$ times the composition A_1ch, where A_1 is the operator of orthogonal projection from $L^2(\mathbb{Q}_p)$ on the space F_0^p of functions supported in $\mathbb{Z}_p$ and invariant under translations by vectors in $q\mathbb{Z}_p$. More generally,

Theorem 7.4.8. *Fixing a prime p and $k = 0, 1, \ldots,$ denote as F_k^p the space of functions on $\mathbb{Q}_p$ supported in $p^{-k}\mathbb{Z}_p$ and invariant under translations by vectors in $qp^k\mathbb{Z}_p$, and denote as P_k the operator of orthogonal projection: $L^2(\mathbb{Q}_p) \to F_k^p$. Then, for $k = 0, 1, \ldots,$ one has $\text{Op}^\bullet(p^{-(k+1)Eul}\Phi_p) = q^{-\frac{1}{2}}p^{-k}P_k$ch and $\text{Op}^\bullet(p^{kEul}\Phi_p) = q^{-\frac{1}{2}}p^{-k}P_k$.*

Proof. Using definition (7.4.14) of the operator $\mathcal{K}$, one first observes that a function $u \in L^2(\mathbb{Q}_p)$ is invariant under translations by vectors in $qp^k\mathbb{Z}_p$ if and only if, given $t \in p^k\mathbb{Z}_p$, the function $x \mapsto 1 - e^{2i\pi\kappa(xt)}$ vanishes on the support of $\mathcal{K}u$: in other words, one has

$$F_k^p = \{u\colon \operatorname{supp}(u) \subset p^{-k}\mathbb{Z}_p \text{ and } \operatorname{supp}(\mathcal{K}u) \subset p^{-k}\mathbb{Z}_p\}. \tag{7.4.43}$$

Next, using (7.4.6) and (7.4.21), one can write, assuming $k \geq 1$ in all that follows,

$$
\begin{aligned}
(\mathrm{Op}^\bullet(p^{-kEul}\Phi_p)u)(x) &= q^{\frac{1}{2}}p^k \int_{\mathbb{Q}_p \times \mathbb{Q}_p} \Phi_p\left(\frac{x+y}{2p^k}, \frac{\eta}{p^k}\right) e^{2i\pi\kappa(\frac{(x-y)\eta}{q})} u(y)\,dy\,d\eta \\
&= q^{\frac{1}{2}}p^k \int_{-x+2p^k\mathbb{Z}_p} u(y)\,dy \int_{p^k\mathbb{Z}_p} e^{2i\pi\kappa(\frac{(x-y)\eta}{q})}\,d\eta \\
&= q^{\frac{1}{2}} \int_{-x+qp^{k-1}\mathbb{Z}_p} u(y)\operatorname{char}(y \in x + qp^{-k}\mathbb{Z}_p)\,dy. \tag{7.4.44}
\end{aligned}
$$

Since, when the sets $-x + qp^{k-1}\mathbb{Z}_p$ and $x + qp^{-k}\mathbb{Z}_p$ intersect, one has $2x \in qp^{-k}\mathbb{Z}_p$, hence $x \in p^{1-k}\mathbb{Z}_p$, one obtains that the integral kernel of the operator $\mathrm{Op}^\bullet(p^{-kEul}\Phi_p)$ is the function

$$(x, y) \mapsto q^{\frac{1}{2}}\operatorname{char}(x \in p^{1-k}\mathbb{Z}_p)\operatorname{char}(x + y \in qp^{k-1}\mathbb{Z}_p), \tag{7.4.45}$$

and the integral kernel of the operator $B_k = \mathrm{Op}^\bullet(p^{-kEul}\Phi_p)\mathrm{ch}$ is the function

$$(x, y) \mapsto q^{\frac{1}{2}}\operatorname{char}(x \in p^{1-k}\mathbb{Z}_p)\operatorname{char}(x - y \in qp^{k-1}\mathbb{Z}_p). \tag{7.4.46}$$

One has $B_k u = q^{-\frac{1}{2}}p^{1-k}u$ if $u \in F_{k-1}^p$: in this case, this certainly agrees with $q^{-\frac{1}{2}}p^{1-k}P_{k-1}u$. On the other hand, from (7.4.43), the space orthogonal to F_{k-1}^p is the algebraic sum of the space of functions $v \in L^2(\mathbb{Q}_p)$ which vanish on $p^{1-k}\mathbb{Z}_p$ and of the space of functions $v \in L^2(\mathbb{Q}_p)$ such that $\mathcal{K}v$ vanishes on $p^{1-k}\mathbb{Z}_p$. What remains to be done in order to show that $B_k = q^{-\frac{1}{2}}p^{1-k}P_{k-1}$ is showing that B_k is zero on each of these two spaces. That it is zero on the first one is obvious in view of its integral kernel. But that it is zero on the second one follows as well, in view of the covariance formula (7.4.8), of the fact that $\mathcal{K}$ is, up to some phase factor, the image under $\mathrm{Met}^\bullet$ of the matrix $\left(\begin{smallmatrix} 0 & 1 \\ -1 & 0 \end{smallmatrix}\right) \in SL(2, \mathbb{Z}_p)$, finally of the fact that $p^{-kEul}\Phi_p$ is invariant under the linear change of coordinates associated to this matrix.

On the other hand, using (7.4.22) and (7.4.23), one obtains

$$
\begin{aligned}
\mathrm{Op}^\bullet(p^{kEul}\Phi_p) &= \mathrm{Op}^\bullet(p^{(k+1)Eul}p^{-Eul}\Phi_p) \\
&= \mathrm{Op}^\bullet(p^{(k+1)Eul}\mathcal{G}_p\Phi_p) \\
&= \mathrm{Op}^\bullet(\mathcal{G}_p p^{-(k+1)Eul}\Phi_p) \\
&= \mathrm{Op}^\bullet(p^{-(k+1)Eul}\Phi_p)\mathrm{ch} \\
&= q^{-\frac{1}{2}}p^{-k}P_k. \tag{7.4.47}
\end{aligned}
$$

□

Remark 7.4.b. For $j, k = 0, 1, \ldots$, one has $P_j P_k = P_{\min(j,k)}$, so that the space of operators with symbols in the linear span of the functions $p^{kEul} \Phi_p$ with $k = 0, 1, \ldots$ makes up a commutative algebra. So does the space of operators with symbols in the linear span of the functions $p^{kEul} \Phi_p$ with $k \in \mathbb{Z}$.

Pseudo-differential analysis of operators acting on (complex-valued) functions on adeles can then be defined by piecing together the pseudo-differential analyses available on $\mathbb{Q}_p$ for all primes. We shall skip this stage and consider instead the same analysis, taking however for symbols distributions on $\mathbb{R}^2$ of a special kind rather than functions on $\mathbb{A}_f \times \mathbb{A}_f$.

Definition 7.4.9. A comb is an $SL(2, \mathbb{Z})$- invariant measure $\mathfrak{S}$ in $\mathbb{R}^2$ supported by $(\mathbb{Z} \times \mathbb{Z}) \backslash \{0\}$. In other words, in view of Bezout's theorem, it is a sum

$$\mathfrak{S}(x, \xi) = \sum_{|j+|k| \neq 0} a((j, k)) \delta(x - j) \delta(\xi - k), \tag{7.4.48}$$

where $a((j, k))$ can be any function of the g.c.d. of j, k. The function a is to be called the indicator of $\mathfrak{S}$. Given an integer $M \geq 1$, it will be said to be M-supported if $a(r) = a((r, M))$ for every $r \geq 1$; it will be said to be finitely supported if it is M-supported for some M.

One can also introduce fractional combs, such an object being just the image of a comb under a transformation $M^{2i\pi\mathcal{E}}$ for some integer $M \geq 1$. A comb with indicator a is a tempered distribution, hence an automorphic distribution, if and only if the function a is bounded by some polynomial: this is always the case when it is finitely supported. One has the identity

$$a(r) = \sum_{d \geq 1} a(d) \left[\operatorname{char}(\{r : d | r\}) - \sum_p \operatorname{char}(\{r : pd | r\}) \right], \tag{7.4.49}$$

where the series reduces to a finite sum for every given $r \geq 1$. As a consequence, the set of functions $r \mapsto \operatorname{char}(\{r : d | r\})$ constitutes a linear basis of the space of indicators, and the same set, taken when d runs through the set of all divisors of M, is a linear basis of the space of M-supported indicators.

Writing the indicator of a comb $\mathfrak{S}$ as

$$a(r) = \sum_{d \geq 1} b(d) \operatorname{char}(\{r : d | r\}), \tag{7.4.50}$$

one can associate to $\mathfrak{S}$ the function $\mathfrak{S}^{\mathbb{A}}$ on $\mathbb{A}_f \times \mathbb{A}_f$ defined as

$$\mathfrak{S}^{\mathbb{A}}(x, \xi) = \sum_{d \geq 1} b(d) \operatorname{char}((x, \xi) \in d\widehat{\mathbb{Z}} \times d\widehat{\mathbb{Z}}), \tag{7.4.51}$$

where $\widehat{\mathbb{Z}} = \prod_p \mathbb{Z}_p$ is the set of adelic integers. The linear map $\mathfrak{S} \to \mathfrak{S}^{\mathbb{A}}$ so defined sends the space of combs with a finitely supported indicator into the space of Schwartz-Bruhat functions on $\mathbb{A}_f \times \mathbb{A}_f$. This comes from the fact that $d\mathbb{Z}_p = \mathbb{Z}_p$ in the case when p does not divide d.

In view of (7.4.49), one has

$$b(d) = a(d) - \sum_{p|d} a\left(\frac{d}{p}\right), \qquad (7.4.52)$$

and (7.4.51) can be rewritten as

$$\mathfrak{S}^{\mathbb{A}}(x,\xi) = \sum_{d \geq 1} a(d)\mathrm{char}((x,\xi)\colon \max(|x_p|_p, |\xi_p|_p) = |d|_p \text{ for every } p). \qquad (7.4.53)$$

Recall from (7.4.21) that the operator p^{-Eul} has been defined, in analysis over p-adic numbers, by the equation $\left(p^{-Eul} f\right)(x,\xi) = pf\left(\frac{x}{p}, \frac{\xi}{p}\right)$, while of course there is no sense given to Eul on its own. Then, (7.4.51) can be rewritten as

$$\mathfrak{S}^{\mathbb{A}} = \sum_{d \geq 1} b(d) \otimes_p \left[p^{-\alpha_p(1+Eul)} \Phi_p\right], \qquad d = \prod_p p^{\alpha_p}, \qquad (7.4.54)$$

where it is assumed that, given d, α_p is defined for every prime p and, of course, zero except for a finite set of primes.

Let us consider the example when $\mathfrak{S} = \mathfrak{T}_N$, as defined in (7.1.23): the indicator of this automorphic distribution is N-supported. In view of (7.1.24), one has, applying (7.4.53),

$$(\mathfrak{T}_N)^{\mathbb{A}}(x,\xi)$$
$$= \mathrm{char}((x,\xi) \in \widehat{\mathbb{Z}} \times \widehat{\mathbb{Z}}) \times \prod\{(1 - p)\colon p \text{ prime}, p|N x_p \in p\mathbb{Z}_p \text{ and } \xi_p \in p\mathbb{Z}_p\}\colon$$
$$(7.4.55)$$

this can be written as $(\mathfrak{T}_N)^{\mathbb{A}} = \otimes_p f_p$, where f_p reduces to Φ_p if p does not divide N while, if it does,

$$f_p(x_p, \xi_p) = \Phi_p(x_p, \xi_p) \times \begin{cases} 1 & \text{if } \max(|x_p|_p, |\xi_p|_p) = 1, \\ 1 - p & \text{if } \max(|x_p|_p, |\xi_p|_p) < 1, \end{cases} \qquad (7.4.56)$$

in other words

$$(\mathfrak{T}_N)^{\mathbb{A}} = \left(\otimes_{p|N} \left[(I - p^{-Eul})\Phi_p\right]\right) \otimes \left(\otimes_{p \nmid N} \Phi_p\right). \qquad (7.4.57)$$

In just the same way,

$$(\mathfrak{T}_\infty)^{\mathbb{A}}(x,\xi)$$
$$= \mathrm{char}((x,\xi) \in \widehat{\mathbb{Z}} \times \widehat{\mathbb{Z}}) \times \prod\{(1 - p)\colon p \text{ prime}, x_p \in p\mathbb{Z}_p \text{ and } \xi_p \in p\mathbb{Z}_p\}$$
$$= \otimes_p \left[(I - p^{-Eul})\Phi_p\right]. \qquad (7.4.58)$$

Theorem 7.4.8, makes the operator, in analysis over $\mathbb{Q}_p$, with symbol Φ_p, as well as that with symbol $(I - p^{-Eul})\Phi_p$, explicit. Whether this may help figuring out any useful topological vector space structure on an appropriate space of combs (or fractional combs), going beyond what can be obtained from a classical distribution viewpoint, remains to be seen. Let us just observe that, starting from the identification of the Dirac comb in $\mathbb{R}^n$ with the function $\Phi = \otimes_p \Phi_p$ on $\mathbb{A}_f^n$, and transferring as can be done in a canonical way the basic operations which make it possible to generate the Schwartz-Bruhat space $\mathcal{S}(\mathbb{A}_f^n)$ from the sole function Φ to analogues in a classical distribution setting, one may come to the conclusion that everything one could do with adeles could also be done within Archimedean analysis. But the adelic point of view may suggest, for instance, completely new Hilbert space structures on spaces of combs or fractional combs. Also, if it turns out (as vaguely suggested at the end of Section 7.2) that local quadratic extensions are useful, the usual distribution point of view would no longer be available in such a context.

Index of Notation

Subject Index

Bibliography

[1] A. Bechata, *Calcul pseudodifférentiel p-adique*, Ann. Fac. Sci. Toulouse (6) **13**, 2 (2004), 179–240.

[2] F.A. Berezin, *A connection between co- and contravariant symbols of operators on classical complex symmetric spaces*, Soviet Math. Dokl. **19** (1978), 786–789.

[3] R.W. Bruggeman, *Fourier coefficients of cusp-forms*, Inv. Mat. **45** (1978), 1–18.

[4] D. Bump, *Automorphic Forms and Representations*, Cambridge Series in Adv. Math. **55**, Cambridge, 1996.

[5] D. Bump, *Spectral Theory and the Trace Formula*, in *An introduction to the Langlands program* (J. Bernstein, S. Gelbart, eds), Birkhäuser, Basel–Boston–Berlin, 2004.

[6] A. Connes, *Trace formulas in non-commutative geometry and the zeros of the Riemann zeta function*, Selecta Mat. (N.S.) **5** (1) (1999), 29–106.

[7] J.M. Deshouillers, H. Iwaniec, *Kloosterman sums and Fourier coefficients of cusp forms*, Inv. Math. **70** (1982), 219–288.

[8] J. Faraut, *Analyse harmonique sur les paires de Guelfand et les espaces hyperboliques*, in *Analyse harmonique*, Les Cours du C.I.M.P.A., Nice, 1982.

[9] P.B. Garrett, *Decomposition of Eisenstein series; Rankin triple products*, Annals of Math. **125** (1987), 209–235.

[10] S. Gelbart, *Automorphic forms on adele groups*, Ann. of Math.Studies **83**, Princeton Univ. Press, Princeton, 1975.

[11] I.M. Gelfand, M.I. Graev, I.I. Pyatetskii–Shapiro, *Representation theory and automorphic functions*, W.B. Saunders Co, Philadelphia–London–Toronto, 1969.

[12] D. Goldfeld, P. Sarnak, *Sums of Kloosterman sums*, Inv. Math. **71** (2) (1983), 243–250.

[13] I.S. Gradstein, I.M. Ryshik, *Tables of Series, Products and Integrals*, vol. 2, Verlag Harri Deutsch, Thun–Frankfurt/M, 1981.

[14] S. Haran, *Quantization and symbolic calculus over the p-adic numbers*, Ann. Inst. Fourier **43**, 4 (1993), 997–1053.

[15] E. Hecke, *Über die Kroneckersche Grenzformel für reelle quadratische Körper und die Klassenzahl relativ-abelscher Körper*, Verhandl. d. Naturforschenden Gesell. i. Basel **28** (1917), 363–372.

[16] D.A. Hejhal, *Some observations concerning eigenvalues of the Laplacian and Dirichlet L-series*, in *Recent Progress in Analytic Number Theory* (H. Halberstam and C. Hooley, eds), vol. 2, Acad. Press, London–New York (1981), 95–110.

[17] S. Helgason, *Groups and Geometric Analysis*, Acad. Press, New York, 1984.

[18] S. Helgason, *Geometric Analysis on Symmetric Spaces*, A.M.S. Math.Surveys and Monographs **39**, Providence, 1994.

[19] L. Hörmander, *The analysis of partial differential operators, III, Pseudodifferential operators*, Springer Verlag, Berlin, 1985.

[20] A. Ichino, *Trilinear forms and the central values of triple product L-functions*, Duke Math. J. **143**, 2 (2008), 281–307.

[21] H. Iwaniec, *Introduction to the spectral theory of automorphic forms*, Revista Matemática Iberoamericana, Madrid, 1995.

[22] H. Iwaniec, *Topics in Classical Automorphic Forms*, Graduate Studies in Math. **17**, A.M.S., Providence, 1997.

[23] H. Iwaniec, E. Kowalski, *Analytic Number Theory*, Am. Math. Soc. Coll. Pub, **53**, A.M.S., Providence, 2004.

[24] H. Iwaniec, P. Sarnak, *Perspectives on the analytic theory of L-functions*, Geom. Funct. Anal. special volume GAFA 2000, Tel-Aviv, Birkhäuser, Basel (2000), 705–741.

[25] N.M. Katz, P. Sarnak, *Random matrices, Frobenius eigenvalues, and monodromy*, Am. Math. Soc. Coll. Pub. **45**, A.M.S., Providence, 1999.

[26] A.W. Knapp, *Representation Theory of Semi-Simple Groups*, Princeton Univ. Press, Princeton, 1986.

[27] T. Kobayashi, B. Ørsted, M. Pevzner, A. Unterberger, *Composition formulas in Weyl calculus*, J. Funct. Anal. **257** (2009), 948–991.

[28] E. Kowalski, *Classical automorphic forms*, in *An introduction to the Langlands program* (J. Bernstein, S. Gelbart, eds), Birkhäuser, Basel–Boston–Berlin, 2004.

[29] S.S. Kudla, *Tate's thesis* and *From modular forms to automorphic representations*, in *An Introduction to the Langlands Program* (J. Bernstein, S. Gelbart, eds), Birkhäuser, Boston–Basel–Berlin, 2004.

[30] H. Kumano-go, *Pseudo-differential operators.* The MIT Press, Cambridge, MA, 1982.

[31] N.V. Kuznetsov, *The Petersson conjecture for cusp forms of weight zero and the Linnik conjecture*, Mat. Sb. (N.S.) **111** (1980), 334–383; Math. USSR-Sb **39** (1981), 299–342.

[32] S. Lang, $SL(2, \mathbb{R})$, Addison-Wesley, Reading, MA, 1975.

[33] S. Lang, *Algebraic Number Theory*, Addison-Wesley, Reading, MA, 1970.

[34] P.D. Lax, R.S. Phillips, *Scattering Theory for Automorphic Functions*, Ann. Math.Studies **87**, Princeton Univ. Press, 1976.

[35] N. Lerner, *Metrics on the phase space and non self-adjoint pseudo-differential operators*, Pseudodifferential Operators **3**, Birkhäuser, Basel, 2008.

[36] W. Magnus, F. Oberhettinger, R.P. Soni, *Formulas and theorems for the special functions of mathematical physics*, 3rd edition, Springer Verlag, Berlin, 1966.

[37] S.D. Miller, W. Schmid, The Rankin-Selberg method for automorphic distributions. *Representation theory and automorphic forms*, pp. 111–150, Progr. Math., **255**, Birkhäuser, Boston, 2008.

[38] C.J. Moreno, *The spectral decomposition of a product of automorphic forms*, Proc. of the A.M.S. **88** (3) (1983), 399–403.

[39] M.R. Murty, *Introduction to p-adic Analytic Number Theory*, Studies in Adv. Math. **27**, A.M.S., Providence, 2002.

[40] L. Pukanszky, *The Plancherel formula for the universal covering group of* $SL(2, \mathbb{R})$, Math. Annal. **156** (1964), 96–143.

[41] H. Rademacher, *Topics in Analytic Number Theory*, Springer Verlag, Berlin, 1973.

[42] M. Reed, B. Simon, *Fourier Analysis, Self-adjointness*, Ac.Press, New York-London, 1975.

[43] M. Riesz, *L'intégrale de Riemann-Liouville et le problème de Cauchy*, Acta Math. **81** (1949), 1–223.

[44] L. Schwartz, *Théorie des distributions*, Hermann, Paris, 1959.

[45] A. Selberg, *On the Estimation of Fourier Coefficients of Modular Forms*, Proc. Symp. Pure Math. **8** (1963), 1–15.

[46] G. Shimura, *Modular Forms of half-integral weight*, Lecture Notes in Math. **320**, Springer Verlag, Berlin–Heidelberg–New York, 1973.

[47] M. Shubin, *Pseudodifferential operators and spectral theory*, Springer Verlag, Berlin, 1987.

[48] C.L. Siegel, *Lectures on advanced analytic numer theory*, Tata Inst. of Fundamental Research, Bombay, 1961.

[49] R.A. Smith, *The L^2-norm of Maass wave functions*, Proc. Am. Math. Soc. **82**, 2 (1981), 179–182.

[50] E.M. Stein, G. Weiss, *Introduction to Fourier analysis in Euclidean spaces*, Princeton Univ. Press, Princeton, 1971.

[51] S. Sternberg, *Group Theory and Physics*, Cambridge Univ. Press, Cambridge, 1994.

[52] J.T. Tate, *Fourier analysis in number fields and Hecke's zeta function*, in *Algebraic Number Theory*, (J.W.S. Cassels, A. Fröhlich, eds), Acad. Press, London, 1967.

[53] M.E. Taylor, *Pseudodifferential operators*, Princeton Math. Series **34**, Princeton Univ. Press, Princeton, 1981.

[54] G. Tenenbaum, *Introduction à la théorie analytique et probabiliste des nombres*, Cours spécialisés Soc. Math. France, Paris, 1995.

[55] A. Terras, *Harmonic analysis on symmetric spaces and applications* I, Springer Verlag, New York–Berlin–Heidelberg, 1985.

[56] F. Treves, *Introduction to pseudodifferential and Fourier integral operators: I, Pseudodifferential operators*, Plenum Press, New York–London, 1980.

[57] A. Unterberger, *Oscillateur harmonique et opérateurs pseudodifférentiels*, Ann. Inst. Fourier Grenoble **29**, 3 (1979), 201–221.

[58] A. Unterberger, J. Unterberger, *La série discrète de $SL(2, \mathbb{R})$ et les opérateurs pseudo-différentiels sur une demi-droite*, Ann. Sci. Ecole Norm. Sup. **17** (1984), 83–116.

[59] A. Unterberger, *The calculus of pseudodifferential operators of Fuchs type*, Comm. Part. Diff. Equ. **9** (1984), 1179–1236.

[60] A. Unterberger, *Quantization and non-holomorphic modular forms*, Lecture Notes in Math. **1742**, Springer Verlag, Berlin–Heidelberg, 2000.

[61] A. Unterberger, *Automorphic pseudodifferential analysis and higher-level Weyl calculi*, Progress in Math. **209**, Birkhäuser, Basel–Boston–Berlin, 2002.

[62] A. Unterberger, *A spectral analysis of automorphic distributions and Poisson formulas*, Ann. Inst. Fourier **54** (5) (2004), 1151–1196.

[63] A. Unterberger, *Quantization and arithmetic*, Pseudodifferential Operators **1**, Birkhäuser, Basel, 2008.

[64] G. Valiron, *Théorie des Fonctions*, Masson, Paris, 1942.

[65] T. Watson, *Central value of the Rankin triple L-functions for unramified cusp-forms*, in preparation.

[66] A. Weil, *Sur certains groupes d'opérateurs unitaires*, Acta Math. **111** (1964), 143–211.

[67] A. Weil, *On some exponential sums*, Proc. Nat. Acad. Sci. USA **34** (1948), 204–207.

[68] D. Zagier, *Introduction to modular forms*, in *From Number Theory to Physics* (M. Waldschmidt, P. Moussa, J.M. Lück and C. Itzykson, eds), Springer Verlag, Berlin (1992), 238–291.

[69] D. Zagier, *Eisenstein series and the Riemann zeta-function*, in *Automorphic Forms, Representation Theory and Arithmetic*, Bombay Coll. 1979, Tata Inst. of Fundamental Research, New Delhi, 1981.

[70] D. Zagier, *The Rankin-Selberg method for automorphic functions which are not of rapid decay*, J. Fac. Sci. Univ. Tokyo, Sect. IA Math. **28** (3) (1982), 415–437.

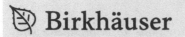 **Birkhäuser** | **www.birkhauser-science.com**

Pseudo-Differential Operators (PDO) Theory and Applications

This series is devoted to the publication of current research in operator theory, with particular emphasis on applications to classical analysis and the theory of integral equations, as well as to numerical analysis, mathematical physics and mathematical methods in electrical engineering.

Edited by
M. W. Wong, York University, Canada
In cooperation with an international editorial board

■ **PDO 7: de Gosson, M.**, Symplectic Methods in Harmonic Analysis and in Mathematical Physics (2011).
ISBN 978-3-7643-9991-7

The aim of this book is to give a rigorous and complete treatment of various topics from harmonic analysis with a strong emphasis on symplectic invariance properties, which are often ignored or underestimated in the time-frequency literature. The topics that are addressed include (but are not limited to) the theory of the Wigner transform, the uncertainty principle (from the point of view of symplectic topology), Weyl calculus and its symplectic covariance, Shubin's global theory of pseudo-differential operators, and Feichtinger's theory of modulation spaces. Several applications to time-frequency analysis and quantum mechanics are given, many of them concurrent with ongoing research.
This book is primarily directed towards students or researchers in harmonic analysis (in the broad sense) and towards mathematical physicists working in quantum mechanics. It can also be read with profit by researchers in time-frequency analysis, providing a valuable complement to the existing literature on the topic. A certain familiarity with Fourier analysis and introductory functional analysis (e.g. the elementary theory of distributions) is assumed. Otherwise, the book is largely self-contained and includes an extensive list of references.

■ **PDO 6: Gupur, G.**, Functional Analysis Methods for Reliability Models (2011).
ISBN 978-3-0348-0100-3

The main goal of this book is to introduce readers to functional analysis methods, in particular, time dependent analysis, for reliability models. Understanding the concept of reliability is of key importance – schedule delays, inconvenience, customer dissatisfaction, and loss of prestige and even weakening of national security are common examples of results that are caused by unreliability of systems and individuals. *Functional Analysis Methods for Reliability Models* is an excellent reference for graduate students and researchers in operations research, applied mathematics and systems engineering.

■ **PDO 5: Wong, M.W.**, Discrete Fourier Analysis (2011).
ISBN 978-3-0348-0115-7

This textbook presents basic notions and techniques of Fourier analysis in discrete settings. Written in a concise style, it is interlaced with remarks, discussions and motivations from signal analysis.
The first part is dedicated to topics related to the Fourier transform, including discrete time-frequency analysis and discrete wavelet analysis. Basic knowledge of linear algebra and calculus is the only prerequisite. The second part is built on Hilbert spaces and Fourier series and culminates in a section on pseudo-differential operators, providing a lucid introduction to this advanced topic in analysis. Some measure theory language is used, although most of this part is accessible to students familiar with an undergraduate course in real analysis.
Discrete Fourier Analysis is aimed at advanced undergraduate and graduate students in mathematics and applied mathematics. Enhanced with exercises, it will be an excellent resource for the classroom as well as for self-study.